RELIGION AND THE PHILOSOPHY OF LIFE

Religion and the Philosophy of Life

GAVIN FLOOD

UNIVERSITY PRESS

Great Clarendon Street, Oxford, OX2 6DP,
United Kingdom

Oxford University Press is a department of the University of Oxford. It furthers the University's objective of excellence in research, scholarship, and education by publishing worldwide. Oxford is a registered trade mark of Oxford University Press in the UK and in certain other countries

First Edition published in 2019

Impression: 1

Published in the United States of America by Oxford University Press
198 Madison Avenue, New York, NY 10016, United States of America

British Library Cataloguing in Publication Data
Data available

Library of Congress Control Number: 2018952832

ISBN 978–0–19–883612–4

Printed and bound in Great Britain by
Clays Ltd, Elcograf S.p.A.

To my wife Kwan

Preface

In William Golding's novel *The Spire*, the protagonist, the Dean Jocelyn, sets out to build the tallest spire of any cathedral hitherto constructed, but there are grave concerns that the foundations will not be deep enough, and the spire will topple. Jocelyn's vision for a new architecture in a new age and his aspiration at the edge of human imagination seems an appropriate image for current human aspiration for a better future realized through technology; we are living through a period of rapid change in understanding the nature of the human and in shaping the future in response to challenges from the environment, from new technology, from population density, and from global economic remodelling; and all this set within the ongoing complexity of, as always, lamentable human politics. In this new context, there is a reshaping of the intellectual landscape with rapid progress in the hard sciences that to some extent is influencing the humanities, with scientists reflecting on their work in the light of humanistic concerns. While each has its distinct sphere of operation, there is a case for humanist disciplines themselves, including the social sciences, to move beyond what we might call the linguistic paradigm and social constructivism towards building models and accounts of the human based on knowledge gleaned from the hard sciences but read through the lens of humanism, in the broad sense, or the *humanum*. This is not to give way to any kind of hegemony of natural science over the humanities, but rather to use knowledge in the service of understanding (and we might even say in the service of *phronesis*).

With this background in mind, this book seeks both to address a problem and to tell a story. The problem is the relation of life to religion, or the explanation of religion as it has developed in the history of civilizations, within the framework of understanding life itself. This is a problem, first, because the nature of religion is much contested, with some scholars even questioning whether it is a viable category; and second, because the nature of life, its place in the universe, and whether we can even speak of life itself are much contested. There are many stakeholders in these debates from those who wish to make claims about a particular religious perspective and see value in religion, to those who wish to see the demise of religion as a human good on ethical and political grounds. This book addresses the issue of religion by offering an explanatory account grounded in the nature of the kind of beings that we are, in our biology and specifically in the idea of life itself, while at the same time recognizing complexity and the necessity of a humanistic account. The story the book tells is the way life itself has been conceptualized and understood in the history of religions and the philosophies of life that have driven civilizations. The double nature of this project is intimately connected, as 'explanation' shows us that human beings have an innate pro-sociality that can be traced deep into our hominin past, while the 'narrative' of religions shows us that human beings have developed institutions and forms of inwardness that support human pro-sociality. I present a combination of explanation and narrative because of the necessary acceptance of the complexity of human reality and of life itself, and as such this work is not an eliminative reductionism. Philosophies of life have developed within civilizations to account for such complexity, in particular advancing the theme of life itself as the foundation of human reality.

The book considers how religion as the source of civilization transforms human bio-sociology through language and through the somatic exploration of ritual, meditation, and prayer. After outlining an account of current ideas about the theory of life (biology) and the philosophy of life, the book goes on to present a narrative of the ways in which life itself has been understood in the histories of three civilizational blocks—India, China, and Europe/the Middle East—particularly as they came to formation in the medieval period. It then develops the story into secular philosophy up to the present, integrating this with an argument for the explanation of religion in terms of life itself, rooted in the bio-sociology of human life: an integration of classical, religious worldviews with modern natural science. This is important because religions remain central to human life in much of the world and have been the creative resources for civilizations, as well as destructive powers that have threatened the human future.

Writing this book has necessitated that I venture into areas where I have relied entirely on the scholarship of others—such as Chinese civilization—but in all of these fields, I have presented a narrative and offered description in the service of my argument about life itself as the wellspring and resource for these histories. While I regard this project as a humanistic account of religion and its place in human life, I wish to absorb what current science tells us about the world and what current knowledge shows us about ourselves. Such an account accepts the integrity and relevance of personal and interpersonal understandings of meaning as well as impersonal or third-person explanations of reality. As such, some may regard it as too conciliatory to religious understandings of person and world, while for others it may be regarded as too conciliatory to scientific accounts, thereby missing the importance of theological understanding. But I regard this position to be a strength, as understanding the meanings of human life must continue to absorb current knowledge, while still operating within a humanistic discourse. The existential primacy of human freedom as one frame for understanding life itself needs to be integrated with the frame provided by the sciences: life in the singular seen in two ways. Such a view is not antithetical to theology, although the theologies of religions will need to absorb current knowledge of person and world if they are to remain relevant.

Indeed, although not my task, this is imperative in order to avoid destructive fundamentalisms (both religious and secular) that denude human life.

The opening quotations reflect the assertion of life, as articulated by Blake, that must be balanced by the reality of death, as articulated in the responsory of John Sheppard. That life entails death and vice versa is a startlingly simple truth that must always govern what we say about life itself.

At one level, the book presents a methodological agnosticism regarding the referents of the traditions it discusses, the gods and supernatural forces of the religions, while at another level it implicitly takes the view that the contemporary sciences offer descriptions of the universe more adequate to the data. Furthermore, the book wishes to develop a kind of abductive argument concerning life itself as the animating force of civilizations. We know more now about the universe we live in than our ancestors, but perhaps we don't know much more than they about how to conduct our lives, and, as they have always been, the metaphysical structures we inhabit are tenuous and subject to revision.

This book therefore offers an interdisciplinary link between religion, science, and philosophy. It is intended to speak to a number of audiences: to those within the religion and science debate, including theologians; to those in the study of religions, including scholars involved in the comparative history of traditions; and to philosophers and intellectual historians of life. This is a broad range of people and I hope that those involved in these disciplines will find some contribution to debate in the book's pages. At the end of the day, we are creatures orientated towards a future that always exceeds us, constrained by the limitations of time and place, while, like Jocelyn's spire, defying gravity.

Campion Hall, Oxford

25 June 2018

Acknowledgements

In a project such as this there are many intellectual debts. The seeds of this work were sown long ago as a student of John Bowker's when his graduate class at Lancaster University discussed sociobiology and we waited with bated breath for the just-published *On Human Nature* by Edward Wilson to arrive. Since then, so much has happened in evolutionary science, the humanities, and wider society. Although interest in the theme of this book goes back a long way, I am indebted more immediately to family, friends, and colleagues. Special thanks to my wife Kwan, for many conversations on this and related topics, with discussion ranging from Gilles Deleuze and Judith Butler to contemporary Western and Chinese art. Her own thinking on these topics has influenced mine. My thanks also to Oliver Davies for many discussions about religion and social cognition, on which he himself has published so insightfully; he offered many invaluable thoughts on the book itself. His own work, from which I have learned much, has been pathfinding in the science and humanities interface. A colloquium, 'Explaining Religion', at Yale-NUS College in Singapore stimulated my thinking and I would like to thank the attendees at that event, including Oliver, Agustin Fuentes, Gary Bente, Kai Vogeley, Johannes Bronkhorst, Joe Alter, Adam Zeman, and Donovan Schaefer. I would also like to thank Yale-NUS College for generously funding the conference. At Oxford I would like to thank the support and collegiality of colleagues at Campion Hall, particularly the Master James Hanvey SJ, whose life and work has been a great inspiration to me, and also to Philip Kennedy, John Barton, Graham Ward, Nick King SJ, Peter Davidson, Joel Rasmussen, and Pat Riordan SJ. Many thanks go to the team at the Oxford Centre for Hindu Studies, including Bjarne Wernicke Olesen, Jessica Frazier, Rembert Lutjeharms, Lucian Wong, and Shaunaka Rishi Das, for stimulating conversation and collegiality. At Singapore I would like to thank colleagues in the Human and Natural Sciences including all the philosophers, Rajeev Patke, Tan Tai Yong, Andrew Hui, Emanuel Mayer, Stuart Strange, Ajay Mathuru, Philip Johns, Terry Nardin, Naoko Shimazu, and Scott Cook for his invaluable help with Chinese terms and texts. I would also like to acknowledge conversations over the years with Francis Clooney SJ and Julius Lipner. My graduate students have always stimulated my thinking and the seminar 'Readings in Phenomenology' has directly influenced this project. A deep engagement of many years with Luke Hopkins has impacted my thought; he introduced me, too long ago, to some eclectic reading from Solovyev, to von Baader, to Norman Brown. Thanks also to my family, Leela, Claire and Adam, Dominic and Emma, and Clare. At OUP I would like to thank the encouragement and help of Tom Perridge and Karen Raith, and thanks to the anonymous readers for their extremely useful reports.

Contents

Introduction 1

Part I. The Theory of Life Itself

1. The Theory and Philosophy of Life 39
2. The Emergence of Religion 66
3. Sacrifice 94

Part II. Religious Civilizations

4. The Sacrificial Imaginary: Indic Traditions 119
5. Earth under Heaven: The Chinese Traditions 175
6. Transforming Life: The Greek and Abrahamic Traditions 218

Part III. Philosophies of Life Itself

7. Philosophies of Life 275
8. The Philosophy of Life as the Field of Immanence 299
9. The Phenomenology of Life 322
10. Bare Life and the Resurrection of the Body 344
11. Religion and the Bio-Sociology of Transformation 365

Epilogue: Modernity and the Life of Holiness 388

Bibliography 395
Index 429

'Energy is the only life, and is from the body; and reason is the bound or outward circumference of energy. Energy is eternal delight.'

William Blake, *The Marriage of Heaven and Hell*

Media vita in morte sumus

John Sheppard

Introduction

LIFE, RELIGION, AND CIVILIZATION

It has almost become a cliché that religions are concerned with the meaning of life; that they claim to offer answers to imponderable questions and prescribe ways of living that enhance the human good. But while a secular irony might be dismissive of the claims of religion, and even of the coherence of the question about the meaning of life, it is the case that for millennia religions have offered hope to individuals and communities, and there is a sense among practitioners that religions offer not only an understanding of life but also a way of living: a way of living rightly in harmony with others in the community, and that may have post-mortem consequence. Indeed, the question of life is inseparable from the question of death and reflection on life is also reflection on death. Such reflection on life's meaning and the importance of behaving in particular ways, whether through correct forms of prayer or correct comportment towards others, has been central to human formation throughout history. Religions have offered philosophies of life that seek to articulate what life itself is and to develop ways of living that enhance life. If ways of living are fundamental to what religion is, then a philosophy of life seeks to establish reasons and justifications for what we might call a religious, somatic orientation towards the world. Well-winnowed ritual and meditation practices along with moral actions have been justified through philosophies of life. But religions also operate beyond individual somatic orientation, at a civilizational level over generations. Religions have been the driving force of civilizations in pre-modernity, fundamental to their formation and development, building resources through art, literature, architecture, social institutions, and sacramental ritual.

There is a deep connection between the ideas of life, religion, and civilization: life has impelled religion, which has driven civilization. Although the history of religions has often been characterized by oppression and violence, this history also shows that they have promoted human wellbeing and enhancement; the completion or fulfilment of life variously conceived. Such enhancement has been regarded as correcting an error or fault in the human condition: ignorance in Buddhism and Hinduism, sin or disobedience to God in Abrahamic religions, and lack of harmony or balance in Chinese religions. The narratives of civilizations respond to life and their religions seek to repair not only a broken humanity but sometimes the cosmos itself; to heal the wound and transform the living. One way of speaking about them might be to say that religions attempt to resolve the tragic.

The terms 'religion' and 'civilization' are, of course, contested—I shall offer definitional reflection in due course—but arguably it is religion that has given rise to civilizations, and while not all religions have produced civilizations, all civilizations have their roots in religions, even though contemporary global civilization may have distanced itself from them. Art and music too have been important and probably preceded religion, but they alone are not sufficient to produce what we call civilization; for that, religions are necessary to create social formation and project hope for the future. Civilizations have emerged in human history at least since urbanization in the Neolithic period, and when we ask why this is, what led to the development of these historical formations over such long periods of time (at least long by human standards), we are inevitably led to ideas of human flourishing, striving towards a future condition of completion and fulfilling desire for life. This striving for fulfilment and desire for life we can read as seeking human self-repair; the correction of a collective distortion of life that narratives of civilization sometimes locate in a specific past, such as a fall (in Abrahamic religions), or fragmentation of a cosmic body (in Vedic religion), or a beginningless distortion of perception (in Buddhism). Such self-repair is intimately connected with human freedom as the capacity to wholly or partially enact it.

To justify such a claim, we have to turn to two sources: first, the history of civilizations themselves and, in particular, the religions that drive them that offer accounts of the relationship between life itself and the living; second, we need to turn to human evolution and to read civilizations in terms of niche construction, the environmental structure that promotes the flourishing of particular life forms. The first account locates explanation in the history of civilizations; the second locates it in the evolutionary and cognitive sciences. Both are necessary for adequate description and explanation of human understandings of life along with the human desire for life and its expression. That is, there is universality to human biology articulated in and controlling the history of civilizations; a biological universality that underlies a human 'moral ontology', to use Charles Taylor's phrase.[1] It is this universality that impels my project and I hope to account for the ways in which religions as the driving forces of civilizations have understood life in relation to the living and how they have offered eschatological hope. This account is rooted in evolution.

[1] Charles Taylor, *Sources of the Self: The Making of Modern Identity* (Cambridge: Cambridge University Press, 1989), p. 8.

Both the forces of history and its biological roots over-determine the history of civilizations. I do not offer a reductionist account because of my acceptance of complexity and the necessity of humanistic accounts of person, but rather demonstrate an integration of the biological and human sciences.[2] Such integration explains the ways in which life itself has impelled religions (almost as an agent) for its self-preservation and explains its modes of articulation through the structures of civilization. This is not to concede to what Taylor calls the 'epistemological construal', the notion that we can understand the world insofar as it fits with mechanistic science,[3] but rather it accepts the philosophical requirement of the primacy of the lifeworld, the hidden backdrop to thought, the generally unarticulated life in which we can make 'our way around rooms, streets, and gardens' and manipulate 'the objects we use'.[4] This backdrop of the lifeworld, which I will have more to say about in due course, is itself closely linked to human desire for a full life, thereby bringing life into the realm of freedom, and to aspiration to a higher purpose or attraction: to verticality or elevation as a human good.

But addressing life itself needs to be approached indirectly through the civilizational route, and I wish to read history with a purpose as this history reflects what we are now as an emergent global society. The story I will tell about life in relation to the living in the history of civilizations is relevant to life lived in the present. Religion and civilization go together, and they generate narratives of life that are being told in this book. In our current age of hybridization and cosmopolitanism it could be that we are entering a new kind of civilization with its own narrative of life, and we also have species-wide scientific knowledge that will be incorporated into that narrative; a narrative that must be grounded in the nature of the kinds of beings that we are. Some have argued that with modernity we have witnessed a clash of civilizations—that conflict zones are always on the edge of civilizational plate-tectonics—and this shows few signs of abating, while others offer more optimistic assessment.[5] Although extremely cautious of making predictions, I nevertheless deal with possible future directions in the Epilogue.

[2] In this, I am in agreement with Vasquez's 'taking biology seriously in religious studies'. Manuel A. Vasquez, *More Than Belief: A Materialist Theory of Religion* (Oxford: Oxford University Press, 2011), p. 173.

[3] Charles Taylor, 'Overcoming Epistemology', in *Philosophical Arguments* (Cambridge, MA: Harvard University Press, 1995), pp. 1–19.

[4] Ibid., p. 6.

[5] For the former view the classic study is Samuel P. Huntington, *The Clash of Civilizations and the Remaking of our World Order* (London: Simon and Schuster, 1996). For an optimistic view of progress in civilization see Steven Pinker, *Enlightenment Now: The Case for Reason, Science, Humanism and Progress* (London: Allen Lane, 2018). But for a compelling critique of the Huntingdon thesis see Peter van der Veer, *The Value of Comparison* (Durham, NC: Duke University Press, 2016), pp. 63–4.

To get at these questions of life itself along with human desire for life and the way civilizations have developed philosophies of life to account for it, I need to demonstrate the relationship between civilization, religion, and life. This is a complex relationship: religion drives civilization in its task of enhancing human life and life comes to be articulated in religion as philosophy of life. Such philosophy deals with the relation between life itself and the living, but understanding this relationship can best be done through the specificity of civilizations. Another way of saying this might be that civilizations express human desire for life that is an anticipation of completeness. This is not without ambiguity because the desire for life has been understood in religions as both positive (in Daoism, for example) and negative (in Buddhism and Gnosticism), with most religions having mixed evaluation.

The story I wish to tell is therefore how religion as the source of civilization transforms the fundamental bio-sociology of *Homo sapiens* through language and through the somatic exploration of religious ritual and prayer. I wish to offer an integrative account of the nature of the human, based on what contemporary scientists tell us, as well as through the history of civilizations. Our moral ontology is supported or even necessitated by our biological natures. This is not a reductionist claim, but rather a claim that what we now know about the evolution of human pro-sociality is a necessary condition for the development of higher-order institutions and civilizations. Life itself seeks articulation and preservation through civilizational structures, particularly through language in the broadest sense of modes of communication. Both nature and culture have formed who we are in a process of coevolution or human niche construction that shows a deep continuity between human beings and other life forms: we share an evolutionary lineage that is the broad context within which to discuss human nature.[6] Religion on this account is a natural and inevitable outgrowth of the biological realm[7] and questions the very distinction between nature and culture. Indeed, in consonance with other work in the humanities, we cannot speak of culture as something separate from the realm of nature.[8] We might even offer a stronger thesis or abductive argument that life itself brings about religion and thereby civilizations: life impels cultural forms for its self-preservation through time

[6] Agustin Fuentes, 'Evolutionary Perspectives and Transdisciplinary Intersections: A Roadmap to Generative Areas of Overlap in Discussing Human Nature', *Theology and Science*, vol. 11 (2), 2013, pp. 106–29.

[7] Thanks to the anonymous reader who re-described the project in these terms.

[8] Tim Ingold, 'Prospect', pp. 9–14, in Tim Ingold and Gisli Palsson (eds.), *Biosocial Becomings: Integrating Social and Biological Anthropology* (Cambridge: Cambridge University Press, 2013), pp. 1–21; Bruno Latour, *We Have Never Been Modern*, trans. Catherine Porter (Cambridge, MA: Harvard University Press, 1993), pp. 103–9. On the way cultural attitudes impact upon tree growth in Japan, see Anna Lowenhaupt Tsing, *The Mushroom at the End of the World: On the Possibility of Life in Capitalist Ruins* (Princeton, NJ: Princeton University Press, 2015), pp. 182–4.

and for its enhancement. The evolutionary part of the story is that with the development of advanced linguistic capability facilitated through an increased brain size, especially the development the neo-cortex, *Homo sapiens* went beyond an earlier social cognition system that promoted bonding between members of the group. This advanced linguistic consciousness simultaneously reduced the epistemic complexity of information through the senses—it may have evolved in part to facilitate practical survival—while at the same time allowing for the development of social complexity because of its ability to discriminate groups and to develop philosophies of life and practices of life. Human linguistic consciousness that is discriminatory—and so develops the ability both to name one group as friend and another as foe—came to modify the all-embracing social brain or the social cognition system.[9] As with other higher primates, our social cognition allows bonding with the group to maximize survival potential, yet unlike other primates our linguistic capacity controls social cognition in the interests of a higher-order sociality that has served to increase group numbers, which led in time to human civilizations. I will briefly sketch these two linked components, evolution and civilization.

First, on this evolutionary account, civilizations are kinds of niche construction that have developed to enhance human life. The social cognition system in the deeper brain allows us to bond with others while advanced linguistic consciousness developed judgement, discrimination, and allows for a more complex social organization. On the evolutionary view, civilizations are thus forms of niche construction that have successfully propagated the species. Furthermore, civilizations have addressed the human need and quest for meaning. That is, civilization has entailed our separation from our pre-linguistic past (even a separation of the linguistic brain from the older social brain), but religions connect these two modes through somatic techniques of ritual, prayer, and asceticism that access non-linguistic ways of being refracted through language. Civilizations might be understood as corporate attempts at human enhancement through the integration of pre-linguistic modes of being, the pro-sociality of our hominin past, with our advanced linguistic and

[9] On social cognition I have learned in particular from Leonard Schilbach, Bert Timmermans, Vasudevi Reddy, Alan Costall, Gary Bente, Tobias Schlicht, and Kai Vogeley, 'Towards a Second Person Neuroscience', *Behavioural and Brain Sciences*, vol. 36 (4), 2013, pp. 393–414. Oliver Davies introduced me to this article and on the cultural application of this science I have learned from his Oliver Davies, 'Niche Construction, Social Cognition, and Language: Hypothesizing the Human as the Production of Place', *Culture and Brain*, vol. 3 (2), 2016, pp. 87–112. On agency and brain see Patrick Haggard, 'Sense of Agency in the Human Brain', *Nature Reviews: Neuroscience*, vol. 18, 2017, pp. 197–208. Also see related issues in Chris D. Frith, *Making up the Mind: How the Brain Creates our Mental World* (Oxford: Blackwell, 2007) and Ethan Kross and Kevin K. Ochsner, 'Integrating Research on Self-Control across Multiple Levels of Analysis: Insights from Social Cognitive and Affective Neuroscience', in Ran R. Hassin, Kevin N. Ochsner, and Yaacov Trope (eds.), *Self Control in Society, Mind, and Brain* (Oxford: Oxford University Press, 2010), pp. 76–92.

therefore reflective capacity. Language itself is a biocultural niche construction and a kind of evolutionary artefact,[10] and through linguistic consciousness *Homo sapiens* can assert their freedom and reflect on the kind of life they have. Language enables us to define who we are and, in one sense, to stand outside of ourselves while at the same time co-constituting the world together. In examining civilizations, we see that we have become alienated from our own bodies because we are primates with advanced linguistic consciousness. Religions have attempted to overcome this through an aspiration towards human freedom in which the body becomes self-aware and language can reflexively articulate that awareness.

One of the effects of religion is to reconnect human communities with their biological roots through the disenabling of language in techniques of the body, through ritual, shamanic trance, and meditation, while at the same time developing narratives that place human beings in a cosmic context with a view to an end time, a collective eschatology, or an individual soteriology. The disenabling of language through somatic techniques is never total because those methods are refracted through the lens of language that gives accounts of them: asceticism, prayer, and meditation are placed within an eschatological or soteriological narrative. Religions, including Asian religions, tend to be eschatological in giving what Reisebrodt has called 'the promise of salvation'[11] in the face of often tragic accounts of the human. The eschatology of religious civilizations links them to life itself. A major function of religion is thus to create structures and practices that draw on a pre-linguistic social cognition, thereby accessing the deeper biological rhythms of the species, the rhythms of life. But while it is important to explain religions with reference to our biology, they are not simply regressions to an earlier development in the human brain. Rather, they are crucial to civilizations in providing the transformation through linguistic consciousness of cooperative biological proclivities. This is not so much a regression to a pre-linguistic past as an integration of linguistic consciousness with the social cognition system and a channelling of life into human imagination. Religions as emergent properties allow us to get in touch with deeper sources of life itself, articulate the desire for life, transform this desire through the imagination allowed by advanced linguistic consciousness, and thereby attempt to heal human alienation from life itself and to fulfil human desire for life. Religions are pathways to life, as well as expressions of life, that aim at integration and transformation. They have not always achieved this, however, and the history of religions has also been a history of strife, but at their best religions bring communities into a vitality and fullness of life,

[10] Chris Sinha, 'Language and Other Artifacts: Socio-Cultural Dynamics of Niche Construction', *Frontiers in Psychology*, vol. 6, 2015, p. 1601.

[11] Martin Reisebrodt, *The Promise of Salvation: A Theory of Religion* (Chicago, IL: University of Chicago Press, 2014).

addressing fundamental human needs in giving eschatological hope and individual salvation, and offering resources for self-repair.

Second, the story involves not only an explanation of civilization in terms of evolutionary niche construction, but also that religions themselves tell of the relationship between the living and life itself, between the desire for life and its fulfilment. Religions offer philosophies of life that explicate this relationship and furthermore specify how eschatological hope can be realized through practices that promote wellbeing and transformation: religions embody philosophies of life that seek to repair the human. While it may be the case that an evolutionary explanation goes against traditional religious accounts, I am convinced that contemporary science offers ways of understanding human life that are relevant to the explication and exposition of religions and therefore of civilizations.[12] Religions advocate a particular way of being in the world and comportment towards others, often (although not always) characterized as compassion or love, making room for the other and advocating the embrace of the stranger. While it may be an exaggeration to say that the religions of the world that have survived promote the golden rule, not to do unto others what you would not have them do unto you[13] (*quod tibi fieri nos vis, alteri ne feceris*),[14] they all promote some degree of universalism: that the claims they make and practice they advocate are binding on at least some human communities because true (this is the case with even restrictive, initiatory traditions).

Indeed, comportment towards others exemplifies an attitude to life that religions have often understood as a kind of love. In one of his last interviews, the French philosopher Jacques Derrida expressed a view about the politics of love, that love is a force linked to civilizations and the future of civilization is contingent upon the development of such a politics.[15] This claim, which

[12] On an interesting account of integrating science with the humanities see Edward Slingerland, *What Science Offers the Humanities: Integrating Body and Culture* (Cambridge: Cambridge University Press, 2008). On the generation of knowledge and interdisciplinary nature of the humanities that incorporates science see Clause Emmeche, David Budtz Pedtesen, and Frederik Stjernfelt (eds.), *Mapping Frontier Research in the Humanities* (London: Bloomsbury, 2017).

[13] Karen Armstrong, 'The Golden Rule across Faiths', Resource on Faith, Ethics, and Public Life, 17 November 2008, Georgetown University Center for Religion, Peace, and World Affairs, https://berkleycenter.georgetown.edu/quotes/karen-armstrong-on-the-golden-rule-across-faiths. Accessed December 2016.

[14] Thomas Hobbes, *Leviathan*, edited with notes by Edwin Curley (Indianapolis, IN and Cambridge: Hackett, 1994), pp. 81, 99, echoing Matthew 7.12: 'Do to others what you would have them do to you'. Cf. Confucius, *Analects* 15.23: 'What you would not want done to you, do not do to others'.

[15] Mustapha Chérif, *Islam and the West: A Conversation with Jacques Derrida*, trans. Theresa Lavender Fagan (Chicago, IL: Chicago University Press, 2008), pp. 19–26. This is close to Martha Nussbaum's idea of compassion. Martha Nussbaum, 'Compassion, the Basic Social Emotion', *Social Philosophy and Policy*, vol. 13 (1), 1996, pp. 27–58.

inevitably remained undeveloped in Derrida's work (he would pass away the following year), contains an insight about the nature of civilization that is intimately linked to the nature of life and of human beings. Derrida's word 'love' implies a fullness of life and ways of being in the world that enhance human flourishing in a civilizational context. We might contend that the relationship between life itself and civilization is crucial for our understanding of human life and the way in which millions of people have inhabited the world throughout history. People's desire for life has been expressed through civilizations whose religions have sought to address that desire by its fulfilment or transformation. Certainly, there is great diversity in the history of civilizations but as the historian David Lord Smail observes, 'it is the similarities that are most startling, the thing that continually reminds us of our common humanity'.[16]

Coevolution

This double story of civilization, that of human evolution and that of religious accounts of life, maps on to coevolution theory, that human beings are the product of gene-culture coevolution over long time periods.[17] In the human case, coevolution is responsible for how we regard others, our fundamental human sociality, our sense of fairness, our capacity to empathize and to develop morality and virtue. I shall address this as an explanatory narrative in Chapter 2 and return to it in the final chapter, but for now, briefly, coevolution is a form of niche construction in which a genetically encoded species constructs a habitable, reproducible world that in turn affects the genetic base.

Central to human bio-sociology are neurological systems, in particular social cognition rooted in the brain below the neo-cortex that social neuroscientists call the social brain, which allows the development of pro-social emotions, so crucial to the success of *Homo sapiens*. I shall address this in more detail in Chapter 2, but to sketch out the terrain we might say that social cognition is the ability of the group to bond and to have a shared

[16] David Lord Smail, *On Deep History and the Brain* (Berkeley and Los Angeles, CA: University of California Press, 2008), p. 199.

[17] For a good summary see Herbert Gintis, 'Gene-Culture Coevolution and the Nature of Human Sociality', *Philosophical Transactions of the Royal Society B*, vol. 366, 2011, pp. 878–88. On the question of the new evolutionary synthesis, whether natural selection or evolutionary development (evo-devo) that includes learned behaviour as affecting the genetic base, see Michael Ruse, 'Evo-Devo: A New Evolutionary Paradigm?', in Anthony O'Hear (ed.), *Philosophy, Biology, and Life* (Cambridge: Cambridge University Press, 2015), pp. 105–24. Ruse discusses the paradigms of 'function' and 'structure', arguing that the tension between these generates first-rate research in biology.

intentionality, as primatologists such as Tomasello have shown, and which is pre-linguistic (in the narrower sense of language). Social cognition is operative in our primate cousins and is developed in *Homo sapiens* to an extraordinary degree, far beyond other higher primates. Indeed, the neurological basis of the experience of community is probably 'the default mode network' of brain function, first described by the neurologist Marc Raichle and his colleagues in 2001[18] that established the priority of our pro-sociality at a neurological level. This means that the default mode of the brain is in fact the mode of its pro-sociality and is active from a very early stage in foetal development.[19] This is a very ancient system stretching back into our hominin past.

But advanced social cognition is not enough to make the human being what it is; we also need language. Language may have developed, at least partly, from tool use fairly late in evolutionary development and is largely, as Davies observes, physically located in the neo-cortex, the layer in which words as codes are stored in physical space.[20] Linguistic ability allows for heightened cooperation including the ability to abstract from a present situation into the future. The origin of language that was at first contextual—language was ostensive—becomes abstracted from immediate context. Along with advanced language comes the ability to make distinctions and judgements, especially concerning social relationships. This advanced linguistic ability combined with the emergence of larger social groups and eventually the emergence of farming and the beginnings of urban communities in the Neolithic. Language capacity implies naming: naming the domesticated dog, certainly, but also the identification of those inside the group from those outside. Language allows for advanced cooperation but also the objectification and demonizing of the other, relegating the other to the non-human in a denial of their humanity.[21] With language ability we have the emergence of religion, political organization, and the monopoly of a particular language with its ability to tell the story of the group: how and why we came to be as we are and the capacity to narrate tragedy and hope.

An early account of this development was given by Merlin Donald,[22] which Robert Bellah has used as a schema in his magisterial work on religion and

[18] M.E. Raichle et al., 'A Default Mode of Brain Function', *Proceedings of the National Academy of Science USA*, vol. 98 (2), 2001, pp. 676–82.

[19] On adult and infant default mode networks see S. Seshami, A.I. Blazejewska, S. Mckown, J. Caucutt, M. Dighe, C. Gatenby, and C. Studholme, 'Detecting Default Mode Networks in Utero by Integrated 4D fMRI Reconstruction and Analysis', *Hum Brain Mapp*, vol. 37 (11), 2016, pp. 4158–78.

[20] For a good summary relevant to our theme see Davies, 'Niche Construction, Social Cognition, and Language'. Also see R.E. Passingham and S.P. Wise, *The Neurobiology of the Prefrontal Cortex: Anatomy, Evolution, and the Origin of Insight* (Oxford: Oxford University Press, 2014).

[21] Davies, 'Niche Construction', p. 5, citing Vittorio Gallese.

[22] Merlin Donald, *The Origins of the Modern Mind: Three Stages in the Evolution of Culture and Cognition* (Cambridge, MA: Harvard University Press, 1991).

evolution.[23] Donald argues that there were three stages in the evolution of human culture: the mimetic, mythic, and theoretic. The mimetic culture is itself based on an earlier episodic culture that humans share with other primates: the ability to recognize at what state or stage another creature is in and to respond appropriately. The episodic culture of higher apes is characterized by their living entirely in the present, for where humans 'have abstract symbolic memory representations, apes are bound to the concrete situation of episode; and their social behaviour reflects this situational limitation'.[24] This situational limitation does not mean lack of intelligence as apes can perceive complex life events, but in this perception episodes are not linked together in a larger narrative. Human mimetic culture, by contrast, does link episodes into a wider autobiographical memory and homo erectus began to mimic past and future events some two million years ago. This mimesis is foundational to human culture; even, as Donald points out, to literary imagination.[25] Mimetic culture especially involved the face and imitation of facial expression[26] and developed into language, which in turn developed into the ability to tell stories (mythic culture) and thence to abstraction and theoretical culture.

While this is a crude sketch of Donald's thesis, what is significant is that this developmental sequence maps onto the development of the brain: advances in cognition go hand in hand with human cultural development. Indeed, Donald's highlighting of mimesis as a fundamental feature of the species *homo* foreshadows findings that support this view. I need to say more about this in Chapter 2, but for now the point is that cultural evolution goes together with genetic evolution and the development of the brain and cognition. There is coevolution of genes and culture that is encapsulated in the idea of niche construction. Advances in cognition were driven by natural selection and mimetic skill that built upon episodic culture and ability. For the development of more complex social organization characteristic of the human species, there needed to be more advanced cognitive capacity and imitative skill in order to produce the niche within which human beings could flourish. The human niche leads to complex culture and civilization.

Civilization is the result of coevolution and its origins can be located in early social, urban formations that are contingent upon a developed neo-cortex and advanced linguistic ability. This creates a disjunction within the human species between the older social cognition system and advanced language, on the one hand, and on the other the possible development of strong social and political bonds through the linguistic control of social cognition. We witness

[23] Robert Bellah, *Religion in Human Evolution: From the Paleolithic to the Axial Age* (Cambridge, MA and London: The Belknap Press of Harvard University Press, 2011), pp. xviii–xix.

[24] Donald, *The Origins of the Modern Mind*, p. 149.

[25] Ibid., p. 170.

[26] Ibid., pp. 180–1.

here the development of law and the emergence of religion as a system of prohibition and injunction, topics I will return to in Chapter 2.

The Emergence of Civilization

This book thus addresses an issue and tells a neglected story. The issue is as broad and deep as they come, namely the explanation of civilization in the context of human evolution and life itself, and the neglected story is the way life has been understood in the history of religious civilizations as necessary for human transformation, a story that necessitates the explication of the relationship between life itself and the living. In the coming pages we will see how life as a category has been a theme, sometimes a hidden theme, in the histories of three civilizational blocks—the Indic, the Chinese, and the European/Middle Eastern—and we will see how life itself can become part of an explanatory paradigm that gives a contemporary account of those civilizations as expressing human desire for life. Why these three? Because they are large scale, ancient, and literate with large bodies of written literature. Peter van der Veer has cautioned that we need to be hesitant about essentializing civilizational units—particularly the West, India, and China—because this risks neglecting the fragmented nature of their histories, but we can, I think, avoid essentialization that plays into the hands of reifying national identity. Van der Veer suggests we can do this through recognizing narratives that go against the idea of unity,[27] but we also need to recognize the distinctive linguistic nature of these civilizational units. Through complex histories with their discrete stories, parallel patterns emerge of life viewed as an animating force. There are other civilizations in history that could have been selected, such as in Zimbabwe or Central America, but these three ancient civilizations, the Indic, Chinese, and European/Middle Eastern, were intertwined with each other, have a massive impact upon global modernity, and are strongly literate with long histories of the written word preserved through the generations.

Such a claim is clearly controversial in the sense that the history of civilizations might be seen not as a history of structures that facilitate human transformation but as structures that have facilitated exploitation and the development of strong social hierarchies for the benefit of the few. Civilization on this view is always built on the subjugation of others: the dominant tribe, the dominant language, the dominant technology repress the minority and the alternative. This is indeed borne out through the archaeological and later historical record. But while the explanation of civilization can be in terms of emergent social hierarchies developing as a consequence of new agrarian

[27] Van der Veer, *The Value of Comparison*, pp. 65–7.

practices, a deeper explanation in terms of coevolution offers a powerful explanatory paradigm and shows that the data of the fossil record refutes 'the notion that our proclivity to organized violence and to waging war was a core, and early, adaptive outcome of human evolution'.[28] To begin to develop this idea, we will need to offer historical description and some definition of terms. This description is also the argument in the sense that an account of the emergence of civilizations from earlier social formation is an account of a shift towards linguistic consciousness and the ordering of the world. A civilization is world ordering over a long period of time.

Civilization is a vague category without a precise definition. There is a case for understanding it in terms of degrees of technology; thus history might speak of a stone age, bronze age, and iron age that link civilization to the human ability to harness resources from the ambient environment: on this view, the more energy we harness, the more civilized we become. But it could also be argued that other factors such as language or moral foundation are even more crucial in the development of civilization. Gandhi, critical of the British, defined civilization as a way of life based on duty or dharma and that pursues the purposes or objects of life.[29] He thought British civilization had emerged since the Industrial Revolution and was based on 'bodily welfare' rather than the higher goals or 'objects of life'.[30] His friend Jawaharlal Nehru likewise considered dharma important and identified a long continuity of Indian civilization from the Indus Valley that he understood in terms of 'some powerful impulse, some tremendous urge, or idea of the significance of life'.[31] More recently, Katzenstein has argued that modern civilizations are linked to states with multiple centres of culture, so we can speak of China, India, and America as 'civilizational states' in which religion has been central to their respective projects of Sinicization, Indicization, and Americanization.[32] I have sympathy with this view, but along with moral foundation and technological development, civilization is a category that necessitates a dominant language—thus we might speak of Arabic civilization or Sanskrit civilization or Chinese civilization—and a dominance of cultural forms and ways of living in the world such as kinds of literature, art, music, and even the

[28] Fuentes, *The Creative Spark*, p. 162.

[29] M.K. Gandhi, *'Hind Swaraj' and Other Writings*, edited by Anthony J. Parel (Cambridge: Cambridge University Press, 1997), p. 65.

[30] Ibid., p. 34. Gandhi is here referring to the purposes of life in Hinduism, the puruṣārthas of duty (*dharma*), prosperity (*artha*), and pleasure (*kāma*) along with the highest good, liberation (*mokṣa*).

[31] Jawaharlal Nehru, *The Discovery of India* (Delhi: Oxford University Press, 1989 [1946]), p. 143.

[32] Peter Katzenstein, 'Civilizational States, Secularisms, and Religions', in Craig Calhoun, Mark Juergensmeyer, and Jonathan Van Antwerpen (eds.), *Rethinking Secularism* (New York: Oxford University Press, 2011), pp. 145–65.

regulation of conduct through manners and civility that Elias brought our attention to.[33]

Religions are integral to this ordering, and the emergence of what we understand by religion is coterminous with the advent of advanced linguistic consciousness. Philosophies of life come out of these religions and in turn exceed them such that we have philosophies of life developing independently of the institutional base from which they originally arose. Put simply, proto-religion emerges in hunter-gatherer communities, then religion first emerges in agrarian cultures as part of an evolutionary niche construction, and civilization emerges from religion along with philosophies of life that express the desire for life.

Long before philosophy in a formal sense, at the end of an extensive period of time in which few technological innovations occurred, a new order began to emerge based on a new way of sustaining life. Rather than hunting and gathering, innovative practices of food production that involved tilling fields, planting, and harvesting sustained increasing populations. Agriculture began to replace foraging around 10,000 kya and this introduction of farming marked off more settled communities from their ambulatory predecessors in what the Marxist archaeologist Gordon Childe referred to as the 'Neolithic Revolution'.[34] For most of the thousands of years that *Homo sapiens* had lived, not much had happened but suddenly, thought Childe, there was 'radical change... fraught with revolutionary consequences for the whole species'.[35] This was a massive event for humankind and transformed small-scale cultures into large-scale civilizations.

Both hunter-gatherer and farming communities may have lived alongside one another for some time, but agriculture won out in the end, and although more labour intensive, it was advantageous in sustaining a growing population. We see this in the increase in numbers of agrarian communities in contrast to the slower development of Palaeolithic habitation.[36] Farming communities initially developed in the Levant and what became known as the Fertile Crescent, an area of high land that runs through parts of modern Israel, Jordan, Lebanon, and Syria, down to Iran along the edge of the Zagros Mountains, a crescent between the mountains of Turkey and harsh desert of Syria.[37] But it was not a straight line from foraging to farming. In this region

[33] Norbert Elias, *The Civilizing Process* (Oxford: Blackwell, 1978).

[34] Gordon W. Childe, *Man Makes Himself* (London: Watts, 1936), pp. 66–104. Some have since disputed the claim on the grounds that the 'revolution' took place over several thousand years, but nevertheless, as Christian observes, it was of such importance that the title is deserved. David Christian, *Maps of Time: An Introduction to Big History* (Berkeley, CA: University of California Press, 2004), pp. 209–10.

[35] Childe, *Man Makes Himself*, p. 74.

[36] Christian, *Maps of Time*, p. 209.

[37] O. Bar-Yosef and R.H. Meadow, 'The Origins of Agriculture in the Near East', in T.D. Price and A.B. Gebauer (eds.), *Last Hunters-First Farmers: New Perspectives on the Prehistoric*

hunter-gatherers created settlements of stone huts for several hundred people, the Natufian culture (13,000–10,000 BCE),[38] but due to climate change that brought a severe drop in temperature, the villages were abandoned only to be re-inhabited a thousand years later when warmer weather returned. Here we have the beginnings of farming. Wild grains were cultivated, and people lived in mud-brick villages, but with no pottery (Pre-Pottery Neolithic A, 10,000–8500 BCE), communities that then grew into complex multi-room homes but still without pottery (Pre-Pottery Neolithic B, 8500–7000 BCE) and which display evidence of religion; bull horns and skulls of ancestors on the walls. Farming also seems to have developed independently in other parts of the world.[39] The size of the communities increased from groups of about eighteen in Natufian culture to about nine hundred in Pre-Pottery Neolithic B.[40]

But there was some hesitation about unreservedly taking up farming—indeed, in early farming communities the average height of *Homo sapiens* fell beneath that of their foraging forebears for a period of time[41]—but in due course cultivation became the norm for the majority of human societies. Agriculture brought with it an eventual increase in numbers and stable populations living in one place. In time these communities grew in size and extended networks of trade and political power to other centres of growth, and we had the beginnings of urban living and networks of exchange (trade, technology, and ideas) that eventually became states. While it is difficult to precisely define a civilization, the cultivation of an agrarian economy marked a shift to a much larger cohesive unit, characterized by a common language and eventually the emergence of institutionalized religion and law.

David Christian presents a model of dense agriculture producing a centre that has outreach to 'hubs' and an exchange of information occurring through these lines of connectivity. Hubs in particular are important for linking different geographical zones; thus Mesoamerica and the corridor joining Mesopotamia with Egypt became important hubs of information exchange that in turn became centres of gravity. These urban centres had

Transition to Agriculture (Santa Fe, NM: School of American Research, 1995), pp. 39–94; Bruce D. Smith, *The Emergence of Agriculture* (New York: Scientific American, 1995), pp. 50–2.

[38] O. Bar-Yosef, 'The Natufian Culture in the Levant: Threshold to the Origins of Agriculture', *Evolutionary Anthropology*, vol. 31, 1998, pp. 159–77.

[39] Christian, *Maps of Time*, p. 223.

[40] Eleni Asouti and Dorian Q. Fuller, 'From Foraging to Farming in the Southern Levant: The Development of Epipalaeolithic and Pre-Pottery Neolithic Plant Management Strategies', *Vegetation History and Archaeobotany*, vol. 21, 2012, pp. 149–62.

[41] Basak Koca Özer, Mehmet Sagir, and Ismail Özer, 'Secular Changes in the Height of the Inhabitants of Anatolia (Turkey) from the 10th Millennium B.C. to the 20th Century A.D.', *Economics and Human Biology*, vol. 9, 2011, pp. 211–19; P.L. Walker and J.T. Eng, 'Long Bone Dimensions as an Index of Socio-Economic Change in Ancient Asian Population', *American Journal of Physical Anthropology*, supplement 42, 2007, p. 241.

high populations and rapid collective learning.[42] Even at this early period we have the beginning of world systems interconnected through networks of trade, technology, religion, and even kinship.[43] With farming and the growth of towns, more complex social structures emerge, with hierarchical societies where workers till the land for the socially superior. Jericho is perhaps the oldest of such communities (9600–7500 BCE). Urban centres of gravity developed into what we can call states with some degree of centralized political power, the earliest being in Sumer around 3200 BCE, followed by Egypt (3000 BCE), the Indus Valley (2500 BCE), and China (2000 BCE). We can call these emergent societies 'states', based on towns with networks of interaction and some centralized government. The earliest law that has come down to us is from the Babylonian king Hammurabi (1792 BCE) that provides evidence of a bureaucratic and legal structure.[44] Early state formations grew out of towns that were conglomerations of activity as a result of the intensification of agriculture, which resulted from the adoption of new ways of food production facilitated through the increased cranial capacity of *Homo sapiens* and the development of advanced linguistic ability.

Civilization is thus a vague category. I think we can speak of civilizations before the rise of states, especially a Neolithic agrarian civilization possibly characterized by a shared language that articulates a religion, perhaps regarded as a sacred language, and characterized by centres of ritual which were probably also centres for trade networks, such as Göbekli Tepe in modern-day Turkey, the world's earliest temples built at the beginning of cultivation. Here twenty temples were constructed over a period from 9600 to 8200 BCE, significantly older than Avebury and Stonehenge. Archaeologists have some idea of how the temples were made—with stone from limestone quarries—but little idea of what went on inside them. We have animal carvings and close by a life-size human sculpture, the earliest known dated to at least 8000 BCE. Indeed, the T-shaped pillars of the temple may represent figures—perhaps deified ancestors.[45] Similarly, in northern Europe from Orkney and the Outer Hebrides to Spain we have stone monuments, the megalithic tombs such as the West Kennet long barrow, and the development of a sacred landscape of linked temple sites such as Avebury, Stonehenge, Woodhenge, and Durrington Walls.[46]

[42] Christian, *Maps of Time*, pp. 291–3. [43] Ibid., pp. 249–50. [44] Ibid., p. 295.

[45] Klaus Schmidt, *Göbekli Tepe: A Stone Age Sanctuary in South-Eastern Anatolia* (Berlin: Ex Oriente e.V., 2012).

[46] H.A.W. Burl, 'Henges: Internal Features and Regional Groups', *The Archaeological Journal*, vol. 126 (1), 1969, pp. 1–28. Things have moved on from Burl's study; see Julian Thomas, 'Ritual and Religion in the Neolithic', in Timothy Insoll (ed.), *The Oxford Handbook of the Archaeology of Ritual and Religion* (Oxford: Oxford University Press, 2011), pp. 371–86. For an interesting account of the meaning of these monuments see Vicki Cummings, 'What Lies beneath: Thinking

We know so little about the worldview of these early people. They were anatomically modern so we can be sure they had advanced linguistic capacity and intelligence as great as modern people, specialization, and an early cosmopolitanism that connected people from a wide geographical area.[47] Religion, social structure, and a general way of life were of apiece and while there must have been a plurality of languages, I suspect that a lingua franca allowed these early people to move across a wide geographical area. Indeed, language became the hallmark of civilization and we can later speak of civilizations united through Latin, Greek, Arabic, Sanskrit, and Chinese that embrace multi-ethnicities, regions, religions, and vernacular languages.

Religions are inseparable from early civilizations and, I would argue, their driving force in the sense that the systems of practice and worldviews that comprise religions facilitated the development of the Neolithic agrarian civilization through ordering space and time and developing eschatological expectation, which we can infer from the building of these early Neolithic temples. While they may have functioned as technology to help farmers predict when to sow the crops, evidence suggests that the Neolithic temples were focused on mortuary rites and secondary burial of the dead; that is, they were eschatological in orientation towards the future.

Farming emerged independently in different regions, although the Fertile Crescent discovery of cultivation is probably the earliest. Technologies spread quite rapidly—tools, farming methods, knowledge of crops—and civilizations linked to political power emerge first in Sumer but then most famously in Egypt where the northern and southern kingdoms unite to form a long-lasting and powerful political, religious, and cultural formation. In China a civilization arose that was to remain in place for thousands of years. In Iran Zoroastrian civilization developed, closely linked to its Indian cousin, Vedic civilization, which replaced the Indus Valley; its effects are still being felt today. These great structures moving through time were characterized by language as a uniting force, law, and religion, in particular ritual patterns that pay attention to respect and honour for ancestors. Both China and India shared an idea of ritual propriety and ritual behaviour which counted above all as correct comportment to the world.

During the first millennium BCE new forms of religion independently arose across civilizations that Karl Jaspers referred to as the Axial Age. Although some have been sceptical, scholars generally accept this to be a description of a shared type of event in the last millennium BCE in which traditional

about the Qualities and Essences of Stone and Wood in the Chambered Tomb Architecture of Neolithic Britain and Ireland', *Journal of Social Archaeology*, vol. 12 (1), 2011, pp. 29–50.

[47] Maura Pellegrini, John Pouncett, M. Jay et al., 'Tooth Enamel Oxygen "Isoscapes" Show a High Degree of Human Mobility in Prehistoric Britain', *Scientific Reports*, 6 (34986), 2016, pp. 1–9.

understandings of human flourishing and social order were brought into question and challenged. In pre-Axial communities there was a sense of human 'embeddedness' in social order, cosmos, and human good that came to be overturned. Charles Taylor writes:

> The surprising feature of the Axial religions, compared with what went before, what would in other words have made them hard to predict beforehand, is that they initiate a break in all three dimensions of embeddedness: social order, cosmos, human good. Not in all cases and all at once: perhaps in some ways Buddhism is the most far-reaching, because it radically undercuts the second dimension: the order of the world itself is called into question, because the wheel of rebirth means suffering. In Christianity there is something analogous: our world is disordered and must be made anew. But some post-Axial outlooks keep the sense of relation to an ordered cosmos, as we see in very different ways with Confucius and Plato; however, they mark a distinction between this and the actual highly imperfect social order, so that the close link to the cosmos through collective religious life is made problematic.[48]

Taylor expresses the essence of the matter: Axial religions and philosophies challenge the older worldviews, and we have a general shift from a ritual system in which human good is seen in terms of conformity to social norms and ritual practices to a questioning of and shift of understanding the human good in terms of ethics, and so a consequent stress on human freedom. Conformity to social and cosmic order—that we see in ancient Vedic, Chinese, Greek, and Judaic religion—is no longer enough to satisfy the human desire for life; the older traditions, the Axial thinkers thought, could no longer repair human life. New civilizational orders were required to address the need for human transformation and to address human desire for life itself. For Buddhism this desire was negatively evaluated in favour of a transcendence of life, at least life seen as a cycle of rebirth and suffering; for Vedic religion the sages of the Upaniṣads rejected the efficacy of ritual sacrifice, that the affirmation of life comes through deeply knowing that life is identical with absolute reality; in China the desire for life was positively evaluated through Confucian ethic and the flow of life expressed by Laozi; and in Judaism the desire for life was reaffirmed in response to the ethical demands of a wholly other God. The post-Axial religions work out the practical consequences of this revolution, and by the time of what we call the medieval period, they had clarified their position, often in relation to each other, and Christianity and Islam had arisen which re-evaluated the desire for life and offered new solutions for eschatological hope. Part II of this book is dedicated to showing how these traditions thematized life in relation to the living and offered solutions to the problem of

[48] Charles Taylor, 'What Was the Axial Revolution?', pp. 34–5, in Robert N. Bellah and Hans Joas (eds.), *The Axial Age and its Consequences* (Cambridge, MA: Harvard University Press, 2012), pp. 30–46.

human alienation, discontent, and longing for completion; how they addressed human desire for life.

Religion and Civilization

Religion was the beating heart of these early civilizations. There has been much debate over the category 'religion' and some scholars have argued that we cannot justifiably project onto non-Western cultures and other times a category that has so evidently developed in European languages and has been closely linked with colonialism and Western hegemony.[49] There is a strong argument that religion is a European, emic category, developed especially in relation to secularization and to John Locke's idea of separating the public from the private realms, relegating religion to the private and governance to the public.[50] This distinction is not justified in other places where religion is inseparably linked to state power. While it is, of course, true that all categories of analysis and areas of discourse arise in particular historical locations, I would argue that religion is no more a socially and politically constructed category than others such as 'politics', 'society', or 'culture'. In one sense there was no politics before the ancient Greeks, and while we can see the force of this argument in the historical specificity of the category's arising, what it refers to—the power of groups within a society, the ordering of communities, the behaviour of societies towards each other—is clearly endemic to human reality. Comparative politics, comparative law, and comparative religion show that there are patterns of human organization and history that cut across regional differences. As regards religion, I agree with the critics who say we cannot define religion in terms of belief. Long ago the genius philosopher Ludwig Wittgenstein critiqued the British intellectualist tradition that had done so. In his remarks on Frazer, he made the point that man is 'a ceremonial animal'[51] and it is quite unjustified to define religion simply as a mistaken belief; religion is primarily action, a comportment, something we simply do as humans rather than an explanation of the world,[52] an approach that was taken up by anthropologists such as Godfrey Lienhardt.[53]

[49] For this argument see, for example, Timothy Fitzgerald, *Discourse on Civility and Barbarity: A Critical History of Religion and Related Categories* (Oxford: Oxford University Press, 2007); T. Masuzawa, *The Invention of World Religions or How European Universalism Was Preserved in the Language of Pluralism* (Chicago, IL: Chicago University Press, 2005). I have engaged with this debate elsewhere: Gavin Flood, *The Importance of Religion: Meaning and Action in our Strange World* (Oxford: Wiley-Blackwell, 2012), pp. 8–16.

[50] See Paul J. Griffiths, *Problems of Religious Diversity* (Oxford: Blackwell, 2001), pp. 103–11.

[51] Ludwig Wittgenstein, *Remarks on Frazer's Golden Bough*, trans. A.C. Miles, rev. and ed. Rush Rhees (Nottingham: Brynmill Press Ltd, 1979), pp. 1e–7e.

[52] Flood, *The Importance of Religion*, pp. 14–23.

[53] Godfrey Lienhardt, *Divinity and Experience: Religion among the Dinka* (Oxford: The Clarendon Press, 1987).

There is force to Wittgenstein's claim, as what we refer to by religions are above all ways of life and ways of acting. Anticipating this approach, Emile Durkheim also criticized the intellectualists. Durkheim offered a famous definition of religion as the sacred, things set apart, and he distinguished beliefs from rites, thinking that rites depend upon beliefs.[54] Robert Bellah's adaptation is as follows:

> Religion is a system of beliefs and practices relative to the sacred that unite those who adhere to them in a moral community.[55]

This certainly captures the spirit of much of what we understand by religion, but some have argued that the emphasis on belief and text is a distortion of wider religious practice that is grounded in a materiality (materials and place), which has been neglected in the study of religions. Manuel Vasquez offers a fresh analysis of religions in terms of social actors embedded and embodied in material, social, and ecological networks.[56] Although my own focus is very much on text, Vasquez's reorientation is a welcome corrective. Other scholars argue that the word 'religion' needs to be analysed into its components as its seeming unity covers quite distinct areas of human life. A good example of the analytic approach is from the Indologist Frits Staal. Staal begins with the succinctness of Durkheim's definition but extends his account by distinguishing three categories—rites, mystical experiences, and beliefs—which are fundamentally independent of each other.[57] Rites are the most independent on Staal's account because they are rooted in our evolutionary past as ritualization occurs in non-human species. Mystical experience can be generated by practices that some religions promote such as prayer, fasting, breathing techniques, and meditation. It is only in the three monotheistic religions of the West, claims Staal, that there is an emphasis on doctrine and belief, but this is not the case outside the West where in Buddhism and Jainism there is no belief in a transcendent God, in Daoism immortality is physically conceived, and in Yoga ideas of deity are subordinated to practice and a theory of practice in which deity is unimportant. The emphasis in Asia, he claims, has rather been on ritual and mysticism.[58] Staal certainly has a point. The emphasis on religion as comprising belief has arisen within Western religions; because of colonialism, the traditions

[54] Emile Durkheim, *Elementary Forms of the Religious Life*, trans. Carol Cosman (Oxford: Oxford University Press, 2001), p. 46: 'a religion is a unified system of beliefs and practices relative to sacred things, that is to say, things set apart and surrounded by prohibitions—beliefs and practices that unite its adherents in a single moral community called a church'.

[55] Robert Bellah, *Religion in Human Evolution: From the Paleolithic to the Axial Age* (Cambridge, MA: Harvard University Press, 2011), p. 1.

[56] Manuel A. Vasquez, *More Than Belief: A Materialist Theory of Religion* (Oxford: Oxford University Press, 2011). See also Donovan O. Schaefer, *Religious Affects: Animality, Evolution, and Power* (Durham, NC and London: Duke University Press, 2015).

[57] Frits Staal, *Rules without Meaning: Ritual, Mantras, and the Human Sciences* (New York: Peter Lang, 1989), p. 388.

[58] Ibid., p. 390.

of Asia and Africa were viewed through the lens of religion, and religions such as 'Hinduism' and 'Daoism' were written about by mostly Western scholars using a model of religion that laid stress on doctrine and belief.[59] I would not wish to claim that religions do not exist or that religion is the product of the scholar's imagination,[60] but rather that what comprises religion varies across the globe and the emphasis on doctrine and language, the product of advanced linguistic consciousness, rests upon much deeper and older patterns of human modes of being, what we call ritual behaviour and, out of the ordinary states of mind, Staal's 'mystical experience'. Staal does make the controversial claim that ritual is meaningless, action for its own sake rather than for the sake of something else, which has a structure or syntax but no semantics: meaning is projected onto ritual post factum. Again, there is truth in this insofar as ritual probably arose before the advent of language, although this does not mean there is no explanation or purpose to ritual behaviour (I shall defer a discussion of the origins of religion to Chapter 2). These are complex issues made more complex by what we now know about human evolution and niche construction, but generally I concur with Staal, and indeed Bellah and Durkheim, that we cannot define religion simply in terms of belief.

To return to Bellah's definition mediating Durkheim, that beliefs and practices combine to form a moral community: ritual clearly has an important function in bonding communities together; shared, participative action contributes to the social glue that makes community cohere. The term 'sacred' is important in Bellah's definition, denoting a human experience of something outside the usual parameters of the everyday. Bellah gives a fine account of this in terms of contemporary experience. We spend much time, he observes, not in the everyday world but in an alternative reality: asleep, watching television or the movies, or playing games. These activities allow us to become absorbed in them such that the everyday world of work is momentarily switched off. Nobody, suggests Bellah, 'can stand to live in it [the everyday world] all the time'.[61] So the experience of a 'non-ordinary' or 'participative' reality is not uncommon. Bellah cites Alfred Schutz, the phenomenological sociologist, who spoke of multiple realities or different provinces of meaning, all of which, for Schutz, are socially constructed.[62] Schutz himself drew on William James for whom there are multiple realities in human experience and also on Bergson's

[59] For an account of the arising of comparative religion in a colonial context see David Chidester, *Savage Systems: Colonialism and Comparative Religion in Southern Africa* (Charlottesville, VA: University of Virginia Press, 1996). For a fine history of the academic study of religion see Guy G. Stroumsa, *A New Science: The Discovery of Religion in the Age of Reason* (Cambridge, MA: Harvard University Press, 2010). See also references in Flood, *The Importance of Religion*, pp. 30–2, note 40.

[60] Jonathan Z. Smith, *Imagining Religion, from Babylon to Jonestown* (Chicago, IL: University of Chicago Press, 1982), p. xi.

[61] Ibid., p. 3. [62] Ibid., pp. xv, 1, 3–4.

notion of 'attention to life' (*attention à la vie*), an idea that points to a tension in consciousness. There are different qualities of attention to life and the tension in consciousness of an everyday mode of being can be contrasted with a different kind of consciousness exemplified by the dream. Schutz writes:

> Activity is united with the highest tension of consciousness and manifests the strongest interest for encountering reality, while the dream is linked with complete lack of such interest and presents the lowest degree of the tension of consciousness. This interest is the fundamental regulative principle of our conscious life. It defines the province of the world that is relevant for us. It motivates us so that we merge into our present lived experiences and are directed immediately to their objects. Or it motivates us to turn our attention to our past (perhaps also our just past) lived experiences and interrogate them concerning their meaning, or rather to devote ourselves, in a corresponding attitude, to the project of future acts.[63]

Ordinary waking consciousness associated with the day-to-day world has a high degree of tension as it is focused on immediate action or directed to future activity in contrast to a dream state that is the opposite. Carl Gustav Jung observed the same thing when he distinguished between directed consciousness made with effort and effortless symbolic consciousness associated with dream and daydream.[64] Religions often cultivate non-directional thinking or rather non-rational thinking outside of language through repetition of chants, prayer, mantra, music, and the development of trance-like states in meditation or the use of drugs.

Bronkhorst has rightly observed how theories of religion have neglected to incorporate mystical states of consciousness and has drawn our attention to the importance of these states of mind. Awareness is a key part of religion and much focus on religious discourse and practice has been on awareness and in particular in developing states of absorption. Awareness, observes Bronkhorst, is variable; awareness of a chair is different to that of a table, and so on, and there is a further dimension of awareness with variable density. That is: 'The same table, at the same place and in the same light conditions, can be cognized differently depending on the density of the web of associations that feed into that awareness'.[65] So once the web of associations is reduced, we experience

[63] Alfred Schutz, *The Structure of the Life-World*, vol. 1, trans Richard M. Zaner and Tristram Engelhardt Jr (Evanston, IL: Northwestern University Press, 1973), pp. 25–6.

[64] Carl Gustav Jung, *Symbols of Transformation: An Analysis of the Prelude to a Case of Schizophrenia*, trans R.E.C. Hull (London and New York: Routledge, 1956), pp. 7–33.

[65] Johannes Bronkhorst, 'Can Religion Be Explained? The Role of Absorption in Various Religious Phenomena', p. 6, *Method and Theory in the Study of Religion*, 2016, pp. 1–30. This theory of mystical states might also be understood in terms of Giulio Tonino's Integrated Information Theory (IIT) in which the properties of a particular state of consciousness are related to causal properties of the systems in the brain. Thus an experience, such as a mystical experience, is an intrinsic property of a complex mechanism. Giulio Tonino, Melanie Boly,

the world differently; the banknote becomes just the piece of paper. Along this spectrum of awareness are mystical states that come upon people or are cultivated through prolonged practice, which reduce the associations of normal experience. Bronkhorst writes:

> Awareness of the world outside us will be *less* interpreted than it normally is, and will rightly be experienced as more direct and therefore more real than ordinary awareness.[66]

Bronkhorst, following Indian meditation traditions, refers to these mystical states as absorption, states that those who experience them claim take us closer to reality. Absorption can come through conscious effort or it can be effortless and spontaneous (when it is sometimes referred to as grace).

In these mystical states cultivated through practices of asceticism and meditation, arguably the brain closes off directed linguistic consciousness, thereby accessing deeper, pre-linguistic layers. This awareness outside of language is an intensification of life. While civilizations rely on hard work on a mass scale to reach their achievements—the cultivation of food, the development of technology—they have also cultivated states of mind that open communities to a wider experience of reality through shutting off directed, linguistic thinking. This has been important in civilizations' recognizing a limit to linguistic thinking and recognizing alternate modes of awareness. Indeed, sometimes these have been foundational, as we see with the Buddha whose enlightenment experience, outside of directional thinking, gave rise to a religion that played a central role in Indian, Chinese, and now arguably Western civilizations.

It is worth dwelling on this for a moment. If, on the one hand, some religious practices enhance a sense of life by cultivating direct access to deeper levels of awareness than linguistic consciousness, then, on the other, religions also cultivate sophisticated levels of linguistic consciousness by integrating and articulating at the level of language these deeper layers of the brain. Such articulation comes to form philosophies of life that broaden or universalize pre-linguistic social cognition in the affirmation and acceptance of the complexity of the other. That is, the social cognition system through which humans are in contact with an interconnectivity grounded in brain structure is expressed through language and broadened out to appeal to a wider audience; human pro-sociality when channelled through the linguistic brain becomes a philosophy of life that embraces a wide range of human actors. It turns strangers into brothers and sisters and becomes articulated as law. Religious law is important in that it regulates human interactions, those within

Marcello Massimini, and Kristof Koch, 'Integrated Information Theory: From Consciousness to its Physical Substrate', *Nature Reviews Neuroscience*, vol. 17 (7), 2016, pp. 450–61.

[66] Ibid., p. 7.

a community and outside of it, which means that it seeks to control the social cognition system. Religious law tends to define who is within the group and prescribe comportment towards others, especially through the control of sexuality (concern about marriage) and food (food laws and rules of commensality). Religions became a driving force operating in people's lives and ordering events of daily life as well as providing opportunity for accessing deeper, pre-linguistic levels of awareness. This is almost a Hegelian *Aufhebung*, 'elevation', the transformation of the human proclivity to pro-sociality into linguistic practices of law and social regulation.

Another way of looking at this process is in terms of desire or drive. Human pro-sociality is desire for life both as affirmation of the other and also as desire for assimilation into the other. Later philosophies of life in the nineteenth century beginning with Hegel began to give shape to the idea that subjectivity has a history, and this is a history of desire. Desire for life is a desire for fullness and completion articulated as subjectivity in relation to the other. If I might generalize, in phenomenology since Hegel the self has an orientation towards the world that later thinkers such as Husserl understood as intentionality (that consciousness has an object) and Heidegger understood as human being-in-the-world, that he called Dasein. Desire is the force that orientates the self to the world and the constant search for the satisfaction of desire is the longing for life itself that civilization seeks to express and address. Desire for life is the desire for satisfaction that is future orientated and, in the end, never finally achieved but constantly renewed. The theme of desire in human life has been central to psychoanalysis since Freud, and while this is not the discourse I situate my own study within, it is worth seeing what it has to say about the desire for life in relation to civilization. Freud's theory was fundamentally about the repression of desire in the unconscious along with resistance to the assertion of unconscious force. This twofold mechanism of repression and resistance ensures the management of unconscious desire—that for the earlier Freud was exclusively sexual—and social control that allows for a regulated society. Repression and resistance are never wholly successful because of the force of desire, which becomes expressed for Freud through the edifice of civilization on a collective level and in the individual person as the symptoms of neurosis. For the later Freud not only desire for life, which he called Eros, but also desire for death, which he called Thanatos, are forces in the unconscious that are channelled into culture. This is sublimation: the transformation of unconscious desire into cultural forms such as art. Indeed, civilization itself is the transformation or sublimation of the instincts.[67] Thus Freud has a tragic sense of life because, on the one hand, people are unhappy and suffer neurosis

[67] Sigmund Freud, *Civilization and its Discontents*, trans. Joan Riviere (London: Hogarth Press, 1930).

due to the repression of the instincts yet, on the other, we have civilization because of that very repression.[68] While Freud's concrete accounts of religion and its origins (the primal horde, Totemism, the son killing and consuming the father, and so on) are highly speculative, his observations on human motivation and the tragic nature of life are insightful. Even for Freud, then, civilization was an attempt to heal never-fulfilled and frustrated desire. Civilization is the salve that covers unhappiness due to the impossibility of fulfilling human appetite.

Without having to buy into the whole psychoanalytic account of human formation, and Freud's imaginative description of the origins of religion cannot be taken in any literal sense, there is an important insight here. Desire for life is the affirming impulse of the world of life and yet a force that is necessarily indirect, channelled, and transformed through cultural expression. Norman Brown picks up on this theme and develops an interesting angle that sees the history of civilization as a history of conflict between the transcendence of Eros in religious quest and the bodily affirmation of Eros in a transformed state that he calls 'the resurrection of the body'. For Brown the Christian vision of life transformed through resurrection, which has traditionally been understood as achieved through the transformation of the instincts in ascetic practice, is compatible with the psychoanalytic account of the whole person, both mind and bodily or material nature. Brown proposes that to move away from the victory of death, future hope lies in the abolition of repression that in Christianity is called the resurrection of the body. This body, claims Brown, is a body reconciled with death. Generally the theme of resurrection has been neglected by modern theologians (Brown is thinking specifically of Barth and Tillich), and within the Christian tradition we must take note of the mystic Jacob Boehme who laid emphasis on the resurrected body, a theme developed by the Russian existential theologian Berdyaev, and an influence felt on the Romantics Blake, Novalis, and Hegel, even through to Freud.[69] Psychoanalysis has much in common with this mystical tradition—but it is not a mysticism of flight, which Brown calls Apollonian, but a Dionysian affirmation of the body that we find in Boehme and that must accept death rather than flee from it. This acceptance of death is recognition that death is not nothing, but a dialectical force unified with life; the resurrected body for Brown is a symbol of such integration of the desire for life with the reality of death, for only by doing so is there hope for a bright future. In Brown's analysis of the modern world an increased Eros, the desire for life, is needed to counterbalance the force of death that is also the affirmation of the

[68] Peter Gay, *Freud: A Life for our Times* (London: J.M. Dent and Sons Ltd, 1988), p. 548.

[69] Norman O. Brown, *Life against Death: The Psychoanalytical Meaning of History* (London: Routledge and Kegan Paul, 1959), pp. 309–10.

importance of play[70] in spontaneous, directionless, and joyful affirmation of life. Bellah too has highlighted play as important in human development, a kind of 'offline' activity in contrast to 'online' pressure: 'Online is the world of fighting, fleeing, procreating, and the other things that all creatures must do to survive. Offline is when those pressures are off and there are other things at work'.[71] Like sleep and other altered states of consciousness, play is important, although little understood, and part of the panoply of behaviours mediated through culture that wholly or partly bypasses linguistic consciousness and directed thinking.

Let us take stock for a moment. If civilizations are collective endeavours sustained over long periods of time that articulate human aspiration and assert desire for and affirmation of life, then religions (even if they can be analysed into component parts) give form to such aspiration and affirmation. Religions have created a certain homogeneity across diverse communities, creating ideals and patterns of behaviour, especially comportment towards others, that have been fundamental to the formation of civilizations. So far, I have presented a view of civilization as a product of coevolution, driven by religion, that expresses human affirmation of and desire for life and which provides a niche for human development. Furthermore, this niche can facilitate human transformation in two ways: through developing practices of life that bypass directional, linguistic consciousness in non-directional, non-linguistic states of mind (absorption) and through developing the philosophy of life that gives expression to human desire for life and thereby to human freedom. At the level of everyday activity, religions pattern life but they also operate at a higher level beyond the individual or community as systems; they operate through big history over long periods of time.

It is their sustained persistence through extended passages of time that links religions to civilizations. Because of religions, civilizations are teleological and eschatological in providing a structure or niche within which human beings as individuals and social groups can flourish. They provide a niche, a structure coterminous with language, religion, social structure, and state formation within which human communities express eschatological hope and seek happiness and a good life, free from hardship and alienation that separation from our social cognition has brought. I would not wish to reduce human longing or civilizational achievement simply to a shift in brain orientation from the deep brain to the neo-cortex, but this is probably a corollary to civilizational achievement. Civilization as a niche for human development works independently through the generations and affects the biology of who we are. We might see civilizations as niche constructions that cultivate values expressing the ideals, goals, and longing of a community sustained over

[70] Ibid., p. 318. [71] Bellah, *Religion and Human Evolution*, p. xx.

generations. Through civilizations we try to become who we are. Thus, civilizations are not simply expressive; they are formative too.

Life Itself, the World of Life, and Transformation

In the course of this book I will often use the terms 'life itself' and 'the world of life'. By 'life itself' I mean the idea of a principle that animates the living or a force that underlies the world. Such an idea is found across civilizations and has been expressed by a number of terms. In Indic civilization we have the Sanskrit neuter noun *jīvana* (nominative singular *jīvanam*) and its instantiation in the particular living being as the masculine noun *jīva* (nominative singular *jīvaḥ*), usually translated as 'self' or 'soul'. These nouns are derived from the verbal root *jīv*, 'to live'. A related concept is the idea of subtle energy or breath that pervades the living body, the masculine noun *prāṇa* (nominative singular *prāṇaḥ*), sometimes identified with the power or energy of God, and the feminine noun *śakti* (nominative singular *śaktiḥ*). In China a similar idea occurs of an animating subtle power in the body known as *qi*, a quality that can be intensified and developed through a range of practices. In ancient Greece we have a similar idea in the neuter noun *pneuma*, whose primary designation is breath but whose semantic range includes subtle energy that pervades the body and in this is closely related to the feminine noun *psuchē*, usually translated as 'soul' but that, like *jīva*, can legitimately be rendered into English as 'life force'. This is true also for the Latin feminine noun *anima* that translates *psuche*. Furthermore, in Greek we have the feminine noun *zoē* and the masculine *bios*, used for example by Aristotle to designate the life of the living. *Pneuma* translates the Hebrew feminine noun *ruah*, the breath that God breathes into living beings. Life itself can be a rendering of these terms although analysis reveals differences in the semantic range of each. Accompanying these terms is the idea of degrees of intensity of life that fills the living. In China, as we will see, *qi* is a force of variable quality, and in ancient Greece, Homer's heroes are filled with a force of energy from the gods[72] that can become depleted.

This idea of life itself has been thematized to varying degrees in philosophy and religion, as we will see in Part II of this book. Life itself also survives in secular philosophy as vitalism, the idea of an animating principle underlying the living that in modern times has been contrasted and opposed to mechanism, with which we can account for the living in mechanistic rather than vitalistic terms. The twentieth-century French philosopher Henri Bergson expresses a strong vitalist philosophy that is taken up in more recent

[72] E.R. Dodds, *Greeks and the Irrational* (Berkeley, CA: University of California Press, 1951), pp. 8–10.

philosophical interest such as in the work of Gilles Deleuze and others, including Luce Irigaray, whose work on the concept of breath as the source of ethics and human connectivity draws on Indic traditions.[73]

These philosophies of life, both religious and secular, have maintained a view that life itself animates the living. The living comprises the world of life, the world of living and dying, coming to form and passing away. The human world of life was called the 'lifeworld' (*Lebenswelt*) by phenomenologists such as Husserl and indicated the world that forms human reality, the world in which we have our being.[74] The lifeworld is thus the human way in which the world of life is experienced, and so the distinction might approximately map onto that between nature and culture. Much of this book will be concerned with the lifeworld in that here, in the human world of living and dying, we see the connection between life itself and human living, a relation that the early philosophy of Heidegger was concerned with but that he later rendered as the relation between Being and beings. Yet the book also wishes to maintain the relevance of truths that the world of science brings.

If I might generalize, the human world of life, the lifeworld, is characterized by desire for life that is simultaneously a desire for fullness, completion, meaning, and the articulation of human freedom. The relationship between the lifeworld and the world that science reveals is a difficult issue, but it is science that is rooted in the lifeworld, and although the scientific paradigm is so powerful because of the success of its accounts of reality, the irreducibility of the lifeworld is an equally important human truth. Life itself is thus accessed through the lifeworld, through living, as well as through the sciences of life. This lifeworld in contrast to the sciences of life roughly maps onto a distinction between 'living' and 'life'. Living is the way someone is in the world and resists objectification or reduction, whereas life is a category that can be apprehended and discussed almost as an object. François Jullien characterizes the distinction as life being understood as a state of representation and objectification, in contrast to living which does not 'authorize exteriority; upon it no backward step is possible'.[75] Living is something we enter into and which resists direct explication—perhaps only through poetry or literature does it come into view—in contrast to life that can be understood and thematized. Living presupposes life and if we wish to understand what human living means, we need to understand conceptualizations of life through the

[73] Luce Irigaray, 'The Age of Breath', Key Writings (London: Continuum, 2004), pp. 165–70; Luce Irigaray, *Between East and West: From Singularity to Community*, trans. Stephen Pluhácek (New York: Columbia University Press, 2003); also see Lenart Skof, *Breath of Proximity: Intersubjectivity, Ethics, and Peace* (Dordrecht: Springer, 2015).

[74] On the origins of the term see Claude Romano, *The Heart of Reason*, trans. Michael B. Smith and Claude Romano (Evanston, IL: Northwestern University Press, 2015), p. 505.

[75] François Jullien, *The Philosophy of Living*, trans. Michael Richardson and Kryzysztof Fijalkowski (London, New York, Calcutta: Seagull Books, 2016), p. 161.

history of civilizations and the explanatory paradigms of life in the harder sciences.

The problem is well articulated by Claude Romano. How, he asks, 'are we to conceive of the relation between the truths of science and the certainties of life?'[76] There have been two solutions, Romano says: one is to ascribe truth to science and confine the lifeworld to *doxa* and the other is to privilege human experience as the true world. But these are not in competition and while the truth of science—in this book, evolutionary anthropology in particular—offers causal explanations, the meanings of life cannot be simply subsumed under this because such an account does not adequately account for the 'data' of life; the texture of living and dying. As I hope to illustrate throughout this book, to understand life we do not have to choose between Proust or Darwin but can work with both humanist and evolutionary paradigms.

Across the civilizations we will encounter in the coming pages, this issue of life itself is recognized in religions that attempt to address it through philosophies and theologies that try to explain it and through practices that seek to transform it. Human desire for life is recognition that life is generally unfulfilled and characterized by existential concern about what life is; for concern about meaning in the face of my annihilation in death. This is a concern that humans have as language-bearing primates; we are the kind of creatures, as Heidegger says, for whom our existence is an issue. We might express this desire for life as an alienation from life itself, an alienation in need of transformation. This is, of course, a metaphor, but a metaphor that has force in the history of religions that offer solutions to human alienation that we can interpret as alienation from life itself. Religions seek to transform the human condition by bringing us closer to the source of life through practices that seek to bypass linguistic consciousness and through philosophies that seek to integrate human reality with a deeper reality of life. It is principally through religions that civilizations have sought human transformation, but art, music, and literature might also be seen in these terms too, although religions differ from these other cultural forms in offering and prescribing practices and ways of living that art, music, and literature cannot do. There is clearly some sense of redemption in, say, Beethoven's A-minor quartet but it cannot constitute a way of life except, perhaps, for those who play it. Religions address the whole of human life in their concern with comportment to others, with alimentation and sexuality, with death, and with practices of prayer and ritual that transform the desire for life in preparation for death.

[76] Romano, *The Heart of Reason*, p. 519.

Human Freedom

One last theme that will be intertwined through the following pages remains to be put in place, namely human freedom entailed by desire for life. Human freedom as the ability to articulate desire has been set against biological impulse. Kojève, for example, regards freedom as the transcendence of biology, and biological life cannot constitute the meaning of human life.[77] But if freedom is integral to human life, we might ask whether there are general features of an anthropology (in the broader sense) entailed by coevolution. If coevolution has produced the human being as its most complex event (the human brain being the most complex entity that scientists have knowledge of), then this means that human life is coterminous with freedom because freedom must be within the paradigm of evolution. Freedom is thus connected with the vitality and fullness of life in the sense that freedom is creativity, which itself is linked to life overflowing with possibility.[78] We might even speak of degrees of freedom depending upon the complexity of organisms, with less complex organisms being therefore more restricted than more complex entities: the greater the degree of constraint, the greater the freedom; the greater the amount of information encoded within the organism, the greater its ability to responded unpredictably. The biosemiotician Jesper Hoffmeyer observes that while human free choice is different to that of the amoeba, nevertheless even the amoeba, he claims, is a subject in a restricted sense as 'a temporal being capable of distinguishing and acting upon selective features of its surroundings and participating in the evolutionary incorporation of the present into the future'.[79] This is because the contribution of organisms to the future is semiotic, the transfer of information (such as DNA that encodes a digital description of the organism), and so the degree of freedom means the depth of meaning a creature is capable of communicating. Such a process culminated in the human as a being within an environment or Umwelt that contained a conception of itself as a builder of that environment.[80]

[77] A. Kojève, *Introduction to the Reading of Hegel*, trans. James H. Nichols (Ithaca, NY: Cornell University Press, 1980), p. 4. Also see the discussion by Judith Butler, *Subjects of Desire: Hegelian Reflections in Twentieth-Century France* (New York: Columbia University Press, 1987), pp. 66–7.

[78] In a critique of neoliberalism, William E. Connolly places freedom as an aspiration to vitality in contrast to pursuit of a failed transcendence or a fullness, *The Fragility of Things: Self-Organizing Processes, Neoliberal Fantasies, and Democratic Activism* (Durham, NC: Duke University Press, 2013), pp. 77–80. But in the histories of religions arguably vitality, fullness, and transcendence are not incompatible goals and are linked.

[79] Jesper Hoffmeyer, 'Some Semiotic Aspects of the Psycho-Physical Relation: The Endo-Exosemiotic Boundary', p. 103, in Thomas A. Sebeok and Jean Umiker-Sebeok (eds.), *Biosemiotics: The Semiotic Web 1991* (Berlin and New York: Mouton de Gruyter, 1992), pp. 101–23.

[80] Ibid., p. 111.

Hoffmeyer has a point. Freedom must be contained within the evolutionary paradigm, although in the human case it is fundamental to the development of language and so to subjectivity, and so we might speak of it as the realization of potential implicit within earlier organic forms, realized to a degree unrecognizable in relation to earlier creatures. Freedom I take along Kantian lines to be independent of determination as rational agency. The difficulty with the Kantian position is that on the one hand, if freedom maps on to his timeless noumenal realm, the realm of the thing in itself outside of nature, then it can have no effects in the world, yet if freedom has effects in the world it cannot be within a timeless noumenal realm.[81] The point of evoking this Kantian problematic is that one way Kant attempts to circumvent the issue is that freedom can be viewed from both the noumenal and phenomenal perspectives. While I do not propose to offer a solution to the dilemma, I suggest that, in a similar way, we might say that human desire for life characterized by freedom as rational agency, as well as naturalistic causal determinism, can be viewed within two frames: life as genetic inheritance and freedom as independence of the human person. As we will see, descriptions of the human person in first-person terms in contrast to third-person terms are a feature of an approach that attempts to integrate both humanities and hard science. On this view, the human being is a 'compound hypostasis',[82] a single event comprising first and third-person accounts. Such a view entails the idea of the world from within subjectivity along with a view of the world from the perspective of nature, from material causation. The general anthropology implied by the coevolution model is thus one of complexity; that the singular-event character of human life can be framed both by the reality of rational agency or freedom and also by genetic inheritance. The single event that is human life can be perceived within both registers.

Framing the singular-event character of human life in terms of freedom entails the recognition of complexity and human response to it, as an act of will and as a rational agent. Framing human life in terms of freedom also entails self-awareness and the possession of values or that there are significances in life; in Taylor's terms, the strong recognition of goods which are 'intrinsically worthy'.[83] We might even say that complexity developed through

[81] The issue is articulated well by Henry E. Allison, *Kant's Theory of Freedom* (Cambridge: Cambridge University Press, 1990), p. 2: 'Either freedom is located in some timeless noumenal realm, in which case it may be reconciled with the causality of nature, but only at the cost of making the concept both virtually unintelligible and irrelevant to the understanding of human agency, or, alternatively, freedom is thought to make a difference in the world, in which case both the notion of its timeless, noumenal status and the unrestricted scope within nature of the causal principle must be abandoned'.

[82] I take this phrase from Rowan Williams, lecture in Oxford, 5 March 2018.

[83] Charles Taylor, 'The Person', p. 266, in Michael Carrithers, Steven Lukes, and Steven Collins (eds.), *The Category of the Person* (Cambridge: Cambridge University Press, 1985), pp. 257–81.

coevolution generates freedom in the sense that the greater the complexity, as in the human brain, the greater the degree of freedom. Rational agency entails a high degree of constraint both internally within the brain/body and externally within natural as well as social and political conditions. In the contemporary, global world, human freedom might entail a rights discourse, although the frame of freedom is not contingent upon external political conditions. This has especially been the case with religions. For example, the voluntary restriction of asceticism implies a greater intensity of subjectivity and so a deeper understanding of human nature as freedom. Within the joint frame of evolutionary anthropology and social cognition, this freedom is the way in which persons become fine-tuned to the social cognition system. That is, asceticism, meditation, and prayer, archetypal practices of Indic, Chinese, and Abrahamic religious traditions, are ways of accessing social cognition beyond language and fine-tuning persons to deeper resonances, although mediated through language in theologies and using language in prayer and mantra to transform the person. Such an anthropology is not the personalism that modernity has produced but rather a particularism that recognizes the unique event of personhood as fundamentally cosmological and as such brought into the fullness of life. On the one hand, civilization is the frame for human freedom, while on the other evolution is the frame for an anthropology rooted in human biological nature. But the person as characterized by freedom cannot itself be simply reduced to the vital, as Taylor considers sociobiology,[84] but civilization and biology/evolution are two frames or descriptions under which a single event of human life can be considered.

VARIABLE NARRATIVES

In the coming pages I will principally address the religions that drive civilizations characterized by a unifying language and literature (Sanskrit, Chinese, Greek, Latin, Arabic, and Hebrew). Cutting across the regional divisions of India, China, Europe, and the Middle East, there are different ways in which religions have philosophized the need for human transformation that also philosophize the relationship between life itself and the living. Although my coming text will organize civilization principally in terms of geography and language, I do think there are common narratives that address the question of life itself, human desire for life, and the transformation of human life entailed. We might call this a typology of narratives of life that we can abstract from their historical specificity and that are shared by different religions. Three of

[84] Taylor, 'The Person', p. 270.

these are ancient—the Gnostic, the non-dualist, and the theistic—while the secular narrative is modern.

1. The Gnostic narrative. In this account human longing is a desire for return to a home from which we have been exiled. On this view desire for life is really a desire to leave life as experienced in the world to a finer life in a non-material realm beyond the world. Plato articulated such a view, as did Plotinus, and it influences the Middle Eastern religions of Judaism and Christianity. Odysseus' journey home is the soul's journey back to its place of origin. This kind of dualism, which Hans Jonas so expertly described, has a generally negative evaluation of the world of life ordinarily experienced, which the Gnostics regard as our prison. Dualism between a spiritual substance that is the soul and a material substance that is the body is found not only in Gnosticism but also in Indic traditions; as we will see, particularly Saṃkhya dualism between self (*puruṣa*) and matter (*prakṛti*) and also in Jainism where the soul's true purpose is freedom from the material constraints that cover and pull it down. Theravāda Buddhism is really a Gnostic narrative too, although without the idea of a permanent self; nevertheless, the goal of *nibbāna* is freedom from the cycle of life as suffering and known as the further shore. Generally, in the Gnostic narrative there is a moral evaluation overlaid onto this dualism such that the spiritual self, the immortal soul, is positively evaluated in contrast to matter negatively evaluated or even seen as inherently evil. There is a spectrum of views here, with Marcion's Christian Gnosticism regarding matter itself as evil to Buddhist views of material life not so much as evil but simply as suffering due to causes and conditions. The Jains, almost paradoxically, have a positive attitude towards life in a deep reverence of living beings, even though the ultimate purpose is to transcend life—and to exit life with grace and dignity in a final, ascetic, ritual death.
2. The non-dualist narrative. In contrast to the negative evaluation of materiality and the desire for life, we have philosophical positions that reject a distinction between a spiritual essence and a material body. Within this there is a range of views that can be described as idealist (the claim that the one reality is consciousness), as materialist (the claim that the one reality is matter), or as emanationist (the claim that material reality is an emanation or transformation of a spiritual substance). Life itself is not distinct from the living and from the material world the living inhabit, as we will encounter for example in Abhinavagupta's non-dualism. Chinese Daoism is close to this in maintaining the flow of life that is essentially a single river, and even the French philosopher Deleuze's univocity of being, echoing Duns Scotus, is a form of non-dualism in which desire for life becomes life's reflexivity.

3. The theistic narrative. This has been the dominant story of the Abrahamic traditions in which life itself has positive moral value because created by a transcendent deity who is invested in it. The God of the Hebrew Bible, the New Testament, and the God of the Qur'an has investment in the world he has created and human desire for life is desire for redemption and healing. The Abrahamic traditions thematize the idea of error, that something went wrong in the created order, in Christianity because of human sin or in Jewish mysticism because creation became shattered and humanity seeks the redemption of life: a reconciliation between the human and divine. The story of human life is thus one of fall that generates the need for repair, a journey through history, and final redemption. But also, in India we have the development of a theistic narrative with the same themes of worship and grace, of salvation from the world of life seen as a cycle of suffering, to a world of grace in which the life force (*jīva*) is reconciled with the Lord (*īśvara*) as the underlying power of the universe, yet who is also beyond it. In contrast to the Gnostic narrative, the theistic story generally has a positive evaluation of the world of life as a reflection of God—even as the body of God for Rāmānuja—and often has a political agenda of world completion or the formation of a sanctified, redeemed world order. The main differences between the theistic narratives of India and the Abrahamic religions is that for the former there is no end or beginning to the 'created' order, no telos other than the salvation of souls, in contrast to a cosmos that has a beginning and an end and in which there is a collective purpose, albeit an inscrutable one known only to God.
4. The secular narrative. I am not sure that we can speak of a single secular narrative of life, but generally the secular world that is the official policy of many national governments, partly in the desire for religious toleration, has a positive evaluation of life. This positive evaluation has not always been benign and biopolitics developed, arguably from the seventeenth century, in which the control of life itself comes to be the purpose of politics, resulting in the dark vitalism of the twentieth century that turns desire for life into desire for death. But the positive, secular evaluation has also produced a human rights discourse. Indeed, the kind of world order now emerging is rooted in a secular worldview. In particular this emergent world order arguably has the idea of human rights as a foundation, a positive evaluation of the world of life and recognition of human desire for life and the legitimate pursuit of human goals such as material prosperity, the right to family life, and the right to good health. This narrative—or series of narratives that Taylor has shown—draws on a positive evaluation of life and on values of knowledge derived from the development of science. An earlier Darwinian perspective that viewed life as bloody conflict and the survival of the fittest driven by the genes has

been replaced by the coevolution view of niche construction in which cooperation has dividends for life. But the secular narrative is not unchallenged, and evolution is not incompatible with religious worldviews. We must not think of an inexorable march to the erosion and final elimination of religion from the world stage. Although traditional religions in the 'North Atlantic'[85] may become eroded and some may even disappear, statistical studies show that religions are not in decline on a global scale.[86] The secular narrative of life is dominant in the North Atlantic, China, and probably South-East Asia, but religions are also becoming important on the world stage, especially global Islam, Hindu fundamentalism, and Christian evangelicalism. Furthermore, there is a burgeoning of what Paul Heelas has called 'spiritualities of life' that resist explanation in purely secularist, political terms as a kind of epiphenomenon of capitalism.[87]

It could be that these different accounts that come to full articulation in Axial and post-Axial eras of civilizations derive ultimately from a common human source. Michael Witzel has argued that all the major myths of the world can be traced back to two fundamental narratives that originated with *Homo sapiens* in Africa: one he identifies as Laurasian which came to dominate Europe and North Asia; the other he calls Gondwana which dominates Africa and Polynesia, and even these are derived from an earlier single source. Witzel makes a case for the Laurasian complex going back to the later Palaeolithic some 40,000 years ago on the grounds that certain cave paintings reflect this system. This is a very early date for advanced language and Witzel claims that the Gondwana system is older as a small group of people ventured out of Africa to travel down the coastline of the Indian Ocean some 65,000 years ago. But comparative linguistics shows common features in what linguists call Nostratic and Proto-Nostratic at about 12,000 BP.[88] I cannot critically assess Witzel's work here—we shall have call to return to it in Chapter 2—but it seems highly likely that human narratives could derive from the early hominins who came out of Africa with the emergence of advanced linguistic consciousness.

This story of the human understanding of life is, then, a potentially huge project, but we can gain leverage on the issue through the specification of the question of life and its articulation through the history of civilizations.

[85] Charles Taylor, *A Secular Age*, p. 1.

[86] Stanley D. Brunn (ed.), *The Changing World Religion Map: Sacred Places, Identities, Practices, and Politics* (New York: Springer, 2015).

[87] Paul Heelas, *Spiritualities of Life: New Age Romanticism and Consumptive Capitalism* (Oxford: Wiley-Blackwell, 2009).

[88] E.J. Michael Witzel, *The Origins of the World's Mythologies* (Oxford: Oxford University Press, 2012), p. 60.

The question of what life itself is and how it relates to the living is inseparable from the question of human transformation, because to account for how civilizations have addressed the philosophical question, we must account for how they have addressed the eschatological question. 'What is life?' is both addressed directly in philosophies that have emerged throughout history and has been inseparable from how we can set things right. Even pure philosophy outside of religion in raising the question of life has at the same time raised the question of how we should live: there has been an ethical component, and we must see ancient philosophy not simply as argument and reflection but also as a way of life, as Pierre Hadot has shown. The question of human transformation is eschatological in the sense that it implies a change in both the human person and community. That human reality is in need of repair is part of the narrative of all human cultures, rooted in our biology and the development of higher brain functions, but, for all that, not losing its existential force. Such repair primarily occurs through ritual (such as sacrifice), which through the history of civilizations has controlled our interactions and which has sought repair in the recognition that human order is part of a wider cosmos, an overarching world order, as Rappaport's insightful account of ritual shows.[89] Ritual efficacy is the way in which human life can be restored to that world order.[90]

In presenting these civilizations that have given rise to life-affirming philosophies and practices over long periods of time, I have adopted a hermeneutical phenomenology; phenomenological in that I let traditions speak for themselves, creating the conditions for what shows itself to appear, and hermeneutical in that I interpret such appearance by drawing on contemporary theory from evolutionary anthropology and social cognition to present an explanation of those appearances. In asking 'how do I narrate these civilizations?' I have come to focus on religions as the heart of civilizations and consequent secular philosophies of life. In practice this means taking philosophical readings from the histories of civilizations and reading them in the light of the question of life and so the question of transformation. This book tells the story of civilizations' accounts of what it is to be alive. Telling such a story entails a kind of empathy that reveals my own desire for life to be identical with everyone else's. It is this pre-understanding that precedes my fuller, theoretical explanation that draws on the panoply of scholarly methods, particularly philology, along with developments in evolutionary anthropology and social cognition. This methodological reflection is another way of saying that the story I have told needs to be re-narrated in a different register such

[89] Roy Rappaport, *Ritual and Religion in the Making of Humanity* (Cambridge: Cambridge University Press, 1999).

[90] Charles Taylor, *The Language Animal: The Full Shape of the Human Linguistic Capacity* (Cambridge, MA: Belknap Press, 2016), p. 274.

that the story of civilizations becomes a story within evolutionary anthropology, social cognition, and hermeneutical phenomenology.[91] The narrative of civilization is different from the narrative of evolution, but I hope to have shown how both are intertwined. Furthermore, the historical narrative becomes a present-day constructive one. A third story could have been told from a theological perspective, Clooney's 'deep learning across religious borders,'[92] but that would be a different enterprise to this. Because the topic is so vast, in one sense this narrative can only be my narrative, but a narrative that I hope is convincing because its power partly derives from an evolutionary story that seeks to theorize life through its narration. In a sense we cannot truly or completely theorize life as we can only partially stand outside of it, and we can only do our best to tell true stories.

§

Part I of the book, 'The Theory of Life Itself', deals with fundamental questions and assumptions: the current state of knowledge concerning life itself, the philosophical issues in that understanding, and how we can explain human development within an evolutionary perspective. I also address the question of the emergence of religion and present a related study of sacrifice as fundamental to religions' views about life and its transformation. Part II, 'Religious Civilizations', offers a reading of religions in three civilizational blocks, India, China, and Europe/the Middle East, particularly as they came to formation in the medieval period when the traditions have defined boundaries and the textual authorities have been established. I take up the idea of a life force in Part III, 'Philosophies of Life Itself', traced through to modern times. The final chapter goes back to the beginning and offers more detailed explanation in terms of evolution and social cognition and how these relate to understanding religions as systems. An Epilogue offers contemporary reflection on life itself, self-repair, and the future, especially in relation to biotechnology.

[91] Beyond the scope of the present volume, I regard phenomenology as inherently hermeneutical because of its historical location and because description and narration are necessarily in language. This does not mean that there is no pre-linguistic experience, or that such experience is excluded from philosophical inquiry. Phenomenological hermeneutics entails that interpretation can lay legitimate claim to truth (as in this project), although this will always be conditional and fallible, of course. See Claude Romano, *At the Heart of Reason*, trans. Michael B. Smith and Claude Romano (Evanston, IL: Northwestern University Press, 2015), pp. 485–503.

[92] Francis X. Clooney SJ, *Comparative Theology: Deep Learning across Religious Borders* (Oxford: Wiley-Blackwell, 2010).

Part I

The Theory of Life Itself

1

The Theory and Philosophy of Life

What is life and how can it be understood or explained? These are questions that have been asked in every human society and that have come to systematic articulation in literate civilizations over several thousand years. What we have come to call religions give accounts of life, its origin and its goal, along with prescribing ways of living that are thought to intensify, prolong, or transform it. Civilizations developed empirically focused portrayals of life—now identified as science—that sought to explain it and technologies to enhance it. Bringing together the categories 'life', 'religion', and 'civilization' is an ambitious project but one that is necessary if we are to explain how life itself has been understood along with the linked idea of the kind of life that makes us who we are, the kind of human beings that we have become in global modernity. I want to explore how the notion of life has been conceptualized in the history of religions that have driven civilizations and through that history has shaped our modern world. Furthermore, I propose that we can offer an account of civilizations in terms of life through explicating its place in religion. Life has animated religion and religion has tried to understand and express life. Answers to the question, 'what is life?' have been offered throughout human history, in earlier centuries through philosophies and religions and more recently through science, particularly the biological sciences, each attempting to identify a metaphysical or physical cause.

The idea that there is an animating principle, a life force, that drives the living, that life itself comes to form through the manifold appearances of the world, is very ancient. It can be found in the Greek ideas of *pneuma* and *psuchē*, in the Chinese idea of subtle energy or *qi*, and in the Indic idea of subtle breath (*prāṇa*) and power (*śakti*). I shall examine these concepts in some depth in the coming pages. We also have more recent philosophical arguments that have understood life in terms of a vital principle or essence alongside down-to-earth philosophies that have sought to understand life in the concrete terms of the biological sciences. Philosophies rooted in biology have tended to be sceptical of vitalist philosophies, while vitalist philosophies have rejected eliminative, materialist explanations. And scepticism towards both kinds of reductionism has been found in philosophies of life that have stressed existence and human living, the more phenomenological and existential philosophies of the last 150 years or so. Are we to understand life primarily in terms of human purposes, desires, fears, and hopes, or are we to explain life primarily in terms of impersonal, biological drives?

To begin to frame this question, we might speak in terms of two models: the Galilean Mathematical Model (GMM) and the Kantian Humanist Model (KHM). The former model emphasizes the scientific understanding of life in which life is measurable, structured by mathematical rules—predictions can be made about it, and above all it is explainable without reference to any supernatural order. The GMM has had immense influence on a scientific worldview that seeks explanation as the location of a cause, to which philosophy responds—and we see this developed particularly in eighteenth and nineteenth-century philosophies. Galileo had laid stress on mathematics as an explication of the structure of nature, a view that impacted both on philosophy, as we see with Descartes, and on science. The chemist and physicist Robert Boyle (1627–91) paid attention to controlled experiment and eschewed general hypotheses about nature, and Isaac Newton (1643–1727) laid the foundations of contemporary science, although there were blind alleys such as the Cambridge Platonist Henry More's explanation of gravity in terms of a world spirit. The new science became integrated into philosophy just as today contemporary philosophy responds to the most recent science.

But philosophy nevertheless tried to understand life at a structurally higher level of human experience. In his later lectures on anthropology, Kant exemplifies this approach. What is important in the human case is not mathematical explanation but giving accounts of human life in terms of motives, culture, and language, a theme taken up with some vehemence in the twentieth century by Husserl in his last published work in which he considered the Galilean project of mathematicization to be a fundamental error in the history of science, leading to distortions and misunderstandings in reducing human meaning to a quantifiable entity that is simply incompatible with an appropriate account of human life that takes seriously life lived at that level. Such distortion, thought Husserl, leads to wrong science and, by implication, a moral and political distortion.[1]

One of the first to be aware of a distinction between the natural and human sciences was Kant in his lectures on anthropology, the study of the human being, which he taught alongside physical geography, the study of nature. Published as *Anthropology from a Pragmatic Point of View*, Kant's lectures open by observing that knowledge of the human being can begin from either a physiological or a pragmatic perspective, the former being concerned with

[1] Edmund Husserl, *Crisis in European Sciences and Transcendental Phenomenology*, trans. David Carr (Evanston, IL: Northwestern University Press, 1970).

what nature makes of the human being, the latter with what the human person 'as a free-acting being' makes of himself.[2] Self-understanding that acknowledges human freedom is distinct from natural understanding that does not. The pragmatic perspective for Kant recognizes the embodied nature of the human being and the unity of mind and body in the person when he says, 'I as a thinking being am one and the same subject with myself as a sensing being'.[3] This unity of the person, their embodied and situated nature in space and time, is the basis upon which even scientific judgements are made.

We need therefore to offer non-reductionist accounts of human life that set it in contrast to broader nature, but both the GMM and the KHM react against an earlier model in which life can only be understood in terms of a grander, supernatural order, which is also an ethical order that prescribes the highest human good and the best ways of living. In such a scheme, life forms were graded in a way based on different principles to the GMM, principles in which lower levels of a hierarchical cosmic system were generated from higher levels, and in which human life was seen to be situated within a cosmic order that comprised other, mostly invisible supernatural beings such as angels and demons, and, in Christianity, a transcendent God as Trinity. We might call this the Religious Cosmic Model (RCM) that we find, although with significant variation, in all pre-modern religions and that forms the structure of civilizations. The RCM has provided a justification for polities with a king or emperor understood in divine or semi-divine terms, a clearly demarcated, hierarchical social system, and a worldview in which the social order is integrated with a natural order. C.S. Lewis articulated the principles of such a worldview up to the Renaissance in his *The Discarded Image*, where authors in the history of Western thinking up to the Renaissance reflected on and refined a grand cosmic schema. The model is fundamental to India and a less ornate, but nevertheless hierarchical, structure is found in China too.

The major civilizations covered in this book exemplify the RCM. Generally, we can say that the RCM comes to be replaced by the GMM with the rise of science and the development of new knowledge, with the KHM emerging to highlight the distinctive nature of human life and picking up on its moral uniqueness that ultimately derives from the RCM. A complex historical picture emerges within these orientations.

This debate within which the nebulous category of life has been understood (the religious-cosmic, the scientific, and the human or humanist) indicates the boundaries within which this study is enframed. But although the book

[2] Immanuel Kant, *Anthropology from a Pragmatic Point of View*, 7.19, in Günther Zöller and Robert B. Louden (eds.), *Anthropology, History, and Education* (Cambridge: Cambridge University Press, 2007), pp. 231–429. On freedom in Kant see Henry E. Allison, *Kant's Theory of Freedom* (Cambridge: Cambridge University Press, 1990).

[3] Kant, *Anthropology*, 7.142, p. 254.

presents an account of life as seen in the history of religions and civilizations, it also wishes to develop an explanation of that history in terms of the category 'life'. To do this, we will need conceptual tools that on the one hand give us leverage on what we mean by 'life', and on the other what we mean by 'religion' and 'civilization'. These are vague terms, which is both a strength and a weakness; a strength insofar as precise definitions do not necessarily give us explanations and may be too restrictive, especially at the beginning of an enterprise, and a weakness insofar as vague categories can be so open to interpretation as to disenable explanatory power. So, I need to tell the story of life as seen by religions in their histories and I need to tell the story of religions seen through the lens of the category life; life, or precise ways in which life comes to expression in human communities, animates religions, and is a fundamental level of explanation. Explicating a history of religions, often a perfervid history, and thereby a history of civilizations in terms of life, will comprise Part II, the bulk of the book, although the goal is not simply to tell a story but to show the pathways by which we have arrived at our modern understanding of life itself, especially human life, our life.

The story of human life and the idea of life itself have been intimately connected to the history of religions, which have responded to life in different ways; on the one hand affirming positive value to life as the expression of a higher or divine order, while on the other offering a negative evaluation of life through claiming a greater value to its transcendence. But this has not been a history of choosing either the affirmation or negation of life, as both values have coexisted and have been held together within the imaginaire of all major religions. Alongside the RCM, throughout history there have been quasi or proto-scientific categorizations of life as we find in ancient Greece (with Aristotle and post-Aristotelian thinking), as well as in ancient India (with Buddhist and Vaiśeṣika categories) and in ancient China (with Mencius and later traditions of alchemy). These resist any easy distinction between science and religion. As we will see, we can identify two general strategies for defining life: one in terms of verticality or hierarchy, a chain of being, to use Lovejoy's phrase, with 'higher' forms above 'lower' ones as we find in the RCM; and the other to see life as immanence, as an all-pervasive force equally present in all forms.[4] These two strategies, the vertical and horizontal, are not incompatible and traditions of thought and practice have contained both. Indeed, the

[4] This way of looking at the history of the category life is close to Thacker's identification of two currents of thinking about the ontology of life—post-Aristotelian Scholastic philosophy and life defined by negation, the idea that life can be defined only through what it is not. Although I formulated the two strategies of hierarchy and immanence some time ago (see Gavin Flood, *Consciousness Embodied: Body and Cosmology in Kashmir Śaivism* (San Francisco, CA: Mellen Press, 1993), pp. 85–94, 246–8), Thacker is clearly speaking about the same modes of thought and his book has influenced my own thinking on these matters. Eugene Thacker, *After Life* (Chicago, IL: University of Chicago Press, 2010), pp. xii–xiii.

relation between proto-scientific and religious ways of understanding life has often not been considered to be distinct in the histories of civilization; the developments of Islamic science or Chinese alchemy come to mind. The history of science and religion has only been one of antagonism since the scientific revolution of the seventeenth century with the development of modernity and the retreat of religion from cosmological explanation. Since then, the sciences have generally taken over the explanation of life in our secular age, producing the powerful contemporary paradigms we have today, particularly in the explanation of human behaviour in the neurological and biological sciences. The retreat of religion from cosmology also heralded the disenchantment characteristic of the secular age and a major shift in the popular understanding of life. As scientific knowledge grows and impacts a wider population through education and communication, attitudes to the category 'life' become more complex, embracing both positive attitudes, almost romantic evaluations of nature and eco-system, and negative attitudes in a desire to escape into an electronic, virtual reality that turns its back on 'nature'.

The different ways in which life has been understood through human history, as summarized in the two strategies or styles of affirmation and negation, are expressions of the issue that I am trying to get at. The evaluation of life is inseparably associated with the evaluation of death, and viewing life as a negative value, as we find in some forms of Buddhism and Hinduism, does not entail valuing death; indeed, the negative appraisal of life is due to its inseparability from death and also the appreciation that without death there is no life.[5] Life and death are two sides of the same coin and religions that have rejected the idea that ultimate value lies in this world have affirmed a value that goes beyond life and death in the affirmation of some other realm, some higher world. E.E. Cummings' line, 'there's a hell / of a good universe next door; let's go'[6] is an idea that we find throughout the histories of higher civilizations. So, I want to tell this story of the ways in which religions have both affirmed and negated life, with a view to a broader claim about life, and to explain the history of religions we need an account of life itself based on the ways in which life comes to appear in human communities, especially in our social and political interactions. To read the history of religions through the lens of life and to claim that human sociality is a kind of life at their heart means that we must engage with sociological and philosophical reflection that draw on the sciences of life. This is to enter a minefield of controversy. Philosophical and sociological reflection on life that has drawn from the

[5] John Bowker, *The Meanings of Death* (Cambridge: Cambridge University Press, 1991), pp. 129–205.

[6] E.E. Cummings, 'pity this busy monster, manunkind', *Complete Poems 1904–1962*, edited by George James Firmage (New York: Norton and Co., 2016 [1973]), p. 590.

harder sciences has led to rigorous and sometimes acrimonious debate about, for example, the importance of genes in human development, whether we can speak of a purpose or goal driving evolution, and whether our decisions are truly free if controlled by neurological processes, themselves driven by our genetic code; not to mention the moral and legal consequences of scientific explanations of life such as debates about stem cell research, abortion, euthanasia, and animal rights. All of these contemporary debates are informed by religious commitments of various kinds. The extent to which we are 'reinventing the human in the molecular age', to use Helga Nowotny's phrase, itself has roots in religious ideas about the nature of life, the constraints upon technologies of life that seek to enhance human capacities, and the moral issues thereby entailed.[7] The foundations of where we are today in our attitudes to life itself and the genetic technologies we are developing have their roots not simply in technological innovation but in religious innovations that have developed practices to realize what are considered to be human goods by those religions, goods that have included the idea of immortality and transcending death. Such aspirations have not gone away in science and technology-orientated visions of a human future in which we, as *Homo sapiens*, are transcended by a trans-human species we wish to become.[8]

I could begin this project to examine the history of religions and civilizations through the category 'life' at a number of points within a range of disciplines. There are several stakeholders and contenders. On the one hand, the sciences, particularly the biological sciences, make claims about the nature of life and its origins, while on the other hand philosophy makes claims about life in terms of human freedom and the social institutions within which we find ourselves. These two reflections on the nature of life itself and the ways of life in human institutions have tended to be distinct, reflecting the GMM and KHM. But to understand and explain religions in terms of life, we need to engage with both kinds of discourse. The biological sciences tell us about life itself, cellular life, the origins of life, and the ways life generates itself, while philosophy and other human disciplines raise questions about the nature of human life, the forms or ways of life that the social sciences attempt to explicate. A distinction that will guide my reflection on religions is between life itself and ways of living. This distinction is closely related to others in philosophy, notably necessity and freedom, and bare life and sovereign power. Life itself is an ancient idea that we find in Aristotle and arguably in the Upaniṣads as the animating force of the cosmos, and in modern times it has been given sharper definition in both the biological sciences and physics.

[7] Helga Nowotny and Giuseppe Testa, *Naked Genes: Reinventing the Human in the Molecular Age* (Cambridge, MA: MIT Press, 2010).

[8] Yuval Noah Harari, *Homo Deus: A Brief History of Tomorrow* (London: Harper Collins, 2017).

What interests me here is how life itself relates to or becomes manifest in ways of life. I shall restrict ways of life to human life, to the lifeworld or cultural world as the enterprise through which human beings have sought to understand their location and so imbue life with meaning and value. Thus, when we are speaking about ways of life, we are speaking about the formation of meaning and the value human communities have placed on life and death. But human ways of life do not necessarily exclude other forms of life, and in the contemporary Western or north Atlantic world we find forms of spirituality that seek to enhance human consonance with nature. With the fading of traditional religion in many regions of Europe and North America, we have the affirmation of the natural world by communities that seek to identify or re-identify with nature: deep ecology and neo-animist movements that seek to redress what they regard as being lost, a harmonization of human life with the natural world.[9] Human life is but one manifestation of the life force that can guide us if we open ourselves to it: the Anthropocene gives way to the post-human. Such affirmation of life is complex in its relation to late modernity with these ecological orientations generally resisting governmental and social attempts to promote conformity and resisting managerialism that they regard as anti-life. These movements, although marginal, impact indirectly upon centre-stage, global politics of the environment.

That religions encode ways of life means that they provide frameworks within which we can live our lives meaningfully. Such frameworks entail a moral perspective, a 'moral ontology',[10] that is the necessary background for moral judgements. Understanding religion in terms of life necessitates understanding religions' conceptions of what it is to lead a good life and to pursue the affirmation of life. The respect for life and the sympathy for the other it entails is an ideal that is embedded in many religions even though their bloody histories have often belied such affirmation. That religions provide moral frameworks within which to pursue human goals is important in understanding who we are and how we deal with contemporary human problems, from issues of gender equality in the workplace, to issues in molecular embryology, to world refugee crises. The ability to feel sympathy, to empathize, is arguably central to human perceptions about the meaning of life and the way we understand our

[9] Graham Harvey, *Animism: Respecting the Living World* (New York: Columbia University Press, 2017). An interesting study of animism among the Chachi of Ecuador is by Istvan Praet who argues that a characteristic feature of animism is to see humanity and life as an achievement 'rather than a birthright' (p. 192). Istvan Praet, 'Humanity and Life as the Perpetual Maintenance of Specific Efforts: A Reappraisal of Animism', in T. Ingold and G. Palsson (eds.), *Biosocial Becomings: Integrating Social and Biological Anthropology* (Cambridge: Cambridge University Press, 2013), pp. 191–210. An example of neo-animism in Western societies is the 'Sacred Living Movement' that offers retreats, especially focused on pregnancy and childbirth. See Ann W. Duncan, 'Sacred Pregnancy in the Age of the "Nones"', *Journal of the American Academy of Religion*, vol. 85 (4), 2018, pp. 1089–115.

[10] Charles Taylor, *Sources of the Self* (Cambridge: Cambridge University Press, 1988), p. 8.

goals. Indeed, the origins of religion may partly lie in human susceptibility to empathize, to appreciate the needs of the other, an idea that is reinforced by the human tendency to dehumanize those who we wish to kill. This capacity for empathy is at the root of cooperation and at the root of human communities, articulated in religions as a moral and sometimes legal imperative to care for others, even when 'others' is a narrowly defined category (those within the initiation group, those on the inside, although larger than the kinship group). Taylor's moral ontology is rooted in the biology of who we are as a species. The capacity for empathy evolved in our early hominin past. It clearly had an evolutionary advantage and, as the primatologist Tomasello has argued, gave us a competitive edge over other primates. Rousseau's contention that humans feel natural sympathy for others may well be correct.[11]

According to religions, human beings are capable of maintaining a higher life, a life beyond mere survival in which shared goals can be achieved and in which sympathy for others is at its heart. That religions have sometimes fostered narrow and restrictive notions of human life and thwarted human flourishing does not detract from the point that religions have consistently made claims about the good life and the overall purpose of life, as we will see. This is even true of groups such as Islamic State that wished to establish a 'new caliphate' in a commitment to a literalist reading of Islam that it believed promotes the human good because it conforms to God's will (at least to those who are not enslaved). Such a view is radically opposed to what Taylor has identified as 'the affirmation of everyday life', as a value of central importance to modern civilization.[12]

To offer an account of religion in terms of life or to explain the nature of empathy in terms of natural selection—that empathy selects for optimal species survival—is potentially, *in nunce*, a reductionist or naturalist explanation. Indeed, naturalism might be seen as part of the disenchantment with religion that Weber identified and that has become a defining feature of secularism. But an evolutionary and biological view of human life that shows us how human beings have an innate susceptibility to empathy is surely an important finding, and I will explore its importance for civilizations in due course; such a perspective can enhance a fundamentally humanist view that seeks to affirm human value at its own level. But a different kind of naturalist account, a strong reductionism, maintains that we can explain religions in terms of their survival potential for the species in enhancing our capacity for empathy, a view that clearly goes against traditional self-understandings of religion. This also goes against humanist understandings in relegating religions purely to their function in protecting gene replication through the

[11] Jean-Jacques Rousseau, *Discourse on the Origin of Inequality*, trans. Franklin Philip (Oxford: Oxford University Press, 1994), p. 27.

[12] Taylor, *Sources of the Self*, pp. 13–14.

generations. On this view, explanations of life put forward in the harder sciences, especially biology, dismiss or challenge traditional religious accounts. Such naturalism clothes itself as an atheism that has wide popular appeal; I am thinking of Dawkins' work in particular that explains life principally in terms of the drive of the genes as the operators of natural selection and Coyne's reductionism that sees religions as mistaken scientific theories.[13] The essential argument here is that every religious account of the origin of the world and the nature of life has been superseded and proved false through science, especially Darwinian natural selection. My account is opposed to this kind of naturalism, not on grounds of supporting religious explanations of world formation,[14] but on both scientific and humanist grounds concerning the validity of human experience entailing a moral perception of the world, and on such accounts' misunderstanding of religions as bad or pseudo-scientific theories. I nevertheless do wish to take on board or internalize the knowledge of life that science provides and in particular a broader view of human evolution necessitated by more recent developments in evolutionary anthropology. Scientific explanation at cellular or even sub-atomic levels does not necessarily lead to eliminative reductionism but is one level of explanation: the specification of a particular range of constraints at the cellular level. Causation is not adequate to the data in biological systems because of feedback between environmental and cellular levels. Along with immediate causes, there are always background constraints. For example, Ottoline Leyser's work on plant morphology shows a complex interaction between cells and environment within the plant, especially in the process of photosynthesis, that the 'ability of plants to adapt their growth and development to the prevailing light conditions involves substantial integration of the different classes of information provided by light'.[15] Although genetic factors are of vital importance in morphogenesis, these need to combine with environmental factors to generate behavioural change and so a web of causation is formed.[16] There are multiple layered systems involved in plant growth that constrain the 'event' of the plant

[13] The most famous and rigorous of Dawkins' works is arguably *The Selfish Gene*, 4th edition (Oxford: Oxford University Press, 2016) but also Dawkins, *The Extended Phenotype: The Gene as a Unit of Selection* (San Francisco, CA: Freeman, 2002). Also, Jerry A. Coyne, *Why Evolution Is True* (Oxford: Oxford University Press, 2010). For a compelling refutation of this kind of view see John Bowker, *Is God a Virus? Genes, Culture, and Religion* (London: SPCK, 1995).

[14] For arguments for the compatibility of religion and science, contra the Dawkins position, see Alistair McGrath, *Inventing the Universe: Why We Can't Stop Talking about Science, Faith and God* (London: Hodder and Stoughton, 2015). Also, Keith Ward, *The Big Questions in Science and Religion* (West Conshohocken, PA: Templeton Foundation Press, 2008).

[15] Ottoline Leyser and Stephen Day, *Mechanisms in Plant Development* (Oxford: John Wiley and Sons, 2009), p. 163.

[16] Ibid., p. 230.

into its outcome.[17] The idea of constraint is important here as implying not only a broader field of influences upon an organism, but also, allowing the possibility of self-organizing systems, that a cause does not necessarily have to be external to its effects.[18] The plant's multi-layered systems are homeostatic. The truth is more complicated than crude reductionism or a singular causation model allows.

As regards human evolution, contemporary accounts show how our ancestors acquired skills—such as the control of fire and tool making—that gave us an edge over other primates and eventually led to the phenomenal success of the *homo* species. Not only did the genes and natural selection affect our life, but also our life experience affected our genes. This co-development is called the extended evolutionary synthesis (EES), sometimes referred to as holistic biology, that explains evolution not simply in terms of the genes and natural selection but also in terms of wider environmental factors. One of the implications of this model is that, along with the genes, culture itself evolves. Agustin Fuentes summarizes evolution as follows:

> Mutation (changes in the DNA) introduces genetic variation, which in interaction with the growth and development of the body (from conception until death) produces a range of variations (differences in bodies and behaviour) in organisms. This biological variation can move around within a species by individuals moving in and out of populations (called *gene flow*), and sometimes chance events alter the distribution of variation in a population (called *genetic drift*). Much of this variation can be passed from generation to generation through reproduction and other forms of transmission and inheritance.[19]

This process of evolution is characterized not merely by natural selection but by four processes of inheritance: genetic inheritance, epigenetic inheritance or systems of the body affected by different factors that can be passed on to the next generation, behavioural inheritance, and the uniquely human symbolic

[17] Michael Marder, *Plant Thinking: A Philosophy of Vegetal Life* (New York: Columbia University Press, 2013), pp. 121–2, on the 'end' (telos) of plant life. Marder's interesting concept is that a plant has non-cognitive thinking, whose identity is a non-identity, being one with its environment: 'Vegetal being revolves around non-identity, understood both as the plant's inseparability from the environment wherein it germinates and grows, and as its style of living devoid of a clearly delineated autonomous self' (p. 162). This seems quite distinct from the human case with our ability to narrate our life as distinct from the *Umwelt*, as we find in the histories of civilizations accounted in this book.

[18] Alicia Juarrero, *Dynamics in Action: Intentional Behaviour as a Complex System* (Cambridge, MA: MIT Press, 1999), pp. 132–44 on the specification of types of constraint. Also 'Causality and Constraint' in G. Van de Vijyer (ed.), *Evolutionary Systems: Biological and Epistemological Perspectives in Selection and Self Organization* (Dordrecht: Kluwer, 1998), pp. 233–42.

[19] Agustin Fuentes, *The Creative Spark: How Imagination Made Humans Exceptional* (New York: Dutton, 2017), p. 6. See also Agustin Fuentes, 'Blurring the Biological and Social in Human Becomings', in Tim Ingold and Gisli Palsson (eds.), *Biosocial Becomings: Integrating Biological and Social Anthropology* (Cambridge: Cambridge University Press, 2013), pp. 42–58.

inheritance. On top of these four processes there is distinctively human cooperation and niche construction.

Many years ago, sociobiology claimed that the explanation of religion could be made genetically on the grounds that the genes control all of life. While the genetic explanation of human beings is important, a necessary condition for understanding what we are, it specifies constraint at far too low a level to have total explanatory power and, indeed, whether the genotype or the phenotype is the dominant evolutionary drive is itself hotly contested in the biological sciences. The anthropologist Tim Ingold, for example, has presented a strong argument that questions the very coherence of the idea of genetic and cultural coevolution. On the coevolution model, biological evolution is the evolution of the 'genotype' while cultural evolution is the evolution of the 'culture-type', but these are not found in the real world; rather they are models constructed in retrospective analysis, which misses the fundamental point that 'at the heart' of the process of evolution is ontogenesis. This, thinks Ingold, is the 'Achilles heel of the entire neo-Darwinian paradigm': ontogenesis on this view is simply the transcription of a prefigured material substrate of the genotype, simply a modern version of Aristotelean hylomorphism.[20] Rather, what needs to be done is to replace a problematic paradigm of human beings having sovereignty over nature mediated by culture with an appreciation of 'the continuum of organic life', and to see ourselves 'not as *beings* but as *becomings*—that is, not as discrete and preformed entities but as trajectories of movements and growth'.[21] Ingold has a point. What emerges from the critique of the ontogeny/phylogeny distinction is the idea of persons as co-creators of their reality, mutually constitutive as 'biosocial' creatures.[22]

Although this new orientation seems basically correct, we need not only an account of ontogenesis but higher-level explanations that specify a wider range of constraints, such as the historical, economic, and social, as important factors in human formation.[23] This is also important for the explication of human institutions, particularly religions. That is, we need an account of religions at a macro-historical level, 'big history', at an interpersonal or intersubjective level (the existential realm of religious experience), as well as the micro-biological and neurological level. The downward reduction of the macro-historical and intersubjective levels to the cellular is inadequate as an explanatory paradigm because it cannot account for the thick texture of everyday human life that forms our dominant reality. An adequate philosophy of life, as Hans Jonas once observed, entails not only a philosophy of the

[20] Tim Ingold, 'Prospect', pp. 6–7, in Ingold and Palsson (eds.), *Biosocial Becomings*, pp. 1–21.
[21] Ibid., p. 8. [22] Ibid., p. 13.
[23] I am strongly influenced in my understanding of constraint by John Bowker's work. See, for example, his *Why Religions Matter* (Cambridge: Cambridge University Press, 2015), pp. 133–64.

organism but also a philosophy of mind.[24] We need to bring together knowledge and understanding, the biology of life with an ontology of life that accepts its existential and ethical quality, the social dynamic of where we find ourselves.

In a nutshell, I wish to relate a narrative that religion has come about in evolution as a particular or characteristically human development that draws on three primary sources: our ability to empathize and form social groups, in other words our pro-sociality; our ability to speak and to develop language; and our ability to order space and time in ritual. Language allows for the development of imagination and opens up the symbolic realm of non-ostensive reference particularly focused on the nature of life itself, and draws on our pre-linguistic ability to form social groups and proclivity to ritual behaviour. Religion expresses life through the transformation of human biological proclivities and reflects on life through the creation of cosmic meaning. The history of religions can be read as a particular kind of evolutionary development fundamentally concerned with the creation of meaning and reinforcing the social bonds of community in relation to broader life understood in a cosmic sense. This is to see religion in terms of the philosophy of life. Religions establish existential meanings for communities, and these meanings at the level of culture might be seen as expressions of life itself or transformations of bio-energy rooted in the evolution of human interpersonal communication or, more technically, constrained by face-to-face social cognition that is specific to human niche construction. The religious imagination creates a metaphorical space for the exploration of life and might be seen as a way in which life reflects itself. Human agency that is part of life reflects back on itself through religious imagination and the practice it entails. Such a thesis involves an *empirical* claim that the origins of religion can be explained in terms of human evolution and the specificity of the human niche as communication and interactivity, what we call social cognition that allows for our pro-sociality; a *historical* claim to show how philosophies of life have operated in the history of religions; and a *philosophical* claim to develop an account of religion in terms of a realist ontology of life, drawing on a hermeneutical phenomenology. But I am racing ahead of myself.

To begin to formulate a thesis about the importance of life in the history of civilizations and to explain that history in terms of life means that I need to say something about our knowledge of life itself—to identify fundamental empirical claims, and how these relate to philosophical claims about the nature of life. I need to delineate the limits of the discussion, on the one hand bounded by biology, the theory of life, on the other bounded by philosophy, the interpretation of life's meaning. What life is and who we are, the biological

[24] Hans Jonas, *The Phenomenon of Life: Towards a Philosophical Biology* (Evanston, IL: Northwestern University Press, 2001 [1966]), pp. 1–6.

and the existential, form the borders of our topic if we are to explain and understand religions and their impact upon contemporary human life. The scientific and philosophical accounts of life provide the basis of the idea of religion as the transformation of the bio-energy of social cognition through language. I will say something of this here but will develop the idea of social cognition and the shift to symbolic thinking through language in Chapter 2. There is in fact a history of this move towards the explication of religion in terms of science and the placing of language as central to human development in Kant's work on anthropology and Ricoeur's work on language. But first we need to begin with life.

THE THEORY OF LIFE (BIOLOGY)

There are many competing claims on the category 'life'. On the one hand, we have formal, scientific definitions in terms of its proposed properties such as reproducibility and metabolism; on the other, we have existential accounts identified with human experience as such. The former kind of third-person account is inadequate to address existential, human concerns, while the latter is weak in terms of explanation and runs the risk of seeing all of life in human terms, the risk of anthropomorphism. Both kinds of accounts—what might be called the reductionist explanation of life and the non-reductionist understanding—operate with different notions of meaning, one in which meaning is explanation (how has life arisen? What is the distinction between living and non-living?), the other in which meaning is identified with the purpose of life or more specifically the purpose of *a* life or of *my* life, which has traditionally been to locate oneself within a cosmos. Both approaches present difficulties. Although my aim is to offer an account of religions in terms of life, I do need to describe different approaches in order to propose a route to explaining religions that goes between the purely reductionist explanation and a humanistic resistance to science. It is possible to present an account of religions as the expression of life that is neither biological reductionism nor a transcendental mysticism.

But is the question about the meaning of life itself even meaningful and, if it is, what could possibly count as an answer? And even if the question could be answered, how would that affect our lives? Religions have come up with different answers to it, often claiming that the answer is not propositional but experiential, and part of this book's task is to present those accounts as they have occurred throughout history. In the first instance we need some idea of what we are looking for; some idea of life, how it is currently understood in the human sciences, and what the most important accounts of life have been. What is life?

At one level we all know what life is through simply being alive, through experiencing life and through feeling the blood flowing through our veins. We also know experientially degrees of life, of feeling full or depleted, energetic or tired, and we have a moral sensibility that makes judgements on our actions. But to understand the nature of life or find a consensual definition is a difficult task. When we raise questions about the difference between living and non-living matter, whether there is a continuum or a disjunction between the living and the non-living, whether life begins at a certain point in time, how life emerges from non-life, whether life has intrinsic value, and so on, we are faced with a quagmire of philosophical and scientific difficulties. While there are prototypical forms of life that we can all recognize, what about the difficult cases that fall between the living and the non-living, such as viruses?[25] Responding to these questions is often a matter of where we begin and from which disciplinary base we are looking.[26] Our intuitions about the nature of life, as we are living beings, are certainly important and take us some way in understanding life, but the explanation of life is different and needs to draw on the rich and varied disciplines that human beings have developed to understand and explain ourselves and the world we live in.

The emergence of life depends upon a very finely tuned universe with a physical basis—atomic and sub-atomic particles—that need to have very precise properties, such that, for example, if the masses of the fundamental particles that make up matter, along with the spin of electrons and quarks, were to be slightly different, 'the physics and chemistry of our universe would not exist'[27] and so there would be no life. Although the chemistry of life is extremely complex, the molecules of life are built from only ninety-two atomic elements, each of which in turn is built from only three particles—protons, neutrons, and electrons, the standard model of particle physics. Electrons are fundamental particles that so far have resisted being broken down further, but protons and neutrons can be broken down into quarks that, like electrons, cannot be further reduced and that are classified as 'up' and 'down' depending

[25] Geraint F. Lewis and Luke A. Barnes, *A Fortunate Universe: Life in a Finely Tuned Cosmos* (Cambridge: Cambridge University Press, 2016), pp. 10–11.

[26] For example, biosemiotics regards the distinction between living and non-living to be whether an entity is 'semiologically active'. Thomas Sebeok cited in Kalevi Kull, Claus Emmeche, and Jasper Hoffmeyer, 'Why Biosemiotics? An Introduction to our View on the Biology of Life Itself', in Kalevi Kull and Claus Emmeche (eds.), *Towards a Semiotic Biology: Life Is the Action of Signs* (London: Imperial College Press, 2014), pp. 1–21. For a history of biosemiotics see Wendy Wheeler, *Expecting the Earth: Life, Culture, Biosemiotics* (London: Lawrence and Wishart, 2016). For an early acceptance of complexity in biology see Robert Rosen, *Life Itself: A Comprehensive Inquiry into the Nature, Origin, and Fabrication of Life* (New York: Columbia University Press, 1991). The basic argument is that to be alive is not to be a machine and that evolution is a corollary to the living: 'it is easy to conceive of life, and hence biology, without evolution. But not of evolution without life' (p. 255).

[27] Ibid., p. 63.

on their electric charge. Thus, in the words of Lewis and Barnes, '[t]he electron has −1 unit of electric charge, the up quark has $+^{2}/_{3}$ units, and the down quark $-^{1}/_{3}$ unit'.[28] These three particles make up all the matter of the universe we can see. If the properties of these particles were to be slightly different—say, a unit or two of difference in the mass of the up and down quarks—then the universe would not sustain life.[29] Some physicists have argued a position that the properties of the universe, its fine-tuning, have been such as to produce observers of it, the anthropic cosmological argument,[30] although this is controversial and inevitably remains speculative.

Philosophers and scientists have approached the question, 'What is life?' in a number of ways, usually trying to detect defining characteristics or properties. Bedau identifies three properties by which life has been defined: the ability to reproduce, the ability to undergo Darwinian evolution, and the ability to metabolize or be a self-sustaining, chemical system.[31] The ability to reproduce is not, as Bedau observes, a universal property as there are living entities that do not reproduce (such as mules) and non-living entities that do (such as some robots), but all living populations undergo evolution and have metabolisms. In an interesting article outlining the field, Bedau argues for a concept of life in terms of a Program-Metabolism-Container (PMC) model. He takes what he calls an Aristotelian approach that tries to explain the characteristic phenomena of life, focusing on organisms in the natural context, in contrast to a Cartesian approach that tries to explain the essential properties of life from a large set of paradigm cases of living organisms abstracted from their contexts.[32] This approach entails giving an account of borderline cases between life and non-life such as 'viruses and prions, which replicate and spread even though they lack a metabolism'.[33] In order to account for minimal chemical life, we need a hypothesis about what life is that an astrobiologist might use, for example, in looking for life in the lakes of liquid methane on Titan, a moon of Saturn, or on Mars. Such a hypothesis is the PMC model that concentrates on chemical functions rather than chemical materials. In Bedau's own terms, the program 'controls cellular processes with replicable and inheritable combinatorial information (the functional analog of genes)', the metabolism 'extracts useable energy and resources from the environment', and the container 'keeps the whole system together'.[34] This model, Bedau argues, explains the hallmarks of life—such as metabolism and reproducibility—and borderline cases between life and non-life, and is the best model to account for the

[28] Ibid., p. 43.

[29] In a fascinating account, Lewis and Barnes describe the different possible scenarios of changing the mass of particles, none of which would result in life. Ibid., pp. 50–3.

[30] Ibid., pp. 15–21.

[31] Mark A. Bedau 'The Nature of Life', p. 14, in Steven Luper (ed.), *Life and Death* (Cambridge: Cambridge University Press, 2014), pp. 13–29.

[32] Ibid., p. 20. [33] Ibid., p. 20. [34] Ibid., p. 23.

characteristics of 'living worlds'. It shows that what is important is not so much the material out of which life is constructed but rather 'the form in which that material is arranged and organized'.[35] While there is a vast difference between humans and bacteria, both share 'the functional integration of various complex capacities',[36] including self-sustainability and the ability to reproduce. The field of biosemiotics would wish to add to metabolism and reproducibility the ability to convey information, so life is composed of molecules that manifest themselves as signs.[37]

There has been much discussion about the origins of life, how organic life could have evolved, and the relation between living organisms and non-living matter. In the first place the universe has to be such a place in which life could evolve. This entails fine-tuning to an extraordinary degree. One theory developed through observations of one of the moons of Saturn is that moving water over rock produces a chemical reaction and this is the first life.[38] But any introductory textbook on biology will describe the biological classification called the three-domain system that divides cellular life into archaea, bacteria, and eukaryote domains.[39] These have a common ancestor in what is called the progenote. This group of self-replicating molecules gave rise to two forms of life called bacteria and prokaryotes, single-celled organisms, which in turn subdivided into archaea and eukaryotes that joined together to form more complex organisms and the myriad of phyla that exist in our world. The main difference between prokaryotes and eukaryotes is that the genes of the former are not enclosed in a membrane-bound nucleus, whereas those of the eukaryotes are: the result of symbiosis between earlier free-living prokaryotes.[40] Prokaryotes that share a similar structure to modern bacteria comprising a single cell developed three billion years ago, with eukaryotes developing two billion years ago and comprising different compartments including a nucleus and cytoskeleton. The transition from prokaryotic cells to eukaryotes was, in the words of Kirschner and Gerhart, 'a major and

[35] Ibid., p. 27. [36] Ibid., p. 29.

[37] Jesper Hoffmeyer, *Biosemiotics: An Examination into the Signs of Life and the Life of Signs*, trans. Jesper Hoffmeyer and Donald Favaresceau (Scranton, PA and London: University of Scranton Press, 2008), p. 15.

[38] A. Le Gill, C. Leyrat, M.A. Janssen, G.C. Choblet, G. Tobie, O. Bourgeois, A. Lucas, C. Satin, R. Howett, R. Kirk, R.D. Lorenz, R.D. West, A. Stoizenbach, M. Massé, A.H. Hayes, L. Bonnefoy, G. Veyssière, and F. Paganelli, 'Thermally Anomalous Features in the Subsurface of Enceladus's South Polar Terrain', *Nature Astronomy*, vol. 1, 0063, 2017.

[39] There is, however, still some controversy here and some biologists claim the three domains arose independently. See William Martin, 'Woe Is the Tree of Life', in Jan Sapp (ed.), *Microbial Phylogeny and Evolution: Concepts and Controversies* (Oxford: Oxford University Press, 2005), p. 139.

[40] Lyne Margulis, *Symbiosis in Cell Evolution: Life and its Environment on the Early Earth* (San Francisco, CA: Freeman, 1981). For a sociological extension of this thesis see Myra Hird, *The Origins of Sociable Life: Evolution after Science Studies* (New York: Palgrave Macmillan, 2009).

enduring accomplishment',[41] because the eukaryotes are so much more complex and larger than their predecessors. And because of the greater degree of constraint, there is a greater degree of freedom of possibility. Indeed, according to the biosemiotician Jesper Hoffmeyer, the transition from prokaryote to eukaryote exemplifies a general principle that 'in emergent process, freedom of possibility will always be constrained at the simpler level in order to allow an altogether new kind of freedom to appear and unfold at a more complex level'.[42] The higher the level of complexity for an organism, the greater the degree of freedom, so more options are open to the eukaryote than the prokaryote, a principle that operates within all living organisms: multicellular life eventually affords greater freedom of movement and operation.

In the four-billion-year history of the earth, the earliest life may have developed as early as within a billion years of the earth forming. The earliest fossil record of colonies of organisms called stromatolites, a kind of eukaryote, are three and a half billion years old and the first multicellular organisms are two billion years old.[43] From multicellularity, body plans emerged about a billion years ago. The origins of life may even be extra-terrestrial, although this just defers the problem of life's origin to a different place, or it may be in archaebacteria that can live well below the surface of the earth in extreme conditions in volcanic fissures under the sea.[44] If life is characterized by reproducibility and metabolism, then some chemical processes seem to possess these properties and 'evolve', thereby implying that the origins of life could be here,[45] but what does seem to be clear is that all life on earth evolved from a single source. While there may have been other sources of life, only one successfully replicated itself to transform into the myriad of living entities that populate the world 'from humans to bananas to sea squirts and amoebae'; all are 'descended from the same (bacterial) ancestor'.[46] The story of life on earth is the story of how organisms have culled energy from their environment—and a particularly important development was photosynthesis, the transformation of energy from the sun—and, as they became more complex, from each other. With the development of organisms that were able to survive within and

[41] Marc W. Kirschner and John C. Gerhart, *The Plausibility of Life: Resolving Darwin's Dilemma* (New Haven, CT and London: Yale University Press, 2005), p. 53.

[42] Hoffmeyer, *Biosemiotics*, p. 258.

[43] David Christian, *Maps of Time: An Introduction to Big History* (Berkeley, CA: University of California Press, 2004), pp. 117–18; John Maynard Smith and Eörs Szathmáry, *The Origins of Life: From the Birth of Life to the Origins of Language* (Oxford: Oxford University Press, 1999), chapter 7. Christian's book contains an excellent survey of the origins of life that clearly explicates the problems involved, as does Robert Bellah, *Religion in Human Evolution: From the Paleolithic to the Axial Age* (Cambridge, MA: Belknap Press, 2011), pp. 57–60.

[44] On the study of this world, see S. Helmreich, *Alien Ocean: Anthropological Voyages in Microbial Seas* (Berkeley, CA: University of California Press, 2009); Christian, *Maps of Time*, pp. 98–100.

[45] Ibid., pp. 94–6. [46] Ibid., p. 92.

transform an oxygen-rich environment, evolution developed at a relatively fast pace to a great proliferation of life forms.

The story of the development of life through natural selection and the other modes of inheritance, from the primeval oceans in our then toxic atmosphere to the thriving of life in the biosphere we know today, is startling for the proliferation of forms that have come and gone, a constant outpouring of life from the evolutionary matrix. The PMC model characterizes the components of all those life forms: program, in particular the genetic code that gives rise to the forms of life; the metabolism that sustains those forms; and the boundary or body, their container. The general consensus has been that the development of simple organisms into complex ones occurs through Darwinian natural selection, the classic view that organisms better adapted to their environment tend to survive and reproduce more successfully, although some biologists think that some deeper law must be involved because the statistical odds of evolution developing through the prokaryote, eukaryote, multicellular organism sequence is simply too slim not to entail some other process.[47] Furthermore, the question remains as to why there was a sudden expansion of life forms; how do we account for the proliferation of life?[48] This remains one of biology's mysteries, although there have been important arguments attempting to answer it. Let us consider two that have relevance for the explanation of religion.

The first argument addresses the question of how diversification or variation occurs when there is such stability and continuity of life forms. How do we explain 'conservation on a cellular level and diversity on an anatomical and physiological level'?[49] Kirschner and Gerhart have argued for a thesis that accounts for this long-term stability alongside sometimes rapid evolutionary change, that organisms are engaged in a higher degree of selective activity than random mutation suggests. Mutations occur only in organisms that are already structures and so core processes preserved through long periods of time facilitate variation such that we have a proliferation of phenotypes. There is a

[47] Walter Fontana, 'Algorithmic Chemistry', in C.G. Langton, C. Taylor, J.D. Farmer, and S. Rasmussen (eds.), *Artificial Life II* (Redwood City, CA: Addison-Wesley, 1992), pp. 159–209; Fontana and Buss, 'What Would Be Conserved if the Tape Were Played Twice?', *Proceedings of the National Academy of Science*, vol. 91, 1994, pp. 757–61; Fontana and Buss, 'The Arrival of the Fittest: Toward a Theory of Biological Organization', *Bulletin of Mathematical Biology*, vol. 56, 1994, pp. 1–64.

[48] In a good layman's guide, George Thomson eloquently puts this thus: 'But what caused these great leaps to occur? If we can believe the fossil record, the simple prokaryotes dominated life on earth for more than two billion years. Why, then, didn't life remain stuck in this limbo? What spurred the prokaryotes to band together into eukaryotes? And why, for that matter, did the eukaryotes later band together to form the multicellular organisms that exist today?' George Thomson, *Fire in the Mind: Science, Faith, and the Search for Order* (London: Penguin, 1997 [1995]), p. 234.

[49] Kirschner and Gerhart, *The Plausibility of Life*, p. 71.

stability of 'core processes'—such as metabolism—but wider 'facilitated variation' which means, as I understand it, that organisms influence their future forms, that organisms constrain 'variation of its phenotype'.[50] The core processes that continue over long periods of time—such as DNA recurrence—facilitate variation without compromising those core processes. Cellular conservation is preserved while allowing for developmental changes due to new environments. Thus, eukaryotic cells develop in much greater complexity from prokaryotes as a consequence of enhanced cell–cell signalling capacities, resulting in basic body plans that are found in the multiplicity of organisms.[51]

Kirschner and Gerhart are not alone in articulating a more complex interaction between gene and organism. Susan Oyama long ago argued that genes 'affect biological processes because they are reactive' and must be seen rather as part of a sequence of events that analysis can begin with, but that of itself 'occupies no privileged energetic position outside the flux of physical interactions that constitutes the natural (and artificial) world'.[52] In her important book she argues that the causes of an organism cannot be identified as genetic or environmental, as alternatives. The opposition or debate between the dominance of nature or nurture has been a drawback in the development of biology, and we need to think about ontogeny in terms of an integration of information and causal functions that produce the organism-environment complex. Biology has failed to appreciate the inseparability of genetic and environmental processes. Rather, what she identifies as three processes of life, constancy, change, and variability, are controlled by both genetic and environmental factors, and we should 'recognize the importance of levels of structure above (and below) that of the genes'.[53] In his foreword to the second edition of Oyama's book, Lewontin makes the point that the language used to account for genes, that they are self-replicating, is simply not true: a parallel would be to speak of manuscripts as 'self-replicating'.[54] Genes respond to environments and environments are affected by genes through organisms, but the claim to gene dominance, as with early sociobiology, is broadly considered untenable in the extended evolutionary synthesis.[55]

[50] Ibid., pp. 220–1.

[51] Ibid., pp. 57–62. For a good explanation related to religion see Bellah, *Religion in Human Evolution*, pp. 60–5.

[52] Susan Oyama, *The Ontogeny of Information: Developmental Systems and Evolution*, 2nd ed. (Durham, NC: Duke University Press, 2000), p. 40.

[53] Ibid., p. 132.

[54] Richard Lewontin, 'Foreword' to Oyama, pp. xii–xiii: 'Genes "do" nothing, they "make" nothing, they cannot be "turned on" or "turned off" like a light or a water tap, because no energy or material is flowing through them. DNA is among the most inert and nonreactive of organic molecules; that is why stretches of DNA can be recovered intact from fossils long after all the proteins have been lost'.

[55] This line of thinking, that genes alone are not responsible for morphology, has been argued for by Samantha Frost who posits that humans are bio-cultural creatures that need to be

A second important development has been formulated by Odling-Smee with Laland and Feldman, which claims to supplement natural selection in a crucial way, namely by showing how organisms change their own and others' environments. These environmental modifications are niche constructions.[56] There had been precursors to the theory in the physicist Edwin Schrödinger and later the biologist Ernst Mayr who spoke of how organisms shift into a new niche or adaptive zone, and Lewontin was an early advocate who formulated equations for niche construction, showing how it differs from standard evolutionary theory.[57] But Odling-Smee, who coined this phrase, and his colleagues have developed a sustained research programme. For them, organisms are the bearers of genes, which survive through organisms that reproduce by natural selection. But organisms 'also interact with environments, take energy and resources from environments, make micro- and macro-habitat choices with respect to environments, construct artefacts, emit detritus and die in environments, and by doing all these things modify at least some of the natural selection pressures present in their own, and each other's, local environments'.[58] That is, organisms choose to create particular environments in which to live, and ways of living in those environments in turn affect natural selection. An organism can actively change a feature of its environment 'either by physically perturbing factors at its current location in space and time or by relocating to a different space-time address'.[59] Odling-Smee and his colleagues' book is replete with examples and opens with the striking account of leaf cutter ants, which cut vegetation, bring it down into their underground nests, and there allow fungus to grow on it that they consume. The fungus, a white mould, is such an abundant source of food that millions of ants thrive within a colony.[60] Here the ants, by developing 'agriculture', have created a particular niche that massively affects the environment—the huge nests can entail the displacement of soil weighing 44 tons—and that provides a space for the population to thrive.

accounted for within a materialist paradigm but 'without falling prey to a biological, environmental, or cultural reductionism'. Samantha Frost, *Biocultural Creatures: Towards a New Theory of the Human* (Durham, NC: Duke University Press, 2016), p. 151. For an account of 'religious naturalism' that parallels Frost's in the sense that it presents a coherent analysis of the sources of life, see Ursula Goodenough, *The Sacred Depths of Nature* (Oxford: Oxford University Press, 1998), pp. 19–25.

[56] F. John Odling-Smee, 'Niche Construction Phenotypes', in H.C. Plotkin (ed.), *The Role of Behaviour in Evolution* (Cambridge, MA: MIT Press, 1988), pp. 73–132.

[57] R.C. Lewontin, 'Gene, Organism, and Environment', in D.S. Bendall (ed.), *Evolution from Molecules to Men* (Cambridge: Cambridge University Press, 1983), pp. 273–85. Edwin Schrödinger, *What is Life?* (Cambridge: Cambridge University Press, 1992).

[58] F. John Odling-Smee, Kevin N. Laland, and Marcus W. Feldman, *Niche Construction: A Neglected Process in Evolution* (Princeton, NJ: Princeton University Press, 2003), p. 1.

[59] Ibid., p. 41.

[60] Ibid., pp. 3–5. This describes the work of Bert Hölldobler and E.O. Wilson. Also see Wilson's earlier work in his fascinating *The Insect Societies* (Cambridge, MA: Harvard University Press, 1971).

Another example would be Galapagos woodpecker-finches, which use cactus spines or twigs as tools to extract insects from underneath bark. Whereas woodpeckers use their beaks to grub, the finches pick up a tool that they have learned to use. The finches have created a niche; they have developed a lifeworld that is not directly the expression of the genes, as the woodpecker's beak is the direct consequence of natural selection, but is behaviour that nevertheless constrains natural selection insofar as finches have evolved short bills appropriate for the manipulation of cactus spines or twigs as tools and perhaps has also selected for increased learning abilities.[61] Knowledge acquired by individuals through ontogenetic development ceases at their death, but the way the environment, the niche, is changed by that behaviour has consequences for future generations. Here niche construction has had an effect on the finches' evolutionary development, on their PMC. The finches reproduce, driven by the genetic code, but they become self-regulating through the 'cultural' niche that in turn has modified their genetic makeup. Niche construction can come to be as important as genetic natural selection: both direct evolution.

The Galapagos finches are a good, simple example that illustrates how biological evolution and learned behaviour can interact. Whether social learning can be called 'culture' is a matter for debate, but social learning does form traditions over several generations in non-human species. The point is that species modify their environment through behaviour such that now 'behaviour is conceptualised in the process of selection'.[62] The interactions between biological evolution and cultural processes in the human case are naturally much more complex than in finches. On the one hand, we have the hard genetic view of sociobiology that the genes are the main driving force that determines the phenotype and in the human case, culture is simply an extension of the phenotype. On the other hand, more recent developments speak about the way the phenotype influences the genotype. Gene-culture coevolutionary theory maintains that culture takes over the evolutionary process and cultural developments are consequences of earlier cultural forms. An extended coevolutionary model adds to the complexity of the niche.[63] This is corroborated by work on human social genomics that has begun to specify 'the neural and molecular mechanisms that mediate the

[61] Odling-Smee, *Niche Construction*, pp. 257–8. However, whether this is 'cultural' transmission is still under debate. See S. Tebbich, M. Taborsky, B. Fessl, and D. Blomqvist, 'Do Woodpecker Finches Acquire Tool-Use by Social Learning?', *Proceedings of the Royal Society B: Biological Sciences*, vol. 268, 2001, pp. 2189–93.

[62] Dorothy M. Fragaszy and Susan Perry, 'Towards a Biology of Traditions', p. 6, in Dorothy M. Fragaszy and Susan Perry (eds.), *The Biology of Traditions: Models and Evidence* (Cambridge: Cambridge University Press, 2003), pp. 1–32. See references here for the debate in the late 1990s as to whether 'culture' is an appropriate term for social learning in non-human animals.

[63] Ibid., pp. 242–6.

effects of social processes on gene expression'.[64] That is, whereas human sociobiology maintains that the human phenotype is only affected by the genotype, the gene pool, as a result of natural selection in particular environments, gene-culture coevolution maintains the independence of a cultural inheritance, and the extended evolutionary synthesis furthermore maintains that in addition there is an ecological inheritance affected by the gene-culture complex. The way in which organisms form their environment is niche construction. In the human case this process has become such that culture and its environmental impact is more important to human evolution than natural selection based on the genes. Culture is a transmission system that provides humans with a non-genetic inheritance. Odling-Smee gives the example of how the domestication of cattle and the development of dairy farming, a culturally inherited practice, gave rise to lactose tolerance in adult populations. Here culture has had a direct impact on genetic inheritance for there are no genes for dairy farming.[65] The persistence of cultural practices over long periods of time can therefore change the genotype; the changed environment can modify biological inheritance. Human genetic inheritance in conjunction with human cultural inheritance provides the basis for gene-culture coevolution[66] and the extended evolutionary synthesis.

If we understand the term 'culture' in a broader sense, not restricted to material culture (tools and nests), then social interactions are important too. It is probable that the large brain in *Homo sapiens* developed over the last 1.7 million years not so much to increase our ability to use tools, but rather due to the complexity of human social living. Humans are part of a group called hominoids (apes and humans) that are in turn members of primates known as anthropoids (monkeys, apes, humans), and primates are a mammalian order.[67] In forming coherent social groups, we need to negotiate with each other to maximize survival capacity in hostile environments.[68] The development of the human social environment has driven genetic evolution of a predisposition for pro-social emotions and even the development of morality,[69] topics to which we will return in Chapter 2. If we have natural selection

[64] George M. Slavich and Steven W. Cole, 'The Emerging Field of Human Social Genomics', *Clinical Psychological Science*, vol. 1 (3), 2013, pp. 331–48.

[65] Ibid., p. 248. P. Gerbaut et al., 'Evolution of Lactase Persistence: An Example of Human Niche Construction', *Philosophical Transactions of the Royal Society Biological Sciences*, vol. 366, 2011, pp. 863–77.

[66] Jeremy Kendal, Jamshid J. Tehrani, and John Odling-Smee, 'Introduction: Human Niche-Construction in Interdisciplinary Focus', *Philosophical Transactions of the Royal Society B: Biological Sciences*, 366, 2011, pp. 785–92.

[67] For a comprehensive account see Fuentes, *The Creative Spark*, p. 15–23.

[68] See Daniel Lord Smail, *On Deep History and the Brain* (Berkeley, Los Angeles, CA: University of California Press, 2008), p. 113.

[69] Herbert Gintis, 'Gene-Culture Co-Evolution and the Nature of Human Sociality', *Philosophical Transactions of the Royal Society B: Biological Sciences*, vol. 366, 2011, pp. 878–88.

in the human case that has been overtaken by cultural inheritance that affects the environment, which itself is then transmitted through the generations, then this has implications for all of human culture. Sterelney has argued that the evolution of modernity has been due to the construction of a developmental niche from the demographic expansion of the Upper Palaeolithic when structured learning environments were developed, allowing apprentice learning and a high fidelity of cultural transmission and skill sets across generations.[70]

Given the importance of niche construction, we need to ask whether religions can be seen as forms of it. As with dairy farming or the use of twigs by Galapagos finches, there are no genes for religion, but we might argue that religions are themselves kinds of niche construction insofar as they are mechanisms for adapting to new environments. Indeed, adaptation, providing resources for group bonding and the formation of meaning, is what religions do. Stripped of particular doctrines, religions are forms of practice that have the effect of uniting a group or facilitating social cognition, as we will see. It is not that the historical complexity of religions can be reduced to nothing but social cognition, rather that a force of deep pro-sociality that is biologically based animates those complex historical entities. Religions provide a sacred cosmos within which human communities find purpose and meaning. Religions were the primary form of social bonding beyond the immediate kinship group or extended social unit, such as the tribe, whose practices of prayer, ritual, and asceticism draw on the wellspring of the animating force of human communities and reflexively seek to control that sociality. Religions form niches for the protection and control of human communities, especially through law that seeks to regulate relationships, narratives that provide meaning through language, and practices that offer meaning through action. It is not as if human beings have much choice in deep sociality: we can choose to disengage with each other, but such disengagement has usually been for the sake of another kind of sociality within a wider cosmos through asceticism or cenobitism or communing with the natural world. Religions construct niches within which humans can and have thrived. The development of non-ostensive, symbolic language has allowed the development of the religious imagination that cultivates human sociality on a cosmic scale. Through religions, through religious language, humans have tried to be at home in the world. Whether religions continue to hold the function of deep sociality into the future is a matter I will return to, but functional analogues of religion that draw on the bio-energy of human sociality will inevitably continue. This is to foreshadow the thesis of social cognition as the basis of religions that I will develop in Chapter 2, whose roots are in the biology of life.

[70] Kim Sterelney, *The Evolved Apprentice: How Evolution Made Humans Unique* (Cambridge, MA: MIT Press, 2012); Peter Hiscock, 'Learning in Lithic Landscapes: A Reconsideration of the Hominid "Toolmaking" Niche', *Biological Theory*, vol. 9, 2014, pp. 27–41.

If life itself, expressed in the fundamental bio-sociality of the human, is one boundary of our project, the second is a philosophy that recognizes complexity and emergence, that the biology of life alone cannot account for religions, and that there needs to be a philosophical account of the existential experience of our life.

THE PHILOSOPHY OF LIFE

The danger of the theory of life is that it can attempt to explain life in a way that rejects human experience as a valid process of understanding. Yet to understand religions, or indeed any aspect of human life, we need to caution against reductionist or 'nothing but' explanations as inadequate to the task of accounting for the complexity of first-person and interpersonal realms.[71] Long ago Husserl argued against scientism, against the GMM, privileging intuition as fundamental to a phenomenological science that distinguished itself from natural science based on objectivism and a purely mathematical explanation of life.[72] While I do not concur with Husserl's complete rejection of the GMM, that Western thinking took a wrong turn with Galileo, his caution that science can lead to the objectification of life—to scientism—has contemporary relevance still. In our explanation of human life, we need to absorb and internalize contemporary knowledge while at the same time maintaining what we might call an existential account of experience and human institutions that accepts the validity of social interaction or rather sees sociality as the basis of even harder scientific accounts: human being-in-the-world precedes reflection, as Ricoeur, echoing Heidegger, maintains.[73]

Hans Jonas, reflecting Kant, claims that the philosophy of life entails a philosophy of the organism and a philosophy of mind, the latter being fundamentally concerned with freedom. Indeed, Jonas argues that freedom

[71] The philosophy of biology is an academic discourse that deals with such problems; the relation of 'reduction' to 'emergence', for example. I cannot review this large body of literature, but Rom Harré's essay clarifies the issues and offers good conceptual boundaries: 'Transcending the Emergence/Reduction Distinction: The Case of Biology', in Anthony O'Hear (ed.), *Philosophy, Biology, and Life* (Cambridge: Cambridge University Press, 2015), pp. 1–20.

[72] Husserl, *Crisis in European Sciences*. For a clear articulation of the distinction between science and scientism, see Ingold, 'Preface', p. 14, in Ingold and Palsson (eds.), *Biosocial Becomings*: 'Science and scientism are quite different. The former is a rich patchwork of knowledge which comes in an astonishing variety of different forms. The latter is a doctrine, or system of beliefs, founded on the assertion that scientific knowledge takes only one form, and that this form has an unrivalled and universal claim to truth'.

[73] Paul Ricoeur, 'Hermeneutics of the Idea of Revelation', p. 143, in *Hermeneutics*, trans. David Pellauer (Cambridge: Polity Press, 2013), pp. 111–70: 'Reflection is never first, never constituting—it arrives unexpectedly like a "crisis" within an experience that bears us, and it constitutes us as the subject of the experience'.

is a quality found in the organic at the level of metabolism,[74] an idea to which we will return, but the point is that the human person is both organism and mind who comes into a pre-existing world. As Ricoeur argues, we are already *in medias res* prior to understanding which is 'never at the beginning or the end. We suddenly arrive, as it were, in the middle of a conversation which has already begun in which we try to orientate ourselves in order to be able to contribute to it'.[75] In an important paper Ricoeur criticizes what he sees as Husserlian idealism, arguing that phenomenology is always hermeneutical even as hermeneutics must presuppose phenomenology. While the Husserlian position is demolished by hermeneutics on the grounds that all understanding is 'always preceded by a relation which supports it',[76] nevertheless both phenomenology and hermeneutics share a mutual belonging in that phenomenology must assume a hermeneutics that claims our lived experience in the world in the modes of fore-having, fore-sight, and fore-conception,[77] while hermeneutics presupposes phenomenology insofar as 'every question concerning any sort of "being" is a question about the meaning of that being'.[78]

In relation to the question of life, therefore, we already find ourselves in its midst, and all inquiry into the nature of life has to be hermeneutical insofar as even the most rigorous scientific investigation already presupposes our modes of being in the world, our human practices. Taking on board the knowledge gained in the life sciences, while not forgetting the Ricoeurian corrective, is central to our explanation and understanding of religions' views about life. The perspective of hermeneutics according to Ricoeur exists within the ontological condition of understanding and this entails, first, a recognition of finitude and, second, a recognition of belonging.[79] Finitude is, of course, the obvious presupposition of all inquiry, but what Ricoeur highlights in belonging is important for our understanding of life and for our understanding of religion. The idea of belonging leads from 'the sphere of objectifying thought' that is characteristic of objectifying science, to 'the ontological condition of belonging, whereby he who questions shares in the very thing about which he questions'.[80] Belonging is equivalent to being-in-the-world as the presupposition of inquiry.

This is an important methodological point about the human sciences in that the very nature of being human facilitates our understanding. In one sense this is obvious, but I think it is particularly relevant in the case of religions. I have

[74] Jonas, *The Phenomenon of Life*, p. 3.

[75] Paul Ricoeur, 'Phenomenology and Hermeneutics', p. 108, in John B. Thompson, ed. and trans., *Paul Ricoeur: Hermeneutics and the Social Sciences* (Cambridge: Cambridge University Press, 1981), pp. 101–28. On fore-having, fore-sight, and fore-conception see Martin Heidegger, *Being and Time*, trans. John Macquarrie and Edward Robinson (Oxford: Blackwell, 1962), p. 191.

[76] Ricoeur, 'Phenomenology and Hermeneutics', p. 105. [77] Ibid., p. 107.

[78] Ibid., p. 114. [79] Ibid., p. 105. [80] Ibid., p. 106.

dealt with this elsewhere[81] but to briefly recapitulate, developing the notion of the formal indication (*die formale Anzeige*) articulated by Heidegger in his lectures on the phenomenology of the religious life, to understand religion we need to appreciate 'the historical' (*das Historische*) which entails the recognition of our humanity in the past, in the texts of the past. Thus, understanding Paul's letter to the Romans requires the formal indication of my being-in-the-world, of my belonging, as a prerequisite.[82]

This is true of both the theory of life and the philosophy of life. On the one hand, the theory of life develops objective explanations through repeatable experiment; it offers an empiricism that develops high corroboration for its claims. On the other hand, the philosophy of life reminds us that we are already *in medias res* and that far from being an obstacle to objective knowledge, this is a prerequisite. This is particularly important for the explanation and understanding of religion in that the theory of life can provide knowledge that explains religions in terms of such phenomena as niche construction, while the philosophy of life can provide understanding that allows us to appreciate the historical, existential, and transformative dimensions of religions and the force of the claims they have made upon people throughout history.

Ricoeur highlights finitude and belonging as being part of hermeneutical understanding. This is also true of religions insofar as they recognize and respond to finitude and also provide a strong sense of belonging. One of the primary functions of religion is to develop in human communities a sense of belonging, a social ontology, that they are able to do through the transformation of human biological proclivities, the pro-social emotions, empathy, and social cognition. Religious imaginations are examples of niche construction, human attempts at being at home in the world projected onto a cosmic scale. If the leaf cutter ants' nest is a home within which the community of ants can thrive, so religions construct a cosmos through a particular kind of imaginaire that allows us to feel at home, to create a meaningful life, although with the secular age there is a loss of the sense of being at home. On the one hand, we can account for, explain, religions in terms of a theory of life that sees religions as niche constructions and the parallel development of genetic and cultural evolution, while on the other we can account for religions in terms of a philosophy of life that sees them as creating a realm of meaning and significance within a cosmic boundary. Religious views on life have always imbued life with meaning through a cosmological imaginaire, which since the

[81] Gavin Flood, *The Truth Within: A History of Inwardness in Christianity, Hinduism, and Buddhism* (Oxford: Oxford University Press, 2013), pp. 15–16.

[82] Martin Heidegger, *The Phenomenology of Religious Life*, trans. Matthias Fritsch and Jennifer Anna Gosetti-Ferecei (Bloomington and Indianapolis, IN: Indiana University Press, 2004), p. 22.

seventeenth century has come under attack from a scientific imaginaire that supports an objective knowledge through empirical means. To understand and explain religions' accounts of life we need, on the one hand, objective science provided by the theory of life, particularly genetics and social neuroscience; while on the other hand we need interpretative, human sciences such as history, philology, sociology, and their philosophical justification in terms of finitude and belonging.

I shall therefore proceed through three stages: the explanation of the emergence of religion and civilization in terms of human biological proclivities—our pro-sociality, our ability to use language, and our somatic ordering of space and time in ritual—that I shall call the bio-sociology of transformation. Here we shall encounter some contemporary accounts of human pro-sociality in terms of social neuroscience that identifies social cognition as fundamental to the human case and how this is relevant to the explanation of religion: how human life is characterized by social cognition that we might call the transformation of bio-energy. I shall then move on to reading the history of religions in terms of the idea of life itself and how life itself comes to expression in the ways of life constitutive of religions that have both positive and negative evaluations enacted through their practices. Finally, I engage with modern philosophies of life along with contemporary practices that seek the human good, thereby situating my study in a broader perspective of concern for the human future.

2

The Emergence of Religion

Robert Bellah offers a Durkheimian definition of religion as 'a system of beliefs and practices relative to the sacred that unite those who adhere to them in a moral community',[1] as we saw in the Introduction. For Durkheim and Bellah, the term 'sacred' is something set apart from mundane life and is an important term in understanding the connection between religion and life. The Indo-European linguist, Émile Benveniste, analyses the sacred in various Indo-European languages showing that while there is no common term for 'religion', there are accounts of the sacred that usually come in pairs, namely in Greek *hieros* and *hagios*, in Latin *sacer* and *sanctus*, and in Avestan *spenta* and *yaozdata*.[2] These terms reflect a double meaning to sacrality; on the one hand, the sacred points to a vital power and state of fullness and prosperity, a state of health, denoted by the first term in the pairs, while on the other, it denotes the sacred as juridical power that controls the vital power of life. In Esposito's reading of Benveniste, these two aspects, what is animated and what is forbidden, come together in the notion of immunity that articulates the function of religion to keep people safe; ultimately religion as the sacred heals life through 'the absorption of something that binds it to its opposite, that draws life from death or includes death in life'.[3] On the one hand, the sacred is celebration of the wellspring of life itself, that which is animated by life, while on the other hand, the sacred is control of life through prohibition and injunction.

This ancient terminology reflects a truth about religion as having roots within life itself or, more specifically, within the biosociology of human evolution, and also controlling life through prohibition and injunction, commandments, and rules. Expression of life and control of life are both part of the function of sacrality or holiness; the sacred is the appearance of life itself along

[1] Robert Bellah, *Religion in Human Evolution: From the Paleolithic to the Axial Age* (Cambridge, MA: Harvard University Press, 2011), p. 1.

[2] Émile Benveniste, *Vocabulaire des institutions indo-européennes* (Paris: Editions de Minuit, 1969), pp. 179–207.

[3] Roberto Esposito, *Immunitas: The Protection and Negation of Life*, trans. Zakiya Hanafi (Cambridge: Polity Press, 2011), p. 53.

with the attempted control of that life. It is the emergence of religion in terms of both expression and control that links religion through the sacred to the bio-energy of life itself that animates human sociality. Put in rather bold terms, it is this source, life itself, that the sacred transforms and attempts to control, to which religion can be traced.

The past is littered with theories of how religions began, with projectionist theories from Feuerbach to Freud, functionalist theories from Durkheim to Bellah, and neurological theories that religion is a by-product of other neurological functions such as moral decision-making. Although many of these theories, some from giant thinkers who have done so much to form the discourse of the human sciences, have great explanatory power, none have gained general acceptance, and it might even be argued that the quest for an origin of religion is doomed to failure. Such pessimism might be based on the idea that we cannot explain religion because the web of causation is so complex that to specify a particular function or mental process is not sufficiently explanatory to do justice to the data. Certainly, a credible explanation of religion needs to be founded on a credible account of the human.

Since the great theorists of the past, things have moved on in the harder sciences and we can now give accounts of human reality supported by bodies of evidence that were previously simply not available. In particular, we now know much more about human development through evolutionary anthropology, we know about genetics and the genetic study of populations, and we know about the development of cognition and neural networking.[4] While the importance of these scientific advances cannot be doubted, the issue arises as to whether we can speak about them in causal terms in relation to religion (which would be the reductionist or naturalist claim). Are the discoveries of neuroscience the *causal* explanation of human behaviour, including religious behaviour? As Bowker reminds us, there are usually not single, causal explanations for complex behaviours such as religious practices,[5] and rather than cause, as we have seen in Chapter 1, we need to speak about the specification of constraints. Rather than locating a cause, these sciences identify a range of constraints that are clearly important in the emergence and development of religions.

If we can account for human reality in terms of evolutionary anthropology and cognitive science, particularly social cognition, then that is where we must

[4] On human development and the out-of-Africa thesis, see the two authoritative volumes from archaeology and paleo-anatomy: Paul Mellars and Chris Stringer (eds.), *The Human Revolution: Behavioural and Biological Perspectives on the Origins of Modern Humans* (Edinburgh: Edinburgh University Press, 1989) and Katherine Boye et al. (eds.), *Rethinking the Human Revolution: New Behavioural and Biological Perspectives on the Origin and Dispersal of Modern Humans* (Cambridge: McDonald Institute for Archaeological Research, 2007).

[5] John Bowker, *Religion Hurts: Do Religions Cause More Harm than Good?* (London: SPCK, 2018), p. 22.

begin to look for the emergence of religion. Indeed, evolutionary anthropology and cognitive neuroscience are strong candidates for the explanation of religion at one level, but even in a discussion of origins, these are not sufficient because they are limited to the third-person perspective. We also need an account of religion in other than naturalistic terms, for naturalism alone does not do justice to the data (to speak in naturalist language) of first and second-person accounts. The two frames within which to perceive the human, the third-person scientific and the first/second-person humanistic, need to both be present to understand evolutionary constraints along with the emergence of human freedom. This makes the situation highly complex because, on the one hand, we have accounts of the origin and development of religion based on scientific methods that are well corroborated and attested within their respective disciplines, while on the other, we have accounts of religion that need to draw on historical data and an inferred analogy between modern and ancient people who have left no literate record. This complexity is compounded by the very problem of locating an origin.

For scientists to understand origins, it is a matter of presenting supporting evidence to back up working hypotheses. This is particularly difficult in relation to the evolutionary past and we need to proceed not only with archaeology but analogically with what we know of contemporary pre-literate or small-scale societies. So, in the current chapter I will discuss what evolutionary anthropology has to tell us about the origins of religion; in particular the development of the pro-social emotions, and the thorny issue of the origin of language along with the development of ritual behaviour. Explaining religion in terms of evolution, in particular gene replication, is not new and there were previous attempts by ethology and sociobiology. The work of Edward Wilson in this field was ground-breaking, if controversial. He developed sociobiology and turned his attention to human life, notably culture and religion, claiming priority to the genetic drive, and that culture and religion are always on the leash of the genes.[6] This one-way, gene-to-culture causation is no longer maintained in the extended evolutionary synthesis. But even an updated evolutionary anthropology finds its limits in the explanation of religion, and we need to introduce a higher-level account in terms of culture and the lifeworld that entails an existential level of 'I–you' language and human aspiration to a vertical ascent or transcendence. While its roots lie in human reality, religion nevertheless

[6] See Edward O. Wilson, *On Human Nature* (Cambridge, MA: Harvard University Press, 1978). For a discussion of Wilson's thesis in relation to religion see John Bowker, *Is God a Virus? Genes, Culture, and Religion* (London: SPCK, 1995), pp. 35–52. For a survey of the debate including Elizabeth Allen's critique ('Against "Sociobiology"', pp. 259–64) and Wilson's defence ('For Sociolobiology', pp. 265–9), see Arthur L. Caplan (ed.), *The Sociobiology Debate: Readings on Ethical and Scientific Issues* (New York: Harper and Row, 1978).

'represents both the means and the product of the human animal's deepest and most extensive search for its own meaning'.[7]

In Chapter 1 we saw the centrality of niche construction along with natural selection as an equally important factor in evolution. To understand religion, we need to understand the kind of niche that is particular in the human case. Arguably what is relevant here are pro-social emotions (empathy and sympathy and their corresponding behaviours, helping, cooperation, and altruism) revealed in social cognition, the development of language (along with symbolic thinking more widely), and ritual action (and so symbolic action). All these were enabled by an increase of brain size, particularly the development of the thin outer layer of the brain, the language bearing neo-cortex, although developments in the sub-cortex were also important as a foundation for language development as well as pro-sociality, and early hominin survival in fact depended more on subcortical enhancement.[8] These are bold claims in that the origin of language is not known (complex language is probably a latecomer in the seven-million-year history of hominin evolution) and the specificity of human ritual in contrast to other animal species is debatable, even whether ritual actually does symbolize. If these constitute human niche construction, then they are foundational to the extraordinary success of *Homo sapiens*, evolutionary developments precipitated by a massive increase in brain size over two million years ago.[9]

I would wish to argue that the communicative practices that comprise religions have their roots in human niche construction. But this is not a cognitivist argument that locates religion in particular regions of the brain (albeit acknowledging the importance of brain development), or a naturalist, biological reductionism that maintains the hegemony of the genes; rather, it contends that forms of communicative practice that are constitutive of religions, while being rooted in human biology, function at a cultural level that has autonomy from the cellular. Religions are niche constructions that create worlds of meaning through imagination within which people can live complete and competent lives and that function eschatologically to facilitate

[7] John Bowker, 'On Being Religiously Human', p. 366, *Zygon*, vol. 16, 1981, pp. 365–82.

[8] Alexandra Maryanski and Jonathan H. Turner, 'The Neurology of Religion: An Explanation from Evolutionary Sociology', p. 114, pp. 133–4, in Rosemary L. Hopcroft (ed.), *The Oxford Handbook of Evolution, Biology, and Society* (Oxford: Oxford University Press, 2018), pp. 113–42.

[9] Brain size probably increased in the *Australopithecus* ancestors of *Homo*, the living human brain having a cranial capacity of 1350–1400 cm^3 in contrast to the *Australopithecus africanus* brain of about 450 cm^3. Brain size along with neurological reorganization gave *Homo* an evolutionary advantage over their ancestors. Prefrontal cortical evolution began with *Australopithecus*, the ancestor that led to *Homo*. See Dean Falk and K.R. Gibson (eds.), *Evolutionary Anatomy of the Primate Cerebral Cortex* (Cambridge: Cambridge University Press, 2001). Also Dean Falk, 'Hominin Paleoneurology: Where Are We Now?', in M. Hoffman and D. Falk (eds.), *Progress in Brain Research*, vol. 195, 2012, pp. 255–72.

self-repair; their roots are in the pro-social emotions, language development, and ritual behaviour.

One of the problems with a quest for origins is, of course, that an origin is simply the consequence of something else, something temporally prior. But this does not necessarily entail a scepticism, and while we need to respect the caution of Derrida regarding the quest for origins,[10] contemporary evolutionary theory along with what we know about human interactive cognition might allow us to speculate in a way that nineteenth- and twentieth-century theorists who posited the origins of religion could not. We now know about human genetic inheritance, the human proclivity to pro-sociality, the coevolution of genes and culture, and the importance of niche construction in the success of a species. The pro-social emotions provide the basis for empathy, law, and ethics; language allows us to tell stories about ourselves, especially where we came from and where we are going; while the repetitive patterns of ritual articulate something deep from our hominin past—socialization processes certainly but also an existential drive for transcendence or elevation and meaning that addresses a perceived lack in human reality. Indeed, the desire for going beyond the conditions of our life, the aspiration to ascend to higher levels, as it were, is a strong factor or compulsion in the expansion of religions, as attested in the histories of religions.

In brief, I would wish to claim that the human proclivities towards pro-sociality, language ability, and the somatic ordering of time into segmented sequences of action that we call ritual are transformed in religion. Specifically, with religion we see the extension of empathy beyond the group in the here and now, particularly articulated as sacred law that attempts to control social cognition[11] and more generally life, sometimes violently; we see language ability developed into meaningful narratives that tell us where we came from, where we are going, and what we can hope for; and somatic ordering of time and space is transformed into ritual that promotes the pro-social emotions as well as altered states of consciousness that take us out of ordinary human transaction. These human proclivities, by which I mean predispositions to behave in certain ways, loosely correspond to human capacities of feeling, thinking, and doing, or emotion, thought, and action (see Figure 2.1).

[10] A consideration of origin consists in 'the consciousness of the implication of another previous, possible, and absolute origin in general'. Jacques Derrida, *Edmund Husserl's 'Origin of Geometry': An Introduction*, trans. John P. Leavey (Lincoln, NE and London: University of Nebraska Press, 1978), p. 152.

[11] An alternative cognitivist thesis to the origins of religion in cooperation is that religion is a by-product of other brain functions, such as moral decision-making. See, for example, Ilkka Pyysiäinen and Marc Hauser, 'The Origins of Religion: Evolved Adaptation or By-Product?', *Trends in Cognitivist Science*, vol. 14 (3), 2009, pp. 104–9. But a causal model such as this is too restrictive in not sufficiently taking heed of the wider web of constraints in order to account for the emergence of religion in evolutionary terms.

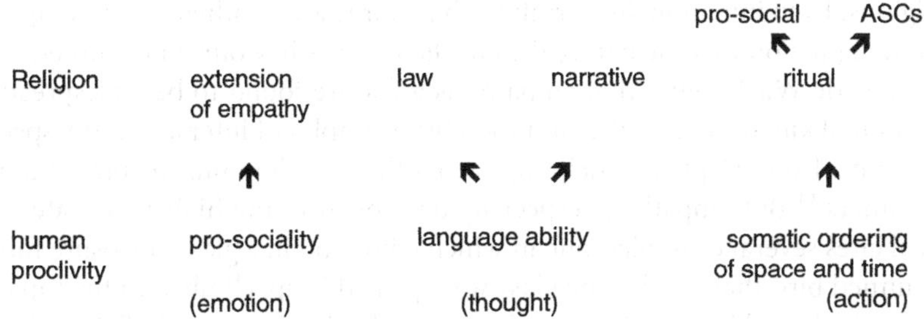

Figure 2.1. The transformation of human proclivities into religion (ASCs: altered states of consciousness)

THE PRO-SOCIAL EMOTIONS

We know the destructive side of human beings simply from reading the news, but the other side of the coin is the ways in which human beings cooperate and go out of their way to help one another, even at the cost of their own safety. We have seen in social cognition how we are deeply affected by the face-to-face and its brain-state correlate, and we have argued that religion can be understood as the transformation of the bio-energy of the face-to-face encounter. While the face-to-face encounter includes aggression, one of the strongest features of human reality is empathy and sometimes sympathy[12] that can become altruism. Religions offer protective boundaries, as systems, and the protection of those boundaries entails control of the face-to-face interaction. One of the key features that we see repeatedly in these systems is the emphasis on the pro-social emotions, particularly empathy that allows us to assess cooperation and belonging, and which is extended in religions beyond the immediacy of communication to a wider community. Many religions urge empathy and more than this, compassion and altruism, that stretch beyond their borders to include those outside—as we see in the Islamic idea of God's compassion and mercy extending beyond the community (*umma*), in the Christian tale of the Good Samaritan, and in Buddhist meditation practice of lovingkindness (*maitri/metta*) extended to all sentient beings. While we will examine how these come to articulation in global civilizational structures and practices in Part II, we simply note here the necessity of empathy in the formation of those communities of imagination that are

[12] On the distinction between empathy and sympathy see Frans de Waal, *The Age of Empathy: Nature's Lessons for a Kinder Society* (London: Souvenir Press, 2009), pp. 88–95. De Waal defines the difference as: 'Empathy is the process by which we gather information about someone else. Sympathy, in contrast, reflects concern about the other and a desire to improve the other's situation' (p. 88).

religions. Even intentionally harmful behaviour towards others entails empathy and imagination in anticipating the effects of our behaviour on the other.[13]

Frans de Waal argues that empathy is a feature found to be widespread in the animal kingdom. He cites innumerable examples of inter and intra-species empathy, from elephants helping each other to dolphins helping human swimmers.[14] But empathy is especially developed in the higher primates. De Waal cites several examples, one in which a bonobo in a glass enclosure found a stunned bird that had hit the glass wall, picked it up, climbed to the top of a tree and released it, spreading its wings to enable it to fly away.[15] This is clearly extraordinary behaviour on the part of the bonobo, one that we immediately recognize as showing compassion for the suffering of another, and in this case of a completely different species. But while empathy and even compassion do occur in other species, they are particularly strong features of *Homo sapiens*.

In many studies Tomasello and his colleagues have shown the dominance of empathy and altruism expressed as cooperation in the human case, and its corollary, helping behaviour. Experimental evidence from studies of children and chimpanzees shows how cooperation and helping occur in both species but altruistic helping behaviour seems to be restricted to the human beings. Through linking the levels of evolution that show the general function of behaviours and psychology that examines the particular mechanisms of cooperation, Warneken and Tomasello demonstrate how helping behaviour is more dominant in children than their ape cousins and while it does not require language, helping does require empathy and the social cognition of understanding the purposes of another, and also entails the motivation to act on the other's behalf.[16] While de Waal's work has emphasized commonalities between humans and other primates, laying stress on what we share, in the field of empathy, Tomasello and his colleagues' experiments have shown marked differences between pre-linguistic children and apes. The children are much more prone to cooperation and helping, which Tomasello calls shared intentionality. Pre-linguistic children up to about eighteen months are generally helpful, generous, and informative, coming to the aid of adults who have dropped pens or cannot gain access to a cupboard when carrying heavy books; the children will see, assess the situation, and without verbal prompting, assist by picking up the dropped pen or opening the cupboard door. Chimps do engage in some helping behaviour but to a much more restricted extent.

[13] Ibid., pp. 210–11. [14] Ibid., pp. 128–9. [15] Ibid., p. 91.

[16] Felix Warneken and Michael Tomasello, 'Altruistic Helping in Human Infants and Young Chimpanzees', *Science*, vol. 311, 2006, pp. 1301–3; Felix Warneken and Michael Tomasello, 'Helping and Cooperation at 14 Months of Age', *Infancy*, vol. 11 (3), 2007, pp. 271–94. See also video of experiments at https://www.youtube.com/watch?v=Z-eU5xZW7cU.

Implicit in Tomasello's work is the idea that individual development (ontogeny) intersects in significant ways with the development of the species (phylogeny). A strong version of this would be that ontogeny recapitulates phylogeny,[17] but a weaker version is that certain features of ontogenetic development are echoes of our early hominin past. The species-specific proclivities that created human niche construction are still nascent in the newborn infant and come to sequential development as the child grows. Tomasello highlights a number of important features about the specificity of human niche construction. As human beings we now live in complex social institutions, some of which cover the globe. These are 'cooperatively organised and agreed upon ways of interacting' that include rules of enforcement for non-cooperators.[18] Social institutions still need to draw on our basic human proclivities that human beings teach one another, which Tomasello calls a form of altruism, and humans imitate one another in the formation of group identity. While imitation certainly occurs in primates and other animals as the basic structure of acquiring non-innate behaviour, in humans this has been elevated to a central feature of our species that allows for the continuity of cultural practices, cultivating cooperative thinking and ways of acting, including the preservation through the generations of successful cultural innovations (such as technologies).

There are innate behaviours expressing altruism, such as cooperation and helping, that human beings are born with and which mark them out, in degree rather than kind, from other primates. Helping others is an important type of behaviour that expresses altruism and is not taught; it is not a product of culture but innate within us. This is an important point because innate altruism is a feature emphasized in religious practices and so this shows the ways in which religions are deeply rooted in human niche construction. Children help from as early as twelve months and one of the key forms of helping is by giving or sharing information. The principal way in which the human infant shares information is through pointing; human infants point to help the other because they have a sense of Tomasello's shared intentionality and belonging to the group. There is some controversy over this. De Waal gives examples of apes pointing. For example, a chimpanzee called Nickie pointed with his head to some bushes where de Waal had, on a previous occasion, thrown some berries. The chimpanzee remembered and pointed for de Waal to retrieve them.[19] Chimps also point with their hands to indicate that they want food. This kind of pointing behaviour, with hand and head, had

[17] This 'law' was first devised by the Darwinist Ernst Haeckel (1834–1919). See W. Coleman, *Biology in the Nineteenth Century: Problems of Form, Function and Transformation* (Cambridge: Cambridge University Press, 1977), p. 47.

[18] Michael Tomasello, *Why We Cooperate* (Cambridge, MA: MIT Press, 2009), p. xii.

[19] De Waal, *The Age of Empathy*, p. 151.

been observed previously by a primatologist, Emil Menzel, and experimentally established by his son Charles Menzel.[20] But even so, as de Waal admits, apes do not point for one another. The uniquely human feature is that prelinguistic children will use pointing to help others with no reward for themselves. Indeed, reward in the form of praise can disrupt helping behaviour; altruism is its own reward. And humans, more than chimps, point with their eyes. This is a significant distinction between *Homo sapiens* and *Pan troglodytes*. While non-human primates have dark eyes with the whites of the eye, the sclera, barely visible, humans have a very noticeable sclera, making the direction of the glance much more visible to others. This indicates the importance of eye direction in gaining advantage for self and other. The white of our eyes 'suggests predominantly cooperative situations in which the individual may rely on others using this information collaboratively or helpfully, not competitively or exploitively'.[21]

The broader evolutionary and cultural implications of this are that in the human case altruistic behaviour was crucial in forming our world. Altruism or mere empathy creates a social bond necessary for the development of shared intentionality that in turn establishes social norms and, in time, institutions. Tomasello's shared intentionality hypothesis involves cognitive representation, inference, and self-monitoring and argues for the importance of humans having shared goals and forming a 'we' as a kind of collective agent. Having identified joint intentionality in human infants, children engaged in a common task and helping each other, Tomasello then goes on to identify shared intentionality as a crucial feature in the development of human culture.

Tomasello's argument may be corroborated by work on early flint knapping; that flint knapping to produce sharp flakes and hammers was done in a public space where the skills could be passed on through observation and imitation.[22] What is characteristic of human tool use is not innovation, the emulative mode or doing something in one's own way having been shown how to do it, as one might have thought, but rather imitation, the mimetic mode. Other primates copy through understanding the basic idea and developing it their own way, but humans copy through precise imitation, as we know from the amount of debris produced in flint tool production (emulation produces

[20] Ibid., p. 154.

[21] Michael Tomasello, *A Natural History of Human Thinking* (Cambridge, MA and London: Harvard University Press, 2014), p. 77; Tomasello, *Why We Cooperate*, pp. 75–6; H. Kobayashi and S. Kohshim, 'Unique Morphology of the Human Eye and its Adaptive Meaning: Comparative Studies on External Morphology of the Primate Eye', *Journal of Human Evolution*, vol. 40, 2001, pp. 419–35.

[22] Peter Hiscock, 'Learning in Lithic Landscapes: A Reconsideration of the Hominid "Toolmaking" Niche', *Biological Theory*, vol. 9, 2014, pp. 27–41.

more debitage than imitation) and from human/ape experimental work.[23] Thus, the form of the handaxe in the Palaeolithic may have depended on high-fidelity cultural transmission,[24] and conchoidal fracture of flints in the Early Stone Age indicates the precise transmission of a skill.[25] Here shared intentionality was focused on the production of technology that had the equally important effect of bonding the group; an instance of social cognition mediated through a task, forming a wide community of attention. By contrast, ancient and contemporary non-human apes engage in individual intentionality characterized by competition along with the evaluation of an action in a particular situation, whether it will be advantageous or not to the individual and whether there are obstacles or opportunities to pursue certain goals. Thus, a chimpanzee might see a tree with bananas and will non-verbally assess the advantages and disadvantages of approaching the tree (such as whether the bananas are reachable, whether there are predators around, whether escape will be easy, and so on).[26] Joint intentionality entails cooperative communication between two agents with joint attention to a task, a joint goal, and the ability to collaboratively achieve that end. Finally, collective intentionality entails group culture or group-mindedness along with conventional forms of communication, characteristic of modern humans.[27]

Tomasello shows how this kind of collective or shared intentionality has been crucial for human development and for culture-gene coevolution. Furthermore, shared intentionality is linked to the development of the pro-social emotions and social norms. Cooperation entails empathy and in addition altruism is natural in the human case. These feed into conformity to the social body and enculturation into culturally specific ways of life that are both genetically inherited and taught. The pro-social emotions form community and create social norms that members of the group adhere to. As Tomasello says:

> The universality of social norms, and their critical role in human evolution, is apparent. All of the well-studied traditional societies incorporate powerful social

[23] David F. Lancy, '*Homo faber juvenalis*: A Multidisciplinary Survey of Children as Tool Makers/Users', *Childhood in the Past*, vol. 10, 2017, p. 7, issue 1, pp. 72–90. Human children have a propensity for imitation; see Claudio Tennie, Joseph Call, and Michael Tomasello, 'Push or Pull: Imitation vs Emulation in Great Apes and Human Children', *Ethology*, vol. 112, 2006, pp. 1159–69. For an excellent survey of tool use development see Fuentes, *The Creative Spark*, pp. 51–67.

[24] Claudio Tennie, David R. Braun, and Shannon P. McPherron, 'The Island Test for Cumulative Culture in the Paleolithic', p. 124, in Miriam N. Haidle, Nicholas J. Conrad, and Michael Bolus (eds.), *The Nature of Culture* (Dordrecht: Springer, 2016), pp. 121–33.

[25] Tetsushi Nonaka, Blandine Bril, and Robert Rein, 'How Do Stone Knappers Predict and Control the Outcome of Flaking? Implications for Understanding Early Stone Tool Technology', *Journal of Human Evolution*, vol. 59, 2010, pp. 155–67.

[26] Tomasello, *A Natural History of Human Thinking*, pp. 10–11.

[27] Ibid., *passim*.

norms about what one can and cannot do, even (or perhaps especially) in the most biologically relevant domains such as food and sex.[28]

Religions tend to reinforce social norms (although sometimes disrupting them) by paying great attention to these biological domains, and it is hardly surprising, given our natural predilection towards altruism, that religions have emphasized the importance of love and compassion in their stated theologies, if not universally then within the group.

Empathy and compassion, which are innate and facilitate niche construction in the human case, are realized through the mechanism of the face-to-face encounter. Joint attention entails the ability to cognize an 'I' and a 'you', which the great apes are able to do, and was arguably a feature of our ancestors. *Homo heidelbergensis*, some 400,000 years ago, engaged in joint attention, speculates Tomasello, which was moving towards collective intentionality. But joint intentionality is only viable while it lasts and once the purpose is over, such as retrieving food, it dissolves. The next 'leap forward' was the institutionalization of joint intentionality into collective intentionality which does not dissolve and establishes norms and institutions.[29] From the realization that there are perspectives, a here and a there, joint attention transforms into collective intention and the ability to abstract perspective from 'I' and 'you' to the whole group, from the second-person to the third-person view. The shift to collectively intentionality, the ability to abstract from the present situation, is produced not merely from the pro-social emotion of empathy, but in the production of language as well.

As human evolution develops, and the pro-social emotions are transformed from a second-person perspective to a third-person perspective, religion emerges as a social force that draws on the bio-energy generated through the pro-social emotional encounter with the other. But shared emotional bond, as posited in affect theory,[30] is insufficient to account for the emergence of religion. Alongside this extension of the face-to-face encounter we have the

[28] Tomasello, *Why We Cooperate*, p. 42.

[29] Tomasello, *A Natural History of Human Thinking*, p. 79. Dunbar speaks of five levels of intentionality and that a fifth level is required for religion to develop. Whether formalized in this precise way, it seems clear that religion does indeed need abstraction and the ability to narrate and project into the space of imagination. See Dunbar, 'Why Are Humans Not Just Great Apes?', in Charles Pasternak (ed.), *What Makes Us Human* (Oxford: One World, 2007), pp. 37–8.

[30] Donovan O. Schaefer, *Religious Affects: Animality, Evolution, and Power* (Durham, NC and London: Duke University Press, 2015). Schaefer's interesting book develops the idea of the inquiry into religions from within a materialist paradigm, through focusing on affect that has been overlooked by textualist scholarship, focusing on 'how power relations touch and move bodies' (p. 34) and that 'maps the deeper embodied formations by which power makes bodies move' (p. 35). As such, 'religious affects' could potentially be seen in non-human animals as well, such as chimps dancing in a waterfall, and 'animal religion' could be characterized by non-linguistic (in the narrow sense) features, particularly bodily movement (pp. 191–2). As we will see, 'ritual' elements of proto-religion may well be non-linguistic, but developed religion entails language.

crucial development of language. It is the combination of the pro-social emotions with language ability in particular that marks out the modern human from our ancestors and from other higher primates.

THE ORIGINS OF LANGUAGE

The origin of language is a problem because we cannot be certain about the point at which it developed in our evolutionary history. Sophisticated linguistic communication may be quite recent in evolutionary terms, as recent as 60,000 years ago, although Barnard suggests 200,000 years ago if we identify language with symbolic thinking,[31] but meaningful vocalizations that express thoughts must go far back far into the past, with what Dunbar calls 'vocal grooming' due to enlarged hominin group size emerging around two million years ago.[32] Increased cognitive skills as a consequence of brain changes, specifically the growth of the neo-cortex that encodes language and stores words in a specific place connected in a network of associations, allowed for the development of language. In particular there has been the discovery of a gene ('forkhead box P2' or 'FOXP2') that enables speech that Neanderthals may have possessed.[33] Perhaps the place to begin is in iconic gestures such as pointing. Now while great apes point, as we have seen, they do not do so for the sake of another. Tomasello writes:

> To comprehend iconic gestures, one must be able to see intentional actions performed outside of their normal instrumental contexts as communication—because they are marked as such by the communicator via various kinds of ostensive signals (e.g. eye contact).[34]

From a young age, before the acquisition of language, human beings know how to point to help out their fellows. The argument is that this shows pointing behaviour, an iconic gesture in Tomasello's terminology, to be innate. The innate capacities of pre-linguistic children must reflect our hominin past because such gestures are prior to learning. Now even if pointing is crucial for early communication, it is a long way from pointing with hands and eyes to the communication of abstract ideas. Language, as Frits Staal observes, consists of phonological and syntactic components but also a semantic

[31] Alan Barnard, *Genesis of Symbolic Thought* (Cambridge: Cambridge University Press, 2012), p. 8.

[32] Robin Dunbar, *Grooming, Gossip, and the Evolution of Language* (London: Faber and Faber, 1996), pp. 114–15.

[33] Alan Barnard, *Genesis of Symbolic Thought* (Cambridge: Cambridge University Press, 2011), p. 83.

[34] Tomasello, *A Natural History of Human Thinking*, p. 60.

component 'that enables language to convey meaning'.[35] Early vocalization would have phonological and syntactic elements but no semantics at first. A story could be told that early pointing became linked with vocalization, but advanced language capacity and the development of semantics are something else entirely, it seems to me.

The key here is the shift from ostensive language, the ability to indicate features of a present situation linked to pointing, to abstract language that entails structure, semantics, imagination, and the ability to think of oneself as another. This is key in the construction of the human niche. The point is that language as part of 'human pragmatic communication is an exaption of—and hence is evolutionarily continuous with—human social intelligence'.[36] Language allows collective intentionality and develops a skill in *Homo sapiens* that far outstrips other primates in their ability to adapt the environment to their needs. But how this shift to complex language came about is a topic of controversy. There are a number of theories about the origins of language, which are not incompatible with each other. Robin Dunbar has put forward the interesting hypothesis that language originated in grooming behaviour to bond the group, in 'gossip'. A larger group entailed more time spent in grooming and so less time in foraging, so language developed as a way of group bonding to supplement grooming activity; speaking becomes 'grooming at a distance' because it allows other activities, particularly foraging, to take place at the same time.[37] Another credible theory is that language originated as singing among Neanderthals in a parallel way to the singing of gibbons as sexual display.[38] This idea that we sang before we talked has appeal. Frits Staal long ago argued that Hindu and Buddhist mantras are left over from the origins of language. Mantras are sound formulas in the Sanskrit language that can be meaningful such as *namaḥ śivāya*, 'Homage to Śiva', or meaningless but used in ritual and meditation such as the Śaiva 'eye' mantra, *oṃ juṃ saḥ*, and in particular what is called the 'seed' (*bīja*) mantras. Indeed, mantras have much in common with music in following a sequence, usually moving from a particular note and returning to it, often via a complex route. Music, like the ritual language of mantras, has syntax, and often a complex structure, but no

[35] Frits Staal, *Rules without Meaning: Ritual, Mantras, and the Human Sciences* (New York, Bern, Frankfurt, Paris: Peter Lang, 1989), p. 188.

[36] Agustin Fuentes, 'Manipulating Materials, Bodies, and Signs: How the Ecology of Creative Problem Solving, Tool Manufacture, and Imaginative Sociality Set the Context for Language in the Later Pleistocene Human Niche', p. 192, in Celia Deane-Drummond and Agustin Fuentes (eds.), *The Evolution of Human Wisdom* (Lanham, MD: Lexington Books, 2017), pp. 191–204.

[37] Dunbar, 'Gossip and the Social Origins of Language', p. 344, in Kathleen R. Gibson and Maggie Tallerman (eds.), *The Oxford Handbook of Language Evolution* (Oxford: Oxford University Press, 2011), pp. 343–5; also see Dunbar, *Grooming, Gossip, and the Evolution of Language*, pp. 78–9.

[38] Steven Mithen, *The Singing Neanderthals: The Origin of Language, Music, Mind and Body* (London: Weidenfeld and Nicolson, 2005).

semantic component (except in the case of singing lyrics). If ritual, as Staal argues, is devoid of meaning but contains a structure (a topic we will return to below), then ritual is associated with the utterance of meaningless sounds, the seed mantras. But while Staal may be correct that mantras are a leftover from meaningless sounds from our hominin past, we still need an account of how we move from pure syntax to semantics; not only how language affects us, but how it conveys meaning, how it communicates. There has to be a transition from singing to communicating semantic content and it is this move that needs to be explained.

Another theory is that language originated through australopithecine mothers babbling to infants and needing to reassure and control them while foraging.[39] Through experiments with apes and mothers with children, Dean Falk argues that *motherese*, musical speech directed towards babies, forms the basis of language. The consequences of an enlarged brain in australopithecines selected for an earlier birth, which gave problems to mothers who needed to find food. Where to park the baby? By adopting new foraging strategies, mothers put their babies down and controlled and silenced their offspring through verbal utterance along with gesture and facial expressions that gradually came to have stable meanings. In contrast to studies that wish to separate prosody from the development of language, Falk argues that language develops from the prosody of mothers to their babies, a view that is compatible with the musical origins of language. A number of scholars have questioned this view, however, such as Fuentes on the problematic assumption that early hominin mothers foraged alone[40] and Bickerton's criticism that this does not account for how structure and symbolism evolved.[41] Indeed, this is a mysterious process, clearly linked to the development of the brain, but necessarily triggered by environmental factors.

Derek Bickerton has presented an interesting theory to account for the advent of developed language: that early hominins were 'power scavengers'. Rather than hunting, most of our food was obtained through scavenging on dead mega-fauna in the Savannah of Africa.[42] When a large animal dies, there is a hierarchy or scavenging chain: first the carnivores such as lions, then hyenas, then vultures. At first *Homo sapiens* was low down in the chain,

[39] Dean Falk, 'Prelinguistic Evolution in Hominins: Whence Motherese?', *Behavioural and Brain Sciences*, vol. 27, 2004, pp. 491–503. This interesting paper argues that language arose through mothers' interaction with offspring, especially in the context of foraging and the need to 'park' the baby. The article is accompanied by twenty-six critical responses that range from the question of whether australopithecine mothers foraged alone (Fuentes) to the criticism that the theory does not account for the development of symbolic thought that language expresses (Bickerton).

[40] Agustin Fuentes, Response, ibid., p. 513.

[41] Derek Bickerton, Response, ibid., pp. 504–6.

[42] Derek Bickerton, *Adam's Tongue: How Humans Made Language, How Language Made Humans* (New York: Hill and Wang, 2007), p. 161.

probably below hyenas. But he had the use of tools—large stone hammers that broke open the bones for their precious marrow. In time, the hominins moved to the top of the food chain. We know this because archaeology has found bones with stone hammer marks over teeth marks, indicating that the hominins were accessing what was left of the carcass after the lions and hyenas. However, in time, this situation is reversed with the teeth marks on top of the blow marks, indicating that humans were accessing the carcass before the carnivores.[43] How did this come about? Bickerton argues that the primary reason is human cooperation that developed due to a new linguistic ability.

Early hominins lived in small groups and language was purely ostensive, concerned with the here and now. To scavenge large animals required collaboration and so the ability of individuals to communicate the existence of a fallen animal some distance away. That is, to say that there is a dead elephant a few hours' walk to the south entails the use of non-ostensive language; it entails a metaphorical space of abstraction and an imagination of the future. What Bickerton calls 'power scavenging' marked homo out from other primate species. A second factor in humans rising to the top of the scavenging chain was the innovation of the Acheulean handaxe, found in abundance, and sharp flakes used for cutting. When a large animal dies, the scavengers have to wait for a period of time before their teeth can penetrate the hide. With sharp tools the women could cut into the hide while the men hurled stone 'axes' at other predators to keep them away. This accounts for *Homo sapiens* becoming the dominant scavenger, even though it might entail fighting off large carnivores or sacrificing some group members while the women worked the carcass. An early kind of non-ostensive language therefore developed in relation to the need for scavenging food, which would entail communicating possible futures to others as well as group cohesion and trust. The survival of the group would depend upon authentic speech because the group would have to trust the word of those who had located the carcass.

This area is fraught with difficulty. Many studies of early hominin evolution, such as those of Tomasello that we have reviewed, are based on contemporary child–ape parallelism, and while it is probable that we can infer back to early hominins from this data, it is not certain. The Bickerton theory has the ring of plausibility about it, but we need to take other factors into account such as the physiology of speech—how and why did we develop the vocal chords to articulate words? Early australopithecines probably had the ability to vocalize words, which requires laryngeal descent and a curved tongue along with 'two tubes', features that distinguish human anatomy.[44] We also need the

[43] Ibid., p. 220.

[44] Anne MacLarnon, 'The Anatomical and Physiological Basis of Human Speech Production: Adaptations and Exaptions', in Dean Falk and K.R. Gibson (eds.), *Evolutionary Anatomy of the Primate Cerebral Cortex* (Cambridge: Cambridge University Press, 2001), pp. 224–35.

hyoid bone to speak and such a bone has been found in a 60,000-year-old Neanderthal skeleton in Israel.[45]

The evolution of language, which has phonology, syntax, and semantics, was crucial for *Homo*'s ability to survive and prosper. It enabled early humans to power scavenge and to communicate future possibilities to others of the group. It was also crucial for the development of religion insofar as religions foster ways of life founded on visions of the world that extend the basic ability of language to project a future. Narratives that develop due to sophisticated, non-ostensive language mould community identities and form eschatologies in which the group sees itself as moving towards a particular horizon of possibility. Of course, this must remain speculation so far back in time, but we can make reasonable inferences from what we know of modern humans while being respectfully cautious. Transcending death is one particularly important theme to emerge with religion that language can allow us to imagine, although religions did not evolve to alleviate fear of dying, as early, textual representations indicate a rather bleak view of death; Sheol in ancient Judaism, a place of darkness, or Hades in the *Odyssey* where the dead are like bats squeaking around Odysseus' head.[46] An aspiration for immortality became important in later religions[47] but is not part of their emergence. How early abstract thinking occurred is difficult to say, but it could be argued that religion in closely linked to the development of language and symbolic thinking that may pre-exist language. Religion in a full sense of a narrative structure that makes sense of the world and gives human communities meaning, along with rules of conduct that ensure group cohesion and conformity, could only have developed with the emergence of complex language. But it may be the case that symbolic thinking pre-dates complex language use and proto-religion developed through behaviours that we might call expressions of symbolic thinking, although without language it is difficult to determine what such behaviours were symbolic of. I am thinking here about the early development of human cultural artefacts, especially those objects we designate as art, some of which may have been simply bodily decoration such as shell beads.

Alan Barnard in a well-considered book argues that symbolic thinking developed around 130,000 to 120,000 years ago before full language, at least before the explosion of the Toba volcano in Indonesia and the findings of the

[45] Barnard, *Genesis of Symbolic Thought*, p. 83.

[46] John Bowker, *The Meanings of Death* (Cambridge: Cambridge University Press, 1991), p. 32.

[47] For an interesting argument that the primary driver of civilization is desire for immortality, see Stephen Cave, *Immortality: The Quest to Live Forever and How It Drives Civilization* (London: Biteback, 2013). The situation is, however, more complex than this presentation because the complexity of religion embodies and accesses the human social cognition system.

Blombos cave in South Africa, dated to around 74,000 years ago.[48] He cites an impressive list of evidence from the Blombos cave, which has yielded etched red ochre and shell beads dated 77,000 or 75,000 Before Present (BP), engraved ostrich egg fragments from southern Namibia (83,000 BP), and possibly earlier uses of ochre in Zambia (270,000 to 170,000 BP). There is also a perforated bear femur from Slovenia produced by non-*Homo sapiens* and called the 'Neanderthal flute', and a possible burial in Spain by *Homo heidelbergensis* (320,000 BP).[49] But certainly with cave art—the earliest from India between 190,000 and 720,000 BP—we have evidence of symbolic thinking. This is, however, to exclude the making of stone tools. The Lower Palaeolithic (2,600,000–100,000 BP) gave rise to Oldowan and Acheulean handaxe industries, which spread out of Africa with early homo migrations. With the middle Palaeolithic (300,000–30,000 BP) we have stone industries named Mousterian and the development of stone-tipped spears with Neanderthals and *Homo sapiens*, and in the Upper Palaeolithic we have bows and arrows, and the full panoply of cultural forms such as rock art and the famous figurines (such as the Venus of Willendorf). During this period too, we have advances in other social practices, particularly the domestication of animals,[50] and probably the development of kinship systems that are arguably still preserved in contemporary hunter-gatherer communities.[51]

So, by the time of the Upper Palaeolithic we have sophisticated stone tools along with what we now would call 'art'. We also may have the domestication of the dog as the first domesticated animal from around 30,000 BP. In the Chauvet cave in France the footprints of a dog have been found accompanying a child dated to around 26,000 BP, much earlier than the established domestication in Natufian culture of the Levant (*c.*12,000 BP).[52] This is particularly significant as evidence of language because we need to use language to control the dog. We need to name the dog and command it (indeed that is what domestication means).[53]

Sophisticated, non-ostensive language must have been in use by around 30,000 BP but there is a long time between this date and the early use of the

[48] Barnard, *Genesis of Symbolic Thought*, pp. 13–24. [49] Ibid., p. 15.

[50] Ibid., pp. 23–7.

[51] Barnard argues that kinship among the San in South Africa may be preserved from our early ancestors because the lifestyle of these hunter-gatherers has not changed. This kinship pattern is universal, which means that members recognize kinship relations for everyone in their society. Barnard, *Genesis of Symbolic Thought*, pp. 53–6. For a different argument that the earliest human kinship system was divided into four groups, see Nick Allen, 'Tetradic Theory and the Origin of Human Kinship', in Nick Allen et al. (eds.), *Early Human Kinship: From Sex to Social Reproduction* (Oxford: Blackwell, 2008), pp. 96–112.

[52] Tamar Dayan, 'Early Domesticated Dogs of the Near East', *Journal of Archaeological Science*, vol. 21 (5), 1994, pp. 633–40.

[53] Mark Derr, *How the Dog Became the Dog: From Wolves to our Best Friends* (New York: Overview Books, 2011), pp. 43–50.

handaxe. During this period rudimentary language of an ostensive nature probably developed, as did, arguably, vocalic music. But by 30,000 BP the genus homo had a large brain and could probably communicate in as complex as any modern language. Indeed, it is perhaps a puzzle that the oldest languages we know about in the Indo-European family are the most complex—Greek, Latin, and Sanskrit—with many grammatical cases and rules, which become simplified into the modern vernaculars (modern Greek, Italian and other romance languages, and Hindi and other Indo-Aryan languages). Language must have become syntactically complex by 2000 BCE and simplified again over the centuries. All this is accompanied by developments in material culture: the use of stone tools, possibly containers for carrying meat, the use of fire for cooking, and the development of what we now call art, from the early scratch marks on red ochre to the stunning sketches and paintings of the Chauvet cave.

RITUAL

I have given an esquisse of some findings and theories about the origins of human empathy and the origins of language. The two may be closely linked insofar as empathy, and by extension altruism, may be related through pointing and pointing may have been accompanied by vocalization that in time developed into the ostensive language of the here and now. How we get from simple phonology and syntax to developed semantics is a problem, and we have surveyed plausible accounts in grooming behaviour, child minding, and power scavenging. Perhaps all played a part. But with pro-social emotions and with language in place, we have the necessary conditions for religion. A third important component of the human niche that is religion is ritual behaviour.

Like 'religion', the word 'ritual' has various definitions and can be specified in a number of ways.[54] I like Rappaport's characterization of ritual as 'the performance of more or less invariant sequences of formal acts and utterances not entirely encoded by the performers',[55] and I have earlier spoken of ritual as the somatic ordering of time and space into segmented sequences of action. Ritual is a patterning of behaviour and the performance of sequences of action that have been received. While I would not wish to promote a

[54] For a survey of theories see Catherine Bell, *Ritual Theory, Ritual Practice* (Oxford: Oxford University Press, 1992).

[55] Roy Rappaport, *Ritual and Religion in the Making of Humanity* (Cambridge: Cambridge University Press, 1999), p. 42. Cited and discussed in Gavin Flood, *The Importance of Religion: Meaning and Action in our Strange World* (Oxford: Wiley-Blackwell, 2012), p. 65.

distinction between nature and culture (because culture is niche construction within nature), we might say that while the origins of ritual may be in innate behaviours genetically driven, what we identify as human ritual is learned and passed through the generations which can add innovations or remove sequences of action. Usually ritual is either preserved trans-generationally with little change or dies out. Ritual in the wider animal kingdom tends to be functional—territorial display to keep others away, sexual display to attract a member of the opposite sex—but human ritual often has no immediately apparent function. It is, in Frits Staal's phrase, action for its own sake.[56] If meaning is restricted to linguistic reference, then Staal's claim that ritual is meaningless, having syntax but no semantics, must evidently be the case in ritual's origin. Ritual precedes language and ritual articulates ways of being that are pre-linguistic.

The functional use of stones for cracking open encased fruit has been observed in non-human primates, with chimpanzees being closest to human stone use. Chimps have also been observed to participate in apparently non-functional stone-throwing ritual in which chimps throw stones at particular trees where the stones accumulate in a pile in the buttress roots or hollows. This accumulative stone throwing has been observed throughout West African chimp populations.[57] Perhaps an extension of aggressive display behaviour, the male chimp actions are reminiscent of human ritual. Throwing a stone against a tree and accumulating the stones in a pile seems to serve no practical function, but we see in this behaviour an emergence of ritual for its own sake without any accompanying linguistic commentary about its meaning. There is a definite pattern to the chimp behaviour: the chimp picks up a rock accompanied by bodily posture of swaying, piloerection, and a bipedal stance, followed by pant hooting and then throwing the stone. The stone is directly thrown at the tree, or thrown into the hollow of the tree, or the tree is repeatedly hit with it. This is followed by pant hooting and drumming with the feet, which ends the performance. Afterwards the chimps travel on their way, perform further displays, or simply wait and listen. Human ritual may have originated in a similar fashion. Behaviour such as breaking open fruit with a stone is used outside of that functional context: the chimp non-functional stone-throwing behaviour is an extension of functional behaviour. The 'sacred trees' of the African chimps may not be religion, but perhaps we can learn from this about the possibilities of religious ritual origin. Were chimps to develop language, there would be some narrative explanation of the stone

[56] Staal, op cit, p. 131: 'Ritual is pure activity, without meaning or goal. Let me briefly digress for a point of terminology. Things are either for their own sake, or for the sake of something else...To say that ritual is for its own sake is to say that it is meaningless, without function, aim or goal'.

[57] H.S. Kühl et al., 'Chimpanzee Accumulative Stone Throwing', *Scientific Reports*, vol. 6, 2016, p. 22219.

accumulation, even though such an account would not be part of the original behaviour. Staal observes that linguistic accounts of ritual are ad hoc and arbitrary; they tell us nothing about ritual acts themselves and while a ritual remains the same, accounts of it in language, accounts of its meaning, change. We seek to attribute meaning to the chimp stone throwing but perhaps there is none; there certainly is no linguistic meaning. Staal's view of ritual as meaningless action for its own sake, with structure and pattern but no semantic component, could perhaps be seen in the chimp example. But even here, the stone-throwing rite would seem to be founded on display behaviour that 'means' or indicates the self-perceived importance of the chimp agent—'I'm the boss' behaviour.

The chimp example has no social function whereas most human ritual seems to have the function of social bonding, of articulating the deep social cognition of our nature, or of establishing social and political relationships. Staal acknowledges that this occurs but argues that it is a by-product of ritual procedure. He writes:

> If ritual is useless this does not imply that it does not have useful side-effects. It is obvious, for example, that ritual creates a bond between the participants, reinforces solidarity, boosts morale and constitutes a link with the ancestors. So do many other institutions and customs. Such side-effects cannot be used to explain the origin of ritual, though they may help to explain its preservation.[58]

For Staal, ritual is a pattern of action that is a remnant of our pre-linguistic past. While I take his point that ritual arose before the development of language, and the emphasis of ritual is on structure and correct performance, it nevertheless comes to be associated with social and transcendent meaning. In the course of time, ritual becomes eschatological and performed within the greater narrative of tradition. Rituals in religions anticipate a goal, prepare participants for the future, and transform them in various ways, often into divine beings. Van Gennep long ago in an important, influential book noted that rituals of passage through life follow a pattern of rites of separation, rites of liminality, and rites of reaggregation.[59] This pattern is identifiable clearly in puberty rites where youngsters are separated from the usual social world, then to be reintegrated into adult life, as Victor Turner has documented among the Ndembu,[60] and even in personal prayer, the act of praying is inextricably

[58] Frits Staal, *Rules without Meaning: Ritual, Mantras, and the Human Sciences* (New York: Peter Lang, 1989), p. 134.

[59] Arnold van Gennep, *Les rites de passage* (Paris: Picard, 1909), not translated into English until 1960 by M.B. Vizedom and G.L. Cafee, *The Rites of Passage* (Chicago, IL: University of Chicago Press, 1960).

[60] Victor Turner, *The Forest of Symbols: Aspects of Ndembu Ritual* (Ithaca, NY: Cornell University Press, 1967).

social.[61] This is a standard pattern of ritual procedure, a threefold structure that allows the transcendence of ordinary social time into a sacred time zone. The *teyyam* dancers of Kerala are transformed into gods for the duration of the performance for an eschatological reason for the community to whom they give the vision of the other world and for whom they offer prediction and advice.[62] Jewish prayer anticipates the coming of the Messiah, and so on. The anticipation of a future that is inherent in ritual structure is meaningful for the community who performs it even if ritual in its origin is in itself prior to semantics. The function of creating bonds between people—in our terminology, the facilitating of social cognition and broadening of the face-to-face encounter—is an important function even if it is a by-product of behaviour that originated in meaningless action.

Religions as autopoietic systems comprise communities of varying degrees of inclusiveness, social cognition with variable extension, along with sacred language set aside in oral or written texts, accompanied by systems of ritual—specific, repeatable, but non-identical actions within a given time frame. Religions as systems are composites that draw from and develop the main building blocks of human niche construction, namely social cognition, language, and ritual. All three may be closely related. Given the evidence of the chimps' 'sacred trees' with their stone-throwing behaviour not linked to any practical outcome, we might argue that ritual that becomes religious is in origin not linked to conscious meaning but becomes harnessed to meaning in religious systems. Any religious ritual from simple Moslem prayer to elaborate Divine Liturgy in the Orthodox Church draws on hominin capacity for ritualization, language set aside, and social cognitive capacity extended to the community.

THE FACES AT ALTAMIRA

Not only does ritual have the consequence of bonding, but it can do this through the important human characteristic of altering the states of mind of the participants. By the time the paintings in the Altamira cave were completed during the Upper Palaeolithic, 45,000 to 10,000 BP, *Homo sapiens* had succeeded *Homo neanderthalensis* and were modern humans probably with language and certainly with curiosity about who they were and what the world is. It is a long journey from the Blombos caves to Altamira, and by the time we

[61] Marcel Mauss, *On Prayer*, trans. Susan Leslie (New York and Oxford: Berghahn Press, 2003), pp. 33–7.

[62] Rich Freeman, 'The Teyyam Tradition of Kerala', in Gavin Flood (ed.), *The Blackwell Companion to Hinduism* (Oxford: Blackwell, 2003), pp. 307–26.

get here, humans had developed the ability to represent the animals they hunted on the walls of caves and to express their visions. At 32,000 BP we have the theriomorphic sculpture of a lion-man found in the Vogelherd cave[63] and small female figures such as the Willendorf 'Venus' common in Europe to central Siberia.[64] At 14,000 BP we have paintings at Lascaux that according to Witzel refer to a late Palaeolithic myth: an ithyphallic figure lies prostrate with a bird-like face, nearby a pole with a duck-like bird on the top and a bison pierced with a spear through its anus and thereby disembowelled. Although somewhat speculative, Witzel connects this scene to early thinking about shamans and hunting that relates these activities to the killing and dissection of a primordial being.[65] Whether this particular example can be so specifically identified or not, the broader point remains that such depictions may have been illustrative of narrative.

Rock art in its long history is probably associated with ritual and a spirit world. David Lewis-Williams has argued that a key to understanding this mysterious art is Shamanism, the name for a kind of tradition in small-scale, hunting and herding societies in which spirits take the form of animals and the shaman, entering a trace state, communicates with them, particularly for the curing of the sick. In the shamanic worldview the shaman goes into a state of trance through drumming and sometimes with the aid of hallucinogens such as the *amanita muscaria* mushroom in Siberia or ayahuasca plant preparation in South America. In Siberian Shamanism he travels to the spirit world and returns with the soul of the sick person that has become lost or abducted by spirits. Lévi-Strauss, years ago, gave a good account of the shamanic trance. The shaman's worldview, or 'fabulation of a reality' in Lévi-Strauss' phrase, consists of various 'procedures and representations'. This fabulation is:

> founded on a three fold experience: first, that of the shaman himself, who, if his calling is a true one (and if it is not, simply by virtue of his practicing it), undergoes specific states of a psycho-somatic nature; second, that of the sick person, who may or may not experience an improvement of his condition; and finally, that of the public, who also participate in the cure, experiencing an enthusiasm and an intellectual and emotional satisfaction which produce collective support, which in turn inaugurates a new cycle.[66]

[63] Ina Wunn, *Die Religionen in vorgeschictlicher Zeit* (Stuttgart: Kohlhammer, 2005), pp. 136–7; Zoya A. Abramova, *L'art paléolithique d'Europe orientale et de Sibérie* (Grenoble: Jérome Millon, 1995), pp. 88–90. On three figurines from the Hohle Fels cave see Nicholas J. Conard, 'Palaeolithic Ivory Sculptures from Southwestern Germany and the Origins of Figurative Art', *Nature*, vol. 426, 2003, pp. 830–2.

[64] Wunn, *Die Religionen in vorgeschictlicher Zeit*, pp. 139–42.

[65] E.J. Michael Witzel, *The Origins of the World's Mythologies* (Oxford: Oxford University Press, 2012), pp. 397–8.

[66] Claude Lévi-Strauss, 'The Sorcerer and his Magic', p. 179 in *Structural Anthropology*, vol. 1, trans. Claire Jackobsen and Brooke G. Schoepf (London: Penguin, 1963), pp. 167–85.

These three are inseparable and there are two poles, the 'intimate experience of the shaman' and the 'group consensus'.[67] The shamanic experience of contact with spirits in the form of animals, of curing the sick, of the community experiencing bonding and the release of tension (Lévi-Strauss calls the shaman a 'professional abreactor') could conceivably have been the kind of experience associated with cave art. The evidence that Lewis-Williams presents is quite convincing, including abstract criss-cross patterns that form part of trance hallucinations in many cultures,[68] the depiction in San rock art in South Africa of dancers apparently in trance, one with a bleeding nose, a symptom of trance, and theriomorphic forms such as an antelope-headed figure.[69]

The journey into the cave would have been a spirit journey into the other world and the kind of rituals associated with it would have been inductive of hypnogogic states. While San art is younger (32,000–10,500 BP) than Altamira or Lascaux, the same ideas are present. What we have with cave art, claims Lewis-Williams, is evidence of a shamanic world in which communities were in contact with the spirit world through the cave and mediated through the shaman. The cave walls were not static but living membranes for the community to make contact with the spirits, an idea supported by finger fluting in soft mud on the walls of a cave in Spain showing that people were touching the walls.[70]

At the end of Altamira in the chamber known as the horse's tail we find the protuberances of the rock turned into two faces by the Palaeolithic artists. As the artists turned to leave the furthest recess of the cave, the two faces would confront them, perhaps animal, perhaps human. It is almost as if the faces looking out at the visitor are those of the cave or the spirits of the cave behind the wall precipitating a face-to-face encounter with the human world.

We have come a long way from chimps throwing stones at trees to the cave art of Lascaux or Altamira, but both activities are within the spectrum of ritual. It is almost inconceivable that the creators of these works did not have language, especially if they depict myth as Witzel thinks, and had a strong drive for somatic exploration and going beyond the world of everyday transaction. The journey into the cave becomes a journey into the self for the shaman artist, which is also a journey for the community. Building on analogues with contemporary shamans, we might say that the cave mediates between the spirit world behind the walls and the community living at the cave's mouth. The faces in the wall claim a face-to-face bond with those who

[67] Ibid.

[68] On spontaneously produced visual imagery or phosphenes see Lorna McDougall, 'Symbols and Somatic Structures', in John Blacking (ed.), *The Anthropology of the Body* (London and New York: Academic Press, 1977), pp. 391–401.

[69] David Lewis-Williams, *The Mind in the Cave* (London: Thames and Hudson, 2002), plate 9.

[70] Ibid., plate 14.

journey there; here social cognition extends to the spirit world. The concern with life for these people is palpable. The life of the community depends upon the herds they hunted, and that life is enhanced and vivified by the spirit world through the cave wall paintings. More than everyday survival, these ancient caves bear witness to human movement and vertical ascent or attraction to a higher, trans-human power or powers.

ON HUMAN NATURE

I have spent some time looking into what the contemporary sciences concerned with evolution have to tell us about the fundamental features of the human condition and how these have evolved from our primate ancestors. In particular I have highlighted three features that it strikes me are crucial for the development of religion, namely our ability to connect through pro-social emotions, our ability to communicate abstractly and narratively through the use of language, and the somatic patterns we create to order our lives and go beyond them. The development of religion as central to the human endeavour has harnessed these capacities and continues to use them, and because these capacities are so fundamental to our nature, religions have thrived through the history of world civilizations. Someone might object that in this list of three features, the crucial development of technology is missing and technology above all defines who we are in contrast to all other species. I certainly would wish to place importance on technology—the million-year or more use of the Palaeolithic handaxe to the development of Levallois flint technology[71]—but I am looking at a prior structural level of the human niche. What are the necessary conditions that create the human niche and therefore that precipitate the development of religion? I have argued that what specifies the human niche and therefore the development of religion are these three proclivities of the pro-social emotions, language, and ritual. Technology fits into this as an extension of the pro-social emotions in that the working on stone tools entails the mimetic teaching of a skill that becomes refined through the generations, and this shared attention on a task entails a bonding of the group; it creates a shared intentionality, to use Tomasello's term. Human practices founded on our pro-social orientation, going out of our way for the other, and our ability to pattern our lives through the ordering of time in a regular, segmented sequence of acts, form the basis of narratives that explain our past and anticipate our future.

[71] Robert Foley and Marta Mirazón Lahr, 'Mode 3 Technologies and the Evolution of Modern Humans', *Cambridge Archaeological Journal*, vol. 7.01 (1997), pp. 3–36.

But I want finally to return to a remark I made at the beginning that the perspective on religion from evolutionary anthropology and social cognition is inevitably limited because restricted to third-person accounts. What I mean by this is that naturalism alone, or reductionism, cannot adequately account for religion because of the human reality of the lifeworld in which, throughout history, humans have experienced a vertical pull or desire for ascent; a height psychology, to use the phrase of Peter Sloterdijk. We need finally to point out the limits of naturalism and to argue for a hermeneutical phenomenology that takes vertical ascent or the human being's deepest search for its own meaning, to use Bowker's phrase, as a feature of human reality that all religions embrace. This deepest search for its own meaning is human self-repair.

In his book *On Human Nature* Edward Wilson argued for the dominance of genetic constraint on human nature over any cultural or environmental factors. We now know this to be wrong because culture provides an equally important constraint articulated by scientists in the ideas of coevolutionary development (EvoDevo) and niche construction; the extended evolutionary synthesis discussed in Chapter 1. What is at issue here is the interpretation, and meaning, of data. I have reviewed important discoveries from social cognition and evolutionary anthropology that provide us with strong evidence for the ways in which human reality, and religion as an essential part of that reality, was generated. The accounts from the perspective of genetics and evolutionary biology have necessarily been what we might call third-person accounts. That is, the objectivity of the data presented through empirical inquiry has necessitated the claims and theory construction in those sciences. This entails as far as possible the eradication of subjectivity through letting the objectivity of the data come through. In the case of human reality, we have seen how that was formed by the pro-social emotions and the development of shared practices among the hominin groups. The meaning of the data is articulated within the boundaries of the discipline (here evolutionary anthropology). But data can be interpreted in different ways and so this entails, as Taylor observed, a distinction between meaning and expression.[72] The expression of the meaning can be articulated in a different way, which entails the basic idea of hermeneutics that something makes sense to someone (for some purpose). This must be true because science can re-visit data and reinterpret it in a new light in the context of new knowledge. We need therefore in addition to the third-person perspective a first/second-person perspective. I say first/second because the first person always entails the other—it is always relational. This is recognized by second-person neuroscience for whom the 'data' is in relation to the inquirer and changes according to the interactive situation.

[72] Charles Taylor, 'Interpretation and the Sciences of Man', pp. 15–16 in *Philosophy and the Human Sciences: Philosophical Papers*, vol. 2 (Cambridge: Cambridge University Press, 1985), pp. 15–57.

So, we need the first/second-person perspective because the third-person view is limited, particularly in the case of human beings and their institutions.

With the origins of religion this is, of course, difficult to achieve because we have no records in language of what those people understood by their own behaviour. Once we arrive at complex religions where people build temples, go on pilgrimage, pray, write texts, protect adherents, and wage wars, we are in a different situation where we are dealing with something that has already interpreted itself. The complexity of explanation by the external observer is increased because of the necessity of absorbing and re-describing something that has already described itself. But given the difficulties of interpreting the ancient materials we have been looking at, the first/second-person account needs to be incorporated, as has been done by Lewis-Williams in his reading of the ancient caves through the lens of the spectrum of consciousness, thereby making sense of data whose meaning would otherwise have remained obscure.

The transformation of the human proclivities for pro-sociality, language ability, and somatic ordering of time and space into religion entails that we need to interpret this in terms of human existential meaning. That is, if we maintain Taylor's distinction between meaning and expression, an important expression of the meaning of religion is in terms of those who lived it. Given that we cannot access those ancient ancestors other than through what they have left behind, the material culture we can trace through archaeology and stories they have told through mythology, we need to use our own pro-sociality, our empathy, to make sense of their lives. This is Heidegger's formal indication: we recognize the humanity of the historical other through the recognition of our own humanity.[73] Our interpretations need to be true to the data given through the sciences yet need to be understood in terms of the kinds of beings we are and the historical reality of our lives, or rather their lives. What I am trying to get at is that third-person explanations of the origins of religion are necessary and have yielded new knowledge, as we have seen, but as part of the explanatory apparatus we can also add the quest for verticality.

By this I mean that the recognition of drive towards meaning and struggle to go beyond the limits of who we are, conceptualized as repair of what we are, is a crucial element in understanding human institutions such as religion. There is a desire for vertical ascent attested throughout the history of religions and arguably present in our early hominin past. When our Palaeolithic ancestors developed Levallois flint technology it was for travelling, to take the tool-kit with them in new exploration or following the herds. This technology is a breakthrough, a transcendence of a limitation. Likewise, the paintings on the cave walls indicate a desire, a movement towards an

[73] Martin Heidegger, *The Phenomenology of the Religious Life*, trans. Mattias Frisch and Jennifer Anna Gosetti-Ferencei (Bloomington and Indianapolis, IN: Indiana University Press, 2004), pp. 42–5.

elevation, a desire for a higher perspective through the art itself and through the human practices of which the art was a part; the shamanistic trance that transcended everyday human reality.

This drive to go beyond the present condition and the development of an array of ascetic disciples to foster such ascent, Peter Sloterdijk's 'height psychology',[74] has proved to be central to the religious imagination and quest. We might look at this in terms of elevation, vertical attraction, or vertical pull.[75] That is, part of human reality is going beyond the limitations of where we find ourselves and we need to understand religions in terms of an attraction towards something expressed in metaphor as verticality. Vertical ascent is attested throughout the history of religions—and we shall have cause to investigate some examples in future pages—but can only be defined or characterized in the instances of its occurrence. That is, we cannot point to something beyond, a kind of giant magnet that pulls at the human world, for verticality is only ever instantiated in the particularity of its occurrence. The mystical ark of Hugh of St Victor or the ladder of divine ascent of St Climacus would be instantiations of vertical attraction, as would be Buddhaghosa's path of purification up through the stages of meditative absorption.[76] Seeking the origins of religion, we might speculate that the existential meaning of early hominin behaviour in terms of ritual and shamanic practices that they undoubtedly had would be in terms of vertical ascent, the going beyond the world of everyday transaction. The interaction with non-human agents, the spirits of the cave, would have been an everyday occurrence and for there to be no such interaction would not be credible within the horizon of their worldview. How this came about is partly a response to mystery and what I have called the strangeness of the world,[77] and partly accounted for in terms of the transformation of the emergent human qualities that we have seen in order to facilitate a healing, a kind of equilibrium.

The question of the origins of religion that I have inquired into here has been in terms of human proclivities highlighted in recent evolutionary anthropology and social cognition. This historical question is inseparable from the question of the sustenance of religion through the generations. What keeps religion alive? One angle on this that I pursue in this book is that life itself comes to articulation in religions, in their accounts of perfectibility, what it is to lead a good life, and how in doing so we aim at self-repair. We will

[74] Peter Sloterdijk, *You Must Change Your Life*, trans. Wieland Hoban (Cambridge: Polity Press, 2013), pp. 111–30.

[75] This useful phrase was coined by Jessica Frazier in the context of discussion in a reading group on phenomenology in Oxford in 2015.

[76] Gavin Flood, *The Truth Within: A History of Inwardness in Christianity, Hinduism, and Buddhism* (Oxford: Oxford University Press, 2014), pp. 69–101.

[77] Gavin Flood, *The Importance of Religion: Meaning and Action in our Strange World* (Oxford: Wiley-Blackwell, 2012), pp. 6–8.

encounter how religions deal with this theme in Part II, but there is one more theme we need to pursue. Tracing our story of religions into later times, from the Neolithic to the Axial age, one common element seems to dominate, and that is the theme of sacrifice—the death-for-life pattern of thought and practice that has been central to religions' self-understanding across all civilizations.

3

Sacrifice

In Chapter 2 I looked at the origins of religion in terms of three proclivities of human nature: pro-sociality, language ability, and the somatic ordering of space and time in ritual. These are the necessary conditions for the emergence of religion, and the practices of parietal and portable art in the Palaeolithic are indications of proto- religious sensibility among *Homo sapiens* in which the affirmation of life was central. Indeed, the affirmation of life through a culture of hunting was a major element in early human practices, and we have here a worldview in which a spirit world was in intimate interaction with the human world, mediated through the animals these people hunted. One important practice emerged that is found across cultures, although mostly excluded from European and American modernity, and that is sacrifice. In this chapter[1] I will move the story on to a stage when human beings are anatomically modern with developed language, technology, and living in probably quite large groups compared to their Neanderthal forebears. With the Neolithic farming revolution, we have new modes of producing food and new kinds of societies that can be much larger, the first urban landscapes beginning to appear in places such as Jericho.[2]

While the origins of sacrifice are obscure, sacrifice is a category central to our understanding of religion and, it could be argued, of human cultural life generally. Sacrifice has had a central place in the history of civilizations as an attempt at human self-repair and bringing people into a fullness of life, attempting to fulfil the desire for life itself and to go beyond death. In this chapter I wish to argue that we need to understand sacrifice not only as an economy of exchange or socio-psychological catharsis, as some major theorists have done, but that behind such exchange, behind catharsis, sacrifice confronts us with the naked situation of human life itself and the attempt to

[1] An earlier version of this chapter was previously published as 'Sacrifice as Refusal' in Julia Meszaros and Johannes Zachhuber (eds.), *Sacrifice and Modern Thought* (Oxford: Oxford University Press, 2013), pp. 115–31.

[2] K.M. Keynton and T.A. Holland (eds.), *Excavation at Jericho: The Architecture and Stratigraphy of the Tel* (London: British School of Archaeology at Jerusalem, 1981).

mend or correct it. We need to understand sacrifice in starker, existential language, in terms of an 'I–you' rather than in third-person terms. While the function of sacrifice can be seen in the context of propitiation of divine powers, expiation of impurity, gift-exchange, communion, the bonding of groups, or even the violent release of pent-up energy, a deeper meaning of sacrifice must be sought in terms of transcending death and the affirmation of life itself against nothingness; the attempt to heal our advanced linguistic consciousness' alienation from life and need to fulfil human desire for life. Theories of sacrifice as gift-exchange and as catharsis are clearly valuable for their explanatory power, but to understand the deeper nature of sacrifice we need to move to a more basic level of human life and to understand it as the refusal of nothingness and death. At this existential level, sacrifice is akin to asceticism or renunciation because it plays out a paradox of the affirmation of life through its destruction. Sacrifice challenges finitude and mortality through the affirmation of life by death and the fundamental revelation or insight that birth in some sense follows death. Moreover, the power of this affirmation lies in the characteristic of sacrifice as refusal: the refusal of death (and time) and therefore of meaninglessness. Developing Heesterman's work,[3] we might say that one of the central elements of sacrifice is refusal. Refusal works at different cultural levels. It can mean refusal of something of human benefit in order to achieve a higher good, and so is akin to asceticism, and it can mean refusal of finitude, refusal of death. If a central theme of religions is the transformation and repair of the human condition, then sacrifice speaks to this theme in various ways but particularly when sacrifice is understood as refusal, which is also an inner transformation. Religious traditions have clearly understood this renunciatory dimension and have developed concepts of sacrifice as metaphor for an internalization of outer practice and recognition of its centrality to human striving for transcendence. Understanding sacrifice in terms of refusal aligns it with renunciation or asceticism, for we can understand both as the refusal of death.

Indeed, this link between sacrifice and renunciation was worked out in a kind of cultural logic in the long history of Indic civilization where renunciation was understood as a response to sacrifice. While I will develop the idea of a sacrificial logic in the Indic historical context in Chapter 4, the current chapter will sketch the idea of sacrifice as refusal, which aligns it with renunciation, and conclude with a methodological claim that a hermeneutical understanding is more appropriate for penetrating this existential layer than a causal explanation.

[3] I am indebted to Heesterman for inspiring these thoughts about 'refusal' and for alerting me to some relevant scholarship. J.C. Heesterman, *The Broken World of Sacrifice: An Essay in Ancient Indian Ritual* (Chicago, IL: University of Chicago Press, 1993), pp. 7–18.

The category of sacrifice in the Judaeo-Christian-Islamic tradition has been a major trope in philosophical and theological reflection, and de Vries suggests that it has been of more philosophical than religious importance in occidental history.[4] We might say the same of the Hindu-Buddhist trajectory where the category of sacrifice—*yajña*, *homa*—has been fundamentally important in reflection on the purposes of life, the nature of the self, and ethics. Although less common as a practice of killing animals, sacrifice has also been a theme in Chinese civilization—the Daoist sacrifice of writing, for example, that carries a person's desires to the next world. Sacrifice is a category that exists across cultures and has been the subject of speculation for several thousand years, from Jewish commentaries on Genesis 22, the story of Abraham and Isaac, to Mīmāṃsā speculation about how sacrifice should not be regarded as violence.[5]

Over the last hundred years or so the idea of sacrifice has played a major role in anthropological theory (particularly Henri Hubert and Marcel Mauss's classic study, but one also thinks of Tylor, Spencer, Robertson-Smith, Evans-Pritchard, and Mary Douglas), psychoanalysis (Freud), philosophy (Bataille, Girard), and theological ethics (Milbank). Leaving aside theological reflection, we might say that theories of sacrifice have seen it primarily in terms of communion (Robertson-Smith), in terms of gift-exchange (Spencer, Tylor), in terms of mediation or communication between the sacred and profane (Hubert and Mauss), and in terms of catharsis (Freud, Girard, Burkert). While it is important to understand sacrifice in functional terms within an economy of exchange that links it, on the one hand, to the gift and, on the other, to a sense of communion following from a cathartic release of tension or simply from sharing a meal, this is not enough.

Some recent scholarship has underplayed the importance of sacrifice: Marcel Detienne has understood ancient Greek sacrifice in political and economic terms involving food distribution[6] and Jonathan Z. Smith has even said of sacrifice, as he said of religion, that it is the creation of the scholar's imagination.[7] These claims, as Heesterman has argued in the context of Vedic material, are exaggerated and there is clearly a body of sacrificial texts and performances that are central to many religions. Less sceptical theorists than Detienne and Smith have understood sacrifice within an economy of exchange, although this emphasis tends to occlude the centrality of sacrifice in ordering religious life and understanding religious action. Sacrifice in the sense

[4] Hent de Vries, *Philosophy and the Turn to Religion* (Baltimore, MD and London: Johns Hopkins University Press, 1999), p. 166.

[5] Wilhelm Halbfass, *Tradition and Reflection* (Albany, NY: SUNY Press, 1988), pp. 87–129.

[6] Marcel Detienne, 'Culinary Practice and the Spirit of Sacrifice', p. 3, in Marcel Detienne and J.P. Vernant (eds.), *The Cuisine of Sacrifice among the Greeks* (Chicago, IL: Chicago University Press, 1989). Political power cannot be exercised without sacrificial practice.

[7] J.Z. Smith, 'The Domestication of Sacrifice', p. 179, in R.G. Hamilton-Kelly (ed.), *Violent Origins* (Stanford, CA: Stanford University Press, 1987), pp. 278–304.

of the violent immolation of living beings has itself been understood as a metaphor for an inner transformation, an *anabasis*, which we might see in terms of an enacted metaphor for fundamental human motivations towards meaning, a meaning that is formed through refusal or renunciation. That is, sacrifice must be understood not only in terms of social function, but also in terms of the formation of religious meaning, in terms of the desire to transcend time and death, to repair the human condition, and in terms of a denial or refusal of death and nothingness. The origins of sacrifice might well be very ancient probably in Palaeolithic hunting traditions, as Burkert argues, supported by the mythological work of Witzel,[8] or in Neolithic farming communities, but irrespective of its origins, sacrifice became a way of ordering the world, expressing hope, and giving life meaning for particular societies through history. This ordering of the world has often, although not only, occurred at times of crisis, at times of the disruption of normal events and the need to re-establish normality within a community. The reinterpretation of sacrifice in Christianity, Judaism, Hinduism, and Buddhism and its transformation into an ethical category is a further articulation of the way sacrifice expresses meaning and orders human relationships to the world, time, and death.[9] Sacrifice is 'a universal component of the human imaginary', as Hedley argues.[10]

WHAT IS SACRIFICE?

The term 'sacrifice' covers a range of practices and cultural themes and we might speak of a 'family of sacrifice'.[11] Although it is useful to restrict the term to the ritual killing of an animal, there are metaphorical extensions quite widely acknowledged in religions, and sacrifice becomes appropriated by narrative; sacrifice is fundamental to the stories we tell about ourselves. The range of practices involves ritual killing, usually an animal and sometimes a human, offering the victim or parts of the victim to non-human powers, and the consumption of parts of the offered animal in the sacrificial meal. Heesterman identifies three major components: ritual killing, destruction, and feast, with

[8] Walter Burkert, *Homo Necans: The Anthropology of Ancient Greek Sacrificial Ritual and Myth*, trans. P. Bing (Berkeley, CA: University of California Press, 1983), pp. 12–22. Michael Witzel, *The Origins of the World's Mythologies* (Oxford: Oxford University Press, 2012), pp. 262–3.

[9] On the Christian and Rabbinic rejection of literal sacrifice see Guy G. Stroumsa, *La fin du sacrifice: les mutations religieuses de l'antiquité tardive* (Paris: Odile Jacob, 2005).

[10] Douglas Hedley, *Sacrifice Imagined: Violence, Atonement, and the Sacred* (London: Continuum, 2011), p. 2.

[11] Burkert cited by Heesterman, *The Broken World of Sacrifice*, p. 9.

destruction as one of the problematic dimensions that does not allow us to see sacrifice in more simple terms of food distribution.[12] The principal meanings evoked by the English term are the violent immolation of a living being for appeasing a wrathful deity, expiation of sin or impurity, and giving up something precious or relinquishing some condition for the sake of another or a higher goal.

Sacrifice is still present in our attempts to avoid misfortune and there are still magical residues in modern cultures in vowing, for example, to perform a certain task or do a certain ascetic practice in return for a divine favour. But the energy release model does not sufficiently account for the ways in which communities face transcendence through sacrifice, or for its narrative and metaphorical appropriations. Of these, the most notable metaphorical extension is war. War in ancient India was understood in terms of sacrifice, and this theme is commonplace today when we speak of soldiers sacrificing their lives for the greater good of the country or cause. Bataille perceives the link between sacrifice and war in his idea of the excess of energy that needs to be wasted to bring the society back to a state of equilibrium, and in contemporary Western society, it is war that best approximates to any residue of sacrificial culture that remains as a way of ordering or making sense of the world. But these 'horizontal' accounts of sacrifice need to be supplemented with 'vertical' accounts that take seriously sacrifice as a site for mediating the human encounter with mystery and confronting life itself through death: the appropriation of death for the affirmation of life.

Hubert and Mauss offered the following definition over a hundred years ago in 1898:

> Sacrifice is a religious act which, through the consecration of a victim, modifies the condition of the moral person who accomplishes it or that of certain objects with which he is concerned.[13]

This is still a relevant understanding and the introduction of the phrase concerning the condition of the moral person is key: literal sacrifice changes the community and metaphorical sacrifice, the relinquishing of personal desire for something or someone else, also brings about change. But the Hubert/Mauss account is really a description of the essential features of sacrifice rather than explanation in functionalist or other terms. Explanations of sacrifice have generally fallen into two general groups: sacrifice as an economy of exchange or consumption, and sacrifice as catharsis. While both

[12] Heesterman, *The Broken World of Sacrifice*, p. 14.

[13] Henri Hubert and Marcel Mauss, *Sacrifice: Its Nature and Function*, trans. W.D. Halls (London: Cohen and West, 1964, p. 13. First published as 'Essai sur la nature et a function du sacrifice', in *L'année sociologique*, vol. 2, 1897, pp. 29–138: *Le sacrifice est un acte religieux qui, par la consécration d'une victime, modifie l'état de la personne morale qui l'accomplit ou de certains objets auxquels elle s'intéresse.*

kinds of theory have much to offer as explanatory accounts, both approach sacrifice in terms of cultural function, but to lay bare the bones of sacrifice, as both naked violence and as metaphor, we need to expose it to the light of a starker existential truth of human life: that we are consumed by death and wish to transcend it. Sacrifice says that death is not final and presents the refusal of death through embracing it. This embracing of death, which is paradoxically the embracing of life, is a refusal of death through renunciation. But to develop this thesis, we need to place it in the context of what we might call the cultural functionalist views.

CULTURAL FUNCTIONALISM

Cultural functionalist explanations of sacrifice in terms of conspicuous consumption or purification through catharsis, the expurgation of impurity or release of pent-up energy, have predominated in the social sciences. Indeed, catharsis has been most significant in theories of sacrifice in the last hundred years. Freud long ago offered a psychoanalytic account of sacrifice through an examination of Totemism. According to Freud, we find in Totemism two prohibitions: against killing the totem except in an annual sacrifice and the rule that marriage must occur outside of the totemic group (an incest taboo). These prohibitions reflect the primal sacrificial act of the sons killing the father and marrying his wives (their mothers). This primal murder is re-enacted in sacrifice.[14] There is, of course, no evidence for the primal murder or the kind of early hominin groups that Freud envisaged, nor is it clear that Totemism is a survival of the earliest kind of religion. But Freud was instigating an important kind of thinking that has been significantly developed by Girard and Burkert, who share Freud's view that the violence in a society becomes channelled into the sacrificial victim, which thereby achieves a return to equilibrium.[15] For Burkert, sacrifice is part of a spectrum of appeasing acts to avert misfortune.[16] For Girard the key point is the channelling of a society's violence into the victim and the subsequent temporary release from the threat of violence, which is thereby dissipated. In Girard's view, the sacrificial victim corresponds to the victim of social scapegoating; both the killing of the sacrificial victim and the scapegoat provide a catharsis that allows

[14] Sigmund Freud, *Totem and Taboo*, trans. James Strachey, reprinted in *The Origins of Religion* (London: Penguin, 1985), pp. 43–224. For a critique see E.E. Evans-Pritchard, *Theories of Primitive Religion* (Oxford: Oxford University Press, 1965), pp. 41–3.

[15] See R.G. Hamerton-Kellyn (ed.), *Violent Origins: Walter Burkert, Rene Girard, and Jonathan Z. Smith on Ritual Killing and Cultural Formation* (Stanford, CA: Stanford University Press, 1987).

[16] Burkert, *Creation of the Sacred*, pp. 40–2.

the re-establishing of the social order. The violence of the sacrifice and the violence of the mob reveal the truth of mimetic desire and the foundations of human culture. In its crudest form, two people want the same thing, become rivals, and slowly mimic each other's aggression until tension builds and spills over into violence. This violence is channelled into the sacrificial victim who absorbs the sins of the community, by which act the community's tension is alleviated. The pattern that Girard has identified is a functionalist argument. Violence builds up in a society, which has to be released through sacrifice or through the scapegoat mechanism. Indeed, a twofold transference occurs: the aggressive transference into the victim followed, once the aggression has dissipated, by a reconciliatory transference which makes the victim sacred. The sacralization of the victim ensures that worship of the victim replaces hatred of the victim.[17] Sacrifice is the mechanism whereby a society cleanses itself of violence through offering itself up as the sacrificial victim or substitute. Sacrifice is only ever partially successful, however, and the cycle of mimetic tension builds up again and the sacrificial pattern has to be repeated.

Girard has tried to demonstrate this formulation in many areas of literature and culture. The simplicity of Girard's model is powerful and functions with success in a number of realms of meaning—especially in literature—but whether it provides an explanation for all aspects of culture is open to question.[18] He recognizes how sacrifice has functioned through history and the negative value that modernity must place upon it in the challenge of a Christian and post-Christian world. For Girard, Christianity disrupts the mimetic, sacrificial model even though at first glance it seems to conform to it. Like Bataille, Girard sees sacrifice as cathartic; it has a purifying or cleansing function for the social group. Like Bataille, it is linked to irrationality and even frenzy, and like Bataille it is more about release of energy and less about an exchange reciprocity. Girard's analysis of sacrifice in terms of mimetic desire is also evaluative in upholding the critique of violence and an optimism that the mimetic process will dissolve through its being revealed by Christianity.

Of course, this summary does not do justice to Girard's thesis, but it seems to me that the hub of his argument is that within pre-industrial economies sacrifice as catharsis fulfils the function of release from tension, and is fundamentally about the violent purification of the tribe. The violence of the sacrifice and the violence of the mob reveal the truth of mimetic desire and the foundations of human culture. But it is Bataille's thesis, inspired by Freud, which combines a cathartic theory of sacrifice with a theory of exchange, and is closer to the existential dimension of sacrifice I wish to bring out.

[17] Rene Girard, *Des choses cachées depuis la fondation du monde* (Paris: Le Livre de poche, 1983), p. 55.

[18] See, for example, critical responses in Michael Dillon and Paul Fletcher (eds.), *Violence, Sacrifice, Desire, Cultural Values*, vol. 4 (2), 2000.

Bataille presents a political economy of what he calls 'the superabundance of biochemical energy' (*la surabondance de l'énergie biochimique*)[19] whose source is ultimately the sun, which is 'the principle of life's exuberant development' (*le principe de son développement exuberant*). In *La Part maudite* that Bataille published in 1949 (revised in 1967 and translated into English as *The Accursed Share* in 1991), he presents an argument that any general economy in a society in time produces an excess and this wealth finds expression in one of two ways: either it is consumed in luxury or destroyed in acts of violence—in the modern context as war and in earlier times as sacrifice—that threaten the social order. The 'accursed share' (*la part maudite*) is the surplus energy expended by a system: if energy cannot be used for growth, it is inevitably and necessarily wasted. Solar radiation that is the source of life is abundant and the sun gives without receiving. Living matter (*la matière vivante*) receives this solar energy and stores it within the boundaries of the space it inhabits. An organism uses this energy for its growth and once that growth has achieved its maximum extension, once it has reached its limit and filled the space allotted to it, then there is an excess produced that needs expression in some way other than growth.[20] An organism that has used all the energy it requires for growth produces excess that needs release. Bataille gives a somewhat prosaic example; thus, he says, the energy of growth in duckweed that fills the space of a pond, once the pond has reached saturation point, needs to be dissipated. Such dissipation occurs in the form of heat. This biological model of energy, build-up and release, also functions in human societies.

From a general discussion of solar energy that accumulates in the biological organism and, indeed, in the total system of organic life, Bataille seamlessly moves into a discussion of human society where the same principles apply. Societies and people reach a limit to growth and the superabundant, unneeded energy finds release in luxury, particularly art and non-reproductive sexual activity, but also in war. Thus labour and techniques (*le travail et les techniques*) have allowed the growth of human societies, and the excess energy thereby produced from 'the pressure of life' (*la pression de la vie*) is consumed (i.e. released) in 'conflagrations befitting the solar origins of its movement' (*des embrasements conformes à l'origine solaire de son mouvement*).[21] If this excess is not consumed in luxury, then it erupts as war and the twentieth century's two devastating wars bear witness to the release of excess energy built up through the labour and economic growth of the nineteenth century.

The excess or superabundance of energy in any system, whose origin is the outpouring of solar energy, is destined for waste, and Bataille sees sacrifice in this context as the outpouring of excessive energy within a society. Excessive

[19] George Bataille, *La Part maudite* (Paris: Les Editions de Minuit, 1967), p. 65.
[20] Ibid., p. 67. [21] Ibid., p. 76.

consumption (as in a potlatch), war, and sacrifice are linked as expressions of energy within a general economy. Indeed, the potlatch is central to Bataille's argument and his discussion of it as the expression of surplus energy is more extensive than his discussion of sacrifice itself. While Bataille's main concern is contemporary society and how human beings can free themselves from this organic inevitability of the release of energy in destructive ways, in order to understand the present, he presents historical analyses, especially of Aztec society. Bataille's Aztecs are the very opposite of us morally ('*se situent moralement à nos antipodes*'); whereas, for us, production is important, for them, consumption is important and they were more concerned with sacrificing than with working.[22] My reading of his somewhat complex argument is as follows.

First of all, Bataille posits a distinction between subjectivity associated with intimacy and the order of things associated with work, production, and the real world. The human sacrifice of Aztec society was the sacrifice of slaves, namely warriors who had become slaves through conquest. As slaves, the victims were bound to labour as their captor's property; the slave becomes a thing, a commodity, he becomes part of the order of things. Drawing on the master–bondsman theme in Hegel, Bataille observes that the slave accepts his situation, which is preferable to dying.[23] The slave prefers to become part of the order of things, prefers to become objectified rather than to die. The owner of the slave becomes the sacrificer of the sacrificial victim who represents the sacrificer. The victim (who is a slave) is degraded to the order of things, made into an object, but at the same time he has the same nature as the subject, the sacrificer, whom he represents. Sacrifice heals the world and 'restores to the sacred world that which servile use had degraded, made profane' ('*restitue au monde sacré ce que l'usage servile a dégradé, rendue profane*').[24] In the distinction between subjectivity linked to intimacy versus objectivity linked to labour, the victim having been made an object through slavery is partially returned to the realm of intimacy through becoming the sacrificial victim. Furthermore, the owner or subject comes to be identified with the victim and, through the restoration of the victim with the sacred realm, is himself restored. Religion is centrally concerned with this restoration and so with repair and healing.

To achieve this restoration, the sacrificial victim must be destroyed; the sacrifice destroys that which it consecrates. This destruction is the destruction of the link with labour and profitable activity and so is an excess, a superabundance. In the process of destruction the victim cannot, of course, be returned to the real order and this refusal of the real order, the world of work, is a principle that 'opens the way of arousal, it releases violence in itself,

[22] Ibid., p. 34. [23] Ibid., p. 94. [24] Ibid., p. 94.

reserving a domain where it rules without equal' (*ouvre la voie au déchaînement, il libère la violence en lui réservant le domaine où elle règne sans partage*).[25] The world of subjectivity, the world of the night for Bataille associated with the dream of reason and with madness, is given violent expression legitimated through the sacrificial victim.

This violence is associated with consumption insofar as it is not connected with work and so is not concerned for the future. The idea of sacrifice therefore entails a particular view of time in which the present is foregrounded. The sacrificer is concerned not with what will be but with what *is* because useless consumption, the consumption of excess, can only occur once concern for the future is relinquished. There is no need to hold anything back if there is no concern for the future, and the person who lives in this way entirely in the present can consume all s/he possesses. Such useless consumption of the kind performed in sacrifice is a way of revealing subjectivity and intimacy to others and is a way in which the subject can connect with others. In sacrifice, for Bataille, we have a return of the thing to the order of intimacy and this return releases violence, but a violence that is contained within the ritual sphere of sacrifice, within the bacchanalian excess of the festival, and, in being given release, preserves and protects those who offer the rites. As for Girard, for Bataille sacrifice protects a community from the excess of violence and from the contagion of violence. Sacrifice stops violence spreading through the community and, through the merging of individuals in the sacred present, allows them to communicate in future-oriented, secular time. In sacrifice we have an attempt to diffuse violence, to continue the status quo, to save a community from ruin. The sacrificial victim is a surplus who is withdrawn from the realm of wealth by destruction: he becomes 'the accursed share' violently consumed who, brought back from the world of things, occupies the realm of intimacy and subjectivity for a time and represents the masters who offer sacrifice, even becoming their equal in the short period before destruction. The victim is both sacred and cursed. Only destruction can rid the victim of his 'thinghood'—that is, of his usefulness—and achieve the release of violence necessary for the energy balance and restoration to a state of equilibrium.

In modern, industrial societies we do not, of course, have sacrifice in anything akin to Aztec society. But Bataille observes that we are still subject to the loss of intimacy in the world of things and that this is indeed a loss. Sacrifice enables a return to intimacy. There is a separation between the sacrificer and world of things and the victim of sacrifice who joins the world of things, returns to the world of intimacy, and thereby restores the immanence between 'man and the world', between subject and object.[26] While he

[25] Ibid., p. 96. [26] Georges Bataille, *Théorie de la religion* (Paris: Gallimard, 1973), p. 59.

does not advocate a return to violent sacrifice, he is sympathetic to the forces released and the irrationality of the intimate, subjective sphere, set apart from the real realm of labour and work and from the future. If sacrifice is a movement from the realm of things, of objectivity, to the realm of intimacy which facilitates the release of violence or tension in a community, then contemporary, Western societies would benefit from it, if not literally then metaphorically. The excess of solar energy, on this view, is a fact about life that we cannot escape, but presumably we do have power over how we respond to the issue. Sacrifice as violent act has largely disappeared from contemporary Western societies but the need for resolution, the need for energy release among organisms that have filled a particular space, has not. The catharsis model of sacrifice does not exhaust its explanation.

CONSUMPTION OR RENUNCIATION?

I would not wish to be over-critical of Bataille's extremely stimulating work; if he is wrong he is wrong in a very creative and interesting way. But there are, it seems to me, a number of problems with Bataille's and other cathartic models of sacrifice. There are historical questions to begin with. It is not clear that Aztec society is prototypical of sacrificial cults; most sacrificial cults have rarely practised human sacrifice. Second, we can take issue with Bataille's claim that sacrifice is necessarily linked to excess. For Bataille sacrifice is an attempted reinvigoration of intimacy through the violent release of energy brought about by the bringing of the sacrificial victim from the realm of work or the real world to the realm of intimacy, luxury, and subjectivity, and so aligns the category of sacrifice with the erotic. The destruction of the victim in Aztec society ensures the return to intimacy, although not completely as the victim who is a slave must remain subordinated to the sacrificer who is the master. The excess of violence channelled into the victim happens because of the victim's retrieval back into the world of intimacy from the realm of work; an event that is inevitably marked by violence as the excess energy in the system seeks release.

Let us then develop the thesis of sacrifice as refusal. My argument is that although sacrifice is a collective enterprise and experience, it is nevertheless deeply connected to subjectivity: the moral subject is transformed through refusal, which aligns sacrifice with asceticism. Both asceticism and sacrifice share in the refusal of death and the affirmation of life and so repair the human condition. In support of this claim, I shall draw on ethnographic and historical research from India before my sustained account of that civilization in Chapter 4. We will also see that the transformation of the moral subject is not individualism but rather sacrifice is conducive to the formation of a shared subjectivity realized in the sacrificial event.

While Bataille and Girard present strong cases for the cathartic and excess models, we need to read sacrifice as human self-repair, in terms of refusal and renunciation.[27] Indeed, this is implicit in common parlance where 'sacrifice' is understood to be giving something up; a mother might sacrifice her career for a child, a soldier sacrifice his life for a country, and so on. This 'commonsense' understanding points to an important dimension of the phenomenon and brings it into alignment with asceticism. There is a dimension of moral transformation in sacrifice insofar as the refusal of a particular kind of wealth (vegetal and animal offerings as symbols of the sacrifier's life) changes the inner condition or the status of the donor.

Could we not argue that a key feature of sacrifice is less cathartic expression of violence, which maintains the cultural status quo, and more the renunciation of material goods that gives expression to the desire for the refusal of death and nothingness? That is, sacrifice is not so much about the consumption of excess, which in such a society would be without hardship, but rather the renunciation of goods or wealth created through labour, through which the community enacts the transcendence of its mortality. On Bataille's account sacrifice is not giving up, not a relinquishing of anything, but the affirmation of luxury through abandoning or destroying excess. We could argue that rather than being about luxury and the intimacy of subjectivity, sacrifice is about suffering and the relinquishing of a self-focused subjectivity in favour of a shared subjectivity in which that which is most precious, a life itself, is given up in order to affirm that very life. Sacrifice is arguably less about the cathartic expelling of excess or surplus energy and more about the renunciation of that which is held to be most valuable and the refusal of that which we seek to overcome. Seeing sacrifice in terms of refusal brings out elements and a deeper, existential level of human repair that remain hidden from purely cultural functionalist views. We therefore need to develop an argument that (i) sacrifice is the refusal of that which is most precious and (ii) there is an analogy between the victim and the sacrificer, which (iii) shows the proximity of sacrifice to renunciation or asceticism and how the indexical-I of the sacrificer is analogous to the indexical-I of the renouncer. This shows (iv) sacrifice to be the refusal of death, nothingness, and meaninglessness; in fact, the refusal of time that is tragic in its necessary failure and in such refusal the affirmation of healing.

[27] Foucault has written extensively on asceticism and, from Mauss, techniques of the body but does not, to my knowledge, address sacrifice directly. See the selection of readings in Jeremy Carrette, *Religion and Culture by Michel Foucault* (Manchester: Manchester University Press, 1999). See also Gavin Flood, *The Ascetic Self: Subjectivity, Memory and Tradition* (Cambridge: Cambridge University Press, 2004), pp. 243–6.

Sacrifice as Refusal

A key dimension of sacrifice that Bataille, and indeed Girard, neglect is refusal: the refusal of goods—animals and other offerings into the fire—that would otherwise be of profit. At a metaphorical level this is the refusal of a person or group for the benefit of others: the hungry mother's relinquishing of her share of bread for her children, the ascetic's refusal to succumb to the passions, or the Christian's turning the other cheek. Sacrifice is perhaps less about profitless consumption and more about refusal or the relinquishing of worldly attachment for a higher goal specified within the tradition, such as the prosperity of a community, the attaining of heaven for the patron of the sacrifice, or the collective bond entailed in sacrificial practice that serves to reinforce social relations. The burnt oblation bears witness to the sacrificer's lack of self-interest; the gift destroyed is the pure gift, the gift that represents the transcendence of present time and space, the here and now. Indeed, rather than the emphasis on present time, as Bataille argues, sacrifice is concerned with the future, with piling up merit in heaven for the Vedic Brahmans, and structuring the universe in an ordered way that relates the future to the present and a higher world to the lower world; it is in fact concerned with vertical ascent.

Contra Bataille, we can read the ethnographic record in a different way: that sacrifice is less the consumption of overabundance and excess and more the refusal of, first, that which is most precious and, second, the refusal of death itself. The giving up of what is good for the community (namely, food or precious substances) is linked to the refusal of death. The ethnographic record seems to support the view that what is regarded as precious is given up to death in the belief that this will somehow renew the sacrificial community. Most sacrificial societies have been agricultural where domestic animals are important for the general economy and livelihood of the people. Evans-Pritchard, for example, writes about this with regard to the Nuer. The Nuer hold cattle in high esteem: they are the topic of most conversation and there is an identification with them in the custom of taking names from the cattle in addition to names given at birth.[28] Cattle are subject to sacrifice, which is a means of communicating with a transcendent reality (Kwoth) and substitutes for the community or particular people who make the sacrifice. This idea of substitution is attested in other areas and times: in Vedic India the patron of the sacrifice is identified with the victim, and even in Christianity where Christ as sacrifice becomes the representative of humanity. That which is most precious is subjected to destruction, which is a symbolic act of substitution of the community or patron for the victim.

[28] E.E. Evans-Pritchard, *Nuer Religion* (Oxford: Oxford University Press, 1956), pp. 248–55.

The Sacrificer and the Victim

This identification of the patron with the victim is an important dimension of sacrifice that Bataille is right to emphasize. The victim of sacrifice is an analogue of the patron of the sacrifice so that instead of the patron dying, the victim dies so that the patron might live in a renewed or refreshed way. The patron is healed or repaired through the victim. Sacrifice is the performance of the ambiguity that killing means death and yet in sacrifice can mean the affirmation of life. The victim dies in order that the patron or community might live and thrive. Sacrifice affirms the desire for life. Death is symbolically transcended through its performance and ritual containment within the boundaries of the sacrificial rite. As Bataille has highlighted, sacrifice faces death and, in a sense, we can see sacrifice as the attempt to control and ultimately transcend death: an attempt that is, of course, doomed to failure and so needs to be repeated in sacrificial societies. Indeed, we might see sacrifice as the ultimate tragic act: tragic in the sense that it attempts to affirm life and transcend death through embracing death, but as we all know, death remains. In sacrifice I meet my death and attempt to transcend it. The substitute dies for me and through that dying, I live. Bataille's Aztecs clearly fall into this pattern where the victim is taken into the world of intimacy and identified with the subjectivity of the sacrificer, as the Nuer is identified with the ox through the personal name. In sacrifice I meet my death, but who is this 'I' of the sacrifice?

Sacrifice and Renunciation

To answer this, we must make a distinction between individualism and subjectivity. Individualism is a value of modernity that highlights the particular agency and creativity of the person over any social collective. What is notable about sacrificial communities is that they are all within what we might call a cosmological worldview. Sacrifice is the performance of a cosmic drama in which the structure of the cosmos is affirmed and the soteriological hope of a community expressed. The 'I' of the sacrifice, the patron who pays for the rite in the Vedic case, while being a particular, individual person, should not be confused with expressing the values of individualism. The sacrificial 'I' is an index of a shared identity that expresses and gives voice to a collectivity or shared subjectivity. In such a cosmological society, such as ancient Vedic India, when I symbolically meet my death in the sacrificial arena through the victim, the 'I' is a particular reference point in the here and now for a broader social identity. Vedic sacrifice is a fine example. Here the patron (*yajamāna*) undergoes a series of ascetic purifications and the sacrifice is performed within the 'sacred canopy' of a cosmology with a fixed order: the three levels of the universe (earth, atmosphere, and sky), the gods who inhabit those layers, and the social order of classes (*varṇa*) that corresponds to the cosmic structure. In such a society, when I perform sacrifice I am standing in for the community and enacting the socio-cosmic order. When I see my death in the victim, it is the community that sees its death through me, which is simultaneously the renewal of life (because I do not die) and the affirmation of the order of the world. The sacrifice transforms and repairs me, who is a sign of the collective identity, into a new life in the world and affirms the order of life. Sacrifice can thus only flourish in a society that is deeply cosmological in its worldview (as Vedic society was).

The identification of the sacrificial 'I' with the victim brings sacrifice into proximity or alignment with renunciation. Indeed, we might say that the 'I' of the sacrificer and the 'I' of the renouncer are analogues. The ideal ethnographic location for this identification is, of course, India, where sacrifice and renunciation are two poles of religious being that we need to say something about in order to develop the more general and widely applicable thesis. There is no space here to develop this history, but suffice it to say that the original, central religious act of early nomadic communities in South Asia was sacrifice, in solemn and complex ritual proceedings accompanied by verses (mantras) from the revealed scripture or Veda.

The oldest, best-documented, and, as it happens, most elaborate sacrificial system we know about is the Brahmanical one. This is attested in the Vedas whose hymns were composed for recitation during the sacrifice. The Vedic ritual sacrifice, which still on occasion takes place, has been well documented.[29] Briefly, one sacrifice called the *agnicayana*, the piling up of fire, takes place over twelve days. An altar in the shape of a giant bird is built of over a thousand bricks in layers within a ritual enclosure. A number of priests perform different functions and recite different mantras. The patron of the sacrifice, the *yajamāna*, along with his wife undergoes purification that involves bathing, fasting, and a 'sweat-lodge' type of enclosure. Goats are sacrificed by being tied to a pole and suffocated, their meat boiled, and the sacred drink soma is consumed. The *yajamāna* after the concluding rites on the twelfth day returns home with three fires which he installs in his home altars and thereafter performs milk oblations into fire for the rest of his life in the evening and morning. In earlier days other animals were sacrificed, particularly the horse (*aśvamedha*), and there is even mention of a human sacrifice (*puruṣamedha*).

These proceedings are highly formalized with great stress on correct performance. There is very little evidence of cathartic release of energy through killing. Indeed, the death of the victims seems simply to be part of

[29] See, for example, Frits Staal, *Rules without Meaning* (New York: Peter Lang, 1989), pp. 65–77.

the procedure along with the recitation of mantras. But what seems to be significant is the change in status of the sacrificer. He is purified through the rite and his faults go into the sacrificial victim who symbolically replaces him. He is healed and repaired through the sacrifice. The sins of the sacrificer are, in a way, symbolically consumed. The focus of sacrifice is not the appeasing of a deity or explicit and overt consumption, but rather the subjective condition of the patron. In this sense, Mauss' definition seems quite correct. Indeed, Mauss probably derived it from the Vedic sacrifice that he was so interested in and that provides the basic example of his text. This idea of purification is later developed in the Upaniṣads where the external sacrifice is interpreted symbolically, or rather the external rite becomes an internal rite.[30] The true sacrifice is the sacrifice of the breath for the higher gnosis of recognizing one's identity with the absolute. The abstinence of the patron in the Vedic sacrifice becomes the asceticism of the ascetic in the Upaniṣads. Sacrifice and renunciation become analogous. This tradition of inner renunciation being identified with sacrifice continues into the medieval period with the Tantras. Here we have the idea of 'yogic suicide' (*utkrānti*) where the yogi intentionally dies through starvation, transferring his consciousness out of his body to a different realm. Here the refusal of life in order to affirm and realize a higher moral condition entails the yogi's own 'sacrifice'.[31] The sacrifice of the yogi himself is not for appeasement or catharsis, but for inner transformation and vertical ascent.

Sacrifice has continued in India through into modernity. At a popular village level, sacrifice is practised among low-caste communities, and particularly linked to possession. In Kerala dance possession rites among the untouchable communities, rites known as *teyyam*, involve a dancer ornately dressed as one of these local deities such as Viṣṇumūrti, becoming possessed by the god, dancing around the sacred compound, parading to the local temple, and upon being refused entry to the realm of the high-caste god, returning to his own shrine. There, offerings are made of alcohol and blood from decapitated chickens. After the blood offering the *teyyam* calms down and the session is complete, almost immediately after which another deity is performed by the same dancer.[32] This is a complex phenomenon that resists any single explanatory framework. There is clearly a strong element of appeasing an angry deity here and there is a cathartic element to the rite in that the goddess (the *teyyam* is usually a goddess in a male dancer) can be

[30] M. Biardeau and C. Malamoud, *Le Sacrifice dans l'Inde ancienne* (Louvain and Paris: Peters, 1996), pp. 65–73.

[31] Somadeva Vasudeva, *The Yoga of the Mālinīvijayottaratantra* (Pondicherry: IFP, EFDE, 2004), pp. 437–45.

[32] There is extensive ethnography on this by Rich Freeman. See for example 'The Dancing of the Teyyams', in Gavin Flood (ed.), *The Blackwell Companion to Hinduism* (Oxford: Blackwell, 2003), pp. 307–26.

violent towards the crowd and is often taunted by groups of boisterous young men. In a *teyyam* that I witnessed with Rich Freeman, the deity struck one of the onlookers so hard that her sacred bow broke in half. Yet while these elements of appeasement and catharsis are present, there is also a renunciatory dimension in that the dancer is in a 'pure' condition to become a vehicle for the deity, and the community itself is purified through the performance. The goddess is perceived to be present among people, giving her 'vision' (*darśanam*) to them as a blessing. She also performs a divinatory and advice-giving function. In a sense it is not the individual who undergoes purification, but the whole community who become touched by the divine. There is a collective subjectivity formed during the rite that is changed as a result of it. Indeed, the *teyyam* songs that are performed by the dancer contain references to tantric terminology of the inner body, the centres of power (*cakra*) that are in a sense exteriorized beyond the individual in the rite.

In all of these cases, even the death of the solitary yogi, the inner transformation entailed by sacrifice, does not entail individualism. Although of course there is a change—a moral change, to use Mauss' vocabulary—occurring in the human person of the sacrifier, there is a sense in which this is occurring in a collective setting that affects the broader community. Although the Vedic sacrifice is individualistic in some ways—it is the patron who is transformed—it is not the affirmation of individualism. Individualism is an entirely modern value associated with self-assertion and a will that would seek to be unique. Sacrifice is clearly not individualistic in this sense as those involved seek a repetition of performance as close as possible to what went before and for those involved to wholly conform to the structures of the pre-existent tradition. The Vedic patron is the performer of a pre-given narrative of tradition that he is following; there is no self-assertion here in Blumenberg's sense as a characteristic of modernity. Indeed, we might even say that sacrifice is antithetical to modernist self-assertion because, on the contrary, it asserts a collective subjectivity and assertion of collective identity. The Vedic patron, the yogic suicide, and the *teyyam* are sharing in a collective performance in which individual will is subsumed beneath a broader, collective will. The *volo* or *icchāmi*, 'I will', becomes a sign for the broader community and the indexical 'I' becomes a carrier of a group identity. Perhaps a better way of putting this is that the performers of the sacrifice participate in a greater narrative such that the story of their own lives, the story of 'my' life, becomes transformed to the narrative of the group.

But even if this is the case, even if we need to understand sacrifice in terms of renunciation or refusal of some good to attain a higher, collective good, then we are still left with the persistent question concerning the link between human finitude and sacrifice. How does the performance of sacrifice and the renunciate subjectivity that it entails address human finitude? Sacrifice in terms of the accursed share does not really address this adequately; it does not address the existential question, or rather it answers it in a way that is inadequate for human finitude. Bataille's understanding is almost a pantheistic sense of continuous consumption or re-cycling of solar energy in an inevitable, life-driven way that leaves no room for subjective sentiment, human narrative, or transcendence. Rather, it might be that through sacrifice, communities confront the inevitability of death and attempt to control it: sacrifice refuses finitude and death and actually means the affirmation of life itself; sacrifice affirms the desire for life and promotes repair of life through the refusal of death. In terms of sacrifice as refusal, it might be that the renunciation of human goods through giving up a life and by extension through asceticism is actually the giving up of death and the affirmation of life. In his actions we can almost hear the sacrifier say: through the death of this animal I am refusing death itself and affirming my own true nature as deathless.

There are different interpretations of what this practice means in the Indian traditions. One school of thinking believed the sacrifice to be so important that it ensured that the sun would rise in the morning and the sacrificial purpose (*kratvartha*) became more important than any human goal (*puruṣārtha*). On this view the community performs the sacrifice because it is injunction (*vidhi*), enjoined in the revealed scripture: the human purpose of performing sacrifice to ensure a place in heaven after death is regarded as secondary.

This indigenous, Vedic view is not a general theory of sacrifice, of course, but we can understand it in terms of refusal: the refusal to let death take over by keeping the universe in process and in affirming the cosmic order, affirming the order of life. Furthermore, the individual patron through ritual procedures that entail his identification with a foetus grows into a new and renewed life through the sacrifice (even if this is not regarded as the main reason for its performance). In later centuries the Upaniṣads have the idea that the sacrifice becomes not an external practice, but an internal practice performed not by the sacrificial patron, but by the celibate, ascetic renouncer (*sannyāsin*). The yajamāna thus becomes analogous to the sannyāsin. The two groups of texts concerned with sacrifice on the one hand and renunciation on the other, namely the Brahmaṇa and Upaniṣad, implicitly recognize the affinity and connection between sacrifice and renunciation. The renouncer has internalized the sacrifice; the sacrifice of the animal becomes the sacrifice of the breath to the cosmic being (*brahman*); the sacrifice of the lower self of empirical experience is sacrificed to the higher self (*ātman*) that is eternal.

The sacrificial patron and the renouncer share the refusal of death, which is the refusal of time and mortality. To attain the immortal (*amṛta*), the flow of time has to be reversed through ascetic practice—fixing the body in a single posture, slowing the breath, stopping the mind—and these practices have their precursor in early sacrifice. The 'I' of the sacrificer and the 'I' of the renouncer are thus linked in the refusal of death, time, and the world of change. For Bataille, the human Aztec victim is brought from the world of things into the world of intimacy where the patron can become identified with the victim: where they share the same substance. For the Vedic sacrificial patron, we might see the external sacrificial rite as brought into the realm of intimacy in the patron's participation, where his asceticism contributes to the success of the procedure. In the renouncer, the 'victim' is the renouncer himself, who sacrifices himself as an act of refusal of time and death. In renunciation, the internalization of the sacrifice, there is thus a distanciation between the indexical-I of the renouncer (the subject of first-person predicates) and the higher self to be realized, which goal is the transcending of death.

This inner sacrifice entails self-reflection and narratability. The renouncer reflects on himself in contemplative practices, turning the self into an 'object' which is analogous to the sacrificial victim. The realization of the timeless self is achieved through the inner process of practice, which is like sacrifice in that it refuses the reality of death. For Bataille, in Aztec sacrifice we have a collapse of objectivity and the world of work into subjectivity, and similarly in Vedic sacrifice we have the collapse of objectivity, the victim as economic commodity, into a shared subjectivity in which the victim comes to represent the patron and thereby the community. Likewise, the Vedic renouncer has withdrawn from the world of work and social productivity to confront the naked aporia of his transient life. But whereas for Bataille the sacrifice plays out a cultural logic in an economy of energy exchange at a purely horizontal level, we must understand both Vedic sacrifice and renunciation as having a transcendent dimension, a dimension that the tradition regards as being outside of death and outside of time, but which is nevertheless expressed through the cosmic order.

The Refusal of Death

Sacrifice is akin to asceticism because of refusal; because it plays out the paradox of the affirmation of life through its destruction. Through death, life is affirmed. The dead animal in the Veda is only sleeping, ready to awaken, and likewise the ritual patron will awaken from his own death to the realm of the ghosts (*pretaloka*) and thence to the realm of the ancestors (*pitṛloka*), where sacrifice will ensure that he does not undergo a second death (*punarmṛtyu*). On the other hand, the renouncer will awaken to the cognition (*jñāna*) that his self is immortal. In both cases there is the recognition of the transcendence of death and of death as a kind of birth: both sacrifice and renunciation affirm the paradoxical idea that birth follows destruction and this immortality is assured for both sacrificer and renouncer. In the end, life overcomes death in sacrifice.

INTERPRETATION AND EXPLANATION

As we see with the example from Vedic India, sacrifice gives order to the world in cosmological societies. The claim about sacrifice as refusal would need to be corroborated or otherwise in other sacrificial societies, but that it seems to be the case in Vedic culture is itself strong support for the claim. Any general theory of sacrifice needs to take the Vedic material into account simply because of the large scale of the literature devoted to sacrifice and its interpretation involved, and because the Indic tradition has reflected upon itself for many hundreds of years on the meaning of the sacrificial practices that were performed. Drawing an analogy with the Aztecs, Bataille could read this material in terms of his thesis, that Vedic sacrifice is the consumption of surplus, of excess, whose violent expression restores order to the group. This is clearly a possibility, but it ignores the indigenous ascetic understanding of sacrifice and allows cultural function (the bringing about of equilibrium, the restoring of energy balance) to take precedence over existential expression. What is important about sacrifice is that it performs the ambiguity of the affirmation of life through its negation in the sacrificial act, and this is the confrontation of a sacrificial community with the bare life of being born and dying. The sacrifice is human self-repair through affirming the desire for life in the face of death and affirming that, in fact, life overcomes death.

Bataille presents a model of sacrifice that is accounted for, on the one hand, in terms of a general economy, and on the other through the cathartic release of energy. Both Bataille and Girard present arguments that conceptualize sacrifice in terms of an economy of consumption and power. But we could argue that sacrifice needs to be understood less in causal terms, explaining one cultural phenomenon by a lower-level one, and more in semantic terms and in terms of transformations undergone in human persons. Sacrifice is not so much energy release as structuring life and ways of seeing the world;[33] it functions to order human reality in relation to a divine realm, to create transcendent meaning, and fosters a social stability that becomes repeated through the generations. If sacrifice orders life and endows meaning in relation to establishing a connection first between human beings and second between a community and a transcendent power, then the emphasis of Girard and Bataille is misplaced. Rather than the process of sacrifice and its consequent energy release that serves to neutralize violence and even give pleasure, what is important is the refusal of material products to meet subjective desires. In this refusal the sacrificial victim becomes a gift both to the god who needs to

[33] For example, Vedic sacrifice provided the structure through which to perceive the cosmos and was fundamental to life ensuring, in some texts, that the sun would rise. The centrality of sacrifice can also be seen in other cultural contexts. See, for example, Evans-Pritchard, *Nuer Religion*, pp. 197–230.

be appeased and to the sacrificial community that needs to be purified and that receives it in the sharing of the sacrificial meal, thereby reinforcing social bonds and the relationship between community and transcendent source. Sacrifice is an effective articulation of social cognition. Through sacrifice we have the formation of new, higher meanings for a community beyond the mundane, the affirmation of verticality, which is the affirmation of life itself. Sacrifice marks a condition and points to a process of transformation and becoming that, within the Indian tradition, for example, moves from a literal enactment to an inner or symbolic enactment within the person.

Gift-exchange and catharsis models of sacrifice are not wrong but operate at one cultural level—a level of energy balance and expression in a community. There is, however, a deeper or more fundamental function of sacrifice as a practice and a metaphor that lays bare the reality of life. Sacrifice brings us into the world through creating a cosmos in which the meaning of human life, of a community, can be located. Sacrifice takes place at a more visceral, cultural level and is the pre-philosophical articulation of problematics that can be articulated philosophically, but they can only be resolved or expressed through action, through ritual. The existential level of sacrifice shows us something about ourselves. It points to mortality and to the human aspiration for transcending mortality and negating death and nothingness.

Because of the existential nature of sacrifice, it requires an approach from the human sciences that accepts its multi-layered complexity. A hermeneutical approach is better suited to gaining a fuller understanding of sacrifice than a scientific approach that seeks causal explanations (such as those offered by cultural functionalist models). Sacrifice is a very ancient human institution and needs to be understood as operating at different cultural levels. Ricoeur offered an interesting approach to social phenomena. If human society displays certain characteristics that are analogous to a text, then we can 'read' human social action in ways analogous to reading a text.[34] Thus, sacrificial practices as social action can be 'read', and the reading I have presented here seeks to understand the complex phenomenon of sacrifice in terms of human existential reality, as self-repair or in terms of the refusal of death and the affirmation of life, concerns that have been fundamental to all human societies throughout history.

The phenomenon of sacrifice is complex and resists explanation in terms of any single paradigm. We can see the sense of theories of the sacrifice that see it in terms of catharsis, or gift-exchange, or even renunciation, but none of them exhaust its meanings. This is why, finally, we need to understand sacrifice in hermeneutical and existential terms. Causal explanations are never sufficient,

[34] Paul Ricoeur, 'The Model of the Text: Meaningful Action Considered as a Text', in *Hermeneutics and the Human Sciences*, trans. John B. Thompson (Cambridge: Cambridge University Press, 1981), pp. 197–221.

and even if we could map with great accuracy where and what is occurring in the central nervous system during the performance of sacrifice, this would tell us little about its meaning, nor would it offer an adequate causal account. It could be that sacrifice resists explanation in terms of causation that social science or brain science could develop, and so we are left with understanding in other terms. Arguably in the study or inquiry into sacrifice, especially across cultures, we first need a phenomenology that allows sacrifice to show itself; as it were, allows what shows itself to be seen. This is essentially an ethnographic or descriptive account (of the kind that I have briefly referred to above). Second, we need a hermeneutical account that generates theory from description or rather uses description to offer particular interpretative angles. In one sense, of course, these kinds of understanding are also explanations, as Ricoeur has pointed out.[35] I have tried to do this here within the general thesis of sacrifice as refusal. Lastly, in the modern Western economies where literal kinds of sacrifice are not performed—unless we count the millions of animals slaughtered in industrial meat production—we have grown unaccustomed to and horrified by it. The world of sacrifice for us is, in Heesterman's phrase, a broken world. We need therefore a kind of human empathy or something like Heidegger's formal indication, using the fact of my being to understand the being of others, to partially step into the world of sacrifice. We can understand sacrifice as renunciation as a feature of all our lives that can become a formal indication of a pre-modern sense of sacrifice. The patron of the sacrifice is a sign for the community and in refusing death is affirming life itself, life for all. The existential *aporia* of human life, facing death, seems to be resolved; although perhaps in truth it is not.

Sacrifice as ritual act is antithetical to contemporary sensibilities and yet is so foundational to cultural life, as I hope to have shown here. With the alignment of sacrifice and renunciation we have externality mapped onto internality and the aspiration for vertical ascent articulated through these two modes. Both sacrifice and renunciation are the affirmation of life itself, the incorporation of death into life, and the refusal of nothingness. This parallel structure of sacrifice-renunciation can be found throughout the history of civilizations as different kinds of transformation of our human proclivities, of our human nature. Religions as autopoietic systems are dominated by the sacrifice-renunciation paradigm, and we might even suggest that extroversive religions, the Abrahamic religions, are dominated by the paradigm of sacrifice, in contrast to which introversive religions, Buddhism, Jainism, and some forms of Hinduism, are dominated by the paradigm of renunciation. Both sacrifice and renunciation entail each other, as I hope to have shown, and both are the

[35] Paul Ricoeur, 'Qu'est-ce qu'un texte?', pp. 153–62 in *Du texte a l'action. Essais hermeneutique II* (Paris: Editions de du Seul, 1986) pp. 153–78.

affirmation of life itself over nothingness and death. We are now in a position to see this pattern of the affirmation of life itself unfolding in the history of civilizations. My theoretical apparatus is now in place to examine the category of life itself in those histories and show how they promote self-repair through affirming desire for life or transforming it.

Part II

Religious Civilizations

4

The Sacrificial Imaginary

Indic Traditions

LIFE'S QUESTS

The history of Indic civilization is a history of the dominance of the Sanskrit language used at first in texts set aside as sacred and regarded as revelation from a higher source, to the classical flowering of epic literature, poetry, law treatises, philosophy, medicine, and so on. This body of literature over a long period of time was composed by a remarkable group of people, the Brahmans or Brahmins, who regarded themselves, and were regarded by others, as a social elite distinguished by their learning and wisdom if not by their power and political authority. Later in the medieval period vernacular languages such as Tamil develop their own literature but modelled on Sanskrit genres and styles, and Persian becomes important with the Mughal dynasty. While political formations have come and gone, the continuity of Sanskrit has been fundamental in the growth and flowering of this civilization, along with a social stability assured by a hierarchical and differentiated social system that has now come to be known as caste. Human self-repair has been a central theme both in the sense of desire for and affirmation of life and in the sense of transcending life for a higher order; all three narratives, the Gnostic, the Non-Dualist, and the Theistic, have had a strong voice in this history. Yet there has also been a naturalist tradition that understood life in terms of a more material causation. While this was not the Galilean Mathematical Model (GMM) that was to develop in the early modern West, it was a naturalism focused on the science of life but tied in to the Religious Cosmic Model (RCM) that saw life in terms of a spiritual, cosmological hierarchy.

Human self-repair has been understood mostly in individualistic terms as a process of purification of the individual soul or life force over a long period of time as it passes through a beginningless series of reincarnations. Although the Brahmans did address the question of life itself in relation to human self-repair, other categories tended to dominate discourse such as Being (*sat*) and absolute reality (*brahman*), along with epistemological categories about how we can know anything (*pramāṇa*). The philosophy of life in Sanskrit culture tended to be replaced by a philosophy of being, although the absolute reality did become identified with life itself, as we will see. The Brahmans also developed spiritual practices—ritual, asceticism, meditation—that articulated an implicit philosophy of life through action. Some traditions, particularly the tantric religion of Śiva, closely integrate practice with a vital philosophy of life that thematized life itself as divine. Other traditions, such as Buddhism, have a more cautious attitude and generally present the Gnostic narrative of wishing to leave the world of life, which is glossed as the world of death, for a higher transcendence or liberation.

In the following pages we will examine the vitalism entailed in Vedic sacrifice, the Axial transformation of that theme in the Upaniṣads, and the philosophies that attempted to categorize and analyse life into specific components. Finally, we will see how a full philosophy of life comes to articulation in the tantric traditions. All these modes of thinking and practice were deeply concerned about offering repair, correcting ignorance and giving relief from the constant suffering entailed by life, and the desire for life that has so often been negatively evaluated in this history. But before we begin, we need to clarify some basic categories within Indic thinking.

Brahmanism distinguishes between Vedic and non-Vedic philosophies,[1] between traditions that accept divine revelation of the Veda and those that do not. But holding too firmly to this distinction does not allow us to see the continuity of discourse (and indeed practice) across that divide. Different traditions of thought developed in the Axial and post-Axial age and by the time of the early Middle Ages, Hindu, Buddhist, and Jain thinkers were in debate across boundaries, and practices of one tradition were even adopted by others: Buddhist meditation influences yoga and Śaiva ritual influences Jainism. Polemic might be just as great between two schools within the Vedic fold, between the Vaiśeṣika and Vedānta schools, for example, as between Vedic and non-Vedic schools. While there is documented tension between the Brahmans, those who followed the Vedic scriptures and tradition of sacrifice,

[1] The *āstika/nāstika* distinction, those who accept the authority of the Vedas (from *asti*, 'it is') and those who do not (*na asti*, 'it is not'), is fairly late. The tenth-century philosopher Jayanta Bhaṭṭa in the *Nyāyamañjari* enumerates six logics (*ṣaṭ-tarki*), the *āstika* as Mīmāṃsā, Nyāya, Sāṃkhya, Jaina, Bauddha, and the *nāstika* as the Cārvāka, and the Jain philosopher Haribhadra (seventh century CE) describes *āstika* theories as five, conflating Vaiśeṣika and Nyāya, with the Cārvāka as the sixth. He defines *āstika* as a system that accepts the existence of another world (*paraloka*), merit (*puṇya*), sin (*pāpa*), and transmigration (*gati*). See Damodara Lal Goswami (ed.), *Ṣaḍdarśanasamuccaya, with a Commentary Called Laghuvṛtti by Maṇibhasra*, by Haribhadra (Benares: Chowkhamba Sanskrit Book Depot, 1905), pp. 98–100. See also Andrew J. Nicholson, *Unifying Hinduism: Philosophy and Identity in Indian Intellectual History* (New York: Columbia University Press, 2010), pp. 154–8, 166–84; Wendy Doniger, *On Hinduism* (Oxford: Oxford University Press, 2014), pp. 45–8.

and Śramanas, ascetics who rejected traditional scriptural authority in the quest for total freedom from entrapment in the world,[2] both shared a continuity of discourse and philosophical terminology. As we begin to explicate this complex history of a civilization, we need to bear in mind our primary focus of inquiry into the category of life and its importance in human self-repair. It is not possible to be exhaustive or even comprehensive, but I will proceed first with reflection on our earliest sources that articulate some understanding of the category of life itself in terms of sacrifice, namely the Vedic scriptures, and what might be called the Axial reinterpretation or internalization of sacrifice. We will then move forward to a time when traditions had become clearly demarcated from each other, taking examples of thinking about life and matter from the first millennium of the common era and tracing this history through the second, to Brahmanism on the eve of modernity. During this long period of time, major shifts in patterns of thinking and practice occurred from a sacrificial worldview, to an Axial transcendental view, to a post-Axial devotional, vitalistic, and scholastic analysis of life, and so to global modernity. That is, we will move from a reflection on Vedic sacrifice to its transformation in the Upaniṣads and the later Vedānta tradition. Second, I shall take up the Axial and post-Axial classification of life in non-Vedic tendencies, and third, examine a confluence of traditions in medieval Tantrism. In a sense, and this is to draw a rough sketch, Vedic sacrifice presents a strong affirmation of life itself which is transformed in the Axial response that emphasizes transcendence. A further response is the non-Vedic rejection of the world of life presented within the sacrificial imaginary, and finally to a further transformation of that imaginary in the tantric tradition from thence into modernity.

These reflections correspond, more or less, to the three major sub-divisions of this chapter, which also correspond, more or less, to ideological developments that were linked to different regions. The mainstream, Vedic tradition focused on sacrifice was centrally located in the homeland of the Noble Ones (*āryavarta*), whose centre of gravity was between the rivers Ganges and Jumna. This was the culture of the Brahmans who spoke the well-formed or perfected language, Sanskrit, maintained strict rules of purity (*dharma*), and were generally theists. To the south-east was the kingdom of Magadha from whence came the ascetic movements of Buddhism and Jainism, which rejected Vedic practice and were atheistic in orientation (at least in the sense of rejecting an omniscient and omnipotent creator). Both had distinct orientations and understandings of life and both come into contact with Brahmanism and influence it. By the medieval period, Buddhism was disappearing from the

[2] S.D. Joshi (ed.), *Patañjali Vyākaraṇa Mahābhāṣya* (Poona: University of Poona, 1968), II.4.9; I.476. Cited in Romila Thapar, *Interpreting Early India* (Oxford: Oxford University Press, 1992), p. 63.

sub-continent, although ascetic ideas and renunciation had been absorbed into Brahmanical culture. The third section reflects this fully developed absorption of the two orientations in medieval Brahmanical learning that we find in the tantric tradition.

This is certainly a shift from a small-scale, rural society, to an urban one focused on kingship and in due course the development of nation in the present era, but not only this. It is also a shift from an understanding of human flourishing in terms of an increase of success in living a prosperous life with abundance of cattle and children in a social order integrated with a cosmos, to an understanding of human prosperity as world transcendence, and then the integration of a sacral order into a political order that was only overturned with the secular politics of empire and nation state. Within this narrative there is a cluster of ideas about ontology or ways of categorizing existence, the transformation of the mind in yoga, and the development of medical science, all of which are concerned with life itself in some way and that feed in to the mainstream pattern of the transformation of the sacrificial imaginary. There are transformations of moral order involved in these developments that I need to chart. The picture is complex insofar as these changes in worldview and social order cannot be understood simply due to a shift from a rural to an urban economy and polity: prior to Vedic society, there was a thriving urban civilization in the Indus Valley parallel to Mesopotamia and Egypt,[3] and Vedic sacrificial culture has never completely died out and persists into the modern world (there is a sense in which nothing ever gets completely left behind in India).

There is a distinct ideological shift in understanding human fullness from an emphasis on everyday social goods to world transcendence and the pull of verticality alongside a horizontal classificatory cultural impulse. Different concepts of self are involved here that Mauss, in his famous essay, described as the beginnings of the notion of person developing in India that was ultimately denied in favour of the negation of the person as a locus of value in traditions of renunciation.[4] While Mauss' perfunctory and surprising treatment of India in this essay leaves much to be desired—he seems to get a little muddled about the doctrines of Sāṃkhya and the *Bhagavad-gītā*—his raising the question about the evolutionary development of the concept of person is germane to our inquiry. Dumont picks up this theme, arguing that the notion of the individual develops in Indian religions with the world renouncer but not with the 'man-in-the-world' who is defined socially.[5] This formulation is itself

[3] Bernard Sergeant, *Genèse de l'Inde* (Paris: Payot, 1997).

[4] Marcel Mauss, 'The Category of the Person', in Michael Carrithers, Steven Collins, and Steven Lukes (eds.), *The Category of the Person: Anthropology, Philosophy, History* (Cambridge: Cambridge University Press, 1985), pp. 1–25.

[5] Louis Dumont, 'World Renunciation and Indian Religion', in *Homo Hierarchicus* (Chicago, IL: Chicago University Press, 1980 [1966]), pp. 267–86.

problematic insofar as the renouncer does not embody individualism as a value in any modern sense because he, and sometimes she, erases social personhood through the act of renunciation. The renouncer gives up name, clothing, and all markers of social belonging in favour of a different kind of identity that emphasizes sameness and being part of tradition.[6] But the renouncer is individual in the sense that he recognizes a truth outside of the human community that he is pulled towards, and that this path is in some sense solitary because it entails the cultivation of inwardness through prayer and meditation. While the quest for the ultimate truth, for *brahman*, might be shared by a religious elite, Weber's *virtuosi*, the experience is particular to the individual, particular to me. Although the value of individualism is absent, the quest for vertical ascent is individual but in a different sense to individualism because it entails the eradication of personal uniqueness.[7]

The two orders are clearly different. On the one hand, we have Vedic social order expressed in the sacrifice in which Vedic society is embedded within the cosmic structure, and which seeks benefits for that society through the magical technology of sacrifice. On the other, we have the renunciate order that places itself outside of mainstream society and seeks a value beyond it, but which it regards as the legitimate goal for all. I can see what Dumont is trying to get at, but in another sense the Vedic sacral order did emphasize the individual as patron of the sacrifice, yet an individuality that was a sign of community. With the Axial shift, value comes to be held in the self's ability to transcend its worldly status, and Dumont highlights a tension throughout the history of Indian religions between the values of renunciation and the values of world affirmation in the householder life. I think there is truth in Dumont's characterization but coming back to our theme, we might say that both sacrificial religion and its renunciation are different affirmations of life through the attempt to transcend death or rather to confront nothingness. If human actions are directed towards goals, and I think on the whole we must affirm this, then the sacrificial act and the renunciatory act, while they may be parallel and address the same issue of life and death, are overtly directed towards different goals. Vedic religion values the prosperity of community that it asserts through the transformation of death in the sacrificial act. Renunciation values spiritual liberation for the renouncer, albeit within a community of renouncers, that it asserts through ascetic gnosis. The shift from a pre-Axial sacrificial economy to an Axial transcendence is a shift from the affirmation of life itself as this-worldly value, to life itself as transcendence of matter and

[6] Gavin Flood, *The Ascetic Self: Subjectivity, Memory, and Tradition* (Cambridge: Cambridge University Press, 2004), pp. 211–34. Also see Sondra Hausner, *Wandering with Sadhus: Ascetics in the Hindu Himalayas* (Bloomington, IN: Indiana University Press, 2007), p. 116.

[7] See my *The Truth Within: A History of Inwardness in Christianity, Hinduism, and Buddhism* (Oxford: Oxford University Press, 2014), pp. 202–8.

nature. In this context human self-repair is understood as the satisfaction of desire for life or going beyond such desire and denying it satisfaction. This shift went hand in hand with the analysis of life and the formation of categories that persist and develop through the long history of Indic religions.

Charles Taylor in a useful essay characterizes the distinction between the Axial and pre-Axial religions as being that the former stresses going beyond the human world, the shift to a new standpoint from which the old can be criticized, and a disembedding in which there is a change from an emphasis on community prosperity through 'feeding the gods' to an emphasis on the human good as a higher ideal. Taylor puts this well:

> What the people ask for when they invoke or placate divinities and powers is prosperity, health, long life, fertility; what they ask to be preserved from is disease, dearth, sterility, premature death. There is a certain understanding of human flourishing here which we can immediately understand, and which, however much we might want to add to it, seems to us quite 'natural'. What there isn't, and what seems central to the later 'higher' religions, is the idea that we have to question radically this ordinary understanding, that we are called in some way to go beyond it.[8]

While sacrificial religion desires community prosperity, renunciation desires community transcendence. It questions the sacrificial order and claims that we must go beyond it through internalizing the sacrifice, which then functions only in imagination; a shift from actual social act to a religious imaginaire. The shift from the pre-Axial to the Axial is a shift in emphasis on what is significant. For the sacrificial order, significance lies in the successful resolution of social or agonistic tension that Bataille took to be the successful circulation of solar energy. For the renunciate order what is significant is transcendence of worldly life, so there is a shift in value from an equilibrium of life gained in the here and now to an eschatological expectation of the transcendence of the world.

This is articulated in the history of Indic institutions in the *āśrama* system, the system of stages on life's way that were originally lifestyle choices, namely the four stages of student, householder, hermit, and renouncer.[9] During the early first millennium BCE communities of renouncers began to live outside of mainstream society, cultivating a meditative way of life in eschatological hope of personal salvation. In time this institution of renunciation became monasticism, which itself had great impact on post-Axial religions, and while there was nothing quite like the dissolution of the monasteries, there certainly was a

[8] Charles Taylor, 'What Was the Axial Revolution?', p. 33 in Robert Bellah and Hans Joas (eds.), *The Axial Age and its Consequences* (Cambridge, MA and London: The Belknap Press of Harvard University Press, 2012), pp. 30–46.

[9] Patrick Olivelle, *The Āśrama System: The History and Hermeneutics of a Religious Institution* (Oxford: Oxford University Press, 1993).

denuding of monastic power and influence in the medieval period, as we see with the Pāśupata order.

With the rise of renunciation as a value and eschatological hope, we have the analysis of life developed in the service of its transcendence or rather in the confidence in a higher life. Renunciate orders saw the goal of this higher life in different ways, as the self's isolation from matter and world (Sāṃkhya), as the realization of its utter transcendence in mystical experience of liberation (Advaita Vedānta), or as the self's annihilation in a vast emptiness (Mahāyāna Buddhism), and these three do not exhaust the possibilities here. We will have cause to examine these notions in the coming pages because these goals are understood as having ultimate significance for our everyday life in the world. I think that Dumont is right in his claim that innovation in Indian religions comes through the renouncers, and our inquiry will inevitably be an inquiry into the literature of the renunciate orders: our texts were composed mostly—possibly only—by world renouncers, although there must have been an infrastructure of householder support, patronage, scribes copying the manuscripts they produced, and the transport of manuscripts to different locations of learning in the sub-continent. We know that manuscripts copied in the South in the medieval period made their way fairly quickly, within thirty years, to the North, and it may be that new ideas, new reflections, spread like wildfire among the intellectual elites in post-Axial India.

While the tradition known as Vedānta is important—some would call it the central tradition of Indian thinking[10]—we now know that medieval India was dominated by the religion of Śiva, beginning from around the fourth century CE and emerging in a great sweep of texts, practices, and royal patronage throughout South and South-East Asia, thereafter to around the thirteenth century.[11] I want therefore to trace a trajectory not only in the Vedic Vedānta tradition but also in a different course from realist philosophies that laid stress on the codification of the world and that feed directly into this explosion of new religious forms. This leads to the vitalism of the tantric traditions, bringing into sharp relief a concern both in terms of understanding and explaining life and of desiring to transcend it to some higher life. I want to follow this path not so much simply as a history but in order to bring out a concern for the nature of life itself throughout this history. We have, then, to

[10] S. Radhakrishnan writes: 'It is said that other scriptures sink into silence when the Vedanta appears, even as foxes do not raise their voices in the forest when the lion appears. All sects of Hinduism attempt to interpret the Vedanta texts in accordance with their own religious views. The Vedanta is not a religion, but religion itself in its most universal and deepest significance'. *The Hindu View of Life* (London: George Allen and Unwin, 1927), p. 23.

[11] See Alexis Sanderson, 'Śaiva Literature', *Journal of Indological Studies* (Kyoto), vols 24 and 25, 2014, pp. 1–113; 'The Śaiva Age: The Rise and Dominance of Śaivism during the Early Medieval Period', in Shingo Einoo (ed.), *Genesis and Development of Tantrism* (Tokyo: Institute of Oriental Culture, University of Tokyo, 2009), pp. 41–350.

speak about sensibilities towards life that can be analytically distinguished as a sacrificial sensibility, renunicatory sensibility, and also a naturalist sensibility that seeks to understand and explain the world almost as a scientific enterprise. These sensibilities are intertwined in that renunciation needs to be understood within the sacrificial imaginary, while naturalist tendencies were not divorced from vertical aspirations to transcendence.

To expedite this discussion, we need to bear in mind the distinctions between sacrifice and renunciation, life and matter, and consciousness and power that echo throughout this narrative. But first I need to pick up the theme from Chapter 3 of sacrifice as refusal, and a suggestion from our discussion of Bataille about the necessity of bringing the sacrificial victim into the world of intimacy from objectivity in order that an identification between victim and patron can be established. We see this pattern in early Indic sacrifice and the irresolvable contradiction of affirming life through death. Jan Heesterman's important work demonstrates the complexity of what we are dealing with and, as Frits Staal also emphasized, the need for any general theory of sacrifice to take on the vast corpus of texts that have sacrifice as one of their central concerns, namely the corpus of Vedic literature from the early Saṃhitās to the Brāhmaṇas.[12] That Hubert and Mauss did so in their important long essay on sacrifice is greatly to their credit. From Vedic sacrifice as the affirmation of life we need to see how the Axial renunciation of sacrifice in the Upaniṣads still functions as the affirmation of life, and how the trope of sacrifice continues to function in the Brahmanical religious imagination throughout the Vedānta literature. I will then be in a position to take up the narrative with more realist classifications that feed in to the Śaiva imaginaire in the early medieval period.

Vedic Sacrifice

Vedic ritual procedure was and remains an elaborate process. In the classical Vedic sacrifice, at the occasion of the *agnicayana* rite, an altar made of bricks in the form of a bird was built, and three fires established within a ritual enclosure. Three priests reciting or singing verses from the three Vedas were in attendance with a fourth priest, called the Brahman, knowing the *Atharva-veda* and in the later ritual overseeing the procedure, although in the *Ṛg-veda* being the poet who spoke the truth.[13] The animal to be slaughtered—a goat or

[12] Jan Heesterman, *The Broken World of Sacrifice: An Essay in Ancient Indian Ritual* (Chicago, IL: Chicago University Press, 1993), p. 9. Frits Staal, *Rules without Meaning: Mantras Ritual and the Human Sciences* (New York: Peter Lang, 1989), pp. 149–52.

[13] Stephanie Jamieson and Joel Brereton, 'Introduction', *The Rig Veda*, vol. 1 (Oxford: Oxford University Press, 2015), p. 28.

cow or even horse—would be tied to a post (*yupa*) outside of the ritual enclosure and suffocated without blood, although in origin decapitation was the most likely *modus operandi*. While the actual killing of the animal eventually becomes de-emphasized, in origin the killing and sharing of the meal were important as the facilitator of social cognition. I have already discussed some theories of sacrifice and offered my own reflections that I shall further support here. Heesterman, following Burkert, speaks of a family of phenomena called sacrifice, a range due to its great antiquity going back to prehistory. Heesterman speaks for three elements—killing, destruction, and food distribution. Killing (the verb *han* 'to kill' is used) is the way the texts express what is done to the animal or even vegetable. It is destroyed in the fire, and distributed as food, thereby becoming 'a primary force in the formation and maintenance of human society'.[14] There was also a fourth component, contest. For Heesterman, the sacrifice was a contest from beginning to end, first in the competition as to who would be the sacrificer, followed by chariot races, dicing games for different parts of the sacrificed cow, and argumentative disputation (*brahmodyas*) in which ritual contestants would challenge each other with riddles.[15] This agonistic dimension was played down over time, but we can see it coming through the ritual or *śrauta* texts. This maintenance of society through sacrifice, in Heesterman's insightful study, has to be repeated because the broken world in which humans live can never be restored to a primal unity, the unity indicated in the famous hymn to the cosmic person whose sacrificed body becomes the universe.[16] So the force behind both cosmogony and sacrifice is desire (*kāma*), the desire for primordial unity and the desire for a healed community, in which death no longer reigns supreme. Heesterman writes:

> In the play of sacrifice man acts out his awareness that he lives in a conflictive world broken by the irreparable rift of his morality. He desperately strives to restore a primordial static unity, fusing life and death into an organic whole. But the dynamics of desire will force him out again to face death and destruction. When he puts together the thousand bricks into the five-layered Vedic fire altar in the form of a bird that will never fly, he attempts to reconstruct the dismembered body of the cosmic man who at the same time is Agni, the fire, his *alter ego*.[17]

Substitution is key to understanding Vedic sacrifice and substitution is also part of a general identification or homology between different elements of the rite and cosmos. Later interpretation in the literature of the Brāhmaṇas and Upaniṣads emphasized these connections or 'bindings' (*bandha*). The victim of sacrifice as economic commodity becomes the substitute for the patron who in turn might be said to represent the whole community. The patron for the

[14] Heesterman, *The Broken World of Sacrifice*, p. 10. [15] Ibid., p. 42.
[16] *Ṛg-veda* 10.90. [17] Heesterman, *The Broken World of Sacrifice*, p. 29.

ritual time frame of the sacrifice has withdrawn from the world of work and productivity to encounter his own death through the victim, and so gain life. This is more than a horizontal economy of exchange, as we saw in Chapter 3; it is rather a cultural imagining where death is transformed into life. The connection (*bandha*) between the patron and victim reinforces life itself not only for the patron as named individual, as the one initiated (*dīkṣita*), but also for the community of whom the *dīkṣita* himself is a sign. The sacrificial victim is the patron who is the community and in a sense all sacrifice is a kind of self-sacrifice but a death that gives life. In Bataille's sense, the victim is shifted from commodity, from the purely functional world of domestic animals to a realm of intimacy in its death. The hymns of the *Ṛg-veda* praise the victim and convey a sense that they are doing the victim a favour by sending it to the world of the gods.[18]

The Brāhmaṇa texts often say, 'man is the sacrifice' (*puriṣo vai yajña*), linking sacrifice to the hymn of the cosmic man, the *puruṣa*, who is immolated by the gods to create the universe. We see this structure of the affirmation of life through sacrificial death in the performance of the rites, the sacrificial arena, and in narratives connected with it, in particular the explicit story of sacrificial substitution in Śunaḥśepa, told in the *Śatapatha Brahmaṇa.* The Brahmin's son Śunaḥśepa (whose infelicitous name refers to a dog's *membrum virile*) is substituted as a sacrificial victim for king Hariścandra's son who had been granted release by Varuṇa on condition that he in due course be sacrificed back to him. Varuṇa agrees to the substitution but Śunaḥśepa skilfully wins his freedom through praising the gods.[19] Sacrifice is implicit substitution for the patron of the sacrifice and thereby for the community. It faces death and attempts to transform death. The myth of Śunaḥśepa says that death is not inevitable and the sacrifice itself affirms life through the immolation of the victim.

This non-inevitability of death is reinforced throughout the sacrifice. The Brahman priest, the most learned fourth priest who recites the *Atharva-veda*, is said to 'heal the sacrifice' which refers to the ritual act of offering a small portion of the sacrificed animal, called the *iḍā* portion, to the Brahman. This symbolizes the wound created by the god Rudra, the excluded feral deity, who shot an arrow and pierced the sacrifice that needs to be symbolically healed by the Brahman priest. Iḍā was the Goddess in the form of a cow who represents the sacrificial meal, and the *iḍā* portion of the sacrificial meal is 'torn apart'. The Brahman priest then heals the wound of the sacrifice and the *iḍā* is even identified with the power of the sacrifice itself, also called *brahman*.

[18] For example, the two hymns praising the horse that is sacrificed. *Ṛg-veda* I.162 and 163.

[19] *Śatapatha Brahmaṇa* III.95.109, trans. Julius Eggeling, *Śatapatha Brahmaṇa according to the Text of the Madhyandina School*, Sacred Books of the East (Oxford: The Clarendon Press, 1900). Heesterman, *Broken World of Sacrifice*, pp. 173–4.

There is an identification of the sacrificial cow, which is food, with the power of *brahman*. Heesterman writes:

> The essential point would seem to be that to become the sustenance of life food has to pass through death, like the immolated *iḍā* cow; and so, in mysterious fashion, has the royal sacrificer—in fact, any sacrifice. *Brahman*, then, appears to be intimately connected with the passage through death. It is the link between life and death. This, the link between life and death, is the riddle of *brahman*.[20]

The sacrifice addresses the bare aporia of human life. The affirmation of life is achieved through strife and through violence. Not only does the patron undergo ascetic purification but they also symbolically take soma and fire, the two sacrificial substances of value, from the demons or Asuras. The demons, the negation of life, are conquered through the act of sacrificial violence that thereby narratively acts out the overcoming of death. Śunaḥśepa through his skill wins out—the victim affirming life and showing how that which is fit for sacrifice gives us life, but we have to keep repeating the act because, tragically, it always fails. And even at physical death this hope is articulated in calling the funeral where the body is burned the last sacrifice (*antyeṣṭi*): the final sacrifice that transforms death into sacrifice and so death into hope of life.

At this early period, there is some philosophical speculation in a few late hymns of the *Ṛg-veda* but the Vedic worldview is articulated primarily through the sacrifice. In a famous hymn the universe is conceptualized as a giant man (*puruṣa*) who is sacrificed by the gods and whose body forms the universe and society.[21] The philosophy of the Mīmāṃsā regarded sacrifice as essential for life itself and for the regular ordering of human life in particular. The universe does not contain something hidden, some truth that it covers, as the Vedānta and renunciate Axial religion believed, but it is what it is, and that is sacrifice. There is nothing beyond the structure of the cosmos and the sacrificial mechanism of life and death that it embodies. To ensure that the sun will rise tomorrow, we need to sacrifice in the never-ending renewal and transformation of energy within it. This is, in fact, a view not far from Bataille. On this worldview Panikkar insightfully writes:

> This ultimate structure is not to be regarded as another or deeper thing or substance; it is in fact sacrifice, the internal dynamism of the universe, universal *ṛta*, cosmic order itself... Sacrifice is the act that makes the universe.[22]

[20] Heesterman, *The Broken World* of *Sacrifice*, pp. 155–6.

[21] *Ṛg-veda* 10.90. Jamieson and Brereton, *Rig Veda*, vol. 3, pp. 1537–40.

[22] Raimundo Panikkar, *The Vedic Experience: Mantramañjari: An Anthology of the Vedas for Modern Man and Contemporary Celebration* (London: Darton, Longman and Todd, 1977), pp. 352–3.

The paradox of life negated by death is played out in the ritual arena and never successfully resolved except as the repeated assertion of refusal. The patron and whole Vedic community symbolically refuse death and affirm shared goods of community and prosperity through the sacrifice, as the hymn to the puruṣa states.

In the last three of the twelve days of the *agnicayana* rite, the building up of the fire, there are final oblations and the patron makes donations or pays the fee to the priests. The patron is anointed like a king and offerings of soma, the juice from particular pressed plants, are made to Indra. Eleven goats are sacrificed, and rites of expiation are made just in case of any ritual omission. The formal relationship between the patron and the priests is dissolved and after a final bath and a final goat sacrifice, the patron with his wife returns to his house accompanied by three fires from the sacrifice that he will keep burning in his home and make offerings to in the morning and evening for the remainder of his life.[23] The long preparation, during which time the patron underwent ascetic purification, abstained from certain foods, and observed celibacy, has paid off with the successful completion. The patron has become parallel to the student of the Veda, the brahmacārin, who likewise is celibate and restricted in dietary regime while he learns the Veda.[24] The hall of recitation is burnt and the ritual arena abandoned back to nature. Any modern performance of Vedic ritual will of course be different simply because of the ambient culture, but being true to textual prescription means that the sequences of act have remained the same for thousands of years.[25]

Vedic ritual knows no permanence. With the completion of the sacrifice, a statement has been made to those present that affirms life itself. The patron became formally connected through ritual identifications with the cosmos and there is undoubtedly a social dimension to the rites in which the status of the patron is enhanced, at least during the longer sacrifice.[26] But more than this, the sacrifice is the attempt to heal the wound, to resolve the irresolvable

[23] This is a condensed version of Frits Staal's description of the *agnicayana* he witnessed in 1976, summarized in *Rules without Meaning: Ritual, Mantras, and the Human Sciences* (New York: Peter Lang, 1989), pp. 76–7.

[24] Heesterman, *The Broken World of Sacrifice*, pp. 165–70.

[25] Michael Witzel remarks that hearing the Veda is like hearing a 3,000-year-old tape recording. Witzel 'Vedas and Upaniṣads', pp. 68–9 in Gavin Flood (ed.), *The Blackwell Companion to Hinduism*, pp. 68–101. In a way, the remnant of Vedic ritual in Kerala among the Nambudri Brahmins is a kind of cultural fossil. One way of accounting for such longevity is to separate form or structure from meaning. This is what Frits Staal does in his explanation in which he argues that Vedic ritual has structure or syntax but no meaning, no semantics. Meaning, such as we find in the Brāhmaṇas and Upaniṣads, is a later ad hoc projection. See his important book *Rules without Meaning*, pp. 131–40.

[26] On the length of these rites see Charles Malamoud, 'Exegesis of Rites, Exegesis of Texts', in *Cooking the World: Ritual and Thought in Ancient India* (Delhi: Oxford University Press, 1996), pp. 226–46.

aporia of life and death through the insistence that death brings life. Life shows itself through ritual act and the ritual actors perform a transformation or repair of life.

The Axial Shift

Once sacrifice ceases to be the dominant, Brahmanical practice and is replaced by renunciatory religion, on the one hand, and devotional religion on the other. But the theme or trope of sacrifice never ceases. There is certainly a shift of values with the development of the Upaniṣads and other renunciatory religions during the first millennium BCE and we enter a new vision of moral order. I am happy to call this an Axial shift in recognition of a general scholarly consensus that this period marked significant changes in different world civilizations. In Indic religions the shift is almost self-conscious in the sense that the Upaniṣads see themselves as a new departure and as the internalization of the sacrifice. But although the practice of sacrifice becomes less widespread, the idea of sacrifice is central to the religious imaginaire and orders ways of understanding life well into the medieval period. The earliest of the Upaniṣads, the *Bṛhadāranyaka*, opens with a traditional account of the connection between the horse sacrifice and the cosmos, with the dawn as the head of the sacrificial horse, the sun as its sight, and so on.[27] While there are certainly some Vedic hymns that display philosophical reflection on the nature and origin of the universe, generally this kind of reflection is absent in the earlier texts. But with the Upaniṣads we have accounts of the origins of life along with the eschatological hope of future life that has transcended the world of appearance. In both cases, in the origin of life and the purpose of life, we have sacrifice as a central theme to explain both. After the opening homology of horse sacrifice with natural phenomena, the text goes into the earliest systematic account of creation. Let us cite the passage in full in Olivelle's translation:

> In the beginning there was nothing here at all. Death alone covered this completely, as did hunger; for what is hunger but death? Then death made up his mind: 'Let me equip myself with a body (*ātman*)'. So he undertook a liturgical recitation (*arc*), and as he was engaged in liturgical recitation water sprang from him. And he thought: 'While I was engaged in liturgical recitation (*arc*) water (*ka*) sprang up from me'. This is what gave the name to and discloses the true nature of recitation (*arka*). Water undoubtedly springs for him who knows the name and nature of recitation in this way. So, recitation is water.

[27] *Bṛhadāraṇyaka-upaniṣad* 1.1. Patrick Olivelle, *The Early Upaniṣads* (Oxford: Oxford University Press, 1998), p. 37.

> Then foam that had gathered on the water solidified and became the earth. Death toiled upon her. When he had become worn out by toil and hot with exertion, his heat—his essence—turned into fire.[28]

This is a very curious and complex text. It goes on to say that Death divided his body into three as the sun, wind, and breath, and the breath he divided into three, although here the text gives a much longer list identifying his head with the eastern quarter, the forequarters with the south-east and north-east, the tail with the west, and so on. That is, from the creation coming out of Death we have the identification of Death with the cosmos and, furthermore, clearly the sacrificial horse is being described. There are thus mercurial shifts in the text that identify absence or nothing with Death, with the horse, and with the cosmos itself. Death is the cosmos along with the sacrificial horse, all of which emerge from the sacrificial liturgical recitation. 'In the beginning there was nothing here' (*naiveha kiṃcanāgra āsīt*), the text says, which is Death (*mṛtyu*).

The paradoxical nature of the text resists easy penetration by the reader. In the beginning there was (the text uses the simple past of the verb *as*, 'to be') nothing. How can nothing 'be'? Of course, it cannot, and yet the text also identifies nothing (*na kiṃcana*) with Death and then Death with hunger (*asanāyā*). We have a short series of substitutions here and we move from nothing, to Death, to hunger. With hunger we have shifted into a different realm because hunger is not nothing, but hunger is desire, although the text does not actually use the word *kāma*. What is this hunger? Is this hunger for life, for being itself, from nothing? From hunger that has come from nothing, Death has the wish to become manifested, to have a body (*ātman*). In a more literal rendering, Death says, 'may I be body-possessing' (*ātmanvī syāmi*), 'may I be embodied'. The word *ātman* that is generally used for 'self' or 'soul' in this early use refers to the living body and is probably derived from the root *an*, to breathe. Indeed, in the *Ṛg-veda*, *ātman* means 'breath', the 'life-breath' of the sacrificial victim such as the horse that flies from the victim like a bird.[29] It is but a small step from breath, to an animate body, to the force that animates it, to the idea of the soul.

This passage, as Olivelle observes, 'is full of word play and phonetic equivalences'.[30] As Olivelle tells us, the verb *arc* means to recite the liturgical text and also 'to shine', and *arka* is a liturgical text that is recited. It also means 'radiance' or 'lightening', as well as being a technical term for a fire used in the

[28] Ibid., 1.2.1–2: *naiveha kiṃcanāgra āsīt | mrtyunaivedam āvṛtam āsid aśanāyayā | aśanāyā hi mṛtyuḥ | tanmano'kurutātmanvī syām iti | so'rcannacarat | tasyārcata āpo'jāyanta | arcate vai me kamabhūditi | tadevārkasyārkatvam | kaṃ ha vā asmai ya evam etad akasyārkatvaṃ veda ||1|| āpo vā arkaḥ | tadyatapāṃ śara āsīt tat samahanyata | sā pṛthivyabhavat | tasyāmaśrāmyat | tasya śrāntasya taptasya tejo raso ninirvartatāgniḥ ||2||.*

[29] *Ṛg-veda* I.162. Jamieson and Brereton, *The Rig Veda*, vol. 1, pp. 344–6. *Ātman* as breath that comes to mean 'soul' is directly parallel to the Greek *pneuma*.

[30] Olivelle, *The Early Upaniṣads*, p. 488.

horse sacrifice. These names indicate both their verbal origin and the thing denoted. In translating such terms Olivelle writes that he uses 'the somewhat awkward and long expression "gave the name to and discloses the true nature of" to convey the pregnant meaning of the simple Sanskrit abstract nouns in these contexts'.[31] I will defer a discussion of naming in relation to giving life for the time being, but there is a semantic density in this passage that plays on the ambiguity of terms.

And at last the text finally admits of desire itself. In a literal rendering: 'He [Death] desired "may a second body be born for me"' (*so'kāmayata dvitīyo ma ātmā jāyeteti*). Then in a further complex series, Death 'with his mind' (*manasā*) copulated (*mithunaṃ samabhavat*) with speech (*vāc*), then with hunger.[32] In a complex sequence that is not entirely clear to me, in which Death's semen becomes the year, Death gives birth to speech (*vāc*) (even though he had just copulated with speech) but then desists from swallowing speech because 'if I kill him, I will only reduce my supply of food'.[33] The masculine pronoun 'him' (*enam*) is used even though speech is feminine. But the text then says that with this speech and with this new, second body, Death 'produced this all' (*idaṃ sarvam asṛjata*) that includes the texts of the Veda, sacrifices, as well as animals and people. The verb *sṛj* is used here that indicates emanation rather than the verb *jan*, to give birth. Then Death began to eat whatever he gave birth to. There are further elaborations with Death himself becoming a corpse, which becomes a sacrificial horse, which is sacrificed to Death! The horse sacrifice and the ritual fire for the oblation together constitute Death, 'merely a single deity' (*ekaiva devatā*).

This is a semantically dense but terribly significant text and indicates the kind of Axial shift I have been speaking of. It is an attempt to interpret the sacrifice and at the same time interpret the world. Here in the first of the Upaniṣads, perhaps about 400 years after the hymns of the *Ṛg-veda* began to be recited, the sacrifice is used as an explanation of the world. It is not simply action that is important, not simply the performance of the rites, but we have to understand what it means. Knowledge rather than action is becoming privileged, not simply as rendering information about the world, but as having soteriological significance. Whoever knows that the horse is the fire and is the sun, whoever understands these equivalences and substitutions, becomes free from the power of death. Death is 'not able to get him' (*nainaṃ mṛtyur āpnoti*) and he avoids 'repeated death' (*punarmṛtyu*). This idea of repeated death probably refers to the early Vedic desire to prevent the second death of the

[31] Ibid.

[32] I follow Olivelle here: *sa manasā vācyaṃ mithunaṃ samabhavad aśanāyāṃ mṛtyuḥ*, 'So, by means of his mind, he copulated with speech, death copulated with hunger' (p. 39). The verb must also govern the second accusative *aśanāyām* rather than 'hunger' being 'death' which grammatically it cannot be. See Olivelle, note, p. 488.

[33] Ibid. 1.2.5: *yadi vā imam abhimaṃsye kanīyo'nnaṃ kariṣya iti.*

ancestors in the next world, the world of the ghosts (*preta loka*) before they transitioned to the world of the fathers (*pitṛ loka*) through post-funerary offerings (*piṇḍa*).[34] Rather than the performance of the sacrifice, knowledge is best. Indeed, the text says that there are three worlds: the world of men (*manuṣyaloka*), the world of the ancestors (*pitṛloka*), and the world of the gods (*devaloka*), and a man can win these worlds respectively through a son, through ritual action, and through knowledge.[35]

We have in this text emergent characteristics of the Axial shift, first in the notion of transcending the particularity of ritual sacrifice, a standing above and looking down that allows us to see the deeper significance of ritual action, and that this knowing distanciation is a soteriological transformation. This verticality is distinct from the sacrificial logic that it reflects on. Second, this knowing is about understanding life itself and such knowing repairs life and brings freedom. It is the beginning of an explanatory orientation; the expression of desire or impulse for knowledge and understanding the world through the development of categories and causal accounts of how life arises. These two modes of reflection, let us call them eschatological hope and classificatory codification, are related insofar as knowing the nature of life is a way of insight into the power that lies hidden within it and gives rise to it. In terms of the two frames of transcendence and immanence, we might say that with the Upaniṣads we have an orientation to transcendence or verticality that is also a recognition of immanence or the horizontal: the hidden power of the sacrifice, which is the sacred word or *brahman*, is actually the hidden power of the universe itself and the animating principle of all appearances. It is also identical to the self. The *brahman/ātman* is the final power that sustains and gives rise to the cosmos; as such it is transcendent and insofar as it penetrates world, it is immanent.[36] Later in the text this power becomes identified with the essence of the self and here the term *ātman* is used not in the sense of body, but in the sense of the life force that sustains the body and all life.

As Olivelle observes, the focus of the Upaniṣads is the human person.[37] The human body along with the bodies of sacrificial beasts, notably the horse,

[34] David M. Knipe, 'Sapiṇḍakaraṇa: The Hindu Rite of Entry into Heaven', in Frank Reynolds and Earle H. Waugh (eds.), *Religious Encounters with Death; Insights from the History and Anthropology of Religions* (Philadelphia, PA: Pennsylvania State University Press, 1977), pp. 111–24. See also Wendy Doniger O'Flaherty, 'Karma and Rebirth in the Vedas and Purāṇas', in O'Flaherty (ed.), *Karma and Rebirth in Classical Indian Traditions* (Berkeley, CA, London: University of California Press, 1980), pp. 1–37.

[35] *Bṛhadāraṇyaka-upaniṣad* 1.5.16.

[36] We will have cause to examine the transcendence/immanence distinction in the Indian context later when we look at the *viśvottirṇa/viśvātma* distinction in the religion of Śiva. For a fine discussion of what the terms 'transcendence' and 'immanence' mean or could mean, see Ingolf U. Dalferth, 'The Idea of Transcendence', in Bellah and Joas, *The Axial Age and its Consequences*, pp. 146–88.

[37] Olivelle, 'Introduction', p. 22, *The Early Upaniṣads*, pp. 3–27. Also see J. Frazier, Hindu Worldviews (London: Bloomsbury, 2016), ch. 3.

have hidden connections to the cosmos, so in understanding the body, we understand the wider universe. We also understand the hidden power that animates it. There is a shift in the use of the term *ātman* as body to *ātman* as the animating principle of the body, as the self or soul, a shift that we see even in the very first chapter of the earliest Upaniṣad, the *Bṛhadāraṇyaka*. The Upaniṣads are deeply interested in what animates the living, what they refer to as the vital energies that pervade body and world. In particular, they analyse the person as having the powers of movement, evacuation, reproduction, breathing, and speaking, along with thinking and the five senses (all of which are systematized in later Sāṃkhya philosophy with mind as the sixth sense). In particular the Upaniṣads privilege breathing, thinking, speaking, seeing, and hearing, all of which are called breath (*prāṇa*). They were created by the creator Prajāpati and competed with each other for dominance, but death takes them all except breath, which it cannot capture, so the faculties became a form (*rūpa*) of breath.[38]

The general term used for 'breath' is *prāṇa*, which is also used for life: a living being is a *prāṇin*, one possessing breath or life. The *Ṛg-veda* mentions the term in the sense of life breath (e.g. 'whomever we hate let breath leave him')[39] and identifies it with life itself (*āyus*) and compares it to one's own son.[40] In a hymn to the Dawn, Uṣas, the derivative *prāṇana* is linked to 'life' (*jīvaṇa*),[41] and *prāṇa* is linked to the wind in a couple of passages, particularly the wind being born from the breath of a sacrificed giant.[42] The oldest systematic identification of breath with wind comes in the *Jaminīya-Upaniṣad-Brāhmaṇa* (3.1–2) and the *Chāndogya Upaniṣad* (4.1–3) in what Bolland calls the oldest level of 'the wind-breath teaching' (der Wind-Atem-Lehre).[43] Linking breath to wind means that breath is the human wind and wind is the cosmic breath. While this general sense always carries on, it comes to be used in the plural (*prāṇāḥ*) to refer to a set of five breaths that animate the body. These are: *apāṇa*, the downward breath of inhalation, *prāṇa*, the outer breath of exhalation, *udāna*, the upward breath from the throat, *vyāna*, the traversing breath between exhalation and inhalation identified with the power of speaking, and *samāna*, the equalizing breath associated with digestion.[44] This is the

[38] *Bṛhadārṇyaka-upaniṣad* 1.5.21.

[39] *Ṛg-veda* 3.53, Jamieson and Brereton, *The Rig Veda*, vol. 1, p. 539.

[40] Ibid. 1.66.1, p. 188. [41] Ibid. 1.48.10, p. 161.

[42] *Ṛg-veda* 10.90.13, Ibid., vol. 3, p. 1540.

[43] Mechtilde Boland, *Die Wind-Atem Lehre in der alteren Upaniṣaden* (Münster: Ugarit-Verlag 1997), p. 15.

[44] *Chāndogya-upaniṣad*, 1.3.3. See Olivelle's succinct summary, 'Introduction', p. 23. For a textual history of the breaths see A.H. Ewing, 'The Hindu Conception of the Functions of Breath: A Study in Early Hindu Psycho-Physics', *Journal of the American Oriental Society*, vol. 22, 1901, pp. 249–308. Also, Peter Connolly, 'The Vitalistic Antecedents of the Ātman-Brahman Concept', in Peter Connolly and Sue Hamilton (eds.), *Indian Insights: Buddhism, Brahmanism, and Bhakti* (London: Luzac Oriental, 1997), pp. 21–38; H.W. Bodewitz, 'Prāṇa, Apāna and Other Prāṇas in

beginning of interesting speculation about the nature of the body as comprising flows of power that we find in later medical, yogic, and tantric literature, although in that later material, the breaths are mapped onto other structures within the body that are understood in a hierarchical sequence of levels. But in this early speculation we have a simpler scheme that identifies the animating force of living beings. Breath as the force of life is also identified with food and ultimately with the force of life itself, the *ātman/brahman*.

In a sequence that develops from the creation story we have looked at above, the *Bṛhadāranyaka* introduces the idea of the creator God Prajāpati who is actually identified with Death as the creator in the opening verses we have seen. Prajāpati, the Lord of creatures, creates demons and then gods who compete for control of the worlds. The gods resolved to overcome the demons during the sacrifice by asking Speech to sing to them the 'High Chant'. As Speech sang, the demons, afraid they would lose their power, rushed at her and 'riddled her with evil'[45] (which is why people say disagreeable things). The gods then ask each of the human faculties to sing to the demons in the same way and the demons rush at each one, filling them with evil. So, they move from speech, to breath, to sight, to hearing, to mind, and finally to 'the breath within the mouth' (*āsanyaṃ prāṇam*). But against this breath they cannot prevail and are smashed 'like a clod of earth hurled against the rock'. The breath in the mouth is the animating force of life against which nothing can prevail, and which keeps death far away. The man who knows this prevails against death. This breath in the mouth also seems to be identified as the central breath of the body, the best of the vital functions. A little later in the text the theologian Ajātaśatru reveals a secret doctrine (*upaniṣad*) to his student Gārgya that just as a spider sends forth thread and as sparks spring from a fire, so all vital functions (*prāṇa*), worlds, gods, and creatures come from the self (*ātman*), which is the truth (*satya*) behind the vital functions.[46]

In an interesting passage that begins to convey this idea, a man who is about to die tells his son that 'you are the brahman, you are the sacrifice, you are the world', and the son repeats this in the first person, 'I am the brahman', and so on. When the man dies, he then enters the son through these that the text calls 'breaths' (*prāṇa*). In this way divine speech, mind, and breath enter the son from the earth and fire (speech), sky and sun (mind), and water and moon (breath). Whoever knows this becomes the self of all beings (*bhūtānām ātmā bhavati*), which is also divine breath (*daivaḥ prāṇaḥ*).[47] As we read, it is almost as if the text is struggling to give articulation to this central idea. The self is the

Vedic Literature', *Adyar Library Bulletin*, vol. 50, 1986, pp. 326–48; K.G. Zysk, 'The Science of Respiration and the Doctrine of Bodily Winds in Ancient India', *Journal of the American Oriental Society*, vol. 113, 1993, pp. 198–213.

[45] Olivelle's translation (p. 41) of *pāpmanāvidhyan*.

[46] *Bṛhadāranyaka-upaniṣad* 2.1.20. [47] Ibid. 1.5.17–20.

essence of the person and the essence of the cosmos and there is a homology between the two.

Now although this self is a cosmic force and the essence of a person, it can also be spoken about as being located within the space (*ākāśa*) of the heart (*hṛdaya*). It is the person (*puruṣa*) that is the inner light of the heart (*hṛdyantarjyoti*)[48] that can be interpreted metaphorically or physically, as Olivelle does in translating these passages in terms of the heart being a cavity surrounded by the pericardium through which the veins of the body emerge.[49] From the space of the heart at death the self leaves the body through the eye or the top of the head and the life breath (*prāṇa*) goes with him.[50] From the body, at death the self goes by one of two paths, through the flames of the funeral pyre to the day, to the fortnight of the waxing moon, from there to the six months when the sun moves north, and so to the world of the gods, thence to the sun, and from there to the lightening and from there to 'the worlds of brahman' (*brahmaloka*, or, more precisely, in locative plural *brahmalokeṣu*), from whence they do not return. Only those with knowledge follow this path. Others who have performed Vedic ritual and austerities without knowledge go through the smoke to the night, to the fortnight of the waning moon, to the six months when the sun moves south, to the world of the fathers, and thence to the moon. Once in the moon they become food for the gods, but once that ordeal ends, they pass into the sky, into the wind, into the rain, and so to the earth where they become food (again) and enter a man's semen and thence to a woman's womb from where they are reborn. And those selves who do not know either of these paths end up as worms, insects, or snakes.[51]

Another creation story in the *Taittirīya-upaniṣad* says that the whole world comes from the self (*ātman*). More specifically, from the self comes space, from space air, then fire, water, earth, plants, food, and man. The text identifies man (*puruṣa*) as the essence (*rasa*) of food (*anna*) and then goes on to praise food as the source of all creatures. Food is the best of all beings from which all beings come into existence and to which they return. Yet within a man at a deeper level is the self that consists of the life breath (*prāṇa*); the breath that gods, people, and animals breathe. This breath is life itself (*āyus*), the life of all beings, and so it is called 'all life' (*sarvāyuṣam*), a term that the text repeats.[52] Here we have a meditation on life itself identified with breath and also with food. Although not always consistent, the text presents a vitalism that identifies self with life and life is articulated as breath and food. The *Kauṣītaki-upaniṣad* likewise identifies breath with the self.[53] Again we are within the sacrificial imaginary in which life passes away and is reborn from death, just as life forms become food that in turn gives rise to life again. This concern and identification of life with breath is an important ingredient in this worldview

[48] Ibid. 4.3.7.
[49] Olivelle's translation, p. 63.
[50] *Bṛhadāranyaka-upaniṣad* 4.4.2.
[51] Ibid. 6.2.15–16.
[52] *Taittirīya-upaniṣad* 2.3.
[53] *Kauṣītaki Upaniṣad* 3.8.

that transforms the Vedic sacrificial imaginary and is taken up and developed later in the science of life (*āyurveda*) that seeks long life and good health as ideal human goods.

We are a long way here from the Vedic hymns that simply praise the gods or from the precise instructions on how to perform the soma sacrifice. An identifiable shift has occurred along two axes; a verticality or transcendence in which the self identified as life itself, as the essence of life, is placed at the heart of the universe as its animating principle, knowing which is transformative or soteriological in the sense that it defeats death, and also a classificatory codification that presents an analysis of the cosmos and in particular a human being and the faculties. In the *Taittirīya-upaniṣad* cited above, for example, we have a typology of beings in the universe along with a typology of the body comprising the five breaths, five senses, and the constituents of the body, namely skin, flesh, sinew, bone, and marrow.[54] This is almost a scientific approach based on observation. These two tendencies develop, on the one hand, into ideas about world transcendence, expressed institutionally in renunciation, and on the other into the classificatory mind-set of the later tradition, the concern with locating entities in their place in the scheme of things. The seeds of the later scholasticism are here. Both the vertically transcendent impulse and the horizontal desire for classificatory knowledge emerge from the same matrix in response to the sacrificial logic of earlier centuries. Furthermore, both sacrificial logic and transcendental detachment are responses to the human existential condition of living and dying.

Although world renunciation and its accompanying ideas of liberation and reincarnation may be from a non-Vedic origin, the Śramaṇa traditions, renunciation was deeply integrated into Brahmanism as the converse of sacrifice. Both are intended to resolve the paradox of life and death: sacrifice through ritual killing that transforms death into life, renunciation through knowledge, or cognition that transforms the self to the truth.

The Theme of Sacrifice

Sacrifice remains a literary theme throughout literature in the Sanskrit tradition. The *Mahābhārata*, a great epic story of an internecine war between two rival factions, the Pāṇḍavas and Kauravas, is itself conceptualized as a sacrifice. Here war is sacrifice and the field of the Kuru, the field of dharma, is the sacrificial arena. The bodies of the heroes become the sacrificial victims and the same theme of affirming life through death occurs again here. Biardeau has traced the theme of sacrifice in the text and right up to the

[54] *Taittirīya-upaniṣad* 1.7.

modern cult of Pottu Rāja.[55] The *Bhagavad-gītā* especially is concerned with the sacrifice-renunciation dynamic. Set on the eve of the battle, the dialogue between the hero Arjuna and Krishna, who turns out to be an incarnation of God, centres on ideas of duty, obligation, and sacrifice. With the retreat of Brahmanical sacrificial ritual, the text bears witness to the rise in dominance of the warrior class who perform sacrifice in battle. And while the *Gītā* extols sacrifice in war, Krishna urges Arjuna to fight because it is his warrior's duty. It also wishes to elevate renunciation. Belonging to or identification with community that war entails is offset by the distance necessitated by renunciation. The conflict between community loyalty and giving up social position for a higher truth, between sacrifice and renunciation, is resolved in the text by a logic that says that Arjuna can be a loyal citizen and be true to his duty by fighting and yet also be a renouncer through inner detachment. The warrior should become detached from the fruits of his action, so long as he acts honourably within the bounds of dharma. Sacrifice is compatible with renunciation and both combined are life-giving forces. So long as the warrior gives up the fruits of his acts to God, the consequences of his action are out of his hands and yet he knows he has acted with impunity.

Sacrifice is the heart of life for the *Bhagavad-gītā*. Both gods and humans owe life to sacrifice and continue living because of it. The gods who sustain the cosmic order of creation are nourished by sacrifice and in return they give the desired enjoyment to the one who sacrifices. Reciprocity is the basis of the cosmic order.[56] The *Bhagavad-gītā* says:

> 3.9 This world is bound by action, save
> for action which is sacrifice;
> therefore, O Son of Kunti, act
> without attachment to your deeds.
> 3.10 When Prajāpati brought forth life,
> he brought forth sacrifice as well,
> saying, 'By this may you produce,
> may this be your wish fulfilling cow'.
> 3.11 Nourish the gods with sacrifice,
> and they will nourish you as well.
> By nourishing each other, you
> will realize the highest good.
> 3.12 Nourished by sacrifice, the gods
> will give the pleasures you desire.
> One who enjoys such gifts without
> repaying them is just a thief.

[55] Madeline Biardeau, *Histoire de Poteaux. Variations védiques autour de la déese hindou* (Paris: Publications de l'Ecole Française d'Extrême Orient, 1989).

[56] Angelika Malinar, *The Bhagavadgītā: Doctrines and Contexts* (Cambridge: Cambridge University Press, 2007), p. 83.

> 3.13 The good, who eat the remains
> from sacrifice, rise up faultless.
> But the wicked, who cook only
> for their own sakes, eat their own filth.[57]

Action binds the world and sacrifice is the only action that does not have binding consequences. Sacrifice in a sense is a free act. Sacrifice is a part of creation by God (the creator Prajāpati who we met earlier in the *Bṛhadāranyaka-upaniṣad* identified with Death) and gives the results of what we desire. Sacrifice is identified with the 'wish fulfilling cow', an idea that echoes the Iḍā cow of sacrificial ritual. The sacrificial order of the universe is the highest good and those who do not participate in it, through eating alone, are wicked (*pāpa*) and impure because they refuse the reciprocal relationship between the human and divine worlds; they refuse to participate in the sacrificial order and only this order is life-giving and life-affirming. Arjuna the warrior must fight and in so doing he acts sacrificially in accepting the bond between the human and divine orders. Arjuna the warrior is like the patron of the sacrifice in the Veda, but whose detachment from the results of his actions is like the renouncer of the Upaniṣads. Killing his foes, he remains free from karmic effects while remaining devoted to God.

It is now that we witness, as Zaehner once said, the entry of theism into Indian religious thinking.[58] Arjuna loves God and God loves humanity represented in Arjuna. Krishna says to Arjuna:

> 18.64 Hear from me now, the supreme word,
> the greatest secret of them all:
> you are indeed my beloved,
> so I will speak for your well-being.
> 18.65 Be mindful of me and devout,
> make sacrifices and revere me,
> and you will surely go to me
> as I promise, for you are dear.
> 18.66 Relinquishing all your duties,
> vow to take refuge just in me!
> I will cause you to be released
> From every evil, do not grieve.[59]

[57] *Bhagavad-gītā* 3.9–13. Flood and Martin, *The Bhagavad Gita*, 2013, pp. 19–20.

[58] R.C. Zaehner, *The Bhagavad Gita* (Oxford: Oxford University Press, 1969), p. 37. This may be the case if the *Śvetāśvatara-upaniṣad* is later than the Gītā, which it probably is, being dated between the first century and 200 AD. See T. Oberlies, 'Die Śvetāśvatara-Upaniṣad: Einleitung-Edition und Übersetzung von Adhyāya I', *Weiner Zeitschrift für die Kunde Südasiens*, vol. 39, 1995, pp. 61–102.

[59] *Bhagavad-gītā* 18.64–6. Flood and Martin, p. 88.

'You are dear to me' (*priyo 'smi me*), says Krishna to Arjuna, a sign of solidarity with Arjuna, reminding us that Arjuna is in fact still the sacrificial victim or identified as the sacrificial victim brought into the world of intimacy. As 'man is the sacrifice' in the early Vedic religion, so here Arjuna is ultimately his own self-sacrifice to God who brings him into the realm of intimacy and love. The final verse quoted here (18.66) is known as the *carama śloka*, the last verse. It is perhaps the most important in the text and has been subject to different interpretations, particularly about agency or grace and effort. Śaṅkara's non-dualistic interpretation takes it to mean that the self has no agency because it is the passive witness and action needs to be abandoned to realize liberation (*jñānayoga*). Rāmānuja, on the other hand, takes the verse to mean that self is indeed an agent and that relinquishing duties refers to expiatory rites as preparation of grace through taking refuge in the Lord. A further development in interpreting the verse followed in Rāmānuja's Śrī Vaiṣṇava school in the thirteenth century, with the northern school claiming that salvation is through the practice of devotion (*bhaktiyoga*) and so takes some effort, in contrast to the southern school claiming that no effort is needed, only taking refuge in the Lord (*śaraṇāgati*), which is a total surrender (*prapatti*).[60] The quid pro quo of sacrificial logic has here been replaced by the total transcendence of a power beyond the human that human action has no control over.

With what has been called the Axial revolution, in the Indic context we have seen a questioning of the values of life affirmation in the sacrifice in favour of a transcendence and the affirmation of an other-worldliness that sees the values of affirming everyday life as being trumped by the higher value of eschatological hope for the ascetic renouncer. This is certainly one response of the Upaniṣads and the trajectory of thinking they engender. The other response is found in the way certain thinkers tried to understand and explain life in terms of horizontal categories; there is a codification that occurs and an analysis of human reality that is scientific and analytical in its method. We have already begun to encounter this in the *Bṛhadāraṇyaka*'s systems of classification of the three realms, the five breaths, and so on. These approaches to reality are linked to Buddhist analytical procedures, to Sāṃkhya classifications of the world, and to Vaiśeṣika categories. While these systems may not have been aware of each other in an early period, they would seem to be drawing on the same conceptual resources that seek to understand life itself through analytical categories. This includes yogic and medical speculation. I now wish to trace these contrasting approaches through presenting examples, beginning with the eschatological hope we find in the Vaiṣṇava theologian Rāmānuja, then retracing a parallel trajectory with a horizontal codification that we find in the

[60] Patricia Mumme, 'Haunted by Śaṅkara's Ghost', in Jeffrey R. Timm (ed.), *Texts in Context: Traditional Hermeneutics in South Asia* (Albany, NY: SUNY Press, 1992), pp. 69–84.

Vaiśeṣika philosophy and in the medical speculation on life, the Āyurveda. These systems of thought resonate with a distinct, non-Vedic, or even anti-Vedic way of thinking we find in the Śramaṇa traditions. It is this second trajectory where I think we find the positive affirmation of life itself in Jainism; the kind of affirmation that reaches its apogee in the religion of Śiva.

Life Here and There

The Vedānta tradition stemming from the Upaniṣads, being articulated in the philosophy of Bādarāyaṇa and the pithy *Brahmasūtra*, is most famously expressed by Śaṅkara (scholarly consensus about dates ranges between 650 and 800 CE).[61] Śaṅkara was the proponent that the correct interpretation of the Vedas is that of non-dualism (*advaita*): that the absolute reality of brahman and its identity with the self (*ātman*) means that our everyday experience of the world as comprising distinct beings and objects is based on ignorance and projection. The everyday experience of the world is illusion (*māyā*). While this is undoubtedly an important philosophy in the development of Vedānta—and Śaṅkara is possibly the most influential Indic philosopher, especially for nineteenth- and twentieth-century Hinduism—a second development of Vedānta has more to say about the category of life itself and so is more pertinent to my question here. This second form of Vedānta is popularly known as 'qualified non-dualism' (*viśiṣṭādvaita-vedānta*) because it rejects the dismissal of everyday experience as projection and accepts the reality of the world, even though that reality must itself be seen in terms of a divine emanation. The most articulate and famous proponent of this view is the South Indian devotee of Viṣṇu, Rāmānuja (traditionally dated between 1017 and 1137).

Like Śaṅkara, Rāmānuja wrote a commentary on the *Brahmasūtra* as a way of establishing the authority of his views. In this he is critical of the non-dualistic Advaita of the kind that Śaṅkara expressed, although his main target of criticism is probably Śaṅkara's disciple, Mandanamiśra. Rāmānuja presents a summary of his thoughts in the *Vedāntasāra*, the Essence of Vedānta, which is a succinct commentary on Bādarāyaṇa's *Brahma-sūtras*, comprising an account of a theology of the relationship between the world and those who live in it, and a wholly transcendent other. Although we might call it a post-Axial theology in the sense that it has abstracted and pushed transcendence even further than we found in the Upaniṣads, it is still within the imaginative power of Vedic sacrifice.[62] The insistence of the sacrificial logic that seeks to

[61] Natalia Isayeva, *Shankara and Indian Philosophy* (Albany, NY: SUNY Press, 1993), pp. 83–7.

[62] Francis X. Clooney, Thinking Ritually: Rediscovering the Pūrva Mīmāitually: Redis (Vienna: De Nobili, 1990).

resolve the irresolvable repeats itself throughout the Vedānta tradition. While Rāmānuja rejects both the Mīmāṃsā insistence on the primacy of Vedic injunction to perform the sacrifice, and the Advaita claim of the illusory nature of life in the world, he is still within the sacrificial imaginary. As with all theists, on the one hand he needs to assert the transcendence of a putative theistic reality and the greater intensity of that reality, while on the other he has to assert that there is value in the experienced world of life. On the one hand, Rāmānuja buys into the Brahmanical view that the cycle of life is one of suffering in which the life principle is reborn repeatedly, yet on the other he must maintain the value of the cycle of life and the manifested universe as the body of God. The world is a real creation that invites our loving response to its creator and sustainer. Indeed, this is an important part of his intellectual agenda because he must reject the Buddhist impersonalists and their wholly negative evaluation of life itself while at the same time rejecting his non-dual predecessors' negative evaluation of life and the world as illusion. He must maintain a realism about life while at the same time adhering to transcendence as validating and giving life meaning. The *Vedāntasāra* is a text that treads this path between the affirmation of a transcendent God outside the universe, distinct from self, alongside the affirmation of life itself as the manifestation of God and something that therefore has value. Alongside this transcendence Rāmānja must operate within the constraints of his revelation, most notably the Upaniṣads, and so still operates within the sacrificial metaphor. Rāmānja must affirm the value of life here while adhering to an eschatological hope of individual human salvation in the future, notably at death with the grace of God. In the same way that Vedic sacrifice wished to deny death and claim that death leads to life, so for Rāmānuja death leads to life in two senses: of eventual reincarnation on the earth for those who have not achieved liberation/salvation and rebirth into the full life of God for those who have. Rāmānuja must affirm the value of life here on earth, but the greater value of life elsewhere, in the transcendence of God and our future life with him.

The concerns of his succinct summary of his theology are how we assert the reality of transcendence and maintain a realism that does not taint that transcendence; how we account for our awareness and the particularity of our life due to our past action, and the nature of salvation. Let us begin with his benedictory or *maṅgala* verse of his text that articulates his theology in a nutshell:

> Homage to God (Viṣṇu), an ocean of pure joy, glorious, who is the essence of all and whose body is all sentient and insentient beings.[63]

[63] Rāmānuja, *Vedāntasāra*, trans. and ed. V. Krishnamacharya and M.B. Narasimha Ayyangar (Adyar: Adyar Library, 1953), 1.1: *samstacidacidvastuśarīrāyākhilātmane/śrīmate nirmalānandodanvate viṣṇave namaḥ//.*

For Rāmānuja God is the essence of existence: the life force that animates the entire universe of sentient and insentient beings (*vastu*) that comprise his body. Rāmānuja is famous for his analogy that as the self is to the body, so the Lord is to the universe comprising sentient and insentient existents (*śarīriśarīrabhāva*). There is a necessary relation between body and soul, which cannot be separately established from each other (*apṛtaksiddha*).[64] This panentheism, as Bartley observes in his illuminating study, 'seeks to accommodate divine transcendence, the qualitative otherness and infinite perfection of deity, and divine immanence in the world'.[65] God in his aspect as brahman is the efficient and material cause of the universe, the manifestation of beings and material reality and who is the inner controller (*antaryāmin*) of all. But there is an aspect of God that is wholly transcendent. Rāmānuja therefore wishes to differentiate his theology from those that maintained the universe to be a transformation of the substance of God, as this compromises the total transcendence of God in his own nature. Indeed, Rāmānuja distinguishes between God in essence or own nature (*svarūpa*) and energy (*vibhuti*), which is also his approachability (*saulabhya*) in which devotees can love God through different models of emotional relation, such as parental affection as between a cow and a calf, friendship, or the affection of lovers. We can know and love God in his aspect of approachability and energy, but we can never know or have access to God in his essence. Human liberation is certainly communion with God in heaven but is participation in God's approachability only. The whole universe and all beings in it are wholly dependent on the transcendent other on whom everything is dependent for its being (a theology not dissimilar to Schleiermacher's idea of the total dependence of creature on creator). Because of the total dependence of the universe and all beings within it on God, all action is ultimately the acts of God, although within this constraint the self is a real agent and so we are responsible for our acts.[66]

From the sacrificial logic of pure immanence in the Vedic world, we have come to full post-Axial transcendence. The universe of conscious and unconscious entities is wholly dependent upon a theistic reality that is its source and hope. God, called Viṣṇu or Nārāyaṇa by Rāmānuja, is the cause of the world, specifically articulated as the cause of the ether/space (*ākāśa*) and life force (*prāṇa*),[67] who should be meditated upon as life and immortality.[68] The person is animated by a life force (*prāṇa*), which is also the self (*ātman*),

[64] See Christopher Bartley, *The Theology of Rāmānuja: Realism and Religion* (London: Curzon, 2002), pp. 79–87.

[65] Ibid., p. 79. Also see C. Ram-Prasad, *Divine Self, Human Self* (London: Bloomsbury, 2013), pp. 77–115.

[66] Martin Ganeri, 'Free-Will, Agency, and Selfhood in Rāmānuja', in Matthew R. Dasti and Edwin F. Bryant (eds.), *Free Will, Agency, and Selfhood in Indian Philosophy* (Oxford: Oxford University Press, 2014), pp. 232–54.

[67] *Vedāntasāra* 1.8.23, 1.9.24.

[68] Ibid. 1.11.29.

who combines with a body in a dependent relationship: the body is only body because of its animation by a force that transcends it. Indeed, Rāmānuja assures us from scripture that with the death of the physical body, the self does not die, being animated by the power of God himself, who appears within the self as the Supreme Person. The self is eternal and not produced but is carried along through time by a stream of act and consequence (*karmapravāhaṇa*) attached to the soul. At death, the soul of a good Brahmanical ritualist who has performed the correct sacrifices and giving to other Brahmans goes to heaven where, although secondary to the gods, he experiences his due enjoyment until that karma is exhausted and he is reborn in this world in the womb of a Brahman or some other caste, in order to experience the remainder of the karma that remains. Rāmānuja writes:

> The self, situated in the bodies of Brahmans and others, performs sacrifice, giving and so on. Then having issued out from that body to enjoy the fruit of that [ritual], he goes forth. Having attained heaven whose form is like fire, he is enveloped by waters in their subtle form and established in bodies mixed together with other elements. He is enveloped by water, which is a transformation as a bodily form made of nectar, and he becomes subservient to the gods. Being there with them he experiences enjoyment, [then] once the effects of those actions come to an end, he comes again into the world of action through the effects of his action that cause him to arrive in the womb of a Brahman and so on, so as to enjoy what remains [of those effects experienced in heaven].[69]

This is a somewhat complex account that is an attempt to explain a passage in scripture. The *Chāndogay-upaniṣad* asks why in the fifth oblation in a particular ritual sequence, water is called 'person' (*puruṣa*). This is Rāmānuja's answer to that question. Water as one of the elements envelops the soul or life force (*jīva*) along with other elements, and thereby comprises the subtle body of the soul through which it experiences the fruits of its actions in heaven. This subtle water is a form of the nectar of immortality. In this subtle body, the soul enjoys the fruits of its action along with the gods, but in due course comes back down to earth. Rāmānuja goes on to explain the process of coming back to earth. The soul enters the clouds and then as rain he enters the earth 'represented as fire' (*agnirūpita*). From the earth, the soul enters the rice and so is transformed into food; from there, through the water of his subtle body, he is transformed into semen; and thence goes into the womb of a woman also imagined as fire, in accordance with or corresponding to the fruits of his acts

[69] Ibid. 3.1.1: *jīvo brāhmaṇādidehastho yāgadānādikarmakṛt tat tatphalabhogāyāsmād dehād uthāya gacchan, etad dehasthābhir bhūtāntarasaṃsṛṣtābhiḥ sūkṣmākārābhiradbhiḥ saṃpariṣvakto 'gnitvena rūpatiṃ dyulokaṃ prāpya, tābhir evāmṛtamayadehākārapariṇatābhir adbhiḥ pariṣvakto devānāṃ śeṣatvam upagamya, taiḥ saha tatratya bhogam anubhūya, karmāvasāne bhuktiśiṣṭabrāhmaṇādiyoniprāpakakarmaṇā saha punarapīṃ karmalokamāgatya.* My translation guided by that of N.B. Narasimha Ayyangar.

(*karmānurūpa*). He is following scripture here in a process that we have seen described in the Upaniṣads. He goes on to say how the self departs from the body in a subtle form along with all the breaths that go with him. At death the soul carries on and, taking the idea of the two paths from the *Bṛhadāraṇyaka-upaniṣad*, Rāmānuja re-describes them through the lens of his own theism. If the soul has knowledge of God, then he goes out of the body into the flames of the funeral pyre and so into the light of God, but if he is only a sacrificial specialist, he goes out through the smoke, ending up in the moon and heaven from whence he returns again to earth through the process described above. The self who worships the gods goes to the gods where he is in the service of the gods (*devopakaraṇatvam*), or he worships the Supreme Person and goes to his service or enjoyment (*paramapuruṣopakaraṇatvam*). Even those who do not perform sacrifices go to the moon after death, while others might end up in one of the seven hells.[70]

But the soul that is liberated goes to the brahman and there leads a full life. Rāmānuja's concept of liberation is not one of impersonal cognition or merging with an absolute; it is rather an individual fullness in relation to others. So, he quotes the *Chāndogya-upaniṣad* that says of the liberated soul: 'this deeply serene one, after he rises up from this body and reaches the highest light, emerges in his own true appearance. He is the highest person. He roams about there, laughing, playing, and enjoying himself with women, carriages, or relatives, without remembering the appendage that is this body. The lifebreath is yoked to this body, as draft animal to a cart'.[71] This, says Rāmānuja, is the fruit of final release.[72] The soul, free from the life of the physical body, goes to a place where there is certainly experience, but experience untrammelled with negative consequence: a life in the fullness of God, in the knowledge of God's powers, and a life complete in the fullness of relationality with others. This release is, Bādārayaṇa's sūtra says, due to the promise of the Lord in scripture ('one is liberated because of the promise', *muktaḥ pratijñānāt*),[73] a liberation, Rāmānuja adds, granted by the promise of removing the concealment of the Lord in the waking state of the world, from whence he does not return here.

We can make a number of observations about this. First, Rāmānuja is still within the sacrificial imaginaire. Even though he places Brahmanical ritual as secondary to knowledge of God, and he includes devotion here as a form of knowledge, it is still an important part of the ambient religious landscape that he cannot ignore. Second, as a careful exegete, Rāmānuja must couch his theology in the language of scripture. Indeed, he speaks through scripture and

[70] Ibid. III.1.13–15; IV.2.1–10.

[71] *Chāndogya-upaniṣad* 8.12.3: *evamevaiṣa saṃprasādo'smāc charīrātsamutthāya paraṃ jyotir upasaṃpadya svena rūpeṇābhiniṣpadyate | sa uttamapuruṣaḥ | sa tatra paryeti jakṣatkrīḍan ramamāṇaḥ strībhir vā jñātibhir vā nopajanaṃ smarannidaṃ śarīram | sa yathā prayogya ācaraṇe yukta evamevāyamasmiñcharīre prāṇo yuktaḥ ||3||*. Olivelle's translation.

[72] *Vedāntasāra* III.3.40.

[73] Ibid. IV.4.3.

his commentary is a string of quotations that he links together through his own theological narrative. Not only is reasoning about the truth important, but also truth revealed through revelation. Third, although Rāmānuja wishes to emphasize the utter transcendence of God, God is nevertheless analogically like a person; indeed he is the Supreme Person, who like a person has a body but unlike a person does not die and is not tainted by impurity. Fourth, the soul is enveloped in a bodily form that comprises the elements, particularly subtle water, and is also surrounded by the breaths. And Rāmānuja uses the term *jīva* rather than *ātman*, a noun that perhaps implies a dynamic understanding of the self as a life force (from the verbal root *jīv*, to live). There is deep connection between the self and the body it has to experience a world. The *jīva* is only *jīva* because it is embodied and this life is but a preliminary to a life elsewhere, a life with God in liberation after death for those who have the devotion and knowledge to attract God's grace. Rāmānuja's theological world is one of eschatological hope alongside a realism that proclaims the value to life lived in the world because the world and the life that dwells within it is the body of God himself.[74]

The contrast between Rāmānuja's world and that of the Vedic ritualists is striking. We are in a different moral landscape here. Although he uses the language of sacrifice, the ethos is of both renunciation and devotion to a putative theistic reality, but a renunciation that is not a negation of the world for a pure transcendence, but one that recognizes the world as within the being of God. There is a sense in which this world is a reflection of a fuller life in God that is to come in liberation.[75]

CLASSIFYING LIFE

I have so far told a story of a sacrificial order reflected in the ancient Vedas in which everyday life of human prosperity, the values of family and sufficient food, along with social status and warrior virtues of strength, prowess, and bravery, were reflected in hymns to gods during the ritual process. The poets in this society articulated social values that affirmed human sociality and whose very language was elevated to divine status. We only know about the articulate levels of this society whose word was preserved in the scriptures, but this representation depicts a social order integrated into a cosmos whose

[74] Julius J. Lipner, *The Face of Truth: A Study of Meaning and Metaphysics in the Vedantic Theology of Ramanuja* (London: Macmillan, 1986), p. 121.

[75] On this theme and the commentarial tradition that follows see Francis X. Clooney, *Hindu God, Christian God: How Reason Helps Break down the Barriers between Religions* (Oxford: Oxford University Press, 2001).

functioning is assured through the life-affirming sacrifice. With fresh winds blowing, this system of values shifted to a new moral order reflected in the Upaniṣads that locates ultimate value in the transcendence of worldly life and an absolute reality, the brahman, which sustains the cosmos and the person. The real meaning of the sacrifice becomes the realization of this reality within the self. A shift has occurred from valuing life in itself lived in a small-scale society where the social and natural orders were integrated, to valuing the negation of those social values in favour or renunciation and the internalization of the sacrifice that leads to liberation. This new value of eschatological hope was the existential understanding of transcendence in the sense of realizing that the self was the essence of life: the sustaining power and goal of human existence. While a political theology of kingship may have dominated the political realm, as we see in the Dharmaśāstra, in ideological terms and textual representation, renunciation rose to the pinnacle of the hierarchy of values.

With this Axial shift there is a transformation in understanding life from the affirmation of everyday values and human flourishing in community, values guaranteed by sacrifice and achieving social equilibrium thereby, to understanding life in terms of transcendence and a new moral order that negated the everyday in favour of what it regarded as a higher purpose. This higher purpose of liberation was reflected in the purposes of human life (*puruṣārtha*) where liberation (*mokṣa*) was added to the three goals, the *trivarga*, of duty (*dharma*), prosperity (*artha*), and pleasure (*kāma*), during the first half of the first millennium BCE.[76]

Alongside the idea of world transcendence, we have a different kind of rising above human sociality in the classification and analysis of life. The renunciate orders not only removed themselves from the community, from the village (*grāma*) to the wilderness (*āraṇya*), but from a detached vantage point tried to understand life through analysis and asking the question of what its essential sustaining power was. We have seen the beginnings of systems of classification in the Upaniṣads that attempt to understand life through categorization and a naturalist impulse that resonates with other traditions of thinking. The sustaining power of the breath (*ātman*) in the early Vedas is the life force that becomes the central claim for the essence of life itself. The *ātman* becomes the essence of the person, identified with the essence of the cosmos (*brahman*). Furthermore, this essence is associated with other systems of classification such as the way the life force sustains the body through five breaths (*prāṇa*) and the analysis of how a person functions. Thus, along with the five breaths we have the operations of locomotion, excretion,

[76] Charles Malamoud, 'Semantics and Rhetoric in the Hindu Hierarchy of the "Aims of Man"', in *Cooking the World: Rituals and Thought in Ancient India* (Delhi: Oxford University Press, 1996), pp. 109–29.

reproduction, and speaking, and the five senses, along with the mind. The quest for the essence of life, for the *ātman/brahman*, through meditation and inwardness, became linked to explaining the way the body and mind work in a scientific analysis.

These early systems of codification that developed probably initially during the first millennium BCE, but whose textual articulation is mostly found in the second, are distinct but related. Alongside the sacrificial and renunciatory sensibilities, we have a naturalist sensibility that seems to have been present at the time of the Buddha (d. *c.*410 BCE) but which only comes to clear articulation roughly between 200 BCE and 200 CE. I now need to trace these two sensibilities to life: the naturalist that seeks to explain life and the renunciatory that seeks to escape from it. I shall deal with them separately, but in the first instance we need to speak briefly of these two sensibilities in tandem as they both arose within the same moral compass that was pulling against and sometimes explicitly rejecting the sacrificial imaginary.

The Axial shift that I have spoken about is reflected not only in the Upaniṣads and the Vedānta but also in renunciate orders that arose during the first millennium BCE, especially Buddhism but also Jainism and other orders such as the Ājīvikas. These ascetic groups were collectively known as Śramaṇas (Samana in Pāli), a term that itself is derived from the root *śram*, to exert oneself or make an effort, and related to the word for the stages of life or lifestyle choice, *āśrama*. The relationship between the Śramaṇas and the Brahmans is controversial. On the one hand, we have the view that Brahmanical renunciation is a development of the Vedic sacrificial order, and on the other, that Brahmanical renunciation is derived from a different worldview, that of the Śramaṇas who are from a tradition quite distinct to the Vedic sacrificial order.

This is a complex historical question and it is not necessary for me to deal with it here, but suffice it to say that there is a convincing case for the origins of Śramaṇas in a different locale to where the Vedas were being composed and the rule of orthodox, Vedic Brahmanism. Johannes Bronkhorst has argued that the origins of this movement are to the south of the centres of Vedic religion in Greater Magadha whose capital was Patna.[77] Ideas of karma and reincarnation, liberation from the cycle of rebirth, and meditation and asceticism come from this quite separate thought-world. We find evidence for this in the Upaniṣads where reincarnation and karma are new doctrines, as well as evidence in the Dharmaśāstras, and in early Buddhist and Jain literature. This is a convincing historical narrative that aligns Brahmanical renunciation and thought about transcendence more closely with Buddhism and Jainism, while yet remaining distinct and defending the scriptures and Brahmanical world

[77] Johannes Bronkhorst, *Greater Magadha: Studies in the Culture of Early India* (Leiden: Brill, 2007).

against the Śramaṇa challenge. The Upaniṣads represent the absorption of new ideas, of a new moral universe, into the Vedic sacrificial world and a reinterpretation of it. As we have seen, the Upaniṣads use renunciation as a way of transforming the sacrificial imaginary while still being within its power such that renunciation becomes a central doctrine and practice of orthodox Brahmanism. This entailed a complex attitude towards life: on the one hand, continuing the positive assessment entailed by the sacrificial imaginary; on the other, adopting a view of life's transcendence for a higher order of being. In time, this higher order of being or eschatological hope comes to integrate the order of life at a lower level in a theism in which life becomes the expression or manifestation of God or even the body of God. All this is happening over a long period of time from around the time of the Buddha who dies around 410 BCE, within the rise of kingship and the Mauryan empire, to its demise and the incursions of Kuṣanas and Scythians, to the establishing in the North of the stable Gupta empire (*c.*320–*c.*500 CE). The sacrificial imaginaire, its renunciatory transformation, and the incursion of a new ideology of renunciation occur within a shifting political landscape that impacts little upon these changes.

Additionally, to this already complex picture we have the Śramaṇa ascetic negation of life that we find in Buddhism and Jainism and the positive attitude towards life entailed in a naturalist philosophy with a strongly empirical dimension. On the one hand, we have sensibility that wishes to leave life, while on the other we have a sensibility that wishes to explain life. The two enterprises are related, and I shall present an account of the latter first.

Life in the Horizontal—Vaiśeṣika Realism

Within this naturalist sensibility, we have a cluster of ideas, movements, and texts in various schools of thinking, including the Sāṃkhya and Vaiśeṣika schools of orthodox philosophy, alongside Yoga, the development of medical science, the Ayurveda, and the complex classificatory system of the Buddhist Abhidharma. It seems to me that these articulate a naturalist sensibility concerned with explaining the world, related to, but distinct from, the sacrificial imaginary in the Vedānta tradition, although Sāṃkhya and Yoga certainly do feed into the Vedānta analysis. It is the Vaiśeṣika school especially that exemplifies the naturalist tendency.

Although technically within the orthodox systems of philosophy, the Vaiśeṣika system displays a style of categorization that echoes the Śramaṇa philosophies. The Vaśeṣika is not directly a philosophy of life in the sense that it seeks a life force or animating principle, but it is an indirect philosophy of life in its concern with ontology or the analysis of categories (*padārthaśāstra*). It is also a system concerned with metaphysics and with soteriology in the sense that eschatological hope is the consequence of knowledge.[78] Although its roots probably go back deep into the first millennium BC, it was likely influenced by Buddhism;[79] the earliest Vaśeṣika source that we have is the mythical Kaṇāda's *Vaiśeṣika-sūtra* composed sometime between 200 BCE and 200 CE and within that range, most probably around 100 CE. The text is commented on by Praśastapada (*c.*500 CE), which itself is commented on by Vyomaśiva (*c.*800 CE) and Śrīdhāra (*c.*990 CE). There is a commentary by Candrānada on the *Vaiśeṣika-sūtra* that pre-dates 1000. This system is closely allied to the tradition of logical reasoning, the Nyāya, and Udayana (*c.*1050–1100 CE) brings the two ways of thinking together, uniting Vaiśeṣika metaphysics with logical analysis.[80] Of all of these, Praśastapada's is the most important as a reworking of the original teachings.[81]

The Vaiśeṣika is a realist system of philosophy primarily concerned with ontology, with the question as to what there is: the enumeration of the categories of existence. It is concerned not so much with life as such but rather with being. The six fundamental categories according to Praśastapada are: substance (*dravya*), quality (*guṇa*), motion (*karman*), universal (*sāmānya*), particularity (*viśeṣa*), and inherence (*samavāya*). Later tradition added a seventh, non-being (*abhāva*),[82] and Candrānanda offers more. These main categories are further subdivided as follows in the words of Halbfass:

> There are nine substances or classes of substances: earth, water, fire, air, ether (*ākāśa*), space, time, souls (*ātman*), and mental organs (*manas*). The first four of these are elemental substances; they consist of indivisible, invisible, and indestructible atoms (*aṇu, paramāṇu*). The atoms form aggregates and constitute those composite and noneternal material things with which we are dealing in our practical and empirical lives. Ether, space, and time are nonatomic, unitary, all-pervasive, indestructible substances. The souls too are omnipresent and eternal, whereas the 'mental organs' (*manas*) are equally eternal, but of atomic dimension, that is, infinitely small.[83]

[78] See Ionut Moise, *The Nature and Function of Vaiśeṣika Soteriology with Particular Reference to Candrānanda's Vṛtti*, DPhil. Oxford University, 2018.

[79] Johannes Bronkhorst, *Buddhism in the Shadow of Brahmanism* (Leiden: Brill, 2011).

[80] There are problems in dating the *Vaiśeṣika-sūtras* and not all commentaries are using the identical text. See Wilhelm Halbfass, *On Being and What There Is: Classical Vaiśeṣika and the History of Indian Ontology* (Albany, NY: SUNY Press, 1992), pp. 79–80. Also see Harunaga Isaacson, 'Notes on the Manuscript Transmission of the *Vaiśeṣikasūtra* and its Earliest Commentaries', *Asiatische Studien: Zeitschrift der Schweizerischen Asiengesellschaft*, vol. 2, 1994, pp. 749–79; Ashok Aklujkar, 'Candrānanda's Date', *Journal of the Oriental Institute Baroda*, vol. 19, 1969/70, pp. 340–1.

[81] Halbfass, *Being and What There Is*, pp. 169–71.

[82] Ibid., pp. 70–1.

[83] Ibid., p. 71. Halbfass goes on to enumerate the remainder. For a clear, well-expressed account also see F. Max Muller, *The Six Systems of Indian Philosophy* (London: Longman, Greens and Co., 1899), pp. 574–602.

Although the Vaiśeṣika's primary concern is the enumeration of the categories of existence, it is relevant to our inquiry because it raises the question of the relationship between being and life.

There has been some debate about the nature of the Vaiśeṣika school, and whether it expresses a purely scientific impulse to explain the world or whether it originally also contained a soteriology that I have called an eschatological hope. On the one hand, some scholars such as Faddegon and Frauwallner see the school purely as a scientific naturalism that became contaminated, as it were, in its later history with the idea of liberation and eventually God from the ambient culture. Frauwallner argues for an original Vaiśeṣika characterized by an atomistic and mechanistic analysis of existence with no room for any religious doctrine of salvation. This was a 'nature philosophy' (*Naturslehre*) giving expression to a purely scientific impulse. According to him, there were four stages in its development: an initial philosophy of nature, a mechanistic view of the world with the account of its atomic constituency, the doctrine of the categories, and a reorganization of the old philosophy of nature in line with the categories. Frauwallner argues that originally there were two streams of thinking in ancient India: one of the early Upaniṣads that is characterized by a doctrine of the world soul, which becomes the basis of Sāṃkhya and Buddhism, and the other characterized by the idea of multiple, individual souls and naturalist philosophy.[84] Frauwallner would seem to oversimplify the picture here, however, in that Sāṃkhya, as well as Jainism, accepts the idea of multiple souls and Buddhism and Jainism arguably have their origins outside of the Vedic world (see Bronkhorst above) and Buddhism rejects the idea of a soul. Jan Houben, by contrast, has argued that the idea of soteriology is there from the very beginning in the Vaiśeṣika sūtras. First, the Vaiśeṣikas align themselves with the Vedic revelation, which does have a soteriological dimension in the Upaniṣads, as we have seen. Second, the Vaśeṣikas are close to Jainism whose idea of the soul in bondage is described in almost physiological terms, not dissimilar to the Vaiśeṣika. Houben argues that the concern for liberation must have been there in the origin of the system because the Vaiśeṣika's 'emulative relation with Jainism and Buddhism would have made it one of the first objectives of the first author

[84] Eric Frauwallner, *Geschichte des indischen Philosophie* (Salzburg: Otto Mullet Verlag, 1953), pp. 192ff., 268; *History of Indian Philosophy*, vol. 1, trans. V.M. Bedekar (Delhi: Motilal Banarsidass, 1973), pp. 152ff., 211. Erich Frauwallner, 'Der ursprünglische Afgang der Vaiśeṣika-Sutrem', in E. Frauwallner, *Nachgelassene Werke* I (Vienna: Osterreichische Akademie der Wissenschaften, 1984), pp. 35–41. Also see Madeleine Biardeau's diametrically opposed view that liberation and the soul were always central concerns of philosophical systems in this period. Madeleine Biardeau, *Théorie de la connaissance et philosophie de la parole dans le brahmanisme classique* (Paris: Mouton, 1964).

of the Sūtra text to assert the Vedic and Brahmanical dharma'.[85] The implications of this are that the naturalist sensibility evinced in the Vaiśeṣika is in the service of a soteriology in the sense that knowing the way the cosmos works, almost mechanically, allows us to unpick our entanglement and to liberate the self from bondage.

The classification of the Vaśeṣika is a horizontal one, by which I mean that it is less concerned with a hierarchy of being, and the value of height in verticality, and more with an analysis of the way things are arranged in the world of experience; with substances encountered in the world and their attributes along with theoretical entities, pairs of atoms, to explain them. The attitude to life here is ambiguous. On the one hand, there is value in explanation and understanding all that there is; I think we can detect a pure impulse to explanation. For example, the *Vaiśeṣika-sūtra* is concerned with biological classifications and considers bodies and how they are sexually or asexually produced within the classificatory scheme of the elements, earth, air, fire, water, and ether. Thus, the body can be classified in two ways, as sexually or non-sexually produced (*yonijam ayonijañ ca*), although this only pertains to the level of the earth where bodies are formed from the comingling of semen and blood. Bodies in the realm of the other elements are non-sexually produced, as we see with sages and gods who are produced through the mind. Sexually produced bodies are also twofold, being egg or womb born. The bodies of trees are considered in the commentary as the realm of experience wherein souls reap the consequences of their previous actions. But some trees, those that grow on the banks of the river Narmada, attain the supreme place (*parama gati*) thereafter because of their contact (*sparśa*) with the holy water.[86]

Yet, on the other hand, there is no value in life itself but only value in the soul's escape from it. This aligns the Vaiśeṣika closely with the impulses of the Śramaṇa traditions of renunciation and parallels what we find in Jain texts, as we will see presently. In its analysis of the person and the mind, the ways in which a human being is entangled in the world, the system resembles Buddhism and Jainism. But in a sense the objectivity of the being of the world is privileged over any subjectivity or even over life itself. The relation between being and life in this system is ambiguous but can be seen more clearly in the concept of the particular (*viśeṣa*), from whence the system derives its name. The particular is the force unique to basic and eternal

[85] Jan Houben, 'Liberation and Natural Philosophy in Early Vaiśeṣika: Some Methodological Problems', p. 732, *Asiatische Studien: Zeitschrift der Schwerizeerischen Asiengesellschaft*, vol. 2, 1994, pp. 711–48.

[86] *Vaiśeṣika-sūtra* 4.5.5 with Śaṅkaramiśra commentary. Nandalal Sinha, *Vaiśeṣika-Sūtra of Kaṇāda (with the Commentary of Shanaka-Miśra and Extracts from the Gloss of Jayanārāyaṇa)*, Sacred Books of the Hindus (Allahabad: The Panini Office Bhuvaneswari Arama, 2nd ed. 1923 [1911]), p. 146.

substances of atoms, ether, time, space, souls, and minds. It is the force that distinguishes each of these simple (i.e. non-composite) substances from each other. With regard to the soul, the particular is what ensures the distinction of one soul from another; the unique force that makes something what it is. In contrast to any notion of life as process or flow, the Vaiśeṣika idea of particularity creates clear boundaries around entities, which in the end are atomistic. The soul that is particular and unique is nevertheless bound within the other categories of the universe but from which it can hope to achieve freedom once it is not affected by conjunction (*saṃyoga*), a link of proximity between substances such as when they touch, as in a heap of threads, but distinct from inherence (*samāvaya*) as threads are related in a whole cloth. Liberation occurs with the cessation of appearance (*pradurbhāva*) of the soul in a new body after death; liberation is an unconscious state.[87] Vaiśeṣika naturalism is here tempered by the Axial transcendence of world and scientific naturalism—the horizontal description of the world—is not distinct from eschatological hope—the vertical transcendence of the world.

The Science of Long Life—The Ayurveda

Within what I have called the naturalist sensibility, I have described one form of ontology, one kind of systematic classification of the categories of existence that also functions as an explanation with an eye to eschatological hope. But alongside Vaiśeṣika naturalism there is also a more vitalistic current that I now want to take up that, like Vaiśeṣika realism, we can take to be a system of horizontal codification. As far back as the *Ṛg-veda* we have found an incipient concern for life itself as an explicit theme that comes to fuller articulation in the Upaniṣads' concern with the breath or vital energy (*prāṇa*), especially as we saw in the *Taittirīya-upaniṣad* that identified the self with breath and breath with life itself (*āyus*). This vitalism expresses a desire for long life—and so is within the sacrificial imaginary. There is therefore some continuity of terminology between the Vedic literature and the Ayurveda[88] and the tradition itself traces such continuity with the gods passing medicine to the human community, but the historical reality may not be so simple as Dominik Wujastyk has pointed out. Emerging in the early centuries of the first millennium BCE, the Ayurveda is 'a system of general medical practice which encompasses both preventive and prescriptive aspects'.[89] This tradition

[87] *Vaiśeṣika-Sūtras* 5.2.19–20. Candrānanda's *Vṛtti* (Baroda: Gaekwad's Oriental Series, 1961). For a good exposition see Moise, *The Nature and Function of Vaiśeṣika Soteriology*.

[88] I have used the Anglicized version of the word *āyurveda*.

[89] Dominik Wujastyk, *The Roots of Ayruveda* (London: Penguin, 1998), p. 3.

emerged 'in an almost fully articulated form' as the medical encyclopaedias of Caraka, Suśruta, and Bhela some 2,000 years ago.[90] Although the idea of the five breaths can be found as early as the *Ṛg-veda*, the Ayurveda represents a distinct development and, while it does in due course incorporate the five breaths, did not originate there. The picture is more complex, as Wujastyk's brilliant studies show. There are certainly links between Caraka's text, the *Carakasaṃhitā*, and the fourth or *Athārva-veda* because that text offers rituals and prayers to prolong life, which, Wujastyk observes, is also the purpose of medicine. But if the roots of the Ayurveda cannot be found in the Veda, where can they be found? Some have argued that its roots go back to ancient Greece and the systems of medicine that would have found their way to the Greek kingdom of Gandhāra in present-day Afghanistan. Indeed, one scholar of Indian medicine, Jean Filliozat, found parallels between the Sanskrit and Greek ideas of breath, of *prāṇa* and *pneuma*.[91] But Wujastyk points out that not one Greek word of a medical nature finds its way into Caraka's text, in contrast to astronomy where Greek terms are borrowed wholesale. There are words, such as *jentāka* or sauna, whose origin is not Sanskrit but nor is it Greek.[92] The earliest evidence for medicine in the Indic context is from Buddhism where the Pāli texts exhort the monks to care for each other. In due course the monasteries included infirmaries to care for the sick. Indeed the Buddha has been compared to a physician, and there is a Mahāyāna sūtra with the name of the 'Medicine Buddha' (*bhaiṣajyaguru*).[93] Wujastyk writes: 'It now seems almost certain that the foundations of classical āyurveda were being laid at the time of early Buddhism in the Buddhist and other ascetic communities'.[94] It is possible, adds Wujastyk, that some early practitioners such as Vāgbhaṭa were indeed Buddhist.

So, the origins of the Ayurveda are not so much textual as rooted in the practical necessity of monks to maintain health and develop treatments for themselves that came to be applied to others, and in empirical investigation of the body and what treatments are effective. In the medical literature, the body is described in some detail as having tissues, ligaments, sinews, bones, muscles, pipes, and so on. Suśruta's *Compendium* describe the fundamental processes of the body, particularly digestion that it sees as a kind of cooking or burning and the transformation of food into bodily tissues, and reproduction that the Ayurveda understands as the union of semen with menstrual blood.[95]

[90] Dominik Wujastyk, 'The Science of Medicine', p. 393, in Gavin Flood (ed.), *The Blackwell Companion to Hinduism* (Oxford: Blackwell, 2003), pp. 393–409.

[91] Jean Filliozat, *The Classical Doctrine of Indian Medicine* (New Delhi: Munshiram Manoharlal, 1964), cited by Wujastyk, 'The Science of Medicine', p. 395.

[92] Wujastyk, 'The Science of Medicine', p. 395.

[93] Ibid., p. 397, citing Kenneth Zysk, *Asceticism and Healing in Ancient India: Medicine in the Buddhist Monastery* (New Delhi: MLBD, 1991).

[94] Wujastyk, 'The Science of Medicine', p. 397.

[95] Ibid., p. 399.

The fundamental process of the body is 'irrigation', the transport of various fluids, such as the humours wind, bile, phlegm, and blood, around the body that occurs through ducts, pipes, and tubes.[96] This scientific literature develops and by the fifteenth century we have the diagnosis of disease by pulse, urine, eyes, face, tongue, faeces, voice, and skin. We also have an analysis of disease aetiology as the imbalance of the humours, but also bad judgements, past actions (karma), or even demonic interference. To cure disease, the Ayurveda recommends a range of procedures including herbal drugs, massage, sauna, exercise, diet, blood-letting, and surgery.[97] There is also reflection on birth and birthing problems.[98] Surgery is one of the notable features of Suśuta's text, which was known by the sixth century AD and whose origins go back to the first half of the first millennium BCE. The Ayurveda also incorporates the soteriological technology of yoga and Caraka offers his own version of the eightfold yoga of Patañjali that leads to the realization of the self as brahman, and the text even cites the *Vaiśeṣika-sūtra*.[99]

The Ayurveda is concerned with life itself, not so much as a philosophical category, but as the lived reality of the body that needs to be understood and analysed in order to produce effective cures for disease. Although there are continuities with the Vedic tradition in the idea of the breaths, and the tradition understands itself in this way, its historical origins are within the Śramaṇa worldview and the horizontal developments of classification and codification so characteristic of this non-Vedic attitude to life. The Ayurveda went on to become very influential in other spheres—on Tibetan Buddhism and on Islamic medicine through Persia. It has also influenced Western medicine and is now part of the repertoire of alternative therapies. But its interest for us lies in the way it seems to come out of the non-Vedic tradition of asceticism. While the ascetic negates the world in favour of a verticality, as we have seen, the Āyurvedin has an implicitly positive attitude to the world in wishing to make life better for people, to cure disease. The drive is far less philosophical and more practical and empirical. It is in this empirical spirit that the tradition has come down to us as a still thriving practice and form of complementary medicine that has attempted to integrate itself into modern medicine,[100] and has now been infused with a vitalist philosophy from Europe, particularly German romanticism, and the science of long life has become a

[96] Ibid. [97] Ibid., p. 404.

[98] Martha Selby, 'Narratives of Conception, Gestation, and Labour in Sanskrit Ayurvedic Texts', *Asian Medicine*, vol. 1 (2), 2005, pp. 254–75; Rahul Peter Das, *The Origin of the Life of a Human Being: Conception and the Female according to Ancient Indian Medical and Sexological Literature* (New Delhi: MLBD, 2003).

[99] Wujastyk, 'The Science of Medicine', p. 406.

[100] Dominik Wujastyk, 'Interpréter l'image du corps humain dans l'Inde pré-moderne', in Véronique Boullier and Gilles Tarabout (eds.), *Images du corps dans le monde hindou* (Paris: CNRS, 2002), pp. 71–99.

philosophy of long life, as Joe Alter has shown,[101] to which we will return presently.

The naturalist sensibility in India developed what we can recognize as science within the medical field and aetiology of disease, as well as a sophisticated science of language far in advance of anything in the world before twentieth-century linguistics in Europe and America. India developed mathematics, architecture, and a science of astronomy.[102] But apart from the Ayurveda that comes closest to it, it did not develop a science of life; it did not develop biology or zoology. This neglect is curious given the empirical thrust of the horizontal classificatory impulse. I think it can be explained by the greater pull of verticality: that thinkers were more concerned with transcendence and understanding or explaining the world as a cosmology that we need to understand to escape from. Seeing the world clearly means that we can avoid its entanglement, but such a view is not conducive to a careful explication of other species or interest in explaining life forms other than in very general terms as born from a womb, an egg, moisture, or from the mind. Two traditions that exemplify this approach of horizontal explication and codification in the service of vertical ascent are both Jainism and Buddhism. The former is the older system with perhaps closer continuities with the Vaiśeṣika and Ayurveda.

The Śramaṇa Rejection

Lastly, we need to consider more closely the counter-tradition to the Brahmanical, Vedic civilization. This was the Śramaṇa culture, following Bronkhorst, that originated in Greater Magadha and rejected the sacrificial cult of the Vedas. As we have seen, from here come ideas of reincarnation and liberation, the verticality that also enters into the Vedic tradition, and from here come ideas of horizontal codification, as we have seen with the Vaiśeṣikas and Ayurveda. Jainism and Buddhism are the archetypal Śramaṇa religions, although there were others including the pure materialism of the Carvakas, the fatalism of the Ājīvikas, and the nihilism of the followers of Kesamkambali.[103] The Carvakas, whose name comes from the verb *carv*, to eat, and the nihilists believed that death was the end of life. On the one hand, these traditions were careful to analyse life and had a classificatory and scientific impulse, yet on the other hand, the main concern of Jainism and Buddhism was escape from life

[101] Joe Alter, *Yoga in Modern India: The Body between Science and Philosophy* (Princeton, NJ: Princeton University Press, 2004).

[102] See Frits Staal, 'The Indian Sciences', in Flood (ed.), *The Blackwell Companion to Hinduism*, pp. 348–409.

[103] Surendranath Dasgupta, *History of Indian Philosophy*, vol. 1 (Delhi: MLBD, 1975 [1922]), pp. 79–80.

in vertical ascent. This was understood differently: for the Jains this liberation or isolation was the freedom of the soul or life force (*jīva*) from the constraints of the material body and karmic residues that weighed it down, so that the soul was free to float to the top of the universe; for the Buddhists this liberation was the extinction of greed, hate, and delusion and the realization that life is suffering, impermanent and without essence or self. Moving forward from the early Ayurvedic texts and the *Vaiśeṣika-sūtra* to the first millennium CE, we reach a time when traditions had clarified their differences, and questions of origin, whether Vedic-Brahmanic or non-Vedic Śramaṇa, had abated to be left with questions of genuine philosophical substance, including questions about the nature of language, the question of what exists, questions about how we know anything at all, and the status of authority in philosophy. In what remains of this section I want to examine the understanding of life in Jainism and Buddhism during the earlier medieval period when the various traditions had achieved a high degree of clarity regarding doctrine and practice, with fixed textual canons, but that also illustrate the porous boundaries of the traditions in the way they absorb new ideas. With the textual illustration of Jainism and Buddhism we will then be in a position to trace these ideas through to the post-Axial medieval synthesis, particularly with the tantric Śaiva tradition.

Like Buddhism, Jain philosophy developed in monastic settings in dialogue with other philosophies. Its founder was said to be Mahāvīra, perhaps an older contemporary of the Buddha, and so is part of the Axial assertion of transcendence. The Jains had a fixed canon of texts in Jaina Prākrit with later texts composed in Jaina Mahārāṣṭrī with a large commentarial literature associated with them.[104] There are also independent treatises on Jain doctrine and practice in Prākrit and Sanskrit which in turn have commentaries written on them. One of these important independent treatises is the first Jain text in Sanskrit by Ācārya Umāsvami (second to fifth century CE) called the *Tattvārthasūtra*, and there are several other independent treatises such as the *Yogaśāstra* by Hemacandra (1088–1172 CE). This text draws on other traditions and re-tools Jainism for his contemporary context in which concerns about liberation through yoga as well as magical manipulation of powers became important; elements which were absorbed from its ambient tantric culture.

The Jain understanding of life deserves our attention because it combines both vitalism that explains animation and also eschatological hope that seeks to transcend ordinary experience. With Jainism, we have a fully fledged philosophy of life alongside the transcendental pull for release. The Jain universe is divided into two broad classes, the living (*jīva*) and the non-living

[104] Ibid., p. 171. Dasgupta provides a clear description of the literature.

(*ajīva*), but even what appears to be non-living is sometimes animated by a life force. This life force, the *jīva*, is distinct from the body but gives life to, and takes the shape of, the body. That is, the *jīva* extends to the length and breadth of the body, which explains why there is sensation throughout the body, and so there is variability of size of the *jīva* depending upon the size of the body. The kind of body, and so the kind of sense perception a being has, constrains the kind of experience it has. Plants, with limited sense perception, experience the world only through touch. Or rather, the soul experiences a tactile world through its body, which is the plant. This principle is true of all bodies in a hierarchy from plants, to worms which possess touch and taste, to ants which have smell in addition, to bees that additionally have vision, up to vertebrates that have five senses. Beyond these are humans, gods, and beings in hell who possess the five senses and also the mind (*manas*) which allows them to think rationally.[105] A life force that is numerically distinct from all others animates the whole of life: there are innumerable *jīvas* who give life to the cosmos and furthermore, although they are distinct from each other, they share an identical quality. The life force of the plant is essentially the same as the life force in the person; the only difference is in the kind of body the soul has and therefore the kind and quality of experience that is available to it. This variability of experience is also due to action in the past that is like a covering over the soul; the soul's experience is therefore constrained by the form of its body and so the world it experiences along with the past karma that has led to that particular form.

The Jain cosmos is hierarchical and populated by innumerable beings in varying degrees of subtlety, born into particular worlds due to their past actions.[106] The Jains produced a sophisticated theory of karmic retribution and their theory of life is integrated into a metaphysics. Umāsvami's *Tattvārthasūtra* describes birth in different realms into three genders: male, female, or neuter (*napuṃsaka*).[107] The entrapped soul lives out its life in the body appropriate to the experience it needs to undergo in that particular life in a particular kind of body. These bodies are listed as five: the gross or physical body, the transformable, the projectable, the luminous, and the karmic,[108] which are progressively more subtle (*sūkṣma*). All beings in the physical universe have physical bodies but beings in other realms have subtle bodies, although the precise designations of 'transformable' and 'projectable' are not entirely clear to me and in a sense all bodies are 'karmic'. But the saint on his or her way to liberation possesses the luminous body. Liberation, then, is the extraction of the life force from this web of entrapment through a process of

[105] *Tattvārthasūtra* by Umāsvāti, trans. with introduction by Nathmal Tatia, *That Which Is* (London: Harper Collins, 1994), 2.21–4.

[106] Ibid., 3.33–4. [107] Ibid., 3.50–2.

[108] Ibid., 3.36: *audārika-vaikriyakāhāraka-taijasakārmaṇāni śarīrāṇi/*.

cleaning the soul of its polluting karmic coverings.[109] This is done through ascetic practice and the non-attracting of new karmic coverings. Thus, icons of the Jain saints depict them standing naked in ascetic practice with vines growing up their legs, indicating the longevity of their asceticism resulting in complete immobility. All action and therefore movement creates consequences, and so the ideal is to achieve immobility of body and mind that allows for the release of the soul from the bonds of matter. This is total freedom of the self in liberation understood as isolation (*kaivalya*): the individual soul realizes its total isolation from matter and, indeed, from other souls. Human freedom is conceptualized not as being within the frame of nature but as being liberated from it. Yet the quest for freedom is enacted within the body through a limited act of freedom, the freedom to end one's life. The path leading to liberation can culminate in ritual death understood as the gradual withdrawal of the life force from the body: a conscious decision that is not regarded as suicide.[110] This is achieved by the Jain ascetic or even householder at the end of life through withdrawal from activity and ultimately through ritual or ascetic death (*sallekhana*).[111] In this process the life force is projected out of the body through the crown of the head.[112] There is a gradual wearing down of all action (*nirjara*) until the ultimate non-action is achieved: the non-act of death.

There is therefore an ambiguity in Jainism to life itself. On the one hand, there is the affirmation of life through an extreme sensitivity to living beings and a great emphasis on non-harm (*ahiṃsā*), so much so that we have the archetypal image of the Jain nun or monk sweeping the floor before them as they walk and wearing a mouth covering so as not to inadvertently swallow any insects. On the other hand, we have a negative evaluation of life as the place of bondage and suffering, with the ideal being the cessation of all action and the achievement of liberation as isolation of soul from the world of life. With Jainism we have a complete rejection of the sacrificial imaginary. This is not only a rejection of sacrifice itself as violence because it reaps negative karmic consequences, but an assessment of life itself as something to be revered and honoured. Yet Jainism seems happy to live with the contradiction or tension between asserting the sanctity of life itself—expressed in respect for and non-harm towards all living beings—and the drive towards transcending life in a liberation in which the life force attains the ultimate fulfilment of its desire, total motionlessness in isolation from its material bondage: human freedom as the realization of the true nature of the life essence juxtaposed to matter.

[109] Robert J. Zydenbos, *Mokṣa in Jainism, according to Umāsvāti* (Wiesbaden: Franz Steiner, 1983).

[110] On *sallekhana* see James Laidlaw, *Riches and Renunciation: Religion, Economy, and Society among the Jains* (Oxford: Clarendon Press, 1995).

[111] *Tattvārthasūtra* 6.22.

[112] *Yogaśāstra* 5.24.

The second Śramaṇa tradition to reject the sacrificial imaginary is, of course, Buddhism. Although the Buddhists and Jains disagreed on the issue of the soul (the Buddhists rejecting such a suggestion) and the degrees of permitted asceticism, both shared a rejection of the Brahmanical ideology of the sacrifice. Within Buddhism, as with Jainism, there were strong tendencies towards the horizontal codification and classification of reality. Yet, like Jainism, there was also a strong verticality that sought the final liberation or *nirvāṇa*, the snuffing out of greed, hate, and delusion. A full engagement with the thought of this world civilization is, of course, not possible in a few pages here, but Buddhism offers deep challenges to philosophies of life and ways of approaching life that have contemporary relevance, if, as Jay Garfield says, Buddhism 'is about the transformation of the way we experience the world'.[113] To use another distinction from Jay Garfield, if we can distinguish between *lectio* (exegesis) and *applicatio* (deployment), then I need here to restrict this chapter to *lectio*.[114]

The fundamental teaching of the Buddha Gautama was summarized in what is traditionally the first discourse of the Buddha, the 'Setting in Motion the Wheel of the Doctrine' (or in Pāli, the *Dhammacakkappavattana-sutta*). This contains the doctrine of four noble truths or, more accurately, the four truths of the noble one (*ariyasaccāni*)[115] that Rupert Gethin describes as the disease, the cause, the cure, and the medicine.[116] This is the central Buddhist doctrine that life is suffering, there is a cause of suffering (which is desire), there can be an end of suffering, and there is a path to the end (the eightfold path). The Buddha taught for many years and established a thriving monastic order that continues to this day. Following his death, a number of councils established the Buddhist canon, versions of which now exist as the scriptural authority of the Theravāda school in Pāli, but also in Chinese and Tibetan versions, some of which may contain material that goes back to the earliest years of the religion. The Pāli canon comprises sayings and doctrines of the Buddha, the Sutta-pitaka, monastic rules (the Vinaya-pitaka), and philosophical elaboration on the teachings (the Abhidhamma-pitaka). Apart from the Theravāda there was an ordination tradition in Kashmir called the Sarvastivāda with its own body of texts preserved in Chinese translation, which was also a philosophical position.[117] This philosophy maintained that the constituents that make up reality, or dharmas, exist in past, present, and future, a

[113] Jay Garfield, *Engaging Buddhism: Why It Matters to Philosophy* (Oxford: Oxford University Press, 2015), p. 179.

[114] Ibid., p. 14.

[115] See Paul Williams' description of K.R. Norman's point on this term: *Buddhist Thought: A Complete Introduction to the Indian Tradition* (London: Routledge, 2000), p. 41.

[116] Rupert Gethin, *The Foundations of Buddhism* (Oxford and New York: Oxford University Press, 1998), pp. 59–84.

[117] Williams, *Buddhist Thought*, pp. 112–28.

position that was refuted by another philosophical school, the Sautrāntika that wished to return to the teaching of the sūtras. Between the end of the first millennium BC and the beginning of the first millennium CE, a new movement developed within Buddhism called the Mahāyāna, the Great Vehicle, distinguishing itself from the earlier schools that it called the Hinayāna, the Lesser Vehicle. This movement claimed that the Hinayāna aspiration was for the enlightenment of the individual whereas the Mahāyāna aspiration was for the enlightenment of all sentient beings.[118]

I refer the reader elsewhere for more detailed developments in this fascinating history,[119] but my concern here is of a different nature and is about the category of life in this religion. In one sense, of course, Buddhism is about life and the tradition sees the Buddha as a physician, offering a cure for the ill that is life. Buddhist practices of the laity—the offering of alms to monks, the veneration of relics, pilgrimage, the development of a good moral life, and the meditation practices of lovingkindness and compassion—are clearly life-affirming. There is a sense in which any tradition, especially one as successful as Buddhism, being the first world religion, must be affirming of life to survive, but Buddhism does not develop life as a category of discourse explicitly. In contrast to Jainism where we have seen a strong philosophy of life in the face of transcendence, Buddhism does not develop the theme of life as a category. Rather, philosophically Buddhism is more concerned with the understanding and cognition of reality in terms of being and non-being. With the Abhidharma we have the analysis of categories that comprise experience, from a classification of the elements of reality to the enumeration of states of mind in specific detail.[120] Developing from this, we have the critique of this approach in the emptiness doctrine of Nāgārjuna and the Madhyamaka School and in the idealism of Yogācāra. Indeed, we might say that the foundations for the whole history of Buddhist thinking lay in the Abhidharma, which constrains later philosophy, even when it disagrees with it. The ontology and anti-ontology of Buddhism is constrained within the Abhidharma trajectory.

But there is one tradition called the womb of enlightenment (*tathāgata-garbha*) that can be seen in terms of a philosophy of life. This tradition maintains that all beings contain the potential for enlightenment, the potential to become a Buddha. Buddhahood (*buddhata*) is an innate propensity within the structure of the world itself. Paul Williams thinks that this tradition falls outside of mainstream Buddhist philosophy because there is little commentarial reflection on its textual sources, except one commentary the *Ratnagotravibhāga*,

[118] Ibid., pp. 131–66.

[119] Heinz Bechert and Richard Gombrich (eds.), *The World of Buddhism* (London: Thames and Hudson, 1984); Paul Williams, *Mahāyāna Buddhism* (London: Routledge, 2009 [1989]).

[120] Rupert Gethin, *The Buddhist Path to Awakening* (Oxford: One World, 2001 [1992]).

and it is mostly taken up in the far east.[121] This doctrine became essential in China where it was transformed into the doctrine of the Buddha nature (*foxing*) and became much closer to a philosophy of life than its Indic equivalent.[122] The *Ratnagotravibhāga*[123] claims that the essence of all beings is this quality of enlightenment that the community of realized beings, the Bodhisattvas, have perceived. In taking refuge in the community of enlightened beings, the aspirant is accessing this innate reality of Buddhahood within them.[124] We are a long way here from the doctrine of emptiness and no-self of the earlier tradition, but it seems to me that the doctrine is not understood as the animating principle of life. Although the *Laṅkāvatāra-sūtra* does introduce the idea of *ātman* that it identifies with the *tathāgatagarbha*, this remains within a Buddhist resistance to reification and seems to be a distinct idea to the *jīva* in Jainism or the Brahmanical *ātman*.

We have come a long way from the sacrificial imaginary of the Vedic roots of Indic culture. Although arguably the Buddha was still responding in some ways to the Brahmanical culture, arguing that the true Brahman was someone who is morally upright and presenting what we might call an Axial re-reading of the Brahmanical tradition, by the time of the first millennium CE that sacrificial imaginary was not operative within Buddhism except in so far as Buddhist philosophers took issue with their Vedic Brahmanical counterparts such as the Vaiśeṣikas, but particularly the Vedic exegetes or Mīmāṃsakas. The Buddhist thinkers responded to their own tradition which remained non-substantialist with a sceptical attitude towards claims about being and God. With Jainism, we have a philosophy of life in which the universe is populated with an infinite number of animating principles that take the form of the body they inhabit, and accompanying this is a deep respect for life itself, even though in the end the ultimate value is transcendence of life and world. With Buddhism we have the importance of transcendence alongside an analysis of existence in order to deconstruct or disentangle attachment and so take the sting out of desire and death, which are both intimately connected. In contrast to Jainism, Buddhist philosophy is orientated to transcendence but without a substantive self that transcends. The moral universe of Buddhism is therefore quite distinct from mainstream Brahmanical tradition, and while there is great emphasis on compassion for all living beings, especially in the Mahāyāna, wisdom lies in understanding that they are empty. If life can be understood as a process of causes and conditions—codified as the twelve links of dependent origination (*pratityasamutpāda*)—then this can be disrupted in penetrating

[121] Williams, *Buddhist Thought*, p. 161.

[122] Sallie B. King, *Buddha Nature* (Albany, NY: SUNY Press, 1991).

[123] For the definitive study see Jikido Takasaki, *A Study of the Ratnagotravibhāga: Being a Treatise on the Tathāgatagarbha Theory of Mahāyāna Buddhism* (Rome: Is.M.E.O., 1964).

[124] Ibid., pp. 177–8.

insight that breaks the link by eliminating craving. The Buddhist analysis of life would claim to be realistic in facing up to death and suffering but is also pessimistic in attributing value to life itself. Value lies in either transcendence in gradualist schools of Buddhism or in awakening to the immediacy of enlightenment here and now.

We might say that while the Śramaṇa traditions of Jainism and Buddhism share an eschatological hope with the Brahmanical tradition of the Upaniṣads, the Vedānta, and even the Vaiśeṣika, this hope is of a different order because the Brahmanical tradition can still invest value in the world of life as the body of God (Rāmānuja), or as the emanation or manifestation of Brahman. It seems to me that this is quite a distinct moral universe in which there is value in transcendence but also value in the world of life: the Brahmanical vision is still within the sacrificial imaginary that the Śramaṇa vision has rejected. Before we leave this world, there is one more important development that we need to consider, namely the tantric revelation that transforms the Brahmanical vision and then in turn affects the Buddhist and Jain worldviews.

AFFIRMING LIFE

With the Axial transformation we saw a major reinterpretation of the sacrificial imaginary in the Upaniṣads and the Vedānta tradition. A further transformation occurs halfway through the first millennium CE and continues until about the thirteenth century and even after. This is the tantric tradition—we might even say tantric revolution—that transforms medieval Hinduism and dominates not only the intellectual elites but the political elites as well. While undoubtedly devotion (*bhakti*) is important—as we see expressed in the mythological treatises of the Purāṇas and in wonderful devotional poetry, such as Jayadeva's *Gītagovinda* or the love poems to Kṛishna of Caṇḍidāsa and Vidyāpati—it is the new revelation of the Tantras that comes to dominate medieval Hinduism in what Alexis Sanderson has called 'the Śaiva Age'.[125] With the tantric traditions we have a new moral compass that was often a threat to mainstream, conservative Brahmanism but that became absorbed into it. Generally, we have a picture of tantric traditions that originate on the edges of Brahmanical society becoming absorbed into the mainstream; we have the Brahmanization of Tantrism. From the fifth century CE tantric traditions focused on the deity Śiva begin to develop, becoming an unstoppable flood of texts and practices by the tenth and eleventh centuries, as is borne witness to by the great amount of manuscripts preserved in Nepal and

[125] Sanderson, 'The Śaiva Age'.

also in South India at Pondicherry.[126] There is also epigraphic material that bears witness to the success of the tantric traditions. But what is Tantra and how is it relevant to a discussion about life itself?[127]

There has been much discussion about the origins of tantric traditions—whether transformations of Brahmanical Hinduism or coming in from another sphere—but I suspect that its roots are in cremation ground asceticism that quickly becomes absorbed and transformed into mainstream Brahmanical culture. As Padoux observes, the origins are on the margins of the Brahmanical world, small groups which were transgressive, visionary, and practising magic that in due course become routinized and absorbed into mainstream Brahmanical society.[128] Intellectually, the roots of this tradition are possibly also in the forms of naturalism that we saw with the Vaiśeṣika philosophy. Indeed, there is a historical connection between the Vaiśeṣikas and the practice of Pāśupata Śaivism.[129] The tantric traditions are characterized by a philosophy of life that gives a positive evaluation to the world of life. While these traditions were rooted in practice, particularly asceticism, they became philosophies in their own right, critically engaging with other systems of thought.

Alexis Sanderson has mapped these traditions of medieval India. To simplify a complex picture by the Middle Ages, say the post-Gupta period after about 620 CE, we have a mainstream Brahmanical tradition that is within what I have called the sacrificial imaginary. This tradition of Smārta Brahmanism, the Brahmans who followed the law books (i.e. the Smṛtis) and Purāṇas, is concerned with ritual purity and living a good life within the constraints of dharma, following prescribed ritual, keeping the household fires, and fulfilling responsibilities to ancestors, sages, and family. The Smārtas could worship different deities and the Śaiva Smārtas were devotees, Śiva Bhaktas, who had faith that at death they would be reborn into Śiva's heaven by his grace. They followed the Smṛtis and Śiva Dharma texts. This tradition remains mainstream, right up to modernity. Alongside this we have another kind of tradition that claimed a higher authority and promulgated the teachings of Śiva (*śivaśāsana*). This was initiatory and itself can be divided into the higher or outer path, the Ati Mārga, and the path of Mantras, the Mantra Mārga. The Ati Mārga was a path to enlightenment of ascetic orders, principally the Pāśupatas, and developed in different stages over time (Sanderson has

[126] A point made by Sanderson, 'Śaiva Literature', p. 1.

[127] For the best introduction available, see André Padoux, *Comprendre le Tantrisme: les sources hindoues* (Paris: Albin Michel, 2010), rewritten for an English-speaking audience as *The Hindu Tantric World: An Overview* (Chicago, IL: Chicago University Press, 2017).

[128] Ibid., pp. 54–5.

[129] George Chemparathy 'The Testimony of the Yuktidīpikā concerning the Īśvara Doctrine of the Pāśupatas and Vaiśeṣikas', *Wiener Zeitschrift für die Kunde Sudasiens*, vol. 9, 1965, pp. 119–46.

identified three main stages). The Mantra Mārga is technically what we refer to as Tantra. This tradition, or group of traditions, revered a body of texts that is regarded as revelation called Tantras. These were usually in the form of a dialogue between Śiva and the Goddess. The religion that developed was called the Śaiva Siddhānta and became the dominant religion in medieval India, or, more precisely, the Śaiva ritual system of initiation and daily and occasional rites became pan-Indian. On top of this mainstream Śaiva religion that regarded itself as orthodox and the Vedic revelation as a legitimate, if lower, expression of divine authority, other kinds of Śaivism developed that went beyond or even rejected the mainstream Siddhānta religion. These non-Saiddhāntika groups may have originated in ascetics living in cremation grounds (and so were polluting to the highly orthodox) and were often concerned with worshipping ferocious Goddesses who demanded to be appeased with offerings of alcohol, blood, and sexual substances in a tradition that became known as the Kulamārga. The point of all this was to gain not simply liberation but power in this world and other worlds that yogis could enter into.

The practice of these groups was intended to go against mainstream orthodox Brahmanism, which they regarded as characterized by inhibition and obsession with ritual purity that was a hindrance to the spontaneous realization of the ubiquity of pure consciousness that these groups believed in. There was, then, an identification between the idea of consciousness, enlightenment, and the Goddess understood as power. These ascetic groups, on the edges of orthodox society, practising sex outside of caste restrictions and dealing with polluting substances, came to be lampooned by orthodox playwrights such as Bhaṭṭa Jayanta,[130] although this reflects the serious concern of elites about the influence of these groups on society. These hard, tantric cults focused on the Goddess and/or a ferocious form of Śiva called Bhairava came to be absorbed by the mainstream, respectable Śaiva establishment, especially in Kashmir where a tradition of householder Śaivism developed that was strongly influenced by these marginal ascetic groups and their scriptures. I refer the reader to Alexis Sanderson's writing for an account of them.[131] These unorthodox groups generally held to a philosophy of non-dualism and were concerned with experience as the fullness of divine life as symbolized in the visage of Bhairava with his bulging eyes coming into the fullness of experience. The philosophical articulation of this school that developed in Kashmir was called the Recognition school (Pratyabhijñā).

[130] Jayanta Bhaṭṭa, *Āgamaḍambara*, trans. Csaba Dezso, *Much Ado about Religion* (New York: New York University Press, 2005).

[131] In particular, Sanderson, 'Purity and Power among the Brahmans of Kashmir', in Michael Carrithers, Steven Collins, and Steven Lukes (eds.), *The Category of the Person: Anthropology, Philosophy, History* (Cambridge: Cambridge University Press, 1985), pp. 190–216.

Non-Dualism as Philosophy of Life

The Pratyabhijñā was mainstream Śaiva philosophy that engaged with other philosophical systems and argued with rival claims to truth. Like the Vedānta, it accepted revelation as a source of knowledge, but unlike the Vedānta, it claimed the tantric revelation as its own.

The Pratyabhijñā developed a philosophy of life based on the Tantras that claimed that the highest value and meaning in life is recognition of the person's innate divinity and opening the eyes to the divine pulsation (*spanda*) that constitutes the essence of life. Beginning with the philosopher Somānanda, whose work the *Vision of Śiva* (*Śivadṛsti*) outlines how Śiva as consciousness manifests world, this tradition followed through a series of teachers, namely Utpaladeva, Lakśmaṇa, Abhinavagupta (975–1025 CE), and Kṣemarāja (*c.*1000–50 CE). Of these Abhinavagupta is the most famous, famous not only for his tantric philosophy but also for his work on aesthetics. After writing a commentary on Ānandavardhana's book on aesthetics, the *Dhvanyalocana*, Abhinvagupta went on to compose learned commentaries based on the *Mālinīvijayottara-tantra*, the root text of the Trika or Threefold religion into which he was initiated. He composed a ritual manual based on this text called the *Illumination of the Tantras* (*Tantrāloka*) and a short summary, *The Essence of the Tantras* (*Tantrasāra*). His most mature and erudite work is the commentary on his grand teacher's work, the *Commentary on the Recognition of the Lord* (*Īśvarapratyabhijñāvimārśinī*). Throughout these works there is a consistent vision of the affirmation of life and a recognition of its innate sanctity.

In the *Essence of the Tantras*, a summary of the *Tantrāloka*, Abhinavagupta summarizes the teachings of the three paths of realization of the pure consciousness of Śiva as expounded in the main scripture, the *Mālinīvijayottara-tantra*, along with an account of a pathless path or non-means to this realization. For Abhinavagupta a person has the capacities of willing, thinking, and acting and these capacities can be harnessed and transformed in the service of a spiritual realization. Each of these is the focus of each of the three paths; thus willing is transformed in the divine means (*śambāvopāya*) through the upsurge of power within the body; thinking is transformed through the energy means (*śāktopāya*) through the development of a pure thought such as 'I am Śiva'; and action is transformed through the individual means (*āṇavopāya*) through bodily practices such as meditation, ritual, and the recitation of mantra. This text is a philosophy of life in the sense that it is a reflection on practices that are thought to bring the practitioner into the fullness of awareness of life as pure consciousness, as light (*prakāśa*) and reflexive awareness (*vimarśa*) personified as the God and Goddess, Śiva and Śakti. Power or energy (*śakti*) conceptualized as the Goddess is at the heart of life and human beings can directly experience this reality through a transformation of awareness.

Through the divine means, the practitioner comes to understand that all appearance is a reflection (*pratibimba*) of pure consciousness, even though this is a way of speaking because in truth there is no distinct or separate original (*bimba*) that is reflected. Awareness is aware of itself and the whole world appears to itself, internally, as it were, within awareness. This awareness is pure and without thought-construction (*vikalpa*), and is God's power of autonomy (*svatantryaśakti*) that surges up to absorb or create an immersion of the person in it. This is the immersion in God (*bhairavasamāveśa*)[132] that is the realization of one's true nature and the essence of the universe itself. The second method, the powerful means (*śāktopāya*), is the cultivation of pure cognition. The practitioner can gradually perfect discursive cognition[133] such that one thought is purified by a subsequent, purer thought, and so on until the ultimate realization of one's identity with God. Because of the power of false or dualistic cognition, people think of themselves as being bound or limited, whereas in truth people are not bound and are unlimited because identical with God who is pure consciousness. This false belief (*abhimāna*) is the cause of keeping beings within the cycle of reincarnation. Abhinavagupta writes:

> This supreme reality alone is ultimately real. It is the ground in which all things are established, it is the energy of the universe by which the universe lives/everything is alive, and I am just that, and therefore I transcend the universe and am immanent within it.[134]

The subject of first-person predicates, the 'I', expresses the essence and energy of the universe. This is the ultimate reality and is identified with God for Abhinavagupta, who is both transcendent and immanent; beyond the universe and the soul of the universe, its life force and energy. Abhinavagupta uses the word *ojas* in this passage which, although a masculine noun, refers to the power (*śakti*) that is the vibrant force of the cosmos itself. It is that which gives life to the living and by which they 'breathe' (*prāṇiti*).

Lastly, the individual means involves all practices focused on the body and on ritual, including mantra repetition and meditation (apart from that described above). Meditation and gnosis bestow the realization of the pure vibrant immanence of power, but also ritual action gives rise to awareness of

[132] *Tantrasāra* by Abhinavagupta, ed. M.S. Kaul (Śrīnagāra: Kashmir Series of Texts and Studies, 1918), p. 10.

[133] *Tantrasāra*, p. 21: 'In this when [the practitioner] gradually perfects discursive cognition in order to enter into the true nature just defined [in the previous chapter] then there is the use of a course of meditation which is based on the teachings of true reasoning, true scripture, and a true teacher'. *Tatra yadā vikalpyaṃ krameṇa saṃskurute samanantaroktasvarūpapraveśāya, tadā bhāvanākramasya sattarkasadāgamasadgurūpadeśapūrvakasya asti upayogaḥ.*

[134] *Tantrasāra*, p. 21: *tad eva ca paramārthaḥ, tat vastuvyavasthāsthānaṃ, tat viśvasya ojaḥ, tena prāṇiti viśvaṃ, tadeva ca aham, ato viśvottīrṇo viśvātmā ca aham iti.*

the fullness of life as pure consciousness. The cultivation of the Śaiva way of life, which entails daily rites and meditation, gives rise to an awareness of non-duality that in due course permeates the practitioner's everyday reality. This awareness evoked through the ritual act extends into everyday action. In a chapter explaining Śaiva ritual, Abhinavagupta writes in the *Tantrāloka*:

> Since here [in the matter of ritual] through worship, there is non-differentiation (*avatirekitva-*) from fullness [or] from fulfilment because of God, of all beings who act, [the practitioner] pervades everything in the ritual system shown to be non-differentiated from God, which is a whole group of actions, even [everyday acts such as] walking and so on. For as a horse mounted in battle, even though freed from the movement of the bridle bit whilst being ridden, does not go beyond his training, even so [the practitioner] having practiced ritual and worship intended to produce transformation into God, abandons the duality of actions even when standing or walking. Thus, for someone having been established in the practice of oneness, this universe spontaneously and forcefully bursts into life like a dancing woman completely filled with the ecstasy of God.[135]

Through ritual, the practitioner becomes one with God (Śiva). This is in fact a formalized process in which there is ritual identification of the practitioner with God through a process of the purification of the elements within the body (*bhūtaśuddhi*), followed by the creation of a divine body through the imposition of mantras (*nyāsa*), followed by worship in the mind (*mānasayāga*) and then externally (*bahyayāga*).[136] To experience this fullness of God in life, the Śaiva religion presented a complex system of ritual and initiation, probably systematizing traditions that were already present. Thus, the teachings of Śiva (*śivasāsana*) demanded that the adept undergo initiation (*dīkṣā*), just as the Vedic patron of the sacrifice had done. The Śaiva initiation performed by both the Śaiva Siddhānta and the non-Saiddhāntika traditions, such as the Goddess cults of the Trika, Kaula, and Krama to which Abhinavagupta belonged, were fundamentally of two kinds. The normative initiation (*samayadīkṣā*) allowed the young adept access to the texts of the tradition while the liberating initiation (*nirvāṇadīkṣā*) guaranteed liberation either at death or within a limited number of lifetimes. Following the *nirvāṇadīkṣā* were two further possibilities: to become a practitioner who sought magical powers and pleasurable experience in heavenly worlds, called a Sādhaka, or to become a teacher in the tradition able to

[135] *Tantrāloka* 15.147–51: *yathaḥ samastabhāvānāṃ śivasiddhimayād atho/pūrṇād avyatirekitvaṃ kārāṇām ihārcayā//147//samasthaṃ kārakavrātaṃ śivābhinnaṃ pradarśitam/pūjodāharaṇe sarvaṃ vyaśnute gamanādyapi//148//yathāhi vāhakaṭakabhramasvātantryam āgataḥ/aśvaḥ saṃgrāmarūḍo'pi tāṃ śikṣāṃ nātivartate//149//tathārcanakriyābhyāsa śivībhāvitakārakaḥ/gacchaṃstiṣṭhannapi dvaitaṃ kārakāṇāṃ vyapojjhati//150//tathikyābhyāsaniṣṭasyākramād viśvam idaṃ hathāt/saṃpūrṇaśivatākṣobhanarīnartadiva sphuret//151.* My translation based on a class by Alexis Sanderson on 15 February 2007. Any errors are, of course, my own.

[136] See Sanderson, 'Meaning in Tantric Ritual'; Gavin Flood, 'The Purification of the Body in Tantric Ritual Representation', *Indo-Iranian Journal*, vol. 45, 2002, pp. 22–43.

initiate others, an Ācārya.[137] These two further consecrations deepened the adept's commitment. Of course, these consecrations that were likened to royal consecrations were not compulsory but options within the system. It this that is alluded to in the above quoted passage. The distinction between fullness (*pūrṇa*) and fulfilment (*siddhi*) refers to the distinction between these two consecrations for one who desires liberation (*mumukṣu*) and one who desires pleasure, experience, and power in higher worlds (*bubhukṣu*).

Although these formal ritual acts result in awareness of the unity of self, God, and universe, this is not only an awareness but also a practice almost like a skill. If we can become aware of our non-differentiation from God through ritual acts in the Śaiva path, then this can also be extended to everyday acts. In a sense, the practitioner's whole life becomes a ritual act and he or she becomes aware of undifferentiated consciousness even in mundane actions such as walking or standing. Through the practice of non-duality, the practitioner comes to realize the reality of non-dualism, which spontaneously bursts into life, like an ecstatic dance, to use Abhinavagupta's striking metaphor.

We are a long way here from Brahmanical purity laws and even the high constraints of Vedic sacrifice. Neither is this the transformation or internalization of the sacrifice through renunciation, nor is it devotion to God and seeing the world as God's body. Rather, we have the realization that universe and self are God; there is a spontaneous overwhelming as the universe itself bursts into life for the practitioner. And this is a skill to be cultivated through repeated practice and discipline. Just as a horse that has been trained does not forget the training even in the midst of a battle, so the practitioner cannot forget the spiritual training that has formed him; a training that produces awareness of life as divine or the spontaneous manifestation of God. This is a world of divine immanence but distinct from the non-dualism of Śaṅkara because it is concerned with the dynamic energy (*śakti*) of pulsating life that the practitioner moves into and experiences in its fullness. This is a strong philosophy of life in the sense that it gives positive value to life and action, in contrast to the ideal of pure transcendence, as in Jainism and the early Upaniṣads, and the affirmation of the everyday through sacrifice.

But we are still within the sacrificial imaginary. For Abhinavagupta, the entire universe is a sacrifice in some sense, a sacrifice to itself in its all self-compassing and self-consuming nature. And sacrifice is still a part of ritual practice. This is because of the influence of Śāktism on Śaivism. Abhinavagupta acknowledges this through inscribing the cult of the Goddess at the

[137] See Hélène Brunner, 'Le Sādhaka, personnage oublié dans le Śivaisme du Sud', *Journal Asiatique*, vol. 263, 1975, pp. 411–16. Also, Brunner, *Somaśambhupaddhati* (Pondicherry: Institut Français d'Indologie, 1977), vol. 3, pp. iii–xxvi.

heart of his practice, and identifying the Goddess as identical with supreme consciousness. The tantric sacrifice in which blood is offered to the Goddess is mapped on to Vedic religion. Thus, in some ways Abhinavagupta's tantric religion is in continuity with the Vedic, although Abhinavagupta consistently claims that the tantric is a higher revelation that teaches the transcendence of Brahmanical inhibition.[138] There are also differences in the sacrifice itself. Whereas Vedic sacrifice sheds no blood, suffocating the victim, Kaula sacrifice involves the decapitation of the goat. The rice beer (*surā*), along with the omentum (*vapāntra*) and heart (*hṛdaya*), are offered as an oblation.[139] Sanderson describes how sacrifice was required in initiation into the form of Śaivism called the Trika, during which the omentum, as in Vedic sacrifice, is offered into the fire.[140]

Abhinavagupta's religion therefore absorbs the old, Brahmanical tradition of sacrifice but qualifies it with the idea that the Śaiva tradition supersedes the Vedic revelation. The Vedic concern with the distinction between purity and impurity, as expressed in topics concerning what should and should not be consumed and in what human practices are pure and impure, is here transcended. Impurity (*aśuddhatā*) is only convention in the sense that the innate nature of something remains the same and is only seen as impure by those who are unenlightened, as taught in the lower scriptures of the Veda. To understand the true nature of God and life, the practitioner must transcend the inhibition of the Brahmanical convention. Thus anything in the world that is within human awareness or that has proximity to consciousness (*āsannaṃ saṃvidaḥ*) is suitable for worship in this system because it can intensify awareness and so is appropriate because 'it is filled with life' (*jīvavat*). So something is pure or impure not because of Vedic prohibition, but depending upon the degree to which consciousness is attracted to or retreats from it. By this principle (how much something is filled with life), 'consciousness goes towards what is proximate or distanced'.[141] In commenting on this passage, Jayaratha (twelfth century) cites a scripture that says 'the substance located in

[138] Alexis Sanderson, 'Meaning in Tantric Ritual', in A.-M. Blondeau and K. Schipper (eds.), *Essais sur le Rituel III: Colloque du Centenaire de la Section des Sciences religieuses de l'École Pratique des Hautes Études*, Bibliothèque de l'École des Hautes Études, Sciences Religieuses, Volume CII (Louvain-Paris: Peeters, 1995), pp. 15–95; J. Hanneder, *Abhinavagupta's Philosophy of Revelation* (Groningen: Egbert Forsten, 1998).

[139] *Tantrāloka* by Abhinavagupta with the *viveka* by Jayaratha, vol. 14, ed. M.S. Kaul (Śrīnagāra: Kashmir Series of Texts and Studies, 1939), 15.172.

[140] Alexis Sanderson, 'Levels of Initiation and Practice in the Śaivism of Abhinavagupta', (1995, unpublished MS); commentary on the first verse of the *Tantrasāra* (Handout, Trinity, 2003).

[141] *Tantrāloka* 15.165cd–166ab: 'By this principle, when consciousness goes towards what is proximate or distanced, then it is shown to be appropriate or inappropriate [for realization]', *anena nayayogena yadāsattividūrate//saṃvid eti tadā tatra yogyāyogyatvam ādiśet.*

one's own body is the pure, best elixir'.[142] We can take this to mean that the body contains the nectar of immortality within it, a belief that located this substance in the head especially in later yogic literature,[143] but we can also take this to mean that natural substances of the body that are regarded as impure in Vedic Brahmanism can be interpreted as pure in the tantric tradition. This is particularly true with regard to sex. Abhinavagupta's religion orientated towards the Goddess traditions demanded erotic worship or sexualized ritual at certain junctures of the year. As Vedic sacrifice required the patron to be accompanied by his wife, so the tantric sacrifice required the practitioner to be accompanied by his partner or 'messenger' (*dūti*) with whom he performed ritualized sex. This was a practice from the more extreme antinomian sects of the Goddess-worshipping Kula tradition that became absorbed into more mainstream Śaivism and later into the Brahmanical Goddess tradition of the South, the Śrīvidyā. Abhinavagupta describes this ritual in which sexual substances (semen and menstrual blood) are offered to the Goddess in a chalice sanctified with mantras. Indeed, the offering of the 'three Ms', meat (*maṃsa*), wine (*madya*), and sex (*maithuna*), was a requirement of this Kula ritual intended to bring the participants into awareness of their innate divinity and identity with God.[144]

What was considered impure in Vedic Brahmanism is considered to be potentially liberating in tantric Brahmanism. For Abhinavagupta's non-dualism there can ultimately be no distinction between the pure and the impure because both are ultimately identical with consciousness. Furthermore, what consciousness is attracted to is potentially liberating. The usual practices of life, particularly sex as intensifying experience, can be used in the service of realizing the true nature of life as filled with divine presence. To become aware of the fullness of life for Abhinavagupta, we need to go beyond the restrictions of tradition that inhibit that realization. His non-dual Śaivism is not simply a rejection of Vedic Brahmanism but a transformation of it. The strictures of the Vedic scriptures are sufficient for a lower level of understanding, but to appreciate the divine immanence and transcendence that comprises the totality of life, we must abandon those limitations and embrace the fullness of life, which is God.

These ideas met with approval in some courtly circles where kings were attracted by the potential for magical enhancement of their powers, but other

[142] *Tantrālokaviveka* 15.166ab: *svadehāvasthitaṃ dravyaṃ rasāyanavaraṃ śubham.*

[143] B. Wernicke-Olesen and S. L. Einarsen, 'Übungswissen in Yoga, Tantra und Asketismus,' in A.-B. Renger and A. Stellmacher (eds.), *Übungswissen in Religion und Philosophie: Produktion, Weitergabe, Wandel* (Berlin: Lit Verlag, 2018), pp. 241–57.

[144] *Tantrāloka* chapter 29. For a translation and analysis see John R. Dupuche, *Abhinavagupta: The Kula Ritual as Elaborated in Chapter 29 of the Tantraloka* (Delhi: MLBD, 2003). Also see Lilian Silburn, *Kuṇḍalinī; the Energy from the Depths*, trans. Jacques Gontier (Albany, NY: SUNY Press, 1988).

more conservative Brahmans were concerned that these practices undermined the social order, as they did, and were a threat to social stability. Thus, the Nyāya Brahman Jayanta Bhaṭṭa was worried about a tantric cult called the 'blue clad' (*nīlāmbhara*) who, he says, practised sex in public places covered only in a blue garment.[145] This potentially undermined the social order in which caste restrictions were so strong, especially the prohibition of sex across caste boundaries. But with Abhinavagupta's Śaivism with Goddess worship at its heart, we have a strong philosophy of life that embraces the Vedic sacrificial affirmation of life but goes beyond it in claiming that Vedic practices actually prevent us from realizing the true potency of God in the world. The universe, the self, and God are in truth a single reality that he calls consciousness: a constant pulsation (*spanda*), the breathing in and out of life itself, that he also calls the opening and closing of the eyes of Śiva. Abhinavagupta tries to make sense of the tantric tradition, bringing it into the field of his substantial learning and interpreting it as enhancing human awareness of life as God.

The message and vision of Abhinavagupta were perhaps too radical for mainstream Brahmanism, and while the tradition does carry on in Kashmir, more or less into the twentieth century, it loses its ritual basis and becomes a purely Gnostic tradition that does not survive without the ritual substructure. The tantric tradition is the closest we come to a vitalism in Indic religions. Life itself is understood here as innately divine. Transformation of human reality comes about through embracing the paradox of death following life, the paradox that the sacrificial imagination tries to resolve in pointing to the dynamism and vibrancy of life here and now. Abhinavagupta's eschatological hope lay in realizing the identity of life itself with consciousness, which he identified with one of Śiva's forms, which is a liberating cognition. He outlines methods to achieve this realization and beyond ritual, this understanding can come through a pure gnosis that disparages all external practices: liberation can be attained purely through awareness.

Let me try to draw together the strands of the history I have described. The sacrificial religion of the Vedas affirmed life in the world while offering hope of heaven after death, and while the gods are clearly important, it is their relation to life in the world mediated through the sacrifice that predominates. The gods have a largely ritual function. Then in partial response to the sacrifice, we have its internalization in the Upaniṣads. The true sacrifice is within the self, of the self to a higher power. We have identified this as the Axial shift that puts forward the idea of vertical ascent. The true purpose of life is to go beyond it and to realize the pure transcendence of absolute reality that is yet identified with the innermost self. But we also have the idea in these texts that this

[145] Bhaṭṭa Jayanta, *Āgamaḍambari*: 'Oh such asceticism (I have never) seen before!' *aho batāpurvam idam tapaḥ*, p. 40. Csaba Dezso, *Much Ado about Religion* (New York: New York University Press, 2005), 2.143.

transcendence is immanent, giving life to the cosmos. The absolute reality is in the breath we breathe and in the food we eat.

While the verticality of the Upaniṣads operates within a Brahmanical logic in relation to sacrifice, there is also an influence from outside of the Vedic arena from the Śramaṇa traditions, whose origin probably lies in Greater Magadha, as Bronkhorst has argued. With these traditions, we have a verticality and eschatological hope that seeks release from life in the world seen as a process of suffering through repeated births and deaths along with a horizontal codification and analysis of the elements that make up the world and human life within it. These two tendencies of vertical transcendence and horizontal codification can be seen in traditions that developed in the wake of the Axial shift, borne witness to in the Upaniṣads. Buddhism is concerned with enumerating the elements that comprise human experience, the Vaiśeṣika tradition is likewise concerned with this kind of enumeration, and the Āyurveda takes this mode of thinking and applies it to the search for health and long life. The Jains absorb these approaches and develop their own distinctive understanding of life as having the potential for liberation in which the soul frees itself from the bonds of action, yet which also has a positive attitude to life in that every living thing contains a life force that pervades it completely. But all remain close to the RCM, even the Ayurveda. Life itself is filled with an animating power and yet in the end the Jain ascetic's goal is to go beyond it, to withdraw the animating power from the world of life. The Upaniṣadic ideal is transformed during the first millennium CE into the non-dualism of Śaṅkara and the qualified non-dualism of Rāmānuja. But we are still within the Vedic sacrificial imaginary as we are with the tantric synthesis. Here the tradition rejects the restriction of the Vedic order while claiming to transcend both it and the Vedānta. Abhinavagupta is a fine example of a vitalism that sees life itself as sacred and which has a generally positive attitude, wishing to maintain the value of the world as the embodiment of life.

There are a number of issues here. Traditionally Indic religions, for example, do not have what we might call political eschatology. They are concerned with transcendence and escape from the world or seeing the world as innately sacred, but until modernity this does not get translated into a social vision or vision of social transformation. The reasons are complex, but this lack of political theology in the history of Indic religions is linked to the manner in which life itself is understood. In contrast to this distinctive history, I need to trace two more trajectories through the history of civilizations: that of Chinese religions and of what we can call Abrahamic.

5

Earth under Heaven

The Chinese Traditions

THE SUSTAINING POWER OF LIFE

With China we are in a very different world to Indic civilization. China did develop the notion of life as a sustaining power of living beings, but in contrast to India, did not pursue the idea of transcendence as its source. Transcendence in the sense of a wholly other realm beyond the world is essentially alien to China. In contrast to any Gnostic view of flight to a higher reality that we find in India or Greece, China throughout its history has maintained a positive evaluation of life, placing human life within a cosmos, but a contained cosmos, less complex in its hierarchy than in India. In particular, pre-modern Chinese civilization emphasized harmony in the relationship between heaven (*tian*), earth (*di*), and humanity (*jen*): there should be a 'unity of heaven and humanity' (*tian ren heyi*). Heaven is the source of all things and generally, with some philosophical exceptions, humanity is thought to be perfectible and by one major philosopher, Mengzi, to be essentially good. There is a continuity between the political and natural orders, thus the terms for 'empire' and 'world' are the same, *tian xia*, under heaven: we might call this a holistic vision. Along with harmony, China has stressed social order in which the emperor is the son of heaven at its centre, with great stress on filial piety and obedience. This means that people need to observe natural limits and that everything has its place: what is good is what is natural. The consequence of this is that China has been less inventive in its cosmological imaginaire than we found in India. Concepts of time, for example, were based on human life and the life span of the dynasty rather than the millions of years of the *kalpa* or *yuga*, thus we find Chinese texts dated precisely in contrast to the difficulty of dating Sanskrit texts that, until the medieval period, were bereft of dates.

To begin our survey, I would like to make some general observations on the differences between Chinese and Indic civilizations that will allow us to appreciate the distinctiveness of China's concept of life itself and China's specific understanding of human self-repair. We can begin with language. Sanskrit is highly inflected in which meanings are changed by the modification of the stem of a word (thus the horse as subject, *aśvaḥ*, can become the horse as object, *aśvam*), and along with this can form long compounds by joining stems together. Chinese by contrast is uninflected and monosyllabic in which word order is important and in which to change meaning, modifiers need to be placed before and/or after a word. This distinction in the nature of language, the one inflected and polysyllabic, the other uninflected and monosyllabic, extends to the written word in which Chinese emphasizes the concrete image and is quite terse in expression in contrast to Sanskrit's prolix abstraction and a tendency towards the hyperbolic.

There is also a contrast in social attitudes, and when Buddhism began to penetrate into China, its otherworldly or Gnostic orientation (true of the Śramaṇa traditions more generally) with its celibate monasticism and asceticism met with strong resistance from Confucian scholars, who saw it as an attack on the very unity of Chinese society. The political values of duty along with the idea that social and political life is essentially to be enjoyed contrasted with the Buddhist ideal of the monk beyond social values and political duty, in which life was seen as suffering to be avoided. The human being as a political entity as located within a broader social fabric contrasted with the renouncer ideal that stressed the analysis of individual psychology: China did not develop an analysis of the mind independently of Buddhism. Partly because of this cultural reticence in which all things had their right place, it took a long time for Buddhism to become established (from its introduction in 65 CE to about 250 CE) and it met strong resistance from Confucianism, although it found accommodation with Daoism, with which there were resonances.

There has been some debate about whether China exhibited a general dualism between mind and body. The prevalent view has been one of holism: that the term for 'mind' (*xin*), often rendered 'heart-mind' as I do here, is regarded as an organ much like any other and so not distinct from the body (*shen, xing, ti*),[1] but this opinion has been questioned by Edward Slingerland and his colleagues in an interesting analysis of a broad range of texts through computer-aided 'distance reading'. Through the analysis of the use of *xin* in a range of historical texts, Slingerland argues that the authors were operating with a (perhaps unconscious) mind–body dualism that challenges the holism

[1] Roger Ames, 'The Meaning of the Body in Classical Chinese Philosophy', in Thomas Kasulis, Roger Ames, and Wimal Dissanayake (eds.), *Self as Body in Asian Theory and Practice* (Albany, NY: SUNY Press, 1993), pp. 157–77; Jane Geaney, On the *Epistemology of the Senses in Early Chinese Thought* (Honolulu, HI: University of Hawaii Press, 1992); François Jullien, *Vital Nourishment: Departing from Happiness*, trans. Arthur Goldhammer (New York: Zone Books, 2007).

consensus.[2] This view in turn has been questioned as it does not explain the strong link between physicality and mentality, especially in Chinese medicine.[3] But even if there is a natural cognitive bias towards mind–body dualism in the Chinese textual corpus, other categories override this in importance and cultural dominance, particularly the ideal of social harmony through good governance. We have to understand the notion of mind not only in relation to body but also to cosmos, as the Chinese imaginaire, like the Indic, is still cosmological in its emphasis on social concord as the expression of cosmological harmony. Indeed, like the Indic, the Chinese conception of life itself is fundamentally cosmological.

To understand the concept of life in the history of Chinese civilization, we need to understand that heaven and earth constitute its totality, where heaven is the ruling principle that governs all things. This is the key to stability and to the good life. Mythologically there is the idea that a primeval oneness of Dao was broken up, resulting in a great sadness according to the last chapter of the Daoist classic, the *Zhungzi*.[4] Human self-repair for Chinese civilization is conceptualized in terms of congruity and striving to establish harmony in one's personal life that is then reflected in the life of state and society. The principles of *yang* and *yin*, the strong and the weak cosmic forces, should be in balance. This important distinction can be traced back to about 1000 BCE: *yang* being positive, active, strong, light, and male, *yin* being negative, passive, weak, dark, and female, and both having equal value. The distinction is central to the ordering of Chinese society and social space: that landscape reflects these two principles and there is a natural geography that can even be expressed in architecture where harmony and balance is sought (that we also see in spatial design called *feng shui*). It is precisely through the fabric of society that the idea of life operates, and it is meaningless to isolate 'religion' from 'society' in the ancient Chinese case, as religion is not formally distinct from social life and the idea of religion giving salvation was an alien concept, although there was the idea of achieving immortality,[5] a theme attested from the earliest

[2] Edward Slingerland, Ryan Nichols, Kristoffer Neilbo, and Carson Logan, 'The Distant Reading of Religious Texts: A "Big Data" Approach to Mind–Body Concepts in Early China', *Journal of the American Academy of Religion*, vol. 85 (4), 2018, pp. 985–1016. See also Edward Slingerland, 'Body and Mind in Early China: An Integrated Humanities-Science Approach', *Journal of the American Academy of Religion*, vol. 81 (1), 2013, pp. 6–55.

[3] Dawei Pan, 'Is Chinese Culture Dualist? An Answer to Edward Slingerland from a Medical Philosophical Viewpoint', *Journal of the American Academy of Religion*, vol. 85 (4), 2018, pp. 1017–31.

[4] Yü Yingshih, *Chinese History and Culture: Sixth Century BCE to Seventeenth Century* (New York: Columbia University Press, 2016), p. 4.

[5] Marcel Granet, *La Pensée chinoise* (Paris: Albin Michel, 1934), p. 426; Henri Maspero, *Daoism and Chinese Religion*, 2nd ed., trans. Frank A. Kierman (Melbourne and Basel: Quirin Press, 2014), pp. 360–1.

period, as Han tombs show.[6] Human repair, at least in mainstream non-Buddhist tradition, was rather understood not as salvation from the world but as living in harmony with it, through training any negative human proclivities into positive social values. The idea of the life force was central to this project as it is this force that drives human virtue, and although life force pervades the universe, it is especially concentrated as ethical power in the human case. Indeed, Daoism that refines the idea of life force has been regarded by some as the inner core of Chinese civilization. 'Yü',[7] The fifth-century text 'Methods of Nourishing the Vital Principle' (*Taiping jing*), an important revealed text, claims that harmony and immortality can be achieved through its method.[8] This is 'to feed one's life' (*yang sheng*), where life means one's vital force or potential.[9]

Civilization in the sense of good governance has been an important ideal in the history of China as being able to affect human nature and facilitate its perfection. These are the general parameters within which the concept of life operates in Chinese civilization. The energy of life is essential for social harmony, which is a corrective to any negative traits of human nature. But perhaps more precisely, the Chinese concept of life serves less in terms of repair and more in terms of flow; that the flow of human life needs to be channelled in morally correct ways as the flow of a river can be guided by channels. In Chinese civilization, to achieve perfection, self-cultivation was necessary through activities such as ritual and music, particularly by virtuous rulers who could guide the populace and develop their intrinsic nature towards harmony. At least this was the ideal. The reality of China's dynastic history has been quite different. Many dynasties such as the Ming operated oppressive feudal regimes with severe punishment for transgressors, a political reality that Chinese philosophers often sought to correct with their concern for the need of virtuous rulers.[10] Huang's critique of the political order in the seventeenth century met with official reprobation.[11] The philosophers' recognition of the need for human self-repair through good governance that cultivated virtue among the population was hardly realized in the history of China. In a way, the message of Chinese philosophy has been that if we neglect virtue, especially those who rule over us, human beings fall away from the way

[6] Michael Loewe, *Ways to Paradise: The Chinese Quest for Immortality* (London: George Allen & Unwin, 1979).

[7] Yü, *Chinese History and Culture: Sixth Century bce to Seventeenth Century*, p. 6.

[8] Maspero, *Daoism and Chinese Religion*, pp. 487–544.

[9] François Jullien, *Vital Nourishment: Departing from Happiness*, trans. Arthur Goldhammer (New York: Zone Books, 2007), p. 13.

[10] Albert Chan, *The Glory and Fall of the Ming Dynasty* (Norman, OK: University of Oklahoma Press, 1982); Charles Hucker, *The Ming Dynasty: Its Origins and Evolving Institutions* (Ann Arbor, MI: University of Michigan Papers in Chinese Studies, 1978).

[11] Huang Tsung-hsi's *Waiting for the Dawn, a Plan for the Prince (Ming-I tai-fang lu)*, trans. Wm Theodore de Bary (New York: Columbia University Press, 1993), 'Introduction', pp. x–xii.

of heaven and this is fundamentally a rejection of the life force, for virtue flows from the life force and good governance flows from virtue. Vitalism became an important political theme for some philosophers in the Confucian tradition, while Daoists sought an understanding of life force in terms that went beyond the political order.

Confucianism

As with the other civilizational blocks, from the sixth to fifth centuries BCE there is an Axial shift in China to a new understanding of tradition and its ethical re-assessment with Confucius or Kongzi (551–479 BCE), whose *Analects* (*Lunyu*) formed the intellectual foundation of Chinese thinking for millennia to come.[12] Kongzi thought that humankind had achieved a golden age in the distant past during the Zhou dynasty (*c.*1045–771 BCE) when the rulers King Wen (d. *c.*1050 BCE), his son King Wu (reigned 1045–1043 BCE), and the Duke of Zhou (reigned *c.*1043–1036 BCE) established a strong connection between humankind and heaven (*tian*) through developing a set of ritual practices (*li*) that controlled human conduct and comportment towards others.[13] Kongzi lived in troubled times when China was governed by local rulers and in his perception, humankind was in need of repair to re-establish ritual comportment that is conducive to social harmony and, in particular, filial piety. If the people were to be guided through ritual by wise rulers, then social harmony would be a consequence.[14] Ritual included sacrifice and unlike the Indian renouncer traditions that rejected it, Kongzi sees sacrifice as part of the overall ritual structure that people need to adhere to; but for Kongzi and his tradition, it is not simply the performance of rites but the ethical human nature formed as a consequence that is the goal. The rites are in the service of moral development. Kongzi understood life to be correct comportment towards others—especially filial piety—and to do this humanity must cultivate morality through

[12] For a translation, along with the Chinese text and contextualization of the *Lunyu*, see Roger T. Ames and Henry Rosemont Jr, *The Analects of Confucius: A Philosophical Translation* (New York: Ballantine Books, 1998), although I found the most readable translation to be the classic of Arthur Waley, *The Analects of Confucius* (New York: Vintage Books, 1989 [1938]). See also the translation in the extremely useful Philip J. Ivanhoe and Bryan W. van Norden, *Readings in Classical Chinese Philosophy* (New York: Hackett, 2001), pp. 1–57.

[13] For an account of Confucius' work as a new, ethical transformation see Heiner Roetz, *Confucian Ethics of the Axial Age* (Albany, NY: SUNY Press, 1993); also Bryan W. van Norden (ed.), *Confucius and the Analects: New Essays* (New York: Oxford University Press, 2001). For an overview of the Confucian tradition see Xinzhong Yao, *Introduction to Confucianism* (Cambridge: Cambridge University Press, 2000). For the historical contextualization of Confucianism see the important three-volume work Ge Zhaoguang, *History of Chinese Thought*, 3 vols (Fudan: Fudan University Press, 2001–12).

[14] Kongzi, *Analects* 1.2; 2.3.

ritual and in this way effect the repair necessary for a harmonious life that is in accord with the way of heaven (*tiandao*).

The category of 'life' and the 'life force' (*qi*) does not play a particularly dominant role in the Confucian oeuvre, although he does use the term,[15] but his follower Mengzi or Mencius (*c.*385–305 BCE) in the *Mengzi*, the works of Mencius, uses an organic image of 'sprouts' to depict the cultivation of moral virtue and *qi* plays a more important role in his philosophy. To improve and transform human beings is to develop the sprouts of benevolence, righteousness, propriety, and wisdom that guarantee the transformation of human nature. The purpose of developing these sprouts is the enhancement of natural human proclivities; without them, without compassion for others, disdain for immorality, deference, and the faculty of approval and disapproval, one is not even human.[16] We might say that for Mengzi, to be human is to be able to practise love as the heart of civilization, which is also to be filled with life. Mengzi illustrates the idea of humans as having hearts that are 'not unfeeling towards others' in an example of seeing a child about to fall into a well; without exception people experience a feeling of alarm and distress and prevent the child from falling; they do so not on rational grounds of currying favour with the child's parents, or to seek the praise of neighbours and friends, or from a dislike of a tarnished reputation at having been unmoved by the situation, but rather due to a natural inclination to help: all human beings have a heart/mind 'which cannot bear to see the suffering of others'.[17] We see the suffering child and we move to help, out of a natural compassion. Mengzi recognizes here an innate human pro-sociality endemic to the human case that for him is the foundation of civilization in the sense of correct comportment and behaviour towards others. When civilization breaks down as, for example, when men do not bury their parents but leave them rotting in a ditch,[18] this is a distortion of human natural proclivities towards compassion and love of parents that Mengzi regards as natural and foundational for human community. Furthermore, such compassion is linked to the force of life as a human animating principle that is not simply an enlivening power but also an innate moral force. Our moral actions increase our life energy. Human nature (*renxing*) is naturally good and inclined towards virtue but a bad environment and neglect of self-cultivation lead to wrong actions and a decline of life force.

The discussion about human nature (*xing*) was initiated by the Yang school of philosophy with its founder Yang Zhu, who advocated human self-interest,

[15] Ibid. 16.7.

[16] James Legge, *The Chinese Classics: A Translation, Critical and Exegetical Notes, Prolegomena, and Copius Indexes*, vol. II, *The Works of Mencius* (London: Trübner and Co., 1861), Book II, part I, ch. 6.4–7, pp. 78–9.

[17] Ibid., Book II, part I, ch. 6.3, p. 78.

[18] Ibid., Book III, part I, ch. 5.4, p. 135.

although no texts of his survive,[19] which is taken up by Mengzi, whose positive attitude that human nature is basically good was not widely accepted until the Song dynasty (960–1279 CE) with the Neo-Confucian reorganization of the canon. Before this period Mengzi was, in fact, criticized for having encouraged one of the rulers he recounts to go to war with the small neighbouring state of Yan, as well as for his understanding of human nature. Xunzi (third century BCE) criticized Mengzi for his optimistic view, thinking human nature to be unformed,[20] a view generally accepted by Han dynasty Confucian texts that regarded human emotions to be in a state of balance rather than goodness, and that therefore need to be cultivated. But the point is, human nature articulated through conduct is perfectible given human cognitive and physiological constraints within and through cultural resources provided by Chinese civilization.

According to Mengzi, once basic needs are met, such as food and shelter, then education must be provided for people in order that they can cultivate virtue and increase the life force within them and so aspire to be perfected or superior. Mengzi even has concrete advice for rulers. To be an enlightened ruler is to regulate people's livelihood and ensure that they have enough to eat and their needs are met through, for example, planting mulberry trees by houses so that 'fifty year olds can wear silk', nurturing livestock, and ensuring cultivation that can feed eight mouths in each household. In this way a man will have sufficient resources to serve his parents and support his wife and children.[21] There is a vision offered here of a sound and harmonious society in which hunger and death are avoided through good planning, and education is provided for the cultivation of moral qualities. A civilization is built upon virtue that increases the life force within humanity, which in turn translates into harmony and correct comportment, meeting the social and moral obligations incumbent on a householder. The modern reader is very conscious of what we might call sexism in Mengzi's language, but writing within the horizon of the fourth century BCE, this kind of language might be expected and should not detract from our appreciation of Mengzi's moral vision and understanding of society as being integrated within the natural order. We do not find resistance in this work, or any Confucian classic, but rather the promotion of conformity and cultivation of normativity. Good governance by a virtuous ruler produces social order and human flourishing. Neglect of the natural order through bad management that leads to famine—as was to

[19] Paul Kjellberg, 'Yangism, "Robber Zhi"', p. 369, in Philip J. Ivanhoe and Bryan W. Van Norden (eds.), *Readings in Classical Chinese Philosophy* (Indianapolis, IN: Hackett Publishing, 2001), pp. 369–75.

[20] Kim-Chong Chong, 'Xunzi's Systematic Critique of Mencius', *Philosophy East and West*, vol. 53 (2), 2003, pp. 215–33.

[21] Legge, *The Chinese Classics*, Book I, part 1, ch. 8.21–4, pp. 24–5.

happen in China, incidentally, 2,500 years later during Mao's rule—can be avoided and this will allow the natural, pro-social proclivities of people to create a harmonious society through the intensification of the life force.

Being filled with life force or *qi*, problematically rendered by James Legge as 'passion-nature', is a quality of being human, but this animating force for Mengzi needs to be controlled by will or resolution (*zhi*). In an interesting and important passage, Mengzi's disciple Gongsun Chou asks the master how he maintains an unperturbed mind/heart (*xin*). He answers that the will is the commander and in maintaining a firm will one should not harm the life force: there is a reciprocal relationship between them.[22] 'When it is the will alone which is active, it moves the life force. When it is the life force alone which is active, it moves the will. For instance now, in the case of a man falling or running; that is from the life force, and yet it moves the mind'.[23] The life force animates the body and affects the mind. This life force is vast and flowing within him, says Mengzi, and furthermore it fills the cosmos between heaven and earth and without it humanity would be in a state of starvation.[24] This metaphorical starvation is life of bare animality, without the cultivation of virtue, without the cultivation of proper behaviour, language, and comportment to others, particularly filial piety. Life force sustains humanity within civilization because it is linked to the accumulation of righteous acts. Righteousness, or more specifically the accumulation of moral acts, produces an increase of life force. Thus, moral nature has to be constantly cultivated at its own pace to increase the life force. Mengzi warns us about the man of Sung who complained that his corn was not growing taller, so he pulled it up. He returned home saying to his people that he had been helping the corn to grow tall, but when his son ran to see for himself, he beheld the corn all withered.[25] Virtue that increases the life force has to be cultivated and allowed to grow while being protected and cared for by the will, as a man weeds the corn rather than pulling it out.

For Mengzi there is an organic relationship between the life force within humanity and civilization. Virtues that are the basis of civil society, namely the four sprouts of benevolence, righteousness, propriety, and wisdom, ensure correct behaviour such as Confucian filial piety through the cultivation of the inner life force. The life force increases with the cultivation of virtue (*de*) through an act of will. That is, the heart-mind is in control of the life force that it needs to nurture through moral decision-making, giving people the strength

[22] Ibid., Book II, part 1, ch. 2.9, p. 64.

[23] Ibid., 2.10, p. 65. I have substituted 'life force' for Legge's 'passion-nature' in order to render a consistent translation of *qi*.

[24] Ibid., 2.12–15, pp. 65–6.

[25] Ibid., 2. *Mengzi* heart-mind is commander of *qi* that it nurtures through moral decisions and gives the mind the strength to do what is right. 16, pp. 66–7.

to do what is right. The general human condition according to Mengzi is one in which humanity is innately good but essentially neglectful or even lazy at cultivating virtue.[26] Human self-repair is through establishing a strong connection with a cosmos filled with life force through an act of will applied to the cultivation of virtue. We might say that Mengzi regards human pro-sociality as a natural condition—the life force that fills heaven and earth—to be cultivated and nurtured by good behaviour and the application of virtue to concrete situations of life, especially in paying attention to correct comportment in human relationships. 'There never has been a man trained to benevolence who neglected his parents. There never has been a man trained to righteousness who made his sovereign an after consideration'.[27] We can even read virtue in the other's face, or specifically in their eyes, says Mengzi, for 'the pupil cannot be used to hide a man's wickedness'.[28] Training in virtue is the highest attribute and a characteristic of the perfected or superior man who constantly helps others in the cultivation of their own virtue. But such cultivation needs to be allowed to develop in concrete economic and social conditions that allow its growth: 'the tranquil order of the empire cannot be secured without a benevolent government'.[29] Civilization depends upon benevolent government and benevolence depends upon self-cultivation: the examining of oneself in all situations. If a man loves others but they do not respond, he should turn inwards to examine himself and his own benevolence.[30]

Mengzi has an organic vision of society expressed in Book IV of his text, in which empire, state, and family form an embedded hierarchy with the root of the empire being the state, the root of the state being the family, and the root of the family the person at its head.[31] On this model, everyone should abide by their duty that is particular to themselves, by 'what is near' rather than 'what is remote'. The work of duty is easy, says Mengzi, and harmony would result if each followed it: 'If each man would love his parents and show the due respect to his elders the whole empire would enjoy tranquillity'.[32] By modern standards this is a deeply conservative view of society, but what Mengzi brings to the table is that there is a natural relationship between the life force within the person and within society and the cosmos that can be cultivated through will or resolution that develops virtue. In concrete terms this means education, ritual, and music and the cultivation of correct comportment: that is, the construction of a civilization that is conducive to social harmony and human perfection and leads away from vice, lethargy, and ignorance.

[26] Ibid., Book IV, part 2, ch. 30.2, p. 213. [27] Ibid., Book I, part 1, ch. 1.5, p. 3.
[28] Ibid., Book IV, part 1, ch. 15, p. 182. [29] Ibid, Book IV, part 1, ch. 1.1, p. 164.
[30] Ibid., ch. 4.1, p. 170. [31] Ibid., ch. 5, p. 171. [32] Ibid., ch. 11, p. 178.

Guodian

The emphasis on social harmony that we find in Kongzi and Mengzi and the emphasis on the natural way of things that we find in Laozi were not incompatible. While there was sometimes hostility between the Confucian and Daoist traditions, the two strands became integrated with Neo-Confucianism in the medieval period. But even long before this, at the roots of Chinese civilization we see the integration of the ideals of social harmony and the natural way. In 1993 a tomb was discovered in the village of Guodian in Hubei Province that contained a cache of bamboo texts. The tomb was of a man of the mid to lower nobility, dated to around 300 BCE, who was buried with 731 bamboo strips containing a number of texts including the *Laozi* and Confucian texts. These were a status symbol for the aristocrat to bring to the realm of the ancestors. Buried during the Warring States period, the strips contain material that was probably read by Mengzi[33] and pushes back the date of the compiling of the Confucian oeuvre. Although it does not mention the term 'the six classics' (*liu jing*), the standard works in the Confucian canon, it does mention the list of all six, thereby establishing the date of this classification at the end of the fourth century BCE.[34] It is significant that the *Laozi* appears here. Although it may be the case that the Guodian texts are not a coherent canon of works but somewhat gathered together for the aristocrat to take with him on his journey, the *Laozi* alongside Confucian classics indicates that there is no tension or conflict generally or popularly perceived between these traditions. Along with the *Laozi*, it inherits two lineages of teaching in the Confucian tradition of Mengzi and Kongzi's grandson Zisi (483–402 BCE).[35] Guodian has been compared by some to the discovery of the Dead Sea Scrolls and even allowing for exaggeration, this is a very important find for Chinese archaeology and the history of Chinese philosophy.[36]

These texts bear witness to a firm belief in the transformative power of the ruler's virtue expressed especially through ritual and music. There is a great urgency for people to cultivate virtue in themselves, for which humans have the capacity and mission to do so. This virtue manifested in the human world has its source in heaven. Thus, social order defined in terms of six relationships between ruler and minister, father and son, and husband and wife reflects a cosmic order or natural way of the universe. One of the Confucian texts, *Chengzi*, says:

[33] Scott Cook, *The Bamboo Texts of Guodian: A Study and Complete Translation*, Part 1 (New York: East Asia Program Cornell University, 2012), p. 106.

[34] Ibid., p. 129. The list of the six classics is: *Shi* (Odes), *Shu* (Documents), *Li* (Ritual), *Yue* (Music), *Yi* (Changes), and *Chunqiu* (Springs and Autumns).

[35] Ibid., p. 110.

[36] Ibid., p. 1. There are sixteen texts in all written in the Chu script, grouped into (1) 1 a–c *Laozi*, 2. *Taiyi sheng shui* ('The Great Unity Gives Birth to Water'), 3. *Ziyi* (BlackRobes), etc.

> Heaven sends down great constancy, so as to bring order to human relations. These are instituted as the proprieties of ruler and minister, manifested as the closeness between father and son, and apportioned as the distinction between husband and wife. Thus the petty man wreaks havoc upon Heaven's constancy so as to violate the great way, [whereas] the noble man brings order to human relations so as to accord with Heaven's virtue.[37]

Heaven's virtue needs to be instituted in human life through cultivating the correct kind of relationships between people. One of the texts, the *Xing zi ming chu*, opens by saying that although everyone possesses human nature, people's heart-mind has no fixed inclination but depends upon external practices to become fixed. By human nature the text means the vital energies of joy, anger, grief, and sorrow, affections that are the human natural endowment that have come down from heaven as heaven's mandate (*tianming*).[38] There is also a connection between the mandate of heaven and human nature expressed through history, especially in the idea of a heavenly ordained dynastic succession in the ancient past that followed from the Zhou dynasty's conquest of the Shang.[39] Thus for these early texts the path of humanity is derived from heaven in its inner nature, the innate proclivities to the emotions of joy, anger, grief, and sorrow, and in its external relationships, particularly the royal lineage that ensures social harmony. Inner human nature and the political order are both derived from heaven, but this heavenly mandate needs to be recognized by cultivating virtue (*de*) and that takes effort and vigilance. As Scott Cook observes, 'the virtues of the human world are none other than the virtues of the cosmos itself'.[40]

But there is a fallibility to human nature and a tendency for people to lose their way and stray from the path of heaven. This fallibility in later Chinese philosophy comes to be articulated as whether humanity is inherently good by nature (*xing shan*) or inherently bad (*xing e*) with Mengzi, and perhaps the more general orientation of Chinese thinking, opting for the positive response and Mozi opting for the negative.[41] This distinction is not known to the Guodian texts, but like the earlier Confucian tradition, they clearly articulate the idea that virtue needs to be cultivated to order or direct human nature in the right way, which means to be in accordance with heaven. Thus, we have the cosmic concepts of heaven (*tian*), mandate (*ming*), and way (*dao*) that need to constrain human nature (*xing*), along with the heart-mind (*xin*), affections (*qing*), and the life force (*qi*). With these texts the basis of the Mencian path is in place, especially the idea of inner contemplation (*si*) and seeking virtue within oneself.

[37] Ibid., pp. 141–2, *Cheng zi* strips 31–3. [38] Ibid., *Xing zi ming chu* strips 1–3.
[39] Ibid. [40] Ibid., p. 142.
[41] Mozi is attributed with composing the *Mozi*, which is now thought to be a composite text (fourth to third century BCE). Ian Johnson, *The Mozi: A Complete Translation* (New York: Columbia University Press, 2010).

Seeking virtue within oneself is central to its cultivation: 'seeking it within oneself (*qiu zhu ji*)' becomes a key theme in the Guodian texts, which is to go to the roots of correct speech and cultivates in the noble man the political efficacy that can influence people.[42] Such inwardness and cultivation of virtue through contemplation is associated with sincerity (*cheng*) and its consequence, bringing things to completion (*cheng zi*), that necessitates the integration of the internal and the external, which we might interpret as the cultivation and guiding of human proclivities along with political institutions we inhabit. Human beings need inner contemplation which, in contrast to the Indic meditation we have encountered in Chapter 4, here means reflection on virtue and one's sincerity. Along with this reflective inwardness, people also need education. These two together, contemplation and education, prevent people from going astray from heaven's mandate and in this regard education, moral persuasion, and tradition are key to cultivating a decent society. There is no Gnostic escape to transcendence here, but rather the anticipation of completion in a social form in which human social order reflects the heavenly order and in which human interactions are governed by cosmological principles. This is understood as the transformation of human nature, of the affections, into cordial relationships characteristic of civilization.

A society therefore needs to be controlled in accordance with natural proclivity. Everything has a way according to its nature, thus the ancient King Yu controlled floods not by damming the rivers but by dredging channels to guide the water in accordance with its natural flow. People need to be channelled in a similar way: governance of the people should therefore be based on virtue (*de*) and guided by ritual (*li*) and music (*yue*), which are the channels to direct human nature, in contrast to governance through punishment and penalty (*xing fa*), which simply obstructs and dams human nature, ultimately to no avail.[43] One text reads:

> In leading the people in the [proper] direction, only virtue is capable of this. The outflow of virtue is even swifter than commands transmitted through postal stations ... Now with virtue there are no greater [stimuli] than ritual and music. They channel happiness and harmonize sorrow, and nurture the heart-mind in compassion and integrity, so that one's faithfulness and trust increase daily without any self-awareness. The people can be made to follow a certain course but cannot be made to understand it. The people can be led, but cannot be coerced ... [If one] honours humanity, holds faithfulness dear, respects solemnity, and makes his home in ritual, and carries all this out without exception, then the people cannot be deluded. To go against this would be insane. Punishments to not pertain to the noble man, and ritual does not pertain to the petty man.[44]

[42] Cook, *The Bamboo Texts of Guodian*, pp. 147–9. [43] Ibid., p. 153.

[44] Ibid., pp. 153–4, *Zun deyi*, strips 28–9l, 31a, 21b–23a, 20b–21a, 31b–32a.

Here in the spirit of the Confucian tradition good governance is held to be not through coercion and punishment but through guidance and culture. The nature of humans—the human way (*rendao*), way of the people (*mindao*), or proper way (*dao*)—is distinct in having affections that can be transformed through cultural forms into socially harmony.[45] Humanity (*ren*) and propriety (*yi*) are conducive to the common good. The human way is identified with the proper way or way of the life force (*qi dao*) and we see again the life force associated with transformation and human repair. Through channelling human nature, the life force within, external governance can produce a harmonious society.

The Guodian texts, as with other thinkers in the Confucian tradition, present a vision of society or rather of what society could be. I hesitate to use the word 'utopian' because the tradition retains a realism about what is possible, but nevertheless there is a clear idea that human nature can be positively transformed and channelled in socially harmonious ways. With correct governance by a virtuous ruler channelled through education and creating the conditions for inner contemplation, a human society can honour relationships between ruler and subject, father and son, and husband and wife, cultivated through channelling the affections by ritual propriety and music. While not all Chinese thinkers agreed with this—the *Mozi* is disdainful of music, for example[46]—most seemed to think that social harmony could be brought about through correct, virtuous governance and the people paying attention to their individual sincerity and correct comportment towards others. Such correct comportment was to be guided through ritual, both performance such as sacrifice and the control of everyday encounters, such as manners and verbal forms of address. It is probably no coincidence that the Guodian texts were compiled during the Warring States period, a time of disruption and war, in the hope of living a decent life without fear from political disruption. There is an anticipation of healing or correcting human proclivities and as a river can be directed through channels, so the life force can be directed through ritual, music, and correct behaviour. Civilization as an edifice built upon a fragile human nature has the power to elevate and transform humanity.

The Confucian tradition continued on a long historical trajectory from these early times through to a post-Maoist revival; a legacy optimistic about human nature yet conservative in its assessment of human possibility and the boundaries of human transformation. Indeed, this restorative normativity that sees life as conformity was to be utterly rejected by revolutionary China.

[45] Ibid., p. 154.

[46] *Mozi*, chapter 32, 'A Condemnation of Musical Performances'. See Lothar van Falkenhausen, *Suspended Music: Chime Bells in the Culture of Bronze Age China* (Berkeley, CA: University of California Press, 1993).

Along with the social harmony of Confucianism, Daoism developed a concept of life itself with the idea of the *dao*, the way. Both Confucianism and Daoism are generally world-affirming traditions in contrast to doctrinal and monastic Buddhism that was to become an important voice in the history of China. We will later see how Buddhism became accommodated to China and the fundamentally Gnostic view of life of early Buddhism was transformed. The strong verticality of Indic religions is absent and, in its place, there is a weak verticality with strong horizontal tendencies towards organic interconnection; nature, society, and the political order need to be in harmony: a pattern exemplified by Daoism.

THE DYNAMIC FLOW OF LIFE

As a brief historical sketch, we might say that Daoism really emerged towards the end of the Han dynasty (202 BCE–220 CE). With the Han collapse many messianic movements arose, including Daoist groups, and Laozi is said to have descended from heaven into Zhang Daoling, the Celestial Master (*tianshi*) who inaugurates the Way of the Celestial Masters which became a highly structured religious community. In a complex history, Daoism spread to other parts of China from the West, introducing new revelations in texts such as the Upper Clarity revelations and Numinous Treasure tradition during the six-dynasties period (220–589 BCE), and developing practices of inner contemplation and alchemy. The tradition became unified during the Tang dynasty (618–906 CE) and developed from there into modern times. There was a particularly important flowering of Daoism during the Song dynasty (960–1279 CE); a tradition called Complete Perfection arose in the Yuan dynasty (1260–1368 CE) and likewise continues into the present.[47] This history bears witness to a stunning amount of textual material, and we might identify a number of themes such as a concern with immortality or at least longevity, cultivated through external and internal alchemy, and a concern with salvation, especially for the ancestors who might be suffering in hell. Throughout this history Daoism absorbs elements from Confucianism and Buddhism, including the development of monasticism, and Confucianism absorbs Daoism in Neo-Confucianism. There is a concern with the power that animates living beings and the potential for harnessing this power for human transformation.

[47] For an excellent summary of this history see Ronan, 'Introduction', pp. 1482–95. Ronan makes interesting observations about the reception of Daoism in the West, particularly how the works of James Legge (1815–97) in the nineteenth century presented a critical, 'Protestant' perspective on Daoism as 'religion' in contrast to the rational agnosticism of Confucianism (pp. 1477–8). See Norman J. Girardot, *The Victorian Translation of China: James Legge's Oriental Pilgrimage* (Berkeley, CA: University of California Press, 2003).

The root text of the Daoist tradition is the *Daodejing* (Wade-Giles—*Tao Te Ching*), the way and its virtue, attributed to Laozi, old master, which articulates a view of the cosmos in which all things follow what is in accord with their nature: humanity needs to follow heaven that itself follows the principle underlying life, called the *dao*. I said earlier that Chinese texts were dated in contrast with Sanskrit texts that were not in the earlier period. This is generally true, especially with historiography, but not with the *Daodejing* which, along with other texts of the Daoist canon, have no dates or authors.[48] Such anonymity is in keeping with the spirit of the tradition in which the sages have no names and are virtually invisible, but this means that dating the text is difficult. While Laozi himself is allotted by tradition to the sixth and fifth centuries BCE, the book attributed to him was compiled in the third century BCE.[49] The tradition developing from Laozi is based on an underlying substratum of popular religion that Schipper calls 'shamanism'[50] and that might be contrasted with the official traditions focused on the emperor and the centralized state. The earliest entry of Daoism into the history of China is during the first empire (221 BCE); in the Han dynasty (206 BCE–22 CE) tombs reflect Daoist ideas of attaining immortality and Daoism was introduced into the court, although under Emperor Wu (140–86 BCE) Confucianism became the state religion and Daoism was excluded. With this bifurcation Daoism becomes popular tradition, on the one hand, in contrast to the official Confucianism of the court, on the other. With the establishing of Confucianism in the court and the excluding of Daoism, Schipper writes:

> With this a deep gulf opened which, despite noticeable variations, was to remain constant throughout Chinese history. On the one hand, there was the state and its administration, the official country, claiming the 'Confucian' tradition for its own; on the other was the real country, the local structures being expressed in regional and unofficial forms of religion. It was then that Daoism consciously assumed its own identity and received its present name.[51]

This distinction runs throughout Chinese history and popular religion as the foundation of Daoism continues to emerge even in modern times. The ritual structure of Daoism provided continuity through the centuries and offered an unofficial articulation of an ideal of bonding the social body along with the quest for immortality, although there have been periods of persecution and even the irredeemable destruction of Daoist texts in 1282.[52] With the decline and collapse of the Han dynasty (third century CE), scholars at the imperial

[48] Kristofer Schipper, *The Taoist Body*, trans. Karen C. Duval (Berkeley, CA: University of California Press, 1993), p. 5.

[49] Ibid. [50] Ibid., p. 6. [51] Ibid., p. 9.

[52] Ibid., p. 14. On the history of Daoism see Schipper, *The Taoist Body*, pp. 5–14. Writing in 1982, he laments that to date 'no serious study yet exists on the history of Daoism through the ages' (p. 5). But one interesting study is Ge Zhaogang, *Taoism and Chinese Culture* (Shanghai: Shanghai People's Press, 1987).

court retreated to read and comment on the Daoist classics, the *Daodejing* and the *Zhuangzi*, in particular Wang Bi (226–49 CE), who presented the texts as metaphysics and created a philosophical orientation to their later reception.[53]

The *Daodejing* was compiled some centuries after its attributed author Laozi. While scholars date the text to the years just before the founding of the first imperial dynasty by Ch'in Shi-huang in 221 BCE,[54] according to legend Laozi gave the teachings to Yin Hsi, a guardian of the pass through the mountains, on his journey to the West to leave this world (in 488 BCE, according to tradition). Although the history of its reception has highlighted the mystical nature of the doctrine—the ineffability of the *dao*—it is nevertheless a philosophical text redacted to strip out mythology. It is also a political text in presenting a social vision of a de-centred state in which people live in small communities and in which there is no law: the world of 'the great peace' (*taiping*), a state of 'natural anarchy and inner harmony'.[55] In this perfect world inhabitants of small communities live out their lives in close proximity but people would grow old and die without ever having visited other villages.[56] In this vision of life, there is an organic relationship between the individual body, the social body, and the cosmos that exist in harmony. The subject of the first-person pronoun, the I, is the sage who contemplates the world from the perspective of the *dao* and who acts without acting, behaving in the world with detachment of a kind that reflects the detached action of the *Bhagavad-gītā*. The sage king rules over the country and ensures that people are bereft of desire through satisfying their need for food—he 'weakens their will and strengthens their bones'—which a commentator observes is parallel to the way the sage rules the individual body, restricting the emission of semen to strengthen the bones.[57]

The Life Force in Daoism

What then is the *dao*? The term designates an underlying principle that in itself is indefinable and only visible through its multiple manifestations. It is a principle that regulates the natural order of the universe, but it is not like the Sanskrit term *brahman* as the ontology of the world. Used in both Daoist and

[53] James Robson, 'Introduction', p. 1475, in Jack Miles (general ed.) and Wendy Doniger, Donald S. Lopez, and James Robson (eds.), *The Norton Anthology of World Religions*, vol. 1 (New York and London: W.W. Norton and Company, 2015), pp. 1473–96. Robson gives a fascinating history not only of Daoism, but also of its reception in the West, tracing the history of translation and evaluation from Jesuit missionaries, particularly Ricci, to nineteenth-century scholars such as James Legge. He notes that the publication or reprinting of the Daoist canon in 1926 inspired important scholars such as Henri Maspero and Kristofer Schipper who did so much to open up Daoism to Western scholarship.

[54] Schipper, *Taoist Body*, p. 184.

[55] Ibid., p. 189.

[56] *Daodejing*, chapter 80, cited by Schipper, *The Taoist Body*, p. 189.

[57] Ibid., p. 191.

Confucian thinking, *dao* has two basic meanings, as road, way, or path and as guidance: a road and a map through life.[58] The idea of the *dao* is pervasive in Chinese civilization, the central concept of Daoism, but is also used in Confucianism. In the Confucian tradition, we can distinguish between the way of heaven (*tiandao*) and the way of humans (*rendao*), the latter being cultivated through ethics, politics, and culture. Dao is a unity that can become plural and might turn back to unity again; it is a creative power (*de*) that generates the forms of the universe and itself is generally conceptualized as female in Daoism. All things are nourished by *de*, a term that designates power and virtue. *Dao* is not simply an abstraction but is enacted through ritual in temples sustained by local communities for thousands of years up to the present, although the formulation of the doctrine is found in Laozi's classic.

For the *Daodejing*, oneness (*yi*) is central to understanding the *dao*. It forms heaven, earth, and humanity into a unity and is the source of all things. This oneness generates twoness, twoness generates threeness, and so on manifesting *yin*, *yang*, and life energy (*qi*) to create harmony.[59] This enigmatic text thus wishes to claim that the world is animated by an ineffable reality and its power, that this is not a mere abstraction, being realized in the body itself and within the ideally peaceful community in which harmony prevails. This almost utopian vision is as much social and political as the caricatured idea of the solitary Daoist sage, but a vision that was turned into an explicit ontology by a later commentator, Wang Pi, who identifies the *dao* with non-being (*wu*) that produces the multiplicity of the world.[60] It is, however, not clear that the text presents an explicit philosophy of life. It does recommend a life in harmony with the *dao*, a life of simplicity without intellectualization with an emphasis on the here and now, 'this here', and in this context we might render the *dao* as life itself, the animating force of the living necessary for harmony and repairing the human tendency to act against the nature of things.

The tradition is replete with images of life, seeing the birth of the universe as parallel to pregnancy and the birth of a child, which reflects the central concern with change and transformation. Wang suggests three primary ways in which we can understand the *dao*, as oneness, as spontaneity, and as the female body.[61] This latter image conveys fecundity and the overflowing, spontaneous production of life, its generation (*sheng*) from the ancestor (*zong*) and the mother (*mu*).[62] Thus the *dao* is the unseen force of all existence, comprising male and female power (*yinyang*) that is associated with ideas of being (*you*) and non-being (*wu*). Although we can render the terms *you* and

[58] Robin R. Wang, *Yinyang: The Way of Heaven and Earth in Chinese Thought and Culture* (Cambridge: Cambridge University Press, 2012), pp. 44–5.

[59] Hans-Georg Moeller, trans., *Daodejing: A Complete Translation and Commentary* (Chicago and La Salle, IL: Open Court, 2007), p. 103.

[60] Schipper, *The Taoist Body*, p. 193. [61] Wang, *Yinyang*, p. 47. [62] Ibid., p. 55.

wu as being and non-being, they more literally mean 'to have' and 'to lack'.[63] The origin of life is in the female emptiness of the *dao*.

Along with *dao* as indicating the source of the living, as life itself, an important related concept is *qi*. *Qi* means breath, but it also indicates life force or life principle. Wang gives a good summary of its semantic range: 'As a shared notion underlying all schools, *qi* is believed to be dynamic, all present, all penetrating, and all-transforming force animating every existence in the universe'.[64] As such it is close to the semantic range of the Sanskrit *prāṇa* and the Greek *pneuma*, although it is articulated in qualitatively different forms such as clear and turbid, dark and bright, the *yin* and the *yang*. *Qi* is manifested in the body as breath and as a subtle force that flows through the central channel (*du*) of the body from the base of the spine (not dissimilar to the Indic view of the body's central channel), and there is a resonance between breathing in and out and the rhythms of nature. Jullien presents the concept succinctly:

> And what is respiration but a continual invitation not to dwell in either of two opposite positions—inhalation or exhalation? Respiration instead allows each to call upon the other in order to renew itself through it, thus establishing the great rhythm of the world's evolution, never absent from the Chinese mind: the alternation of day and night and the succession of the seasons. Thus respiration is not only the symbol, the image or figure, but also the vector of vital nourishment.[65]

Qi is an idea in all traditions whose roots go back to ancient China, being found as a character on the oracle bones of the Shang dynasty (1766–1122 BCE) where it is used as a verb and adjective rather than a noun. Wang observes that the Chinese character, part of which resembles the number three, might derive from observing lines of steam rising from morning dew or from steam arising from cooking rice, as one early text, the *Shuowen Jiezi*, says.[66] In his translation of the *Mengzi*, James Legge makes an observation about the character *qi*: 'Originally it was the same as "cloudy vapour". With the addition of *mi*, rice or *huo*, fire, which was an old form, it should indicate "steam of rice", or "steam" generally'.[67]

Qi is a philosophical idea but is also present in popular religion as a substance that can be increased by acts of reverential devotion such as temple pilgrimage,[68] and so is an active concept in everyday religious life. In a different character, this *qi* with the same sound denotes prayer or alms seeking.[69] The rich semantic range of the concept means that it has been widely used in the

[63] Ibid., p. 56. [64] Ibid., p. 59. [65] Jullien, *Vital Nourishment*, p. 31. [66] Ibid.

[67] Legge, *The Works of Mencius*, p. 64, note 8.

[68] Hsun Chang, *Incense-Offering and Obtaining the Magical Power of Qi: The Matzu (Heavenly Mother) Pilgrimage in Taiwan*, PhD (Berkeley, CA: University of California, 1993). DAI-A 55/07.

[69] Wang, *Yinyang*, p. 59.

Chinese traditions, indicating abstract life force but also having degrees of intensification with tangible results in the body: as the body ages, *qi* wanes until death when it disappears from that particular body.[70] This all-pervading life force was classified into different types (*lei*) such as *yang*, *yin*, wind, rain, dark, and bright in the *Zuozhuan*, the Chronicle of Zuo (fifth century BCE), and becomes identified with *yingyang*, which become two dimensions of *qi*.[71] A work of the Han dynasty, the *Huainanzi* (200 BCE), gives the earliest account of the formation of the universe in which *qi* is its source, bifurcating into *yang*, a diffuse force, and *yin*, a concentrated force. The text reads:

> When heaven and earth were not yet formed, all was ascending and flying, diving and delving. Thus it was called ultimate manifestation (*taizhao*). The Dao began in the nebulous void. The nebulous void generates time-space (*yuzhou*); time-space generates *qi*. *Qi* moves within the border. The light and bright *qi* spreads and ascends to form *tian* (heaven) and the heavy and turbid *qi* congeals and descends to form *di* (earth).[72]

Qi is of central importance in the Chinese conceptual landscape, and landscape is an appropriate metaphor because *qi* is described in two modes of rising up through the spring and culminating in summer, the *yang-qi*, with the *yin-qi* descending at the commencement of autumn and culminating in winter,[73] the sunny side of the hill and the dark side.[74] The cyclic energy of the seasons captures the dynamic movement inherent in the concept. It is the life force that animates the living and causes the constant change from hot energy of *yang* to cool energy of *yin*, from light to dark, and from living to dying. These forces in balance regulate the flow of life expressed in the changing seasons and in the heavenly, earthly, and human realms. An elaborate system of connections articulating the balance of *qi* is developed, linked to five phases or elements of wood, fire, earth, metal, and water, further associated with the directions and the planets.[75] *Qi* is the force that sustains the balance of these properties of reality. Even though the term *qi* may not appear frequently in early sources, it has been taken as the fundamental, distinguishing feature of a Chinese worldview: the basic presupposition and equivalence to a Western atomic conception of the world.[76] Even in literary

[70] On *qi* in the history of Chinese medicine see Paul U. Unschuld, *Medicine in China: A History of Ideas* (Berkeley, CA: University of California Press, 1985), pp. 67–83.

[71] Ibid., p. 61.

[72] H. Roth, trans., *The Huainanzi: A Guide to the Theory and Practice of Government in Early China* (New York: Columbia University Press, 2010), p. 143, cited by Wang with modified translation, *Yinyang*, p. 61.

[73] Wang, *Yingyang*, p. 62.

[74] Robson, 'Introduction', p. 1483.

[75] Ibid.

[76] Roger T. Ames and David L. Hall, *Focussing on the Familiar: A Translation and Philosophical Interpretation of the Zhongyong* (Honolulu, HI: University of Hawaii Press, 2001), p. 24; Chung-ying Cheng, '*Li* and *Ch'i* in the *I Ching*: A Reconsideration of Being and Non-Being in Chinese Philosophy', *Journal of Chinese Philosophy*, vol. 14, 1987, pp. 1–38.

theory *qi* is an important concept, being the force that allows the selection of particular words and treating a literary text itself as a living organism.[77] Daoist ideas of *qi* also pervade Daoist sexual practices—that have some affinity with extreme tantric practice—intended less to experience pleasure and more to develop sexual control in the service of longevity and the preservation of sexual fluids understood as the physical analogue of subtle energy. The Chinese sex manuals—that incidentally are the earliest ancient texts to treat women and men equally in this regard—are concerned with pleasure but also with cultivating a healthy and long life.[78] The life force expressed in health and sex is held to be a precious substance contained within the body in a physical manifestation.

Qi is the force responsible for making up a human being, ably summarized by Schipper. It takes different forms or 'souls' that account for the body and its patterns of life. *Shen* or 'spirit' is the purest, heavenly *qi* in the body, also called the 'cloud soul' (*hun*), in contrast to the earthly *qi* called *jing*, which designates bone marrow, sperm, or menstrual blood.[79] *Shen* resides in the heart, *jing* in the kidneys. There is also the *po*, the bone soul of the skeleton, comprising six souls that are in opposition to *shen*. There is then a tension in the body between the bone souls, regarded as demons (*gui*), which try to return the skeleton to earth and the heavenly soul that tries to control them. For long life we need to cultivate *jing*. The gods and demons in the body are analogues to the gods and demons in the cosmos:[80] the microcosm reflects the macrocosm. Thus, *qi* is the essence of the person and corresponds to the cosmos as a manifestation of the same fundamental force. The life force relates to the living as their essence and more than that as the principle that moves the external cosmos as well. This has practical consequences for Daoists (and Buddhists) in that if a person dies through suicide or accidental death, the heavenly soul cannot become an ancestor but is condemned to wander as an orphaned soul (*gu hun*).[81] These souls are the subject of great ritual attention to appease them with sacrificial paper money and with prayers. Schipper notes that on such ritual occasions Buddhist priests join with their Daoist counterparts to offer prayers to reintegrate these orphans back into the cycle of birth and death.[82] The offerings of money only become real upon being burned: destruction ensures its presence in a more intense life beyond this one.

There is also another word, with a distinct character, pronounced *qi* that means 'vessel'. In the commentarial tradition associated with the ancient

[77] David Pollard, '*Ch'i* in Chinese Literary Theory', in Adele Austin Rickett (ed.), *Chinese Approaches to Literature from Confucius to Liang Chi-chao* (Princeton, NJ: Princeton University Press, 1978), pp. 43–66.

[78] Schipper, *Taoist Body*, p. 126. [79] Ibid., p. 36. [80] Ibid. [81] Ibid., p. 37.

[82] Ibid., p. 38; Terry F. Kleeman, *Celestial Masters: History and Ritual in Early Daoist Communities* (Cambridge, MA: Harvard University Press, 2016).

Book of Changes (*Yijing*),[83] this term designates not only vessels (from water-holding bowls to sacrificial bowls) but becomes a general term for concrete things. The *Xici* commentary says that what is above tangible forms (*xing er shang*) is called *dao* and what is below tangible forms (*xing er xia*) is called 'vessel'.[84] The concrete reality of the world embodies the invisible power of the *dao*. Change is a constant in reality, a perpetual movement from one state to another governed by the principle of the *yinyang*, a term that captures the dynamism of the world of change. We have then a network of closely related concepts. At a higher level we have the abstract *dao* in its mode as life force or *qi* that contains the principle of polarity, the *yinyang*, and manifests the world of concrete reality comprising the unending flow of events. In this worldview what in the West came to be classified as 'dead' matter or what in India was classified as unconscious matter (*jaḍa*) is here part of the dynamic flow of life. The living reality of world is just that: a living reality as a manifestation of the underlying principle that contains an incipient dynamism controlling events into their outcome. This process is captured in the term *yi*, change and constancy; the dynamic process that the *Book of Changes* seeks to capture in its series of sixty-four hexagrams comprising the interpenetration of the two principles of *yin*, softness, and *yang*, hardness.

The *Yijing* and the *Daodejing* provide the textual foundations for Chinese metaphysics and speculation about the underlying principle of reality and its dynamic force that forms the world into its present outcomes. In particular we have the interaction between the passive and active forces inherent in the life force itself. One of the earliest and famous philosophers to reflect on this dynamism was Zhuangzi (late fourth century BCE), whose eponymous text was compiled around 300 CE. Often considered a Daoist, Zhuangzi was in fact active before the formation of the Daoist schools, although he is regarded as a sage within the Daoist tradition.[85] He thinks about *qi* in terms of emptiness and emptiness is the mind of the perfected person that reflects events 'not welcoming things as they come nor escorting them as they go',[86] thus achieving an equanimity and sage-like attitude to life and death. The wise person needs to maintain equanimity between life and death, which are really only

[83] This text, famous in the West through Wilhelm's translation, came to formation over a 700/800-year period from the early formation of the hexagrams in the Western Zhou period (1045–771 BCE) to the commentaries on the text after the Warring States period (403–221 BCE) (Wang, *Yinyang*, p. 64).

[84] Wang, *Yingyang*, p. 66. Wang notes that the term *xingershang* was used by Yan Fu who translated Aristotle into Chinese as 'metaphysics'.

[85] Burton Watson, *The Complete Works of Chuang Tzu* (New York: Columbia University Press, 1968).

[86] *Zhunagzi* ch. 27, p. 243, in Ivanhoe and Van Norden (eds.), *Readings in Classical Chinese Philosophy*.

two aspects of the same process of universal change. At the heart of this change is a force, but a force that is essentially empty.

Qi is the life force animating the living and as such it gives an account of the coming into existence and destruction of 'the myriad things' in its gathering and scattering. But in an interesting article Tu Youguang raises the question about what differentiates the myriad forms that are due to the gathering of *qi*: 'how is it that some come into existence as belonging to the class "tree," and others to the class "human being"?'[87] This is a question about the specification of constraint: what is it that makes something what it is? The philosophy of life force (*qi*) can account for the existence of something as a manifestation of the force but cannot account for its particularity. We need also the idea of principle (*li*) that orders the gathering of *qi* into particular classes. Furthermore, we need to account for the particularity of each example, as one pine tree is distinct from another. In the Daoist example, how are we to account not only for the class 'female aristocrat' as a sub-class of 'human being', but for Black Jade (Lin Daiyu) being different from Precious Clasp (Xue Baochai)?[88] According to Fung, the Daoist response seems to be that individuality is characterized or particularized by virtue.[89] What differentiates the two female aristocrats is degree of virtue. The great scholar of Chinese philosophy, Fung, problematizes the issue but it seems that Daoism needs an account of what differentiates forms animated by *qi*, which it does not adequately provide. The same issue comes up in Latin medium theology with Duns Scotus, as we will see, whose solution, namely univocity, is not made by the Daoist philosophers, other than to say that *qi* is the life force that animates beings into their particularity, and once particularized there can be an ordering into classes.

Outer and Inner Alchemy

Uniquely intense in the history of civilizations, China fostered a deep interest in the practice of long life (*xiu-yang*) and the idea of immortality through an external alchemy that produced elixirs to lengthen life and a corresponding inner alchemy to cultivate meditation for the same result.[90] Alchemy

[87] Tu Youguang, 'Daoism Stresses Individual Objects', p. 49, in *Contemporary Chinese Thought*, vol. 30 (1), 1998, pp. 45–57.

[88] Ibid., p. 49.

[89] Fung Yu-lan, *History of Chinese Philosophy*, vol. 2, *The Period of Classical Learning from the Second Century BC to the Twentieth Century AD*, trans. Derek Bodde (Princeton, NJ: Princeton University Press, 1951–2), pp. 231–6.

[90] Fabrizio Pregasio, 'The Alchemical Body in Daoism', in Manuel Vasquez and Vasudha Narayana (eds.), *The Wiley-Blackwell Companion to Material Religion* (Oxford: Wiley-Blackwell, 2019); 'Which Is the Daoist Immortal Body?' *Micrologus*, vol. 26, 2018, pp. 385–407; I. Robinet, *Taoist Meditation: The Mao-shan Tradition of Great Purity* (Albany, NY: SUNY Press, 1993 [1979]).

originated in China, in experimenting with and producing substances for the prolongation of life.[91] Ironically, many emperors died prematurely as a result of imbibing such elixirs. Both the inner (*neidan*) and outer alchemy (*waidan*) tried to cultivate the animating principle of life that with the inner alchemy is conflated with the soteriological goal of becoming one with the *dao*. A body of literature develops through Daoism's long history and the accompanying institution of monasticism, modelled on Buddhist monasticism, cultivates the inner goal of the *dao*. Along with a focus on immortality and inner practices for Daoist enlightenment, there are rituals for the dead and more than simply appeasing the ancestors, raising them to heaven.

The *Daodejing*, the *Yijing*, and the *Zhuangzi* form the textual basis for the unfolding Daoist tradition through the centuries and the formation of the Daoist canon that includes texts such as the *Huainanzi*, the book of Master Huainan (second century BCE) that stresses the potency of the *dao* as the foundation of all things, the *Inward Training* (*Neiye*) of the fourth century BCE that focuses on the creative force of the *dao* in self-cultivation practices such as contemplative breathing, and the important *The Seal of the Unity of the Three in Accordance with the Book of Changes* (*Zhouyi cantong qi*), a book steeped in symbolism and metaphor that attracted a long commentarial tradition. A composite text, it was composed between the Han and Tang dynasties, commented on by the famous poets Li Bo (701–62 CE) and Bo Juyi (772–846 CE), and contains material on external and internal alchemical traditions;[92] the distinctly Chinese concern with long life and immortality. The *Daodejing* attracts over 700 commentaries in its history from the third century BCE[93] and Daoism grows into a flourishing state religion in the Tang, by which time Laozi had been divinized and identified as the people's saviour[94] along with the growth of a panoply of other deities. For example, in the *Wondrous Scripture of the Upper Chapters of Limitless Salvation* (*Lingbao wuliang duren shangpin miaojing*), Laozi appears as a god along with a pantheon in a pearl suspended in the air from which the scripture emanates. Its first recitation cures deafness, its second blindness and so on, up to the tenth recitation that revives the dead and gives them long life.[95] This tradition continues with purification rituals for the universal salvation (*pudu*) of the dead as we find in the *Purification Rite of Luminous Perfected* (*Mingzhen zhai*).[96]

External alchemy's concern with long life comes to be transformed with inner alchemy to produce the elixir of immortality within the body, called the immortal embryo (*shentai*), and union with the *dao*. There may be the influence of Buddhism on the tradition as it too developed the idea of the inner embryo or womb of enlightenment, as we saw in Chapter 4. A number of texts bear witness to the cultivation of bodily meditations and visualization

[91] Schipper, *The Taoist Body*, p. 174. [92] Robson, pp. 1556, 1561–4.
[93] Ibid., p. 1570. [94] Ibid., p. 1609. [95] Ibid., p. 1734. [96] Ibid., p. 1742.

techniques that bring cosmological forces into the body and circulate those energies within it in order to actualize a transformation of person. This kind of material can be found as early as the fifth century CE with the evocation of the Jade Woman of Great Mystery who comes down to lie with the practitioner after ingesting liquid emissions from the Goddess and from the sun and the moon.[97]

One of the most important texts of inner alchemy is the *Awakening to Reality* (*Wuzhen pian*), composed by Zhang Boduan (987?–1082 CE), the first patriarch of the southern lineage of internal alchemy.[98] This collection of eighty-one poems depicts meditation practices, stressing the importance of cultivating inner vision to gain the goal of awakening, an idea that reflects Buddhist influence. The golden elixir is found within the self, 'the wondrous Reality within Reality, where I depend on myself alone'.[99] Replete with the symbolic language of alchemy, the text speaks of all people having the medicine of long life within them, the elixir that can be cultivated and grows and once ripened 'Gold fills the room'; that is, the body is filled with the golden light of awakening and so then 'what is the point of seeking herbs and learning how to roast the reeds?'[100] The inner alchemy takes over the external alchemy, balancing *yin* and *yang* within the body and cultivating the inner lead, not the physical lead of external alchemy. Cultivating inner practices leads to a vision of 'the bright pearl as round as the moon', and apprehending within the self the three Daoist heavens of the Great Clarity (Taiqing), the Upper Clarity (Shangquing), and the Jade Clarity (Yuqing).[101]

By the medieval period, the Song dynasty (960–1279), the quest for the inner elixir had become well established and texts from this time and the later Yuan (1260–1368) make up half the Daoist canon that was put together during the Ming in 1445.[102] Examples of texts concerned with the inner alchemy of transformation are preserved in the fifteenth-century collection, the *Corpus of Taoist Ritual* (*Daofa huiyaun*) that contains 'Song of the Dark Pearl' composed by Wang Wenquing (1093–1153) along with a commentary. This text comprises a fusion of a magical tradition of gaining power from thunder, the Thunder Ritual, and the tradition of inner alchemy in order to produce the perfected person or infant within the body.

Having established the connection between the planets and the orifices of the body, the planets governing over the body, the text develops the idea of the inner person as an infant or perfected person within. In this text the idea of *qi* plays a crucial role, depicting it as a substance that can be increased or diminished. For example, being over-loquacious will reduce *qi* in the body so the practitioner must focus his *qi* within him through reserve and self-containment. This involves the meditation practice of locking the breath and

[97] Ibid., pp. 1800–1. [98] Ibid., p. 1896. [99] Poem 4, ibid., p. 1898.
[100] Poem 6, ibid., p. 1899. [101] Poem 16, ibid., p. 1902. [102] Ibid., p. 1891.

cultivating inner concentration. The Pearl reads 'so lock your breath and gaze within' with the commentary:

> To 'lock' your breath means that nothing can enter from without, and nothing can exit from within. 'Gaze within' means to not have one single strain of distracted thought. Your eyes look at the Muddy Pellet, your tongue is pointed towards the palate, spirits and qi come and go of their own accord. When you 'lock your breath' your qi will gather; when you 'gaze within' your spirits will concentrate.[103]

This is reminiscent of yoga techniques in India. Breath control facilitates inner concentration without distraction from which arises an inner vision, here identified as the 'muddy pellet', a term for the 'dark pearl' or pearl of enlightenment hidden within the self, the central 'palace' of nine within the head. This is the place where 'myriad spirits' convene and where the Celestial Emperor dwells, realized through cultivating *qi* in an upward movement from the base of the spine, again in a method reminiscent of tantric yoga; *qi* needs to be circulated through the body. This enigmatic text also speaks of the source of *qi* being the Mysterious Female, which is the root of the universe and foundation of one's life. Furthermore, the Mysterious Female is identified with the ancestral *qi* as well as with an 'aperture'. Through closing off 'five *qi*', which we might take as the energies of the five senses, the ancestral *qi* can be cultivated.[104] A complexity of identifications occurs here: the ancestral *qi* that is the root of life is identified with the Mysterious Female, which in turn is identified with the 'aperture', which we might interpret as the inner vision awakened with the stilling or closing off of the outer senses. In other words, a culture-specific description of a meditation process cultivating inwardness.

One last text is worth mentioning in this tradition of inner alchemy: the seventeenth-century *Secret of the Golden Flower* (*Taiyi jinhua zongzhi*) made famous in the West through Richard Wilhelm's 1929 translation and Carl Jung's introduction. The golden flower is a synonym for the golden elixir and inner pearl of the inner alchemy tradition that we have already encountered. The aim is to reverse the outward-flowing life energies, to reverse the process of external distraction, and to return the spirit to the primordial *dao*. This method of reversal or backward-flowing life force is through a technique of circulating light within the body for '(w)hen the light is made to move in a circle, all the energies of heaven and earth, of the light and the dark, are crystallised'. Usually life energy flows downward, says the text, but through this meditation technique it can be reversed and made to flow upward through the body; by closing the outer eyes and opening the inner eye: 'When both eyes are looking at things of the world it is with vision directed outward. Now if one closes the eyes and, reversing the glance, directs it inward and looks at the

[103] Ibid., p. 1924. [104] Ibid., pp. 1925–6.

room of the ancestors, that is the backward flowing method'.[105] The text goes on to refine the method in terms of half-open eyes concentrating on the tip of the nose, which it derives from the teachings of Laozi and the Buddha. The person contains two energies, the feminine *yin* energy identified with the white soul (*po*), linked to the earth, and the masculine *yang* energy identified with cloud soul (*hun*), linked to heaven and heavenly *qi*. There is a conflict between these and if the *yin* energy wins out, at death the white soul flows into the earth and becomes a ghost, but if the yang energy triumphs through the backward-flowing method of this meditation, then the cloud soul becomes a spirit or god (*shen*) and the meditator reaches the highest condition of the golden flower.

If in reality most inner alchemists were men, the practice was intended to be gender neutral, although later during the seventeenth and eighteenth centuries, a tradition of women's inner alchemy emerges, focused exclusively on women and concerned with reversing the flow of menstrual blood, 'cutting the red dragon', regarded as the seat of women's vital energy (*qi*), in a way that parallels the reversal of the flow of semen in men.[106] Some twentieth-century literature—written by men—described precise physical meditation techniques and the method may have gained in popularity with the post-Mao tolerance of ancient religions and the re-establishing of Daoist female monasticism.

The Neo-Confucian Synthesis

The tradition of inner alchemy explicitly thematizes the life force. Chinese alchemy's concern with *qi* is not an obscure interest but is part of the central religious and philosophical tradition throughout China's long history. The Daoist concern with *qi* as part of the way (*dao*) of the flow of life is complemented in Confucianism with a focus on comportment and how to behave, with ethics as the expression of correct social mores. The focus of Confucianism was less on life force (*qi*) and more on correct behaviour and moral uprightness (*ren*), and so with the centralized regulation of human affairs. There needs to be a reciprocal system of social relations in which good actions lead to harmony in both society and nature. The principle of heaven (*li*) is expressed in rites, rules for human conduct (also *li* but a different character), and etiquette. The superior man needs to act in accordance with the principle of heaven (*li*) and humanity (*jen*). Mengzi had claimed that all people are basically good by nature although humans need to be cultivated to live in accordance with the way of things (*dao*). Confucianism thus had a political concern in a way that Daoism did not, a concern reflected in Confucianism's general support of the emperor and sometimes the relegation of Daoism to the

105 Ibid., p. 2043.

106 *Cutting the Red Dragon* (*Duan Honglong*) by Fu Shan (1606–84); ibid., p. 2047.

margins of history. The focus on ethics was about fulfilment of human nature, not a concern with cultivating the inner experiences of alchemy.

But with the Song dynasty (960–1279) during which Daoism flourished, there was a corresponding new impetus to Confucianism and what has become known as Neo-Confucianism.[107] This retains the traditional emphasis on correct behaviour and the ontological basis of morality but is influenced by the life-force tradition of Daoism. Neo-Confucianism saw desire (*yu*) as the reason for the failure of humanity to understand its deeper harmony with heaven and earth, but whereas the Buddhist response was to retreat from the world, the Neo-Confucian response is not to withdraw from political and social responsibility but to move into the world and control desire through learning.[108] Desire is a response to the world that takes place at the level of *qi* but *qi* needs to be guided. With Neo-Confucianism we have *qi* represented not simply as life force but transformed into ethical ideal. Neo-Confucianism, on the one hand, emphasized good conduct, order, or principle embodied in the idea of *li*, while on the other there was an emphasis on *qi*, life force as the animating principle of life. In due course, the *qi* tradition came to be integrated within the *li* tradition (life itself comes to be seen as part of moral order), but it is significant that Neo-Confucianism recognized the centrality of this idea.[109]

Zhu Xi (1130–1200) is largely responsible for the synthesis of ideas that we call Neo-Confucianism. Developing the thought of Cheng Yi, one of two important brothers, he highlighted the idea of *li*, principle, order, or pattern, to be the guiding force of life.[110] Zhu Xi looked to Zhou Dunyi (1017–78) who he claimed to be the greatest sage since Mengzi, who composed a famous text, *The Explanation of the Diagram of the Great Ultimate* (*Taijitu shuo*). This controversial book, controversial because of its reliance on Daoism, brought together ideas of the life force with Confucian concern about right conduct. The diagram comprising nine circles hierarchically arranged can be read from the top down to describe the manifestation of the universe, as the Confucians so read it, or bottom up to describe the reversal of energy back to the *dao*, as the Daoists read it. Beginning with the Daoist heaven, the Great Ultimate (*taiji*) also translated as the Supreme Polarity and also identified with the Great Void (*wuji*), the text says that from here *yang* activity flows which generates *yin*, stillness. These two alternate to produce the elements water, fire, metal, wood, and earth, which are phases of *qi*, the life force that produces the

[107] For a historical survey see Peter K. Bol, *Neo-Confucianism in History* (Cambridge, MA: Harvard University Press, 2008).

[108] Ibid., pp. 170–1.

[109] On Neo-Confucian thinking see Wm Theodore de Bary, *The Message of the Mind in Neo-Confucianism* (New York: Columbia University Press, 1989); de Bary, *Neo-Confucian Orthodoxy and the Learning of the Mind-and-Heart* (New York: Columbia University Press, 1981).

[110] John Berthrong, 'Transmitting the Dao: Chinese Confucianism', in Wonsuk Chan and Leah Kalmanson (eds.), *Confucianism in Context* (Albany, NY: SUNY Press, 2010), pp. 9–13.

seasons. In the realm of becoming and transformation, the elements are controlled by *yin* and *yang*.[111] In turn human beings are produced through the same combination, comprising the four Confucian virtues of righteousness, prosperity, knowledge, and integrity. Thus, from the interplay of *yin* and *yang* the generation of myriad things is produced.[112] The Great Ultimate is the *dao* and the diagram depicts its flow through manifestation. In this worldview there is no metaphysics independent of cosmology and the whole diagram represents the cosmos as governed by stillness and activity, with all change from one to the other resting within *qi*; 'Even though the changes and transformations constantly replace each other, their *qi* originally is one'.[113]

The privileging of *qi* as an ultimate reality that animates all life is further developed by Zhou Dunyi's contemporary Zhang Zai (1020–77). For Zhang Zai *qi* becomes a new way of understanding the world of life, which is subtle to explain yet without which there could be no life. He identifies this life force with the Void for *qi* is 'something and nothing', in its essence ungraspable, and yet he dissociates this idea from anything in Buddhism. Indeed, he is vehemently opposed to Buddhism and sees it as compromising the ethical teachings of Confucianism and being a negative influence on Confucian scholars. What distinguishes Zhang Zai from his Buddhist contemporaries is his insistence that the ultimate reality, identified with the Great Void (*taixu*) in Buddhism, contains *qi*.[114] In contrast to Buddhism, *taixu* has the quality of *qi* and is both its dispersal and condensation. *Qi* produces the myriad things in condensation within *taixu*, but when it dissipates the myriad things dissipate too. Condensing and dispersing (*jusan*), bending and expanding (*qushen*), and motion and rest (*dongjing*) are properties of *yinyang* which itself is a property of *qi*.[115] The condensing and dispersing of *qi* controls the appearance and disappearance of forms in the universe. There is one principle, the principle of life itself, with multiple manifestations. Zai uses an analogy with water to express this:

> The condensing and dissolving of qi from the vast emptiness is like the freezing and melting of ice from water. If it is realized that the vast emptiness is qi, one realizes that there is no nothingness.[116]

[111] Reproduced in Wang, *Yinyang*, p. 75. Also see Robin Wang, 'Zhou Dunyi's Diagram of the Supreme Ulitimate Explained (*Taiji shuo*): A Construction of the Confucian Metaphysics', *Journal of the History of Ideas*, vol. 66 (3), 2005, pp. 307–23; Joseph A. Adler, *Reconstructing the Confucian Dao: Zhu Xi's Appropriation of Zhou Dunyi* (Albany, NY: SUNY Press, 2014).

[112] Ibid., pp. 1905–6. [113] Wang, *Yinyang*, p. 79.

[114] Zhang Zai, *The Complete Collection of Zhang Zai's Work* (Beijing: Chinese Press, 1976), p. 8, cited in Wang, *Yinyang*, p. 79. See Jung-Jeup Kim, *Zhang Zai's Philosophy of Qi: A Practical Understanding* (New York: Lexington Books, 2015).

[115] Wang, *Yinyang*, p. 80.

[116] Zhang Zai, *Zhengmeng*, p. 8, quoted in Jung-Yeup Kim, *Zhang Zai's Philosophy of Qi* (Lanham, MD, Boulder, CO, New York, London: Lexington Books, 2015), p. 32. References to the *Zhengmeng* are from Kim's study, who uses the standard Chinese edition of Zhang Zai's work published in 1978.

This is to elevate *qi* to a very high position indeed: life itself as the essence of the universe that produces the myriad of appearances, both living and apparently non-living. The vast emptiness gathers as *qi*, which becomes the myriad things, which in turn disperse as *qi* and *qi* back into the vast emptiness.[117] There is controversy over precisely what Zhang Zai means, which has been open to various interpretations. On the one hand, he is presented as a substance monist;[118] on the other, as an organic pluralist.[119] The substance monist view has been compatible with a materialist reading of Zhang Zai in keeping with Maoism, although this is arguably to project back a contemporary concern. On the other hand, Kim's organic pluralist view might be at the cost of appreciating Zhang's commitment to a unified field of being that underlies innumerable forms of the cosmos.

But what is unique to Zhang Zai is the way he brings *qi* into the realm of ethical meaning, thereby synthesizing Daoist vitalism with Confucian ethics. Morality is simply another manifestation of *qi*. We see this in his last work, *Rectifying the Ignorant* (*Zhengmeng*), in which he responds to Daoism and Buddhism in the assertion of a Confucian perspective, critiquing Buddhism in particular for being antithetical to the world.[120] Rather than the practice of trying to go beyond the physical world in meditation, we should cultivate ritual that connects us with others; in Kim's words, Zai promotes 'a strictly embodied and shared practice that disposes us to engage with human society, as a part of the phenomenal world of diversity, in terms of creating integrative wholes among ourselves, through forming fitting relationships, which produce affective vitality'.[121] Ritual propriety (*li*) is important as it allows us to go beyond our personal concerns, to develop (*da*) our heart-mind, and to connect with others in a vital harmony. Human beings need to empty the heart-mind and to resonate with the ambient cosmos. He writes: 'The relation between the transformation of bodily dispositions and emptying one's heart-mind... is that of the reciprocity of the outer and inner'.[122] Through traditional Confucian ritual we connect and resonate with others and with nature; what is within comes into harmony with what is without. Thus, Zai finds a vitalism in human ritual action that produces resonances within society and with nature itself. Ritual is the expression of a natural capacity or inner nature (*xing*) within human beings and connects us together and with the wider world. The body expresses *qi* and intensifies *qi* through somatic exploration that is antithetical to Buddhist asceticism.

[117] Wingsit Chan, *A Sourcebook in Chinese Philosophy* (Princeton, NJ: Princeton University Press, 1963), pp. 495–504.

[118] Junyi Tang, 'Chang Tsai's Theory of Mind and its Metaphysical Basis', *Philosophy East and West*, vol. 6, 1956, pp. 113–36.

[119] Kim, *Zhang Zai's Philosophy of Qi*, p. 2.

[120] Ibid., pp. 20–3. References to the *Zhengmeng* are from Kim's study, who uses the standard Chinese edition of Zhang Zai's work published in 1978.

[121] Ibid., p. 27.

[122] Zai, *Zhengmeng*, p. 274, quoted ibid., p. 56.

The affective vitality produced through ritual propriety creates a resonance between the practitioner and wider community that expresses *qi*. The family is of key importance here as it supports and sustains wider social harmony and the processes of the state. But Zai has a wider conception of family and wishes to extend benevolence to a community that embraces nature as well as the wider society. In a striking passage he writes:

> The sky is my father and the earth is my mother. I minutely exist intermingled in their midst. Thus, that which fills up nature I regard as my body, and that which directs nature I consider as my capacity to resonate. All people are my brothers and sisters, and all things are my companions. Those who are exhausted, feebled, crippled, or sick, those who have no brothers or children, wives or husbands, are all my brothers and sisters who are in misery and have no one to turn to... To sustain the heart-mind and cultivate the capacity to resonate is to not be indolent.[123]

This interesting passage shows Zai's expansive conception of life itself as extending beyond the body to the wider social body and cosmic body of sky and earth. The family is the capacity for resonance that goes beyond the immediacy of blood relations and marriage ties. *Qi* is the animating force of life understood not simply as living organisms but as the wider society and the whole of nature. Again we see the emphasis on harmony and resonance so important in both Confucian and Daoist understandings of life, and while Zai displays resistance to mere philosophical abstraction, it is clear that he has a metaphysics of life itself as the animating force of the living and of the wider world; an animating force that is not nothing but pervades the vast emptiness prior to life and that in some sense pervades it. Such a life force cannot be grasped in the abstract but only realized through human action, the ordered life of ritual propriety that sets up resonance between the individual body, social body, and interpersonal body of the cosmos. This explicitly vitalist tradition carries on after Zai,[124] but with his philosophy we see the integration of a Daoist emphasis on life itself as animating principle and the moral order of society and universe; we cannot speak of life itself outside of ethics. All of creation, 'heaven and earth and the myriad things', is a coherent whole, a unity that includes human society within it.[125]

The figure who best articulates this union of life force with ethics is the figure I mentioned at the beginning of this section, Zhu Xi (1127–1279). Born during

[123] Ibid., quoted p. 52.

[124] Ibid., p. 7. After Zhang Zai there are a series of philosophers who emphasize the philosophy of life (*qixue*): Luo Jinshin (1465–1547), Wang Tingxiang (1474–1544), Wang Fuzhi (1619–92), all in the Ming Dynasty, and Dai Zhen (1723–77) in the Qing. On these later developments see Peter Kees Bol, *Neo-Confucianism in History* (Cambridge, MA: Harvard University Asia Centre, 2008).

[125] Bol, *Neo-Confucianism in History*, p. 200.

the Southern Song in an intellectual climate in which there was a strong Daoist tradition and Buddhism had made significant inroads into China, Zhu Xi was somewhat of a prodigy in passing the civil service examination at nineteen when the average age of passing was the early thirties.[126] His particular philosophy came to be known as the Cheng-Zhu school and his commentary on what he identified as the four Confucian classics, the *Collected Commentary on the Four Books* (*Sishu*), became the basis for the civil service exam until its dismantlement in 1905. One of the main problems he inherited was the issue of how to integrate Confucian ethics with its emphasis on filial piety as the individual person's responsibility with a Buddhist metaphysics of the interrelatedness of all things and its denial of the individual person. He did this through two key ideas of principle or pattern (*li*), the structure of the universe that connects everything, and life force (*qi*) that comprises the universe. The pattern is the intelligible way that things fit together and that connects everything in the universe and life force as the matter of the universe is inseparable from this.

The life force differentiates the pattern into particular types, thus there are various patterns as manifestations of the pattern itself, but with different life force. This differing endowment of life force establishes a hierarchy of beings with the clearer or brighter life force as the more sagacious. Conversely, the more turbid the life force, the more the pattern is directed away from the qualities of perfection. The wise have clear life force; other people have various degrees of turbidity.[127] The hierarchy of beings is thus controlled by the degree of density of life force, so some animals are regarded as benevolent because their social life reflects human relationships—'wolves have fathers and sons, bees have rulers and ministers, jackals and otters [leave food behind to] give thanks, geese and swans mark social distinctions'.[128] Life force condenses in the universe to create things and pattern as 'the place where life force condenses' adheres to it.[129] Zhu Xi explains this through a metaphor of a bowl of water. Life force is like water, which is clear, but which changes colour according to the colour of the bowl it is held in.[130]

Zhu Xi thus uses the idea of the pattern and interconnectedness of all things as a concept to differentiate the uniqueness of human nature, because it is in human nature that the pattern is most manifested as the disposition towards benevolence (*ren*), righteousness (*yi*), propriety (*li*), and wisdom (*zhi*), and is the life force that allows varying degrees of clarity for the pattern. In this way Zhu Xi has a distinct interpretation of the tradition, reading classical Confucianism through the lens of life force that is constrained by the pattern. Some later thinkers, such as Wang Yangming (1472–1529), rejected his ideas on the grounds that the pattern is too difficult to find, but Zhu Xi certainly

[126] Wing-tsit Chan, *Chu Hsi, Life and Thought* (Beijing: Chinese University Press, 1987).
[127] Ibid., 14, p. 174. [128] Zhu Xi, 10, p. 173. [129] Zhu Xi, 7, p. 172.
[130] Zhu Xi, 14, p. 174.

established the centrality of the idea of the life force controlled by the pattern within the central Confucian tradition, thereby cementing a fusion of Daoist and Buddhist ideas, and bringing them into the orbit of his Confucian hermeneutic.

LIFE AS BUDDHA NATURE

While Neo-Confucianism rejected Buddhist asceticism in affording positive value to life itself and social life in particular, Buddhism over a long period of time gained a strong foothold in China and had to accommodate and adapt to Daoist and Confucian ideas. Zai's critique might be somewhat of a caricature of Buddhist asceticism, for in some ways Buddhism in China comes to have a more positive assessment of life itself than we found in India, because of the idea that all beings contain the Buddha nature. From its first mention during the Han dynasty (206 BCE–220 CE), Buddhism was regarded as a foreign religion and the process of translating Buddhism to China took several hundred years from the Han dynasty (around 65 CE) to the Tang, by which time it was fully integrated into Chinese culture. After the breakup of the Han dynasty, China split into northern and southern parts, the northern being governed by non-Chinese Huns, the southern by a series of petty dynasties that lacked the centralizing power of the Han.[131] In 316 the Huns sacked Loyang, people were forced South, and the Chinese tried to make sense of this social collapse. From this point onwards, Buddhism becomes a significant influence. In the North, Buddhist missionaries gained a reputation for psychic powers to which rulers were attracted for strengthening state power, while in the South Buddhism appealed more to the literati who had fled the North and were interested less in magical powers than in the perfection of wisdom.[132] The period of disunity (220–311) came to an end with the Sui that reunited China, followed by the Tang during which Buddhism flourished, as did Daoism, as we have seen.

Part of the transition did indeed involve literal translation of Buddhist texts into Chinese and the finding of appropriate equivalents. A monk from central Asia, Kumārajīva (334–413 CE), ran a translation bureau and produced good Chinese translations of texts that had particular appeal to the educated classes. Thus, *nirvāṇa* becomes *wu-wei*, non-activity, dharma and awakening (*bodhi*) become *dao*, morality (*śīla*) becomes filial piety (*xiao*), the Buddhist saint (*arhat*) becomes the immortal or pure man (*chen-ren*), and homage (*vandana*)

[131] Denis Twitchett and Michael Loewe, *The Cambridge History of China*, vol. 1: *The Chin and Han Empires* (Cambridge: Cambridge University Press, 1986).

[132] Peter Harvey, *Introduction to Buddhism*, 2nd ed. (Cambridge: Cambridge University Press, 2013), p. 213.

becomes social etiquette (*li*).[133] This translation inevitably involved a great semantic shift and some ideas, such as no-self, never firmly took root and China interpreted the idea of rebirth to mean the reincarnation of an immortal soul.

Confucianism was hostile to Buddhism. Confucians accused the Buddhists of not mentioning the Confucian classics; they were critical that monks do not work and are parasites on society, nor do they marry and they give up worldly pleasures; cultural values of family and affirmation of worldly life central to Confucianism.[134] Furthermore, Buddhism posed a political threat of being a state within a state; monks, for example, were tried by the monastic community, not by the civil authorities. But in spite of this resistance, Buddhism was adopted by China and we can read this history as the history of the acceptance of the monastic community. Even deep Chinese values such as filial piety came to be absorbed by Buddhism.[135] The literati took to Buddhism, especially in the face of political unrest and invasion in the North after the collapse of the Han.

Buddhism in China, however, was not a unified tradition. On the one hand, we have the emptiness teachings initially promoted by Dao-an (312–65 CE), while on the other we have pure land teachings promoted by Hui-yuan (354–416 CE). In the course of time a number of schools (*zong*) developed, each a clan that traces its origin to a patriarch.[136] The initial forms of Buddhism were direct imports from India: the Mādhyamika emptiness teachings (introduced by Kumārajīva) and the Yogācāra teachings introduced by Xuanzang (Hsüan-tsang, 602–64).[137] Four schools in particular became important: Tiantai (T'ien-t'ai) founded by Zhiyi (Chih-I; 539–97), the Huyan (Hua-yen) founded by Dushun (Tushun; 557–640), the Chan or meditation school, and the Jingtu (Ching't'u) or Pure Land school. Each had a distinctive teaching. The Tiantai, named after mount 'heavenly terrace', the place where it was founded, stressed meditation and the doctrine that all beings have the Buddha nature, in particular revering the *Lotus Sūtra*. Huayan, the flower garland tradition based on the *Avataṃsaka-sūtra*, emphasized the interpenetration of all things. The Pure Land school that became widely popular is a Buddhism of grace founded on the pure land meditation texts, particularly the

[133] Erik Zürcher and Jonathan Silk (eds.), *Buddhism in China: Collected Papers of Erik Zürcher* (Leiden: Brill, 2013). See also, for example, C. Pierce Salguero, *Translating Buddhist Medicine in Medieval China* (Philadelphia, PA: University of Pennsylvania Press, 2014).

[134] For these accusations and the Buddhist defence see Mozi, *The Disposition of Error*, in Kevin Reilly (ed.), *Readings in World Civilizations*, vol. 1 (New York: St Martin's Press, 1994), pp. 165–70. The Confucian rejection of Buddhism continues through the generations. Even in the seventeenth century Huang Tsung-hsi rejects Buddhism not because of its asceticism, but because of its ornate rituals and 'superstition'. Huang Tsung-hsi, *Ming-I Tai-fahng lu*, trans. Wm Theodore de Nary, *Waiting for the Dawn* (New York: Columbia University Press, 1993), p. 169.

[135] Kenneth Kuan Sheng Ch'en, *Chinese Transformation of Buddhism* (Princeton, NJ: Princeton University Press, 1973), pp. 14–50.

[136] Harvey, *Introduction to Buddhism*, p. 213. [137] Ibid., p. 214.

Sukhāvativyūha. Chan, the meditation school based on the idea that teachings were directly transmitted through the mind from the Buddha, was said to be introduced by the possibly legendary Bodhidharma (*c.*470–520 CE). Finally, I should mention a fifth school, Tantric Buddhism (Zhen'yan), which so impacted Tibet but in China did not flourish, although its presence characterized as the repetition of mantras was felt during the Tang dynasty. It is in the context of these schools that we need to locate the notion of life itself and the Buddhist response to a developed and sophisticated notion of *qi* that existed in China prior to Buddhism.

In general terms Buddhism had a less positive attitude to the world of life than Daoism or Confucianism in the sense that the world of rebirth was regarded as a place of suffering. There is a mixed attitude to the world of life in Chinese Buddhism. On the one hand, any negative evaluation of life became transformed in contact with the ambient culture. We see this especially with the idea of the Buddha nature (*foxing*). With the Indian Mahāyāna, the *tathāgatagarbha* ('womb/embryo of the enlightened one') teachings had claimed all beings to have the potential for enlightenment and the *tathāgatagarbha* comes to be conceptualized as the essence of existence, linked to the luminous mind (*prabhāsvara citta*) and all beings' potential for Buddhahood (*buddhata*). From Buddhahood it is but a small step to the Chinese idea of Buddha nature (*foxing*). Yet, on the other hand, a negative evaluation of life develops in the emphasis on transcending the body. In the *Lotus Sūtra*, chapter 23, the 'medicine king' Buddha Bhaiṣajyarāja expresses his dedication to the Buddha through self-immolation. His burning body illuminates the universe for 200,000 years. Some followers of the tradition in China followed this practice of self-immolation out of devotion to the Buddha. *In Praise of the Lotus Sutra* (*Hongzan Fahua Zhuan*) is a text that presents an account of a monk Huiyi during the Song dynasty, who performed a ritual self-immolation before the Emperor who pleads with him not to go through with his severe observance. The monk replies: 'What is worth preserving in this feeble body and worthless life?'[138] The text records more such incidents. Two nuns who were sisters reciting the Lotus Sutra 'held deep loathing for the physical body' and offered themselves in self-immolation to the dharma.[139]

This kind of negative attitude to the life of the body would have been alien to Chinese Confucianism and even Daoism, but the idea that all beings contain the Buddha nature is potentially less hostile to the body. But can we say that the Buddha nature is the power that gives life to the living? Is Buddha nature the Chinese Buddhist equivalent of the *dao*, which itself is associated with *qi*?

[138] In Donald Lopez (ed.), 'Buddhism', p. 1203 in Jack Miles (ed.), *The Norton Anthology of World Religions*, pp. 727–1467.

[139] Ibid., 1203–4.

The term *qi* generally has little importance in Chinese Buddhism, although there are exceptions. There is an eighth-century text, the *Yuan Ren Lun* (*Discussions of the Origin of Humanity*), where *qi* plays an important role in its cosmological structure, a text important for Huayen and Chan Buddhism and that may have been a precursor to the ways in which later Song Confucians theorized *qi*. Here the vital force constitutes the origin of the cosmos, engendering heaven and earth, which in turn produce myriad things, including human beings.[140] But the explanation of life here is less naturalistic and more metaphysical insofar as Buddhism posited the Buddha nature as the underlying essence and support of the world and its inhabitants and in some ways is akin to the idea of *dao*. One text in particular is important in answering the question as to whether they are equivalent ideas, and that is the *Platform Sutra of the Sixth Patriarch* (*Liuzi-tan-jing*).

The Platform Sutra

Hui-neng (638–713) was the sixth patriarch of Chan Buddhism who was born and taught in a town in Southern China. This southern branch of Chan Buddhism was in dispute with the northern branch headed by Shenxiu (600–706) over whether enlightenment was sudden or gradual, the former view winning out in due course when the Emperor chose to support the southern school, subsuming the northern within it.[141] Hui-yen's teaching is minimal and simple. Indeed, the text boasts that he was an illiterate boy working in the kitchen of the monastery. In the story, Shenxiu writes a verse on the monastery wall. The verse reads:

> The body is a Bodhi tree
> The mind is like a standing mirror
> Always try to keep it clean
> Don't let it gather dust.[142]

In response Hui-neng asks his friend to write two further verses on the wall:

> Bodhi [awakening] doesn't have any trees
> this mirror doesn't have a stand
> our Buddha nature is forever pure
> where do you get this dust?
> The mind is the Bodhi tree

[140] Peter N. Gregory, *Inquiry into the Origin of Humanity: An Annotated Translation of Tsung-mi's Yuan jen lun* (Honolulu, HI: University of Hawaii Press, 1995), p. 44.

[141] H. Dumoulin, *Zen Buddhism: A History: India and China*, trans. J.W. Heisig and P. Knitter (Bloomington, IN: World Wisdom: 2005), pp. 107–21.

[142] Hui-neng, *Platform Sutra* 6, in Red Pine, translation and commentary, *The Platform Sutra: The Zen Teaching of Hui-neng* (Berkeley, CA: Counterpoint, 2006), p. 6.

the body is the mirror's stand
the mirror itself is so clear
dust has no place to land.[143]

Upon reading them, the patriarch named Hui-neng as his successor rather than Shenxiu. In fact, the verses illustrate the gradual process of enlightenment with effort in contrast to the sudden awakening without effort. There is a simplicity to Hui-neng's teaching. The human mind and its concepts block us from direct, experiential understanding of the truth of the Buddha, which is to see one's true nature. All of life is empty of any substantial reality and the only truth is the Buddha nature, which is pure like an unsullied mirror. The text describes the Buddha nature as non-duality beyond conceptualization. In a discussion between Hui-neng and a disciple, Yin-tsung, the disciple asks about the authority of Hui-neng who showed him his bowl and robe received from his teacher. Hui-neng describes how the Buddha nature is beyond the dualities of good and bad and beyond either permanence or impermanence.

Chan's emphasis on the direct, wordless experience of reality through meditation excludes developing metaphysics and excludes the imaginative space within which to speculate about life itself and its relation to the living. While the idea of the Buddha nature is the closest Chinese Buddhism gets to thematizing life itself, this is not really conceptualized, as *dao* was, in terms of life energy or the sustaining power of appearances. Chan is anti-metaphysical, concerned with bringing awareness into the present moment as the full reality of enlightenment and the existential realization of the Buddha nature beyond rational understanding. This tradition was said in Chinese sources of the Song period to be from the Buddha himself who raised a flower in a silent sermon enlightening Mahākaśyapa, who transmitted the teachings directly from mind to mind, thereby instigating the Chan tradition.[144] Bodhidharma received the direct transmission and introduced it to China. This foundation myth sets the tone for the Chan emphasis on direct experience beyond words and establishes the school's four principles of the transmission outside of the formal Buddhist teachings, not relying on words, pointing to the mind, and seeing one's true nature as the Buddha. The Buddha's true teaching is found in silence, beyond words (and also therefore in laughter, likewise beyond words).[145] The later edition of Hui-neng's text presents the famous question: 'When you're not thinking of anything good and not thinking of anything bad, at that very moment, what is your original face?',[146] at which point Hui-neng's disciple Hui-ming gained awakening. This kind of apparently nonsensical statement came to have importance in a meditational context, especially

[143] Ibid., sutra 9, p. 8.

[144] Harvey, *Introduction to Buddhism*, p. 222.

[145] See Bernard Faure, *Double Exposure: Cutting across Buddhist and Western Discourse* (Palo Alto, CA: Stanford University Press, 2004), p. 97.

[146] *Platform Sutra*, 11, p. 120.

in Japan where Zen used a dialogue form between master and disciple in which the disciple responded to a riddle (such as 'what is the sound of one hand clapping?') to achieve a sudden awakening (*satori*). The non-dual realization beyond works is the heart of the Chan/Zen tradition; a wordless teaching articulated with some eloquence by Dōgen in the thirteenth century in which 'everyday mind' is the Buddha's awakening and 'the moment is already here'.[147]

This teaching is about awakening to the nature of reality as the reality of enlightenment here and now but does not develop any implications of the doctrine. We seek in vain to find any justification of life other than the stark contrast between the unenlightened and enlightened state and the need to shift from the former to the latter. Indeed, the emphasis on direct experience through meditation (and so there being little need for an elaborate library) meant that Chan survived the late Tang persecution of Buddhism (842–5) when the Emperor Wuzong, a patron of Daoism and strapped for cash after a civil war, targeted the wealthy monasteries.[148] Tiantai and Huayen Buddhism did not recover, nor did Chinese Tantra, but Chan continued because of its minimal requirements and also Pure Land because of its wide popular appeal.

In one sense, the emphasis on direct experience in Chan is a philosophy of life by claiming fullness to experience in the present moment, but life itself as a concept is not developed. The often repeated claim that Buddhism is a philosophy rather than a religion is challenged by Chan in the sense that the meditation tradition is anti-philosophy. But then Chan might be seen as anti-religion as well, if by that we mean reflection on revelation and practices that follow from it, because Chan has a strong iconoclastic tradition that seeks the disruption of organized religion.[149] But we should not over-emphasize this because there is reverence for the tradition of the patriarchs and the preservation of the truth of the teachings of the Buddha through the generations. There are clear affinities with the Daoist Zhuangzi whose writings point to the *dao* as immediate experience, but in contrast to Daoism, Chan rejects conceptualization in emphasizing the need to go beyond the dualities that thought entails. In more contemporary parlance, we might say that Chan attempts to bypass linguistic consciousness and through meditation accesses deeper levels of the brain and the bio-energy of human pro-sociality.

But aside from this Chan non-philosophy, there is a popular, non-monastic version in which people venerated relics of the Buddha and went on

[147] Shobo Genzo Dōgen, ed. Kazuaki Tanahashi, *Treasury of the True Dharma Eye: Zen Master Dogen's Shobo Genzo* (Boston, MA and London: Shambhala, 2012), p. 426.

[148] Harvey, *Introduction to Buddhism*, p. 223; Heinrich Dumoulin, *Zen Buddhism: A History*, vol. 1, *India and China*, trans. James W. Heisig and Paul Knitter (Indianapolis, IN: World Wisdom, 2005), p. 211.

[149] See, for example, the wonderful book by M. Conrad Hyers, *Zen and the Comic Spirit* (London: Rider, 1974).

pilgrimage to gain practical benefits in life, as Bernard Faure has shown.[150] In this popular Buddhism we have thaumaturgic practices, Buddhist saints who are regarded as miracle workers, funeral rites, and devotion to images and relics. Such practices are clearly affirmative of life in the sense that they affirm non-monastic values of family life, reverence for the dead, worldly success, and faith in a higher power, but they are not a philosophy of life itself, even if there is an implicit philosophy within them.

The Pure Land

This popular Chan overlaps with Pure Land Buddhism, which had wide appeal. The Sanskrit Mahāyāna text the *Sukhāvatīvyūha sūtra*, the text displaying the land of happiness, is related to visualization texts of the universe filled with innumerable world systems or Buddha fields (*buddhakṣetra*). This, along with a text unknown in India, the Sutra on the meditation on Amitayus (*Guan Wuliangshou Jing*), became the textual foundation of the tradition in China.[151] The Pure Land teachings, although originating in India, never became established there and unlike the Yogācāra and Mādhyamika schools, there was never a pure land *śāstra* or formal philosophical reflection in India. The situation changes, however, with its introduction into China: the Buddhism of faith came to have wide general appeal and attracted some philosophical reflection, even more so when it enters Japan. All that is required is to repeat the name of Amitābha with faith ten times and he will take you to the pure land, to heaven, at death from where there will be no return to the world of suffering. The simplicity of the teachings is in sharp contrast to the complex philosophy of the Buddhist commentarial traditions and its effortlessness contrasts with the meditative effort of Chan. As the tradition develops in East Asia, even one slight thought of the Buddha Amitābha (Amto Fo in China, Amida in Japan), if it is a pure thought, will guarantee salvation.

Repeating the name became the key practice, which has surface overlap with mantra repetition in tantric or esoteric Buddhism but is conceptually distinct. Whereas mantra yoga is concerned with re-structuring the mind, repeating the name of Amito Fo is devotional. Its origins are in the practice of recollection or mindfulness of the Buddha (*buddhānusmṛti*), an essentially meditative practice that comes to be reconceptualized as a practice of devotion.

[150] Bernard Faure, *The Rhetoric of Immediacy: A Cultural Critique of Chan/Zen Buddhism* (Princeton, NJ: Princeton University Press, 1991).

[151] See J. Foard, M. Solomon, and R.K. Payne (eds.), *The Pure Land Tradition: History and Development* (Berkeley, CA: Institute of Buddhist Studies, 1996); Harvey, *Introduction to Buddhism*, pp. 216–17.

Hui-yuan (334–416) introduced meditation on Amitābha and Tanluan (T'an-luan: 476–542) popularized repeating the name, becoming the first patriarch of the tradition. Inspired by the Daoist quest for immortality, he thought this would be guaranteed by rebirth into Amitābha's heaven. Although it is ultimately Amitābha's grace that saves us, Tanluan advocated recollection of the Buddha or *nianfo*, Jap. *nembutsu*, which is both recollection and calling on the Buddha.[152] The second patriarch Daochuo (Tao-ch'o; 562–645) introduced the idea of the decline of the teachings of the Buddha (*mofa*, Jap. *mappō*) that since people are living in a degenerate age, then devotion is the only path available. Rather than rely on self-power (*zili*), we must rely on other power (*tali*) or the power of Amitābha. The third patriarch, Shandao (Shan-tao; 613–81), consolidated the tradition, writing commentaries and advocating recitation with a calm and unruffled mind.[153] But once the tradition is transplanted into Japan, there is no Chinese reserve and the tradition becomes purely devotional. The Japanese monk Honen (1133–1212) thought that in the degenerate age the only chance of salvation is chanting the name of the Buddha, and his disciple Shinran (1173–1263) took this up and radicalized the teaching to mean that only a moment of faith in Amida is sufficient for salvation; the repetition of the names is pointless other than as praise. From meditative effort we have come to a doctrine of pure, effortless grace. In his book *Tannisho: Lamenting the Deviations*, Shinran develops this doctrine: we have to rely wholly on the other power and the practice of the *nembutsu* is in fact a non-practice, an effortless practice because of this reliance.[154] In complete contrast to the earlier *Sukhāvatīvyūha* that says that wicked people may be reborn in a lotus in the pure land that will open in due course once their sins are purified, Shinran says that 'even a good person attains birth in the Pure Land, how much more so the evil person'.[155] Amida's primal vow ensures universal salvation.

The Pure Land tradition is deeply optimistic about the possibility of salvation in a world characterized by wickedness, but the optimism is for rebirth in the pure and happy land. Like other forms of Buddhism, the world of life is one of suffering with salvation seen as being taken out of it. Like other kinds of Chinese Buddhism, there is minimal metaphysical speculation and while the pure land entails an imaginaire, the Chinese texts offer no description over the translation of the Sanskrit text. Although there is concern for the living and compassion of their suffering, there is no speculation on life itself and the relation between life itself and the living. With the Pure Land we are in a kind of pragmatism in which reliance on the power of that which is wholly other to the self takes precedence over any philosophical reflection.

152 Harvey, *Introduction to Buddhism*, p. 216. 153 Ibid.

154 Shinran, *Tannisho*, VII, p. 1297, trans. Taitetsu Unno in Lopez (ed.), 'Buddhism', pp. 1293–307.

155 Ibid., III, p. 1296.

LIFE FORCE AND VIRTUE

The very long and mostly uninterrupted history of Chinese civilization has demonstrated the great sustaining power of Confucianism, Daoism, and Buddhism. By the time of the Tang (618–906 CE) through to the Ming (1368–1644 CE), which roughly corresponds to the European Middle Ages, the traditions are consolidated, and their textual, institutional, and philosophical boundaries established. There is some commentarial tradition, although nothing as elaborate as we find in India or indeed Europe, but there is a philosophical commitment to working out the implications of doctrine and developing ways of life that express those philosophies.

What is unique to the Chinese situation is the strong connection between life force and virtue. Ethical conduct is linked to the degree or intensity of the life force such that the noble or perfected person has a greater density of *qi*: Mengzi speaks of being filled with flood-like *qi*. Life force is not neutral and is inextricably linked to the very structure of the universe, as we saw with Zhu Xi's idea *qi* and of pattern (*li*) controlling its multiple forms. The political order was thus theorized by the Chinese philosophers on a scale of virtue (and therefore effectiveness) depending upon the degree of life force it articulated. From the earliest times of the Axial shift with Kongzi and Mengzi we witness an understanding of Chinese civilization as a force for repair or guidance of the human community, from following what it regarded as lower human proclivities that led to neglect of social responsibility, particularly filial piety, to following virtue and abiding by correct relationships within the society. There is no transcendence here in a Gnostic vision of otherworldly escape, but rather the affirmation of human life, and therefore political life, as transformation within world. All significant Chinese thinkers have recognized the importance of human pro-social emotions and the ability of the political order to nurture pro-sociality that is thereby to align human reality with cosmic reality. That the political reality of China, as everywhere else, has continued to fail to live up to this aspiration has been a theme in Chinese philosophical critique.

In the early modern period, one text in particular stands out as a critique of the political order and especially of despotism, and that is Huang Tsung-hsi's *Waiting for the Dawn, a Plan for the Prince* (*Ming-I tai-fang lu*). Huang Tsung-hsi (1610–95)—a contemporary of Hobbes and John Locke—offers a strong ethical evaluation of the Chinese political order of his day. His vision emphasized the importance of law in regulating governance. At one time in the distant past, during the period of the Three Dynasties, there was a harmonious political order, but with the rise of the Han and the Song dynasties, this order became corrupted due to the selfishness of the rulers. Huang argues that the main reason for the success of the earlier order and the later decline of that order is due to law. To the end of the Three Dynasties there was law, claims Huang, but after that there has been no law.[156] The early rulers of China, he claims, knew that the people, 'the all under heaven', needed sustenance so gave them fields to cultivate, clothing so gave them hemp and mulberry trees to grow, and education so gave them schools and institutions such as marriage to protect against promiscuity and the military to protect against disorder.[157] This social order was ensured by the presence of law and once rulers go against law, they simply act to preserve their own interests and try to preserve the dynasty into the future without concern for people. 'Only when we have governance by law can we have governance by men', Huang famously says.[158] Huang's idea of self-repair is therefore linked to the necessity of good governance that is for the people. Indeed, he has been regarded as advocating a kind of democracy and as being the intellectual forebear of Sun Yat-sen's revolutionary politics and the establishing of modern China.[159]

Since Huang's book and in later political thinking, the idea of the life force is not an explicit theme, but his idea of law as articulating popular will draws on a Confucian heritage, sketched earlier, that draws on the theme of the life force. For the Confucian scholars Mengzi and Zhu Xi, life force was integral to their political vision as the force that drives ethical behaviour. It is this intellectual inheritance that is behind Huang and although unique in his articulation of a specific political vision, he is part of a long tradition that wished to claim that human self-repair comes through self-cultivation, which is the intensification of life force within a person expressed in civic behaviour.

At the time of the Chinese revolution (1911) at the end of the Qing dynasty,[160] there were those who wished to embrace modernism (and therefore westernization) and to completely reject the past, such as Chen Duxiu, for whom Confucianism was 'an inefficacious idol and a fossil of the past'[161] and for whom any notions of life force are outmoded. What would create China as a modern nation would be the unqualified adoption of Western modernity and technology to take the country out of a dark past and not so much to seek self-repair but rather to create a new and never-before-realized future. Confucianism would be incompatible with such a future and new society. A second school of thinking wished, by contrast, to revive what was good in Confucianism and adopt a rights consciousness from the West that would allow such an awareness to be rediscovered in the ancient tradition. Liang Qichao (1873–1929), for example, wished to retrieve human rights from within Confucianism, who drew on the German legal theorist Rudolph von

156 Huang, *Waiting for the Dawn*, p. 97. 157 Ibid. 158 Ibid., p. 99.

159 De Bary, 'Introduction', pp. 78–9.

160 For a good history see Michael Dillon, *China: A Modern History* (London: I.B. Tauris, 2010), pp. 120–69.

161 Chen Duxiu, 'The Constitution of Confucianism' (1916), in Stephen C. Angle and Marina Svensson (eds.), *The Chinese Human Rights Reader: Documents and Commentary 1900–2000* (Abingdon and New York: Routledge, 2001), pp. 67–76.

Jhering, arguing that rights consciousness needs to be highlighted in a new China and this is indeed a resource within the tradition.[162] But again, the language of life itself is lost in this new discourse, and even the traditionalists who wished to revert to tradition and create a transformed tradition to address the needs of the new China[163] did not revert to the language of life itself. Needless to say, with Mao's revolution and the Great Leap Forward, the language of life itself was rejected as a means of addressing the human good in favour of an imported materialism that saw economic necessity as determining the social good and human nature.

§

We are now coming to the close of this reflection on the life force in Chinese civilization and its implications for human flourishing. While the idea of the life force is important in Confucianism, it is above all Daoism that develops a philosophy of life through conceptualizing life itself as the reality of the Dao, the way, which entails the claim that living beings are animated by a force (*qi*) that is coextensive with them. As *dao*, this *qi* is the totality of all that is, on the one hand, and has variable degrees of intensity, on the other. The practitioner can build up *qi* within the body and, conversely, deplete *qi* through bad living. Death and life are alternate phases in a cycle along with the transition from growth to decay, from *yang* to *yin*, and the Daoist sage is thought to understand this process and become detached from it, thereby attaining wisdom. Wisdom is the refection of life itself on itself and the concomitant understanding that action as a characteristic of living beings is, in the end, non-action. While there is some similarity to Buddhism here in the notion of detachment, the underlying metaphysics is quite different. Buddhism offers a way out of life as suffering but refuses speculation on life itself. Daoism sees life as the expression of life itself and simply as flow; life is what it is, and wisdom is understanding that.

But this does not mean that there is not a right way to live. To live correctly is to be in harmony with the *dao*, such that heaven, earth, and the human resonate with each other. This is no overt political philosophy, but there is an implicit political philosophy here in which each thing has its place and meaning is understood as the location of something within the broader cosmic scheme. It is Confucianism, rather, that has a more developed ethical view and is the central political philosophy of China, although Daoism was favoured

[162] Liang Qichao, 'On Rights Consciousness', pp. 5–15.

[163] Carson Chang et al., *A Manifesto for a Reappraisal of Sinology and Reconstruction of Chinese Culture*, in Carson Chang, *The Development of Neo-Confucian Thought*, vol. 2 (New York: Bookman Associates, 1962), pp. 455–83. See also Stephen C. Angle, *Contemporary Confucian Political Philosophy* (Cambridge: Polity Press, 2012), pp. 61–3.

during the Tang and Song dynasties. For Neo-Confucianism, as we have seen, an ethical worldview articulated through social, political relationships is not enough for the understanding of life, but conversely the *dao* alone without reference to human and therefore moral and political life is not enough either. Combining the Daoist philosophy of life with the Confucian ethics of life, Neo-Confucianism came up with an understanding of the relationship between life itself and the living, which was inherently ethical and political. This view was antithetical to Buddhist monasticism in which life is suffering to be escaped from: a fundamentally Gnostic worldview. For Neo-Confucianism to live in the world is to express the life force, and to live fully in the world as a human being is to express the life force socially through filial piety and family commitment and politically through obedience to state and emperor.

While *qi* has semantic equivalents in other civilizations—*prāṇa* in India, *pneuma* or *psuchē* in Greece[164]—it is the degree to which it becomes central to a civilization's self-definition that is important in understanding the history of life itself. China, through Daoism, has highlighted the theme such that it became a central philosophy and impacted upon the political philosophy of Confucianism. All this was to be rejected. As its complex history shows, the Chinese revolution at the end of the Qing dynasty was influenced by Western ideas of democracy more than ideals of harmony or balance of the life force, and the Maoist revolution, perceiving the failures of earlier regimes to institute social justice, rejected traditional understandings of life itself in favour of an imported materialist Marxism. Transformed into a Chinese context by Mao, the revolution and later cultural revolution rejected traditional concepts as having failed China and attempted to institute ways of life unrelated to any notion of a mystical life force, to seek repair through rejection of the foregoing civilization but in doing so ironically instituting the far worse excesses of oppression than earlier dynasties. Contemporary China again looks not only to economic development but to its older ideals of virtue and balance, wishing to avoid the excesses of Maoist collectivism and the errors of Western individualism. Within the Chinese state controlled by democratic centralism, there is currently little reflection on older ideas of life itself, although the idea of *qi* still has impact on popular culture in the quest for long life and health, if not as an effective political philosophy. The idea of life itself may yet have traction in Chinese intellectual traditions, although mediated through a massive expansion of consumerism and globalization. But although not thematized in this way, bio-sociology that accesses human pro-sociality is just as crucial in China for social and political coherence as for anywhere else in the world. It is the modes of accessing such life in Europe and the Middle East to which we now turn.

[164] For an interesting comparison of *qi* and *pneuma* see Shigehisa Kuriyama, *The Expressiveness of the Body and the Divergence of Greek and Chinese Medicine* (New York: Zone Books, 2002), pp. 242–59.

6

Transforming Life

The Greek and Abrahamic Traditions

MAPPING LIFE

Although the category 'Abrahamic religions' is contentious,[1] Judaism, Christianity, and Islam do form a group of traditions related historically (Christianity emerges as a Jewish sect, Islam from roots in ancient Judaism), conceptually (in their ontological commitments, namely to a theistic conception of world along with eschatological hope), and practically (in their organizing of space/time, namely patterns of ritual, prayer, and pilgrimage). The theme of life itself has been central for them all, especially the sacrificial model of transforming death into life and an eschatology that awaits the end of days. The history of these three religions has, on the whole, been more conflictual than harmonious, almost inseparable from the history of power and the emergence of the idea of the nation and state. We see this clearly with the rise of Athens and the importance of the city's patron Athena, with Yahweh's patronage of the Jewish people as nation, with the adoption of Christianity as the official religion of the Roman Empire, and with Islam where state power was animated by religious force.

A conception of life begins to emerge in these religions from common roots in Jewish and Greek thinking. We can trace three trajectories of interest in life itself that cut across the boundaries of the religions. As with India, we have, first, a sacrificial imaginary at work across the Greek, Semitic, and Christian histories. This is the affirmation of life through death that with Christianity becomes the human/divine sacrifice to end all sacrifice. In a second trajectory, mostly developed within the Greek ethos, we have life understood in terms of verticality; that there is a hierarchy of the cosmos with the purest, most intense

[1] Remi Brague, *The Legend of the Middle Ages: Philosophical Explorations of Medieval Christianity, Judaism, and Islam*, trans. Lydia G. Cochrane (Chicago, IL: Chicago University Press, 2009). Brague argues that the different religions were grounded in different concepts of revelation. For a discussion and support of the category see Guy Stroumsa, *The Making of the Abrahamic Religions in Late Antiquity* (Oxford: Oxford University Press, 2015), ch. 8.

forms of life at the top that generate the lower forms. This verticality was emphasized by theologians with a certain kind of reflection distinct from legal or dogmatic considerations, which in Christianity came to be named mystical theology. Through contemplating the essence of individual life, the soul can ascend back through the hierarchy to the source of life itself. In Chapter 1 I called this the Religious Cosmic Model (RCM), the idea that there is a hierarchy of being within which all life forms find their place. Third, we have a horizontal, scientific drive to understand life; stemming from the Greeks, it finds its way into Islam where the Arabic-medium theologians transform it, from whence into Christian theology and so into the Renaissance and the modern world. In Chapter 1 I called the formulation of this idea in modernity the Galilean Mathematical Model (GMM), which in its application becomes inextricably linked to the advance of technology. This is the objectification of life with all the great success and trouble that it has engendered that ends up being a view of life antithetical to both the sacrificial and hierarchical imaginaries. Indeed, if the hierarchical view of life focuses on the experience of the fullness of life within a person, then the horizontal, scientific account emphasizes an objectivist empiricism with a deep suspicion of mysticism and inner-worldly ascent. This scientific trajectory is intimately linked with rationalism, part of which is critical reflection on what the sciences show us.

We will begin with Athens and Jerusalem, moving on to the Christian synthesis and its unique claim to the transformation of death in the resurrection of Christ, to the medieval Islamic philosophers and their impact on that Christian history. In the end, it is the horizontal view of life, the view that emphasizes the classification of world, the location of the causes of particular lives, and the transformation of life through technologies that has won out and proved the more successful paradigm in human history: Aristotle's main bequest to our modern, global civilization.

Breathing, Naming, Sacrificing

In the Hebrew Bible, the Tanakh that becomes the Christian Old Testament, in the beginning God created the heavens and the earth. He then creates creatures and in particular Adam, who he forms from a clod of earth, breathing life into matter. Adam then gives names to all the animals. Both breathing and naming are ways of giving life: there has to be the first impetus, the life-breath, but then there has to be naming, the bringing living forms into language and categories. Breath, as its precondition, becomes speech. Adam recapitulates God's giving life and through naming the animals he gives them life, bringing them into conceptual order, bringing nature into culture, bringing bare life into the realm of meaning. God breathes into man and man names, thereby creating and becoming God-like or reflecting the divine image in which he has been made.

Speaking depends on breathing and the order of life depends on speaking. A relationship between life, breath, and speech seems to have emerged very early in the history of civilizations. We found it in India and in some of the earliest literary sources of the West; Greek texts and Jewish scriptures. The Bible has precursors in Mesopotamian creation narratives and also from the polytheism transformed into monotheism in Egypt.[2] A pattern emerges of divine breath giving life, human language ordering life, and sacrifice transforming life through preventing death. The history of sacrifice is predicated upon breathing and naming or ordering cosmos. The human community has ordered the world through speech and maintains it through sacrifice. Sacrifice expresses the attempt to have power over death, to bring the meaninglessness of death into the realm of meaning, into the linguistic present of naming.

Ancient religions in the West, Judaism and Greek religion, bear witness to this pattern. Sacrifice was central to the history of Judaism at least up to the destruction of the temple in 70 CE. Even before the great temple in Jerusalem was built to house the Law of Moses, sacrifice symbolized the covenant between God and the Jewish people, incumbent upon individuals and the whole society. Individuals seem to have continued sacrifices after the destruction of the temple and into the second century, but sacrifice as a central pillar of Judaism ended at that point. The Mishnah contains a mass of detail about rules for sacrifice, the kinds of animals that can be killed, on what occasions, and so on, and we have detailed accounts of the temple sacrificial system from Josephus and Philo.[3] Blood offerings were central to Judaism along with a coterie of priests in the temple. During the second temple period, all Jews were expected to worship the One God in the temple and all were expected to bring offerings, particularly at important festivals. Philo gives great detail about animal sacrifice and presents a kind of ethnography of the practice that he was thoroughly familiar with. Furthermore, he allegorized the sacrifice to such an extent that different parts of the animal body come to represent human structures, thereby making concrete the human-sacrificial victim identification.

Of a classification of ten kinds of animal, three are suitable for sacrifice (sheep, oxen, and goats) along with two kinds of bird (turtle dove and pigeon), with seven animals fit for food.[4] These are divided into three types: whole

[2] Jan Assmann, *Moses the Egyptian: The Memory of Egypt in Western Monotheism* (Cambridge, MA: Harvard University Press, 1998).

[3] See Maria-Zoe Petropoulou, *Animal Sacrifice in Greek Religion, Judaism, and Christianity, 100 BC to AD 200* (Oxford: Oxford University Press, 2008), pp. 128–206. Also, Guy Stroumsa, *The End of Sacrifice: Religious Transformations in Late Antiquity* (Chicago, IL: University of Chicago Press, 2009). William K. Gilders, *Blood Ritual in the Hebrew Bible: Meaning and Power* (Baltimore, MD: Johns Hopkins University Press, 2004).

[4] Philo, *The Special Laws*, I.33, pp. 192–3, in *Philo*, vol. VII, *On the Decalogue, On the Special Laws Books* 1–3, with trans. by F.H. Colson, Loeb Classics (Cambridge, MA: Harvard University Press, 1937); Petropoulou, *Animal Sacrifice*, pp. 161–2.

burnt offerings, preservation offerings, and sin offerings,[5] and Philo presents quite a detailed account of each. The first type is for the glory of God, the second for the preservation of human affairs, and the third for the purification of the people from sin.[6] He gives details of the sacrifice and what kinds of animal are permitted and what are not. The animal has to be without blemish, cannot be pregnant, and so on.[7] Indeed there was a raft of ritual specialists inspecting animals and attending the chief priest. With burnt offerings there is a regime of regular sacrifice, a lamb at dawn and dusk daily, and more on the Sabbath accompanied by copious incense: 'Thus the blood offerings serve as thanksgivings for the blood elements in ourselves and the incense offerings for our dominant part, the rational spirit-force (*logikou pneumatos*) within us which was shaped according to the archetypal form of the divine image'.[8] On special occasions more animals are offered: at the corn harvest in the spring two calves, a ram, and seven lambs are to be offered to be consumed by the fire along with two lambs for the priests.[9] On the most holy day of the year, the Day of Atonement, the same number of beasts should be offered along with a further ram and two kids. One of the kids is set free into 'a trackless and desolate wilderness', bearing on its back the transgressions of the law for the community who offer it: the scapegoat bearing the sins of the group.[10] These sacrifices represented the life principle (*psuchē*) comprising reason (*logikoi*) and unreason (*alogikoi*), corresponding to the male and female, says Philo. The blood (*'aima*) is to be poured around the altar because it is the libation (*spondē*) of the life principle (*psuchē*).[11] It is significant that Philo calls this the vital oblation (*psuchikos spondēs*) that gives thanks to God for creation of the universe, both for the totality of the universe and for its parts, which are like the limbs of a living being (*zōon*).[12] Moreover the oblations for the preservation of life are for the mind and the body's good health.[13]

While the practice of sacrifice may not have been theorized in the early period, Philo certainly offers allegorical explanations as to its meaning. The ritual purification of the body entailing prayers and slaughtering the animal in a particular way, depending on the occasion, stands for the cleansing of the soul, Philo tells us, and the sacrificial victim stands in for the offerer in the case of the 'sin oblation'. The sacrifice thereby affirms life through death. By pouring the blood of the victim around the altar, the priest offers life itself back to God who is the giver of life. The oblation affirms the power of creation, preserving the life of the community and absorbing its sins. The breath of God gives life to the living, to the human community—specifically the Jewish

[5] Philo, *On Special Laws*, I.35, pp. 207–11. [6] Ibid., I.36, p. 211.
[7] Ibid., I.34, p. 195.
[8] Ibid.: *τὰ δὲ θυμιάματα ὑπὲρ τοῦ ἡγεμονικοῦ, τοῦ ἐν ἡμῖν λογικοῦ πνεύματος, ὅπερ ἐμορφώθη πρὸς ἀρχέτυπον ἰδέαν εἰκόνος θείας.*
[9] Ibid., I.35, pp. 203–5. [10] Ibid., p. 207. [11] Ibid., I.38, p. 217.
[12] Ibid., p. 219. [13] Ibid., I.40, p. 229.

nation—brings the living into the realm of meaning by naming, by language, and offers life so transformed back to God through living beings in sacrifice. This enacts the overcoming of death and purifies the tribe. Philo theorizes this explicitly through his typology, but we find essentially the same structure in the ambient Pagan sacrifices that Philo would have witnessed.

The ancient Greeks connected life itself with breath, an idea that has deep Indo-European roots, and while it finds philosophical expression in the work of Aristotle, the folk psychology of the idea can be found in epic literature. The *Iliad* is replete with the phrase 'his life breath left him' when heroes die in battle, and the dead whose breath has left them only have minimal life in Odysseus' underworld in the *Odyssey*, where the dead squeak like bats. But even they need sacrifice to survive and live their thin life; they drink the blood of sacrificial victims, becoming momentarily fuller creatures than before. Filled with the blood, although dead, Tiresias' breath is turned into speech for the questing Odysseus.[14] Breath is linked to passion that energizes the warriors and to speech that inspires them. The anger of Achilles is linked to the fullness of his life: his breath fills him and from its source in the *phrenes* animates his anger and warrior's power. The *phrenes* may refer to the lungs or to the kidneys as the seat of power and from this source Achilles speaks his truth and destroys his enemies.[15] Once the black blood flows, his life force leaves him but remains as a trace in his ghost.

Ancient Greek sacrifice was the dominant mode of worship functioning at all levels of society from an ancient period, as we see in Homer, through to the second century CE. While the nature and meaning of Greek sacrifice has been contested,[16] literary and epigraphic sources bear witness to its continued importance. Literary sources provide descriptions of animal sacrifice and the occasions on which it was held. There were various types of Greek sacrifice with specific functions, particularly offering, divination, purification, propitiation, and the sealing of an oath.[17] There were military sacrifices before a battle, oath sacrifices, and purificatory sacrifices. Before battle a ram would be sacrificed as 'the slaughter victim' (*sphagion*)[18] and the same term was used for sacrificing to a river into which the animal's blood was poured. Animal sacrifice was widespread and even the 'revival' of sacrifice in Roman times may have been less of a revival in Greece than a continuity, as Petrapoulou argues.

[14] *Odyssey*, book 11, p. 407 in *Odyssey*, vol. 1, trans. A.T. Murray, Loeb Classical Library 104 (Cambridge, MA: Harvard University Press, 1919), p. 104.

[15] R.B. Onians, *The Origins of European Thought about the Body, the Mind, the Soul, the World, Time and Fate* (Cambridge: Cambridge University Press, 1951), pp. 14–15, 28–30.

[16] For a summary of the debate see Petrapoulou, *Animal Sacrifice*, pp. 1–28.

[17] Petrapoulou, *Animal Sacrifice*, p. 34.

[18] Robert Parker, *On Greek Religion* (Ithaca, NY and London: Cornell University Press, 2011), p. 155.

A range of terms is used for the sacrifice, some indicating particular types. *Thusía* indicates an 'Olympian' sacrifice at which the victim's meat was eaten in contrast to the *enagízein* 'chtonian' or heroic sacrifice that was wholly consumed by the flames. The most common verb to sacrifice is *thuō*[19] and there are verbs associated with sacrifice such as *katárkhesthai*, which in Homer's time meant sprinkling barley on the animal's back and *holokautein*, designating the complete immolation of the victim. The chtonian sacrificial victims were black, sacrificed at night in a special pit with their head pressed downward, whose blood was poured onto the ground,[20] in contrast to the *thusía* offered on an altar (*bōmós*) in the day. There is also *spondē*, libation, that we have already come across with Philo that designated a libation of wine or non-alcoholic fluids, but also the blood of the victim itself poured on the ground, in the river, and on or around the altar.

We know the process of Greek sacrifice in some detail from contemporary sources. The satirist Lucian, although mocking the procedures, gives an account of sacrifice, how different people offer different animals—the farmer an ox, the shepherd a lamb—inspecting them to make sure there are no imperfections (as in Jewish sacrifice), decking the animal with garlands, and leading it to the altar to be slaughtered before the god. Lucian compares the priest, all bloody with the offerings, to the Cyclops and describes how he lights the fire on the altar and offers the goat or the sheep, the smoke rising to heaven.[21] Plutarch offers a description of sacrifice, as does Pausanias in his *Description of Greece*, noting regional variations of ritual practice.[22] One rite described by Pausanias is the festival of Laphria where the people of Patrai annually perform a sacrifice to Artemis in which a large bonfire is lit and many animals thrown onto it, driven back into the flames when they try to flee.[23] The process seems to have contained standard elements with regional variations. Apart from these extreme orgies of death by fire, the animal was often killed with ritual hesitancy, Meuli's 'comedy of innocence' in which there was a pretence that the ox sacrificed was not dead at all, even stuffing its corpse with straw and yoking it to a plough, and by putting the knife or axe that killed it on trial. Sprinkling the animal's head with water ensured that it nodded, thereby assenting to its imminent demise.[24]

But what does it all mean? In contrast to India where there is elaborate and sophisticated discussion about sacrifice—even though in the end completely inconclusive—in Greece there was no accompanying explication; Greece did

[19] Parker, *On Greek Religion*, p. 154. [20] Ibid.

[21] Lucian, *On Sacrifices*, pp. 168–9. In A.M. Harmon (trans.) *Lucian* volume 3, Loeb Classics 130 (Cambridge, MA: Harvard University Press, 2014), pp. 155–71.

[22] Petrapoulou, *Animal Sacrifice*, pp. 103–5.

[23] *Pausanias Description of Greece*, Archaia XVIII, pp. 277–9, trans. W.H.S. Jones, Loeb Classical Library 93 (Cambridge, MA: Harvard University Press, 1918); Parker, *On Greek Religion*, p. 167.

[24] Parker, *On Greek Religion*, p. 129.

not theorize its cultic practices.[25] Theories of sacrifice abound, as we have seen, but for the Greeks the sacrifice was above all a gift of wealth and a gift that permits communication between the god and human.[26] It was also obligatory in ancient Greece. This link between sacrifice and law shows how important civic authorities considered sacrifice to be. There was clearly the function of elevating the status of the citizen who offered sacrifice,[27] but the legal imperative to perform it demonstrates how important it was to establish a connection with the gods and thereby renew the community. Sacrifice then, Parker observes, covers a family resemblance of rites,[28] for we have, on the one hand, the meticulous choosing of the victims yet, on the other, the indiscriminate slaughter of any animal in the Laphria rite.[29] But this sacrificial range is linked by the idea of offering and communicating. For our purposes, we see that this general pattern of ritual killing gives life to a community through the community's participation in the rite (the bonding entailed by a common ritual objective) and affirms that life, through literally feeding it. The exception here is the totally burned offering that feeds the gods who will then, hopefully, reciprocate.

Sacrifice was central to religious practice in the ancient world and was generally antithetical to asceticism practised by some philosophers who may also have been vegetarian.[30] In Greece, sacrifice pervaded the society at all levels, being legally incumbent upon citizens of the city-state of Athens, and even being performed by the poorest through a simple gesture of kissing his own hand.[31] The philosophers simply assumed the ambient sacrificial culture but reflected different tendencies in the society, tendencies towards a way of life that emphasized circumspection and reserved dignity in behaviour, and whose reflection on the nature of life was divorced from cultic sacrifice. The philosophers in Greece reflected on the gods, certainly, but came to reflect on life itself as an abstract idea and how to live: on metaphysics and ethics. In this history, the two most important ways of thinking were embodied by the two giants at the root of the Western philosophical tradition, Plato and Aristotle. In general Plato contains an account of the world as reflection of the higher reality of the forms, a verticality in which meaning in human life is ensured by a transcendent structure. This in time became transformed into the hierarchical cosmos of the Neo-Platonists. I will not address Plato himself explicitly here but rather see him as reflected in the tradition he began, namely Neo-Platonism.[32] Aristotle, on the other hand, is the philosopher whose work opens up a whole tradition of reflection upon nature and the actuality of the living. In his thinking we find both reflection on life itself and reflection on living beings,

[25] Ibid., p. 126. [26] Ibid., p. 139. [27] Ibid., pp. 76–80. [28] Ibid., p. 154.

[29] Ibid., p. 168. [30] Ibid.

[31] Lucian, *On Sacrifices*, in *Lucian*, vol. III, trans. A.M. Harmon, Loeb Classical Library 130 (Cambridge, MA: Harvard University Press, 1921), p. 167.

[32] I do not discuss Plato explicitly because the force of his ideas come through in Neo-Platonism and those philosophers embody and reflect his thinking. When speaking about Plotinus, for example, we are always in the shadow of the master.

going so far as to classify them, introduce experiments, and lay the foundations for the whole of Western science and medicine that followed in the centuries to come.

Aristotle—Mapping the Horizontal

It is probably no exaggeration to say that Aristotle laid the foundations of all contemporary knowledge in the West, composing works in logic, metaphysics, physics, and biology that were to have enormous impact upon Scholastic thinking in Latin, Arabic, and Hebrew and from thence into modernity. Comprehensive in his range, Aristotle offers empirical observation on living beings along with an account of life itself as animating principle and offering definition as to precisely what it is to be alive. As Gareth Matthews observes, Aristotle seems to have been the first to explain life in terms of a list of life functions that he called 'soul powers' or 'psychic powers' (*dunameis tēs psuchēs*).[33] A number of works focus on the study of animals and natural processes, but the one book that is most important in his understanding of the nature of life itself is *De Anima, On the Soul.* It is here that Aristotle attempts a systematic analysis of the 'soul' or 'life force' by seeking to understand its properties and operations. But we are immediately in difficulty before we begin, over the translation of Aristotle's *psuchē* that was translated into Latin as *anima* and that into English as 'soul'. While the Latin theologians wished to bring out the implications of Aristotle's understanding with a view to explicating a Christian vision, it is far from certain that 'soul' is an appropriate rendering of what Aristotle is trying to get at. Indeed, Thacker in his insightful study argues that this is really a mistranslation and we are far better to render the term as 'life itself' or 'vital principle'.[34] By 'soul', as Denys Turner observes, Aristotle simply meant that which accounts for something being alive in a certain kind of way as, say, a spider or a fly.[35] Something having a soul is a way of saying it has a living body.

Having discussed the ideas and theories of various philosophers who precede him in Book I, and in the end dismissing them for incoherence or contradiction, Aristotle then wishes to make a fresh start in determining what the soul is[36] or, depending on how we wish to render *psuchē*, what the life force

[33] Gareth B. Matthews, 'De Anima 2.2–4 and the Meaning of Life', p. 185, in Martha C. Nussbaum and Amélie Oksenberg Rorty (eds.), *Essays on Aristotle's* De Anima (Oxford: Clarendon Press, 1992), pp. 185–93.

[34] Eugene Thacker, *After Life* (Chicago, IL: Chicago University Press, 2010), p. 7.

[35] Denys Turner, *Thomas Aquinas: A Portrait* (New Haven, CT: Yale University Press, 2013), p. 60.

[36] Aristotle, *De Anima* II.1 (412a3), trans. W.S. Hett, *On the Soul; Parva naturalia; On Breath*, Loeb Classics 288 (Cambridge, MA: Harvard University Press, 2014 [1957]). I use the usual Bekker line numbering system.

is that animates living beings from rational animals such as ourselves, to non-rational animals and plants. He begins with a reflection on substance (*ousia*) as a kind of existence, and classifying such substance as matter, which is potentiality, form, which is actuality, and a combination of both. Among substances are bodies, some of which have life (*zōe*) and some of which do not, 'by life we mean the capacity for self sustenance, growth and decay'.[37] Given that there are different kinds of bodies, Aristotle goes on, the soul as 'having life' cannot be the body but 'the soul must be a substance in the sense of being the form of a natural body, which potentially has life'.[38] The substance of the body is actuality (*entelechia*). On one view, then, there is a serially ordered structure of life in which each type of soul is distinct without common definition, while on the other there is the view that there is a shared notion of an animating principle.[39] Now while there are different kinds of bodies, what they have in common is that the soul is the actuality of a natural body that has the potential for life. On this account, the soul and the body are not disunited in the sense that the wax and its shape are one; the soul plus the body constitutes the animal. Aristotle illustrates this idea with a tool: being an axe constitutes the 'soul' of the tool, for without this it would not really be an axe but only an axe in name as, by analogy, the pupil plus the power of sight constitutes the eye.[40] On this account the soul is not separate from the body. If a substance comprises form and matter of which the former is actuality and the latter potentiality, then the soul cannot be without a body, although it cannot be a body: 'for the actuality of each thing is naturally inherent in its potentiality, that is in its own proper matter'.[41] He writes:

> As we have already said, substance is used in three senses, form, matter, and a compound of the two. Of these matter is potentiality, and form actuality; and since the compound is an animate thing, the body cannot be the actuality of a

[37] Ibid., 412a14: *ζωὴν δὲ λέγομεν τὴν δι' αὐτοῦ τροφήν τε καὶ αὔξησιν καὶ φθίσιν.*

[38] Ibid., 412a20–1: *ἀναγκαῖον ἄρα τὴν ψυχὴν οὐσίαν εἶναι ὡς εἶδος σώματος φυσικοῦ δυνάμει ζωὴν ἔχοντος.*

[39] Eli Diamond argues that the truth for Aristotle is between these two positions, although in the end 'the nature of life (and of soul as the principle of life) lies in thought, and most perfectly in the uninterrupted thought of God'. Eli Diamond, *Mortal Imitations of Divine Life: The Nature of the Soul in Aristotle's* De Anima (Evanston, IL: Northwestern University Press, 2015), p. 40. Diamond rejects Fraser's account here that maintains that the nutritive soul in plants is the only one that can exist independently, thereby paralleling the category of substance (*ousia*); Kyle Fraser, 'Seriality and Demonstration in Aristotle's Ontology', *Oxford Studies in Ancient Philosophy*, vol. 22, 2002, pp. 43–82. Also see the collection of essays on this theme: Gerd van Riel and Pierre Destrée (eds.), *Ancient Perspectives on Aristotle's* De Anima, Ancient and Medieval Philosophy Series I 41 (Leuven: Leuven University Press, 2009); Richard Sorabji, 'Body and Soul in Aristotle', *Philosophy*, vol. 29, 1974, pp. 3–89.

[40] Aristotle, *De Anima*, 413a1.

[41] Ibid., 414a26–7: *ἑκάστου γὰρ ἡ ἐντελέχεια ἐν τῷ δυνάμει ὑπάρχοντι καὶ τῇ οἰκείᾳ ὕλῃ πέφυκεν ἐγγίνεσθαι.*

> soul, but the soul is the actuality of some body. For this reason those are right in their view who maintain that the soul cannot exist without the body, but is not itself in any sense a body. It is not a body, it is associated with a body, and therefore resides in a body, and in a body of a particular kind.[42]

Form and matter constitute one entity and *psuchē* is the principle that gives substance actuality. On this view the life force is individuated in a particular form and it would not make sense to speak of it existing independently of that form; the shape of the wax is not different from the wax. We can even say that the soul is the cause or source of the living body, its end, and its essence.[43]

But there is some ambiguity in Aristotle's account in the sense that he leaves off his definition with a problem of whether the soul might be the actuality of its body in the sense that the sailor is the actuality of the ship (and so potentially separable from the ship).[44] He then goes on to reflect on the characteristics of living entities as possessing 'thinking or perception or local movement and rest, or movement in the sense of nutrition, decay and growth'.[45] Of these features of the soul, self-nutrition is the most basic as, for example, in the plant that can grow in all directions and continues to live so long as it can absorb nutrients. This is the only power of the soul that the plant possesses and it is the basis for all other forms of life. This is the first power that allows us to speak of something as living. Other life forms possess further powers; so sensation is a feature of animals, for even living things that do not possess the power of local movement do possess the power of sensation. It is this power that allows us to speak of living things as animals.[46]

To summarize Aristotle's text, we can, then, distinguish living beings according to the powers of the life force that they possess. These powers are primarily the nutritive (*threption*), the sensory (*aisthāesis*), the locomotive (*kinēsis*), and the power of thinking (*dianoētikon*). Plants have only the first whereas animals have the first along with the sensory and the appetitive. Appetite (*orexis*) is the genus with passion (*thumos*) and desire (*epithumia*) as the species. All animals have at least the sense of touch and so the capacity for pleasure and pain and thereby the capacity of appetite, for the sense of touch apprehends nutrients that are hot, cold, dry, or moist. Other animals also possess locomotion and man, and higher beings than him (the gods), possesses all of these plus the power of thinking.[47] These constitute an embedded hierarchy such that perception is never found distinct from

[42] Ibid., 414a14–21: *τριχῶς γὰρ λεγομένης τῆς οὐσίας, καθάπερ εἴπομεν, ὧν τὸ μὲν εἶδος, τὸ δὲ ὕλη, τὸ δὲ ἐξ ἀμφοῖν· τούτων δ᾽ ἡ μὲν ὕλη δύναμις, τὸ δὲ εἶδος ἐντελέχεια· ἐπεὶ δὲ τὸ ἐξ ἀμφοῖν ἔμψυχον, οὐ τὸ σῶμά ἐστιν ἐντελέχεια ψυχῆς, ἀλλ᾽ αὔτη σώματός τινος. καὶ διὰ τοῦτο 20καλῶς ὑπολαμβάνουσιν οἷς δοκεῖ μήτ᾽ ἄνευ σώματος εἶναι μήτε σῶμά τι ἡ ψυχή· σῶμα μὲν γὰρ οὐκ ἔστι, σώματος δέ τι, καὶ διὰ τοῦτο ἐν σώματι ὑπάρχει, καὶ ἐν σώματι τοιούτῳ.*

[43] Ibid., 515b9–12. [44] Ibid., 413a7–9. [45] Ibid., 413a21–5.

[46] Ibid., 413b1–4. [47] Ibid., 414a29–414b19.

nutrition and touch is found alone in some animals but never sight, hearing, or smell. Likewise, thought and imagination assume the existence of the other powers. Aristotle speaks of the different forms that these powers constitute as different types of soul, the nutritive, the sensitive, and the rational; the nutritive being the most basic, particular to plants,[48] with humans having all three. And while these constitute a typology of living beings who share life (*zoē*), which is sweet like a beautiful day (*euēmeria*),[49] beyond mere existence there is human life (*bios*) constituted of ways of life in the polis, namely the philosopher's life of contemplation (*bios theōrētikos*), the life of pleasure (*bios apolaustikos*), and the political life (*bios politikos*).[50]

With Aristotle we have an analysis of types of living beings and an inquiry into what animates them. But what animates bodies cannot be distinguished from the body in the sense that it does not outlive the body. This is quite distinct from Plato's idea of the soul and the later Neo-Platonic doctrine of the soul's journey out of the body back to its source. Aristotle does have the idea that the goal (*telos*) towards which things strive is participation in the eternal and the divine as far as their natures as plants, animals, or humans allows,[51] but the particular soul existing outside of the body is alien to him, although there is a sense in the disembodied intellect (*nous*) surviving death, although this is not survival of the particular person. The *nous* cannot die and so outlives the body,[52] but it remains vague as to its cosmic function independent of the body for the *nous* is not particular to each but general and shared by all. There is little eschatological hope in Aristotle; indeed, he is simply not interested in any such notion but in the analysis of life as we find it, and in developing his theory that became known as hylomorphism, the combination of matter and form.

Whether Aristotle continues to be philosophically relevant is a matter of dispute. On the one hand, philosophers Nussbaum and Putnam have sought in Aristotle a way through the body–mind problem that neither reduces mind to pure physicality nor supports a Platonic and Cartesian dualism. On the other hand, Burnyeat has argued that this understanding cannot be right because Aristotle's physics is no longer credible, and he does not account for how matter and form arise in the first place.[53] The dispute remains unresolved,

[48] Ibid., 416b26.

[49] Aristotle, *Politics*, 1278b, 23–31, *Aristotle XXI*, trans. H. Rackham, Loeb Classics 264 (Cambridge, MA: Harvard University Press, 1932).

[50] Aristotle, *Nicomachean Ethics*, I.4, pp. 13–16, *Aristotle IXX*, trans. H. Rackham, Loeb Classics 73 (Cambridge, MA: Harvard University Press, 1932). On the necessity of biological life for contemplation see Matthew D. Walker, *Aristotle on the Uses of Contemplation* (Cambridge: Cambridge University Press, 2018), pp. 16–18, 42–5.

[51] Aristotle, *De Anima*, 414a25–415b1.

[52] Ibid., 408b18, 429b22.

[53] See the essays, including Burnyeat's, in Martha C. Nussbaum and Amélie Oksenberg Rorty (eds.), *Essays on Aristotle's* De Anima (Oxford: Clarendon Press, 1992).

but it shows clearly the ongoing impact of Aristotle's thinking. My purpose here is to offer a description rather than assessment of his understanding of what it is to be a living being in order to demonstrate his central concern with life itself, but a very different conception of life itself as an animating principle that cannot be separated from what it animates. We might call this a horizontal view of life that disparages a Platonic verticality in stressing that we have to account for the reality of living in broadly naturalist terms. Even given this naturalist orientation, Aristotle's *De Anima* had a massive impact on theological thinking in the Abrahamic traditions of the medieval period (on the Arabic, Hebrew, and Latin theologians), on biological thinking through to the eve of modernity, and even on the philosophy of Franz Brentano and thence into the phenomenological tradition, being an especially important influence on Heidegger's categorization of life.[54] But there was also another tradition that had a massive impact on the history of the West, namely Platonism: equally concerned with life itself and its relation to the living, it develops a quite distinct view of the self and the necessity of transcendence in order to fulfil the purpose of life and understand its nature.

Plotinus—Aspiring to the Vertical

Plotinus (204–70 CE) marks the end of what is called Middle Platonism and the start of Neo-Platonism. With Plotinus we have a return to or rather transformation of Plato's philosophy; a transformation in the sense that Plotinus focuses on the experience of the forms in mystical ascent, the return of the individual life to its life source. There is a simplicity to Plotinus' vision, as Hadot observes,[55] but a vision that can become complex in the unfolding of life into its myriad forms. The general model that Plotinus advocates, a model almost identical to the Indic hierarchical cosmologies we have already encountered,[56] is that pure being that we can understand as life in itself generates forms in a pure realm, uncontaminated by matter, that then manifest the world in gradual degrees of solidification. The forms generate the multiple life of the world. All beings contain a rational principle (*logos*) and are the activity (*energeia*) of a universal soul (*psuchēs pases*). The soul is in

[54] Martin Heidegger, *Phenomenological Interpretations of Aristotle*, trans. Richard Rojcewicz (Bloomington and Indianapolis, IN: Indiana University Press, 1985), pp. 64–97. On the three traditions of interpretation see Thacker, *After Life*, p. 9.

[55] Pierre Hadot, *Plotinus or the Simplicity of Vision*, trans. Michael Chase with an introduction by Arnold L. Davidson (Chicago, IL: Chicago University Press, 1993 [1989]), p. 42. On Plotinus's worldview see in particular Gary M. Gurtler SJ, *Plotinus, the Experience of Unity* (1988).

[56] Others have observed this. See for example Frits Staal, *Advaita and Neoplatonism: A Critical Study in Comparative Philosophy* (Madras: University of Madras, 1961), pp. 170–9.

harmony with its activity and in harmony with other souls and their activities, although they may appear to be radically different beings; even though beings may be opposites, they are part of a common order (*homos suntakseis*) because they come from a unity. All creatures are brought together under a single genus, 'the living creature' (*to zōon genos*).[57] The totality is itself alive: 'a multiplex living thing' (*zōon pole*) with distinct parts.[58] The entire universe is a living being (*zōon de ontos tou pantos*)[59] of which the particular living beings are parts.

For Plotinus a single life flows through all things, life that is identified with light and with the One (*monos*). The One, who is light and life, generates the multiple forms of the universe in a hierarchical sequence, through the mind (*nous*) where we experience the forms, to soul (*psuchē*), to the multiplicity of the world.[60] Life in the higher world is movement, driven by the One, a movement that implies space and extension, although the world of forms is only extension by analogy, for that world is quite distinct from our own. The One might be understood as life in itself, manifesting the forms that descend lower, generating a hierarchy, and becoming condensed and particularized in materiality within which the individual soul is trapped. In a famous passage Plotinus speaks of waking up out of the body in a divine realm, which is also entering into himself, before descending back to the body via discursive reasoning and being puzzled as to how he came there.[61]

Behind the appearance of the material world we have the forms, the source of visible manifestation.[62] These forms—Plato's pure realm of truth, goodness, and beauty—are animated by spirit (*psuchē*) for Plotinus. All manifestations in the material world have their source in the forms. Thus, we might speak of an original plant or animal that finds expression in different types of plant and animal. Aristotle criticized the Platonic forms as being lifeless, but Plotinus in a sense animates them and for him 'Plato's world of Forms has become a complex and dynamic intelligible universe in which unity and plurality, stability, and activity are reconciled'.[63] The entire universe, the whole cosmical hierarchy, is itself a living thing akin to the way an individual body expresses life.[64] The One and the cosmos form an organic unity in

[57] Plotinus, *Enneads*, III.3.1. A.H. Armstrong (trans.), *Plotinus: Porphyry on Plotnus*, Loeb Classical Library 440 (Cambridge, MA: Harvard University Press, 1966).

[58] Ibid. [59] Ibid., III.3.2.6.

[60] On these three hypotheses see Michael Atkinson, *Plotinus, Ennead V.1: On the Three Principal Hypotheses: A Commentary with Translation* (Oxford: Oxford University Press, 1983).

[61] Plotinus, *Enneads*, 4.8.1. [62] Hadot, *Plotinus*, p. 40.

[63] John M. Dillon and Andrew Smith, 'Introduction', p. 6 in Gary M. Gurtler, *Enneads of Plotinus: Ennead 4.30–45 and IV.5: Problems concerning the Soul: Translation, with an Introduction and Commentary* (Las Vegas, NV, Zurich, Athens: Parmenides Publishing, 2015).

[64] Gurtler, *Enneads*, pp. 227–8.

which all the parts are interrelated. Plotinus writes in a passage worth quoting at length:

> First of all we must posit that this All is a 'single living being which encompasses all the living beings that are within it'; it has one soul which extends to all its parts, in so far as each individual thing is a part of it; and each thing in the perceptible All is a part of it, and completely a part of it as regards its body; and in so far as it participates in the soul of the All, it is to this extent a part of it in this way too; and those things which participate in the soul of the All alone are altogether parts, but all those which also participate in another soul are in this way not altogether parts, but none the less are affected by the other parts in so far as they have something of the All, and in a way corresponding to what they have. This one universe is all bound together in shared experience and is like one living creature, and that which is far is really near, just as, in one of the individual living things, a nail or horn or finger or one of the other limbs which is not contiguous: the intermediate part leaves a gap in the experience and is not affected, but that which is not near is affected. For the like parts are not situated next to each other, but are separated by others between, but share their experiences because of their likeness, and it is necessary that something which is done by a part not situated beside it should reach the distant part; and since it is a living thing and all belongs to a unity nothing is so distant in space that it is not close enough to the nature of the one living thing to share experience.[65]

Even though parts of the individual body, or parts of the cosmos, are distant from each other, they affect each other because of their participation in the cosmic entity, the one life of the All. The whole cosmos is an organic unity in sympathy or harmony, the one part affecting the other parts. Life itself 'boils over', to use Hadot's phrase, with abundance.[66] Even as parts of the body that are discontinuous are connected because of the unity of the ensouled body, so all beings are united by the world soul. The multiple appearances of the world are but emanations of the forms, which should be the true object of contemplation. Yet while, on the one hand, Plotinus has a positive evaluation

[65] Plotinus, *Enneads*, IV.4.32, pp. 236–9: *πρῶτον τοίνυν θετέον ζῷον ἓν πάντα τὰ ζῷα τὰ ἐντὸς αὑτοῦ περιέχον τόδε τὸ πᾶν εἶναι, ψυχὴν μίαν ἔχον εἰς πάντα αὐτοῦ μέρη, καθόσον ἐστὶν ἕκαστον αὐτοῦ μέρος· μέρος δὲ ἕκαστόν ἐστι τὸ ἐν τῷ παντὶ αἰσθητῷ, κατὰ μὲν τὸ σῶμα καὶ πάντη, ὅσον δὲ καὶ ψυχῆς τοῦ παντὸς μετέχει, κατὰ τοσοῦτον καὶ ταύτῃ· καὶ τὰ μὲν μόνης ταύτης μετέχοντα κατὰ πᾶν ἐστι μέρη, ὅσα δὲ καὶ ἄλλης, ταύτῃ ἔχει τὸ μὴ μέρη πάντη εἶναι, πάσχει δὲ οὐδὲν ἧττον παρὰ τῶν ἄλλων, καθόσον αὐτοῦ τι ἔχει, καὶ κατ' ἐκεῖνα, ἃ ἔχει. συμπαθὲς δὴ πᾶν τοῦτο τὸ ἕν, καὶ ὡς ζῷον ἕν, καὶ τὸ πόρρω δὴ ἐγγύς, ὥσπερ ἐφ' ἑνὸς τῶν καθέκαστα ὄνυξ καὶ κέρας καὶ δάκτυλος καὶ ἄλλο τι τῶν οὐκ ἐφεξῆς ἀλλὰ διαλείποντος τοῦ μεταξὺ καὶ παθόντος οὐδὲν ἔπαθε τὸ οὐκ ἐγγύς. οὐ γὰρ ἐφεξῆς τῶν ὁμοίων κειμένων, διειλημμένων δὲ ἑτέροις μεταξύ, τῇ δὲ ὁμοιότητι συμπασχόντων, καὶ εἰς τὸ πόρρω ἀφικνεῖσθαι ἀνάγκη τὸ παρὰ τοῦ μὴ παρακειμένου δρώμενον· ζῴου τε ὄντος καὶ εἰς ἓν τελοῦντος οὐδὲν οὕτω πόρρω τόπῳ, ὡς μὴ ἐγγὺς εἶναι τῇ τοῦ ἑνὸς ζῴου πρὸς τὸ συμπαθεῖν φύσει.*

[66] Ibid., p. 50. On these three hypotheses see Michael Atkinson, *Plotinus, Ennead V.1: On the Three Principal Hypotheses: A Commentary with Translation* (Oxford: Oxford University Press, 1983).

of the world as the manifestation of the pure forms—the world is indeed beautiful and marvellous—he also has a negative evaluation of it as the trap or prison of the soul. The animating principle of the body, the soul or *psuchē*, is trapped within this world of multiplicity and needs to ascend back through the hierarchy from whence it came to unite with the source of life, goodness, beauty, and light.

Plotinus describes the cosmic hierarchy in terms of three hypostases: the One (*monos*), the intellect (*nous*), and the soul (*psuchē*).[67] Each of these is a lower emanation of the higher and the soul incarnates or descends into physical or material bodies.[68] The soul mediates between the intellect and the material world, and indeed generates and creates in a parallel, but more limited, way to nature. There is in Plotinus a complex of ideas mapped onto a generative hierarchy from the One. At the level of the intellect we have the forms, and the forms, including nature, generate the multiplicity of the universe, specifically the four elements earth, air, fire, and water, within which the soul acts, being both a life-giving or animating force for the body and having a contemplative nature. Interestingly, Plotinus rejects the Stoic identification of the soul with breath (*pneuma*) on the grounds that the doctrine posits the soul as intelligence but also as body, which he thinks is not coherent.[69] Thus the soul moves into the world through body, enacting the results of its past acts, but has a proclivity to turn away from the world, to face inwards towards its true home and source in the One. The soul that inhabits different bodies is diminished as it moves further from its origin; as the souls hasten towards matter they become less,[70] there is a fading of intensity, and as the soul moves through different bodies it takes with it the results of its actions in previous lives.[71] Indeed souls trapped in the world originally took form as stars, but descended to a lower materiality. Taking his cue from Aristotle, the soul comprises different levels, the vegetative level, growth, sensitive, and rational,[72] and in this way is equipped to interact with the material world and also to develop higher, contemplative functions that allow it to return to its source.

In the end, the beauty of the world is a deception that draws us in and keeps us bound, as Odysseus, through his attachment, was bound to Circe and her island. Plotinus offers a mystical interpretation of Plato in which the forms are also objects of contemplation, which become experiences of the soul. The soul's ascent back through the hierarchy to its source is a reversal of the process. In this the appearance of lower forms shuts out or closes out the higher and conversely in mystical ascent the lower layers of the hierarchy

[67] Plotinus, *Enneads*, V.3.1–2.

[68] Barry Fleet, *Plotinus Ennead IV.8: On the Descent of the Soul into Bodies* (Las Vegas, NV, Zurich, Athens: Parmenides Publishing, 2012).

[69] Gurtler, p. 333 discussing *Enneads*, IV.7.4.

[70] Plotinus, *Enneads*, III.3.2.3.

[71] Ibid., III.3.2.4.

[72] See Gurtler's translation of *Enneads*, IV and his discussion (p. 8).

are closed off as the soul rises. Only from the perspective of the One, who is at the top of the hierarchy and yet who also pervades it and is not different from it, is there a perception of totality.

The ascent of the soul back up through the hierarchy is envisaged as a journey back across the ocean dividing the world from its source of light, beauty, and the good. As Odysseus left the island of Circe to return home, even though the island was enchanting and delightful, so the soul must leave this world to return home to its father, the One, the source of life. But we have no ship to carry us there across a horizontal ocean; rather, we must cultivate contemplation and an inner sight that will transport us back through the cosmos to our true home. Plotinus writes:

> This would be truer advice: 'Let us fly to our dear country'. What then is our way of escape, and how are we to find it? We shall put out to sea, as Odysseus did, from the witch Circe or Calypso—as the poet says (I think with a hidden meaning)—and was not content to stay though he had delights of the eyes and lived among much beauty of sense. Our country from which we came is there, our Father is there. How shall we travel to it, where is our way of escape? We cannot get there on foot; for our feet only carry us everywhere in this world, from one country to another. You must not get ready a carriage, either, or a boat. Let all these things go, and do not look. Shut your eyes and change to and wake another way of seeing which everyone has but few use.[73]

On the inner journey back home, we need to develop a new way of seeing, an inner sight that turns away from the external sight of the eyes onto the world, to an inner world where begins the journey. After the above-quoted passage, Plotinus says that we must go back into ourselves and look for there we find the beautiful; we must cultivate ourselves as a sculptor making a statue cuts away and polishes it—so the soul must cultivate virtue. The cultivation of this inner sight, this inner vision that beholds the beautiful and the good through ascent, is dependent upon virtue to purify the self. To behold the good, one must be good. As Plotinus says, 'no eye ever saw the sun without becoming sun-like nor can a soul see beauty without becoming beautiful. You must become first all godlike and all beautiful if you intend to see God and beauty'.[74]

[73] Plotinus, *Enneads*, I.6.8, 16–28: *Φεύγωμεν δὴ φίλην ἐς πατρίδα, ἀληθέστερον ἄν τις παρακελεύοιτο. Τίς οὖν ἡ φυγὴ καὶ πῶς; ἀναξόμεθα οἷον ἀπὸ μάγου Κίρκης (φησὶν) ἢ Καλυψοῦς Ὀδυσσεὺς (αἰνιττόμενος, δοκεῖ μοι) μεῖναι οὐκ ἀρεσθείς, καίτοι ἔχων ἡδονὰς δι᾽ ὀμμάτων καὶ κάλλει πολλῷ αἰσθητῷ συνών. Πατρὶς δὴ ἡμῖν, ὅθεν παρήλθομεν, καὶ πατὴρ ἐκεῖ. Τίς οὖν ὁ στόλος καὶ ἡ φυγή; Οὐ ποσὶ δεῖ διανύσαι· πανταχοῦ γὰρ φέρουσι πόδες ἐπὶ γῆν ἄλλην ἀπ᾽ ἄλλης· οὐδέ σε δεῖ ἵππων ὄχημα ἤ τι θαλάττιον παρασκευάσαι, ἀλλὰ ταῦτα πάντα ἀφεῖναι δεῖ καὶ μὴ βλέπειν, ἀλλ᾽ οἷον μύσαντα ὄψιν ἄλλην ἀλλάξασθαι καὶ ἀνεγεῖραι, ἣν ἔχει μὲν πᾶς, χρῶνται δὲ ὀλίγοι.*

[74] Ibid., 1.6 [1] 9: *Οὐ γὰρ ἂν πώποτε εἶδεν ὀφθαλμὸς ἥλιον ἡλιοειδὴς μὴ γεγενημένος, οὐδὲ τὸ καλὸν ἂν ἴδοι ψυχὴ μὴ καλὴ γενομένη. Γενέσθω δὴ πρῶτον θεοειδὴς πᾶς καὶ καλὸς πᾶς, εἰ μέλλει θεάσασθαι θεόν τε καὶ καλόν.*

That is, the soul takes on the properties of its objects of perception, so to contemplate the beautiful and the good is to adopt those properties. Perception is a kind of transformation and the transformation or purification is contingent upon the degree to which the self conforms to its objects. So, to become divine one must focus inner sight on the divine and thereby take on its qualities in a gradual process of purification. With such cultivation the soul rises up through the cosmic hierarchy within itself, gradually ascending by way of different stages or levels until it reaches and beholds the One and is united with it. This is mystical transport and although Plotinus does not actually attempt to describe any correlative experience, he does tell us that the One is formless, beyond the intellect, and anything that we can say about manifestation must be denied of it.[75] The self unites with this un-predicable source of life and mystery as a lover rests in the beloved, but yet the soul may be held back in the ascent by some impediment that hinders vision.[76] But if it surpasses these, the soul joins to the One, which is self-sufficient, in need of nothing, and is the one principle of all things. In this dance the soul sees the spring of life (*peges zōes*)[77] and is finally at the end of the journey (*telos poreias*). This, in Plotinus' most famous phrase, is 'the flight of the alone to the alone' (*puge monou pros monon*).[78]

Here horizontal experience is replaced by vertical ascent. The journey of the alone to the alone is a return of a fragment of life back to life itself. To reach the One, our true home, we must leave reflection behind, the rational mind thinking through language, and replace it with contemplation. Contemplation is vision turned inwards so that, as in India, the self ascends within itself, within the body, on a journey back to the One. The soul must wake up from its slumber in the body for this world is like sleep:

> This, then, is our argument against those who place real beings in the class of bodies and find their guarantee of truth in the evidence of pushings and strikings and the apparitions which come by way of sense-perception; they act like people dreaming, who think that the things they see as real actually exist, when they are only dreams. For the activity of sense-perception is that of the soul asleep; for it is the part of the soul that is in the body that sleeps; but the true wakening is a true getting up from the body, not with the body. Getting up with the body is only getting out of one sleep into another, like getting out of one bed into another; but the true rising is a rising altogether away from bodies, which are of the opposite nature to soul and opposed in respect of reality. Their coming into being and flux and perishing, which does not belong to the nature of reality, are evidence of this.[79]

[75] Ibid., 6.9.3. [76] Ibid., 6.9.4. [77] Ibid., 6.9.9.

[78] Ibid., 6.9.11.50. There is some dispute over the translation of this phrase. Armstrong translates it as 'escape in solitude to the solitary', and Hadot as 'to flee alone, towards the Solitary One' (p. 97). The more metaphysically non-dualist rendering here retains a sense of irony that perhaps Plotinus intended.

[79] Ibid., III.6.6: *Ταῦτα μὲν οὖν εἴρηται πρὸς τοὺς ἐν τοῖς σώμασι τιθεμένους τὰ ὄντα τῇ τῶν ὠθισμῶν μαρτυρίᾳ καὶ τοῖς διὰ τῆς αἰσθήσεως φαντάσμασι πίστιν τῆς ἀληθείας λαμβάνοντας, οἳ παραπλήσιον τοῖς ὀνειρώττουσι ποιοῦσι ταῦτα ἐνεργεῖν νομίζουσιν, ἃ ὁρῶσιν εἶναι ἐνύπνια ὄντα.*

As we have seen in some Indic material, this world of appearance is like sleep in which the soul is reborn repeatedly due its past actions. Plotinus adheres to a doctrine that transmigration and suffering is due to the soul's misdeeds in a previous life. Following from Plato, Plotinus tells us that the soul conforms to the body it inhabits depending upon its past deeds. Thus, the rational principle (*logos*) of the soul contains within itself the rational principle of matter (*hule*) to which it corresponds.[80] This is remarkably close, indeed identical, to Indic teachings about reincarnation and karma. While Plotinus' disciple Porphyry tells us that Plotinus joined Gordian's army against the Persians in the hope of getting to India to speak with Brahmins,[81] he never reached there as Gordian was defeated and killed. The extent to which Plotinus was exposed to Indic thinking is unclear but he may have been influenced. It is also vaguely possible that the influence was the other way around; that Greek ideas of reincarnation entered India, although the two cultures may have come up with the idea independently. But the model that Plotinus develops is fundamentally the same as a yogic model of human nature, that the soul is ensnared in the world of passions and needs to withdraw from that world, turn inwards, and contemplate the higher reality which is also a return to its source. The life of the sage is focused on contemplation in contrast to the life of action in world, focused on family and political success.[82]

The journey to the One is more than a metaphor for Plotinus. The inner journey of ascent is an actual experience and he does think the world of the forms to be more real than the world of appearances, yet he also wishes to maintain that life itself can be seen in the myriad, beautiful forms of the world. Although there is a negative evaluation of life because we are trapped by it—we fall under the spell of nature's magic[83]—Plotinus also wished to acknowledge the beauty of the world as a reflection of the forms.

This then is the general picture that Plotinus sketches, although there is some slippage in his understanding and use of the category 'life'. The word that Plotinus uses is *zōe*. He uses the term for actual living beings but also in an

Καὶ γὰρ τὸ τῆς αἰσθήσεως ψυχῆς ἐστιν εὑδούσης· ὅσον γὰρ ἐν σώματι ψυχῆς, τοῦτο εὕδει· ἡ δ' ἀληθινὴ ἐγρήγορσις ἀληθινὴ ἀπὸ σώματος, οὐ μετὰ σώματος, ἀνάστασις. Ἡ μὲν γὰρ μετὰ σώματος μετάστασίς ἐστιν ἐξ ἄλλου εἰς ἄλλον ὕπνον, οἷον ἐξ ἑτέρων δεμνίων· ἡ δ' ἀληθὴς ὅλως ἀπὸ τῶν σωμάτων, ἃ τῆς φύσεως ὄντα τῆς ἐναντίας ψυχῇ τὸ ἐναντίον εἰς οὐσίαν ἔχει. Μαρτυρεῖ δὲ καὶ ἡ γένεσις αὐτῶν καὶ ἡ ῥοὴ καὶ ἡ φθορὰ οὐ τῆς τοῦ ὄντος φύσεως οὖσα.

[80] Ibid., III.3.2.

[81] Porphyry, *On the Life of Plotinus and the Order of his Books*, 2, p. 9, in A.H. Armstrong (trans.), *Plotinus: Porphyry on Plotnus*, Loeb Classical Library 440 (Cambridge, MA: Harvard University Press, 1966).

[82] Plotinus, *Enneads*, IV, 4.44.1–6. See Gurtler's discussion, p. 209. Also see Suzanne Stern-Gillet, 'When Virtue Bids Us Abandon Life (*Ennead* VI 8 [39] 6, 14–26', in Filip Karfik and Euree Song (eds.), *Plato Revived: Essays on Ancient Platonism in Honour of Dominic O'Meara* (Boston, MA and Berlin: Walter de Gruyter, 2013), pp. 182–98.

[83] Plotinus, *Enneads*, ch. 44; Gurtler, p. 217.

abstract sense as a quality that they possess. He does speak about the totality of the All as a living being, as we have seen, but also speaks in terms of the source of life being beyond life. Plotinus' non-dualism is more of an emanationist cosmology in which the One emanates the manifold forms of the cosmos and these forms are life, while the One transcends life: 'So if there is anything prior to actuality, it transcends actuality, so that it also transcends life'.[84] The One is beyond life in the sense that it is the source of living beings but nevertheless the intellect that generates the lower manifestation contains a trace (*ichnos*) of the good, the source,[85] which is then recapitulated in the lower orders. Although the source is in a sense beyond life, it nevertheless contains life (and intellect), otherwise it would be lifeless and mindless and these are the qualities of non-being. It is not possible for a mindless thing to generate mind, or a composite body to generate life.[86] The fullness of the One generates intellect, which generates the rest of manifestation, and all is characterized by life in varying degrees of intensity. The soul is the force that animates the body, as the forms at the level of intellect are the force that animates the forms of the world. Both form and soul are animated by the One.

The world is filled with life, the luminosity of the One, the power that flows through all things, and yet as souls we long to return to our true home beyond even the highest levels. Like Odysseus, we are wanderers and like Odysseus, we must cross an ocean of trouble. To do this we turn our gaze inwards in contemplation, but this is not self-absorption like Narcissus (whom Plotinus contrasts with Odysseus)[87] but a turn towards the One, which is within us: the light and melody of pure transcendence that attracts and pulls the soul out of the material world. As with the Indic material, there is an ambiguity in Plotinus towards the world—it is a reflection of beauty but not itself the wellspring of life that is our soul's destiny—with both a positive and negative appraisal of the material world. And yet his vision of the interconnectedness of all things is clearly an ideal with contemporary resonance and indeed according to Hadot, Plotinus is one of the sources for Bergson's 'organic solidarity' (*totalités organiques*) and 'the immediate' (*l'immédiat*).[88] Plotinus is an important thinker who we can understand in terms of the philosophy of life because he has a strong sense of life overflowing from a single source of luminosity, but like many non-dualists he has to balance the idea and goal of transcendence with the immanence of life itself, a balance that Plotinus manages to achieve.

Plotinus' emphasis on contemplation, inwardness, and mystical ascent is a rejection of externality and the sacrificial religion of his day. He refused to participate in external ritual and, rather than go to the sacrifice, says that the

[84] Plotinus, *Enneads*, VI.7.17: *εἴ τι τοίνυν ἐνεργείας πρότερον, ἐπέκεινα ἐνεργείας, ὥστε καὶ ἐπέκεινα ζωῆς.*

[85] Ibid. [86] Ibid., IV.7.2. [87] Ibid., I.6.8. [88] Hadot, *Plotinus*, p. 41.

gods should come to him. Like other mystics, Plotinus had an ambivalent attitude to the world that allows different readings of his work. The emphasis on mystical ascent is a more solitary reading that stresses the particularity of the soul and its journey, while the emphasis on the world as a single living being of which we are parts is a more communal reading that stresses the interrelationship and connection between all things. This latter reading resonates more with contemporary sensibilities and is open to claiming Plotinus for a contemporary non-dualism that stresses the organic unity of life. The former kind of reading influenced the Christian mystical tradition, the Neo-Platonic tradition through to Ficino and Western esotericism, while the latter kind of reading found resonance in modern thinkers, in Goethe and in Bergson. For Plotinus life is good and beautiful, but at the same time it can be a trap that ensnares us from which we must flee back to our true home, the source of life. But although Plotinus advocated the separation of the soul from the body in mystical ascent, his attitude towards the body was nevertheless one of gentleness. The body housed the soul and is to be respected but calmed through contemplation. This is no Gnosticism where flight from the body implies a negative attitude towards it, and Plotinus' life bears witness to the affirmation of the world in his dealings with disciples, his compassion towards orphans, and his desire to propagate the way of the philosopher.[89]

In terms of the concerns of Part I of this book, Plotinus offers an example of an extended pro-sociality that encompasses the totality of life. In offering an account of the interconnectedness of all things, Plotinus illustrates a way in which social cognition can move beyond the human group to encompass the wider world. It is true that Plotinus' teaching was probably intended for the esoteric few, but it nevertheless necessarily embraces everyone—including all living beings—in its vision. In this vision the One pervades the universe as its source and so is accessible from all points of it, yet through vertical ascent is realized in a more intense form. But as Plotinus says, the One is experienced in the small things of life as well as the great.

Ancient Philosophy and Christianity

Saint Paul laid the foundations of the Christian world with his high-energy input into fledgling Christian communities, laying the theological foundations of doctrine. While being influenced by Jewish Merkavah mysticism[90] of vertical ascent and operating in the highly articulate world of Pagan philosophy and the Platonic orientation towards transcendence, he is nonetheless

[89] Hadot, *Plotinus*, pp. 93–6.

[90] Christopher Rowland and C.R.A. Morray-Jones, *The Mystery of God: Early Jewish Mysticism and the New Testament* (Leiden: Brill, 2009), pp. 137–66.

affirmative of the body and sees the necessity of operating successfully in the social and political realm. If Christianity is to flourish, it needs to operate strategically within the social and political realities of the time. Paul was remarkably successful in his transformation of the Jewish sect into the world religion it became, even though there is a certain precariousness about Christianity's early years and genuine questions about whether it could survive at all in a climate of disdain and suspicion in the Roman world.[91] Although Paul's concerns were arguably more political than philosophical, he nevertheless had to locate his Christian theology in the intellectual world in which he found himself. This was Platonism to be sure, which develops into the Plotinian tradition we have just seen, but more important were Stoic and Epicurean philosophy in Paul's ambient intellectual environment, a philosophical force that influences Jewish, Hellenistic philosophy, especially in Alexandria and the philosophy of Philo. Alongside this he is influenced by apocalyptic Judaism that anticipates the coming of the Messiah.[92]

Stoicism was arguably the dominant philosophy in Rome in the years prior to Christianity that Paul encounters and needs to engage with. On the whole, Stoicism has an affirmative philosophy of life that, while being hierarchical in its ontology, nevertheless gives positive value to the dynamism that, for the Stoics, invigorates the world. Cicero, living under the Republic, expresses the Stoic philosophy that is out of consonance with Platonism in its materialism and affirmation of life. Thus, for example, the Stoics believed in a providential God who cared for the cosmos but is immanent within it, in contrast to the Platonic deity outside of cosmos who leaves its running to the Demiurge (*dēmiourgos*), and the Epicurean deity who, having created, does not intervene.[93] A fine articulation of Stoic philosophy is Cicero's *De Natura Deorum*, written about 50 BCE. For Cicero there is a 'natural scale' (*scala naturae*) of life with the four elements dominating this world and forming non-organic entities, plants, sentient but non-rational animals, and human beings as rational animals. Above the four elements was the ether (*aetheris*) which

[91] For an engaging account of the unlikely rise of Christianity see Larry W. Hurtado, *Destroyer of the Gods: Early Christian Distinctiveness in the Roman World* (Baltimore, MD: Baylor University Press, 2016).

[92] For a definitive study see Christopher Rowland, *The Open Heaven: A Study of Apocalyptic Judaism and Early Christianity* (London: SPCK, 1982).

[93] Thomas Bénatouil, 'How Industrious Can Zeus Be?', in Richard Salles (ed.), *God and Cosmos in Stoicism* (Oxford: Oxford University Press, 2009), chapter 1. This volume is an excellent up-to-date source on Stoic cosmology. Also see Michael Lapidge, 'Stoic Cosmology', in J.M. Rist (ed.), *The Stoics* (Berkeley and Los Angeles, CA: University of California Press, 1978), pp. 161–85; M.W. Wright, *Cosmology in Antiquity* (London and New York: Routledge, 1995), pp. 47–8, 84–6; Robert B. Todd, 'The Stoics and their Cosmology in the First and Second Centuries AD', in W. Haase (ed.), *Aufstieg und Niedergang der Römischen Welt* (Berlin: de Gruyter, 1989), pp. 1365–78; David E. Hahm, *The Origins of Stoic Cosmology* (Columbia, OH: Ohio State University Press, 1977).

constituted the matter of the heavens—the sun, moon, stars—and which affected the lower levels where it is transformed into intelligence (*intelligentia*) and perception (*sensus*) in the human realm. This ether is a substance akin to fire, yet unlike worldly fire it is not destructive but rather animates the bodies of living creatures. This fire is 'the life and health of the body' (*corporeus vitalis et salutaris*).[94] The bodies of the stars, moon, and sun are made of ether and it is the fire that gives them intelligence, for they are conscious beings for Cicero; that is, gods.[95] This designing fire spreads downwards from the ether as breath or life force (*pneuma*) that animates the bodies beneath the heavenly bodies. In human beings it becomes the soul (*psuchē*) constituted by reason (*logos*) or mind (*nous*), but in lower animals it is only soul. Descending lower, the life force becomes nature (*physis*) and inorganic states of being (*hexis*).[96] This pneuma is the life of the world, the power (*dunamis*) that animates the entire universe, controlling everything from rocks to plants to human intelligence and political affairs. Nothing that is inanimate and irrational can give birth to the rational and animate, says Cicero, quoting Zeno, therefore the world is animate and rational.[97] This animating force is indicated by the presence of heat. Cicero writes:

> Every living thing therefore, whether animal or vegetable, owes its vitality to the heat contained within it. From this it must be inferred that this element of heat possesses in itself a vital force that pervades the whole world.[98]

This heat manifests itself in male and female and so is responsible for generation that itself is within a hierarchy of scale with higher principles governing lower. Thus the 'ruling principle', and Cicero cites the Greek term *hēgemonikon* here, has supremacy over that which it governs. The ruling principle over the whole of nature is therefore the highest. By this ruling principle, 'the glowing heat of the world is purer, more brilliant, and more mobile' (*mundi ille fervor purior perlucidior mobiliorque multo*) than the ordinary heat we experience.[99] Movement, of animals for example, is generated by this heat. Indeed, the entire world is held together by this principle which is a divine nature (*natura divina*) and we can therefore say that the world is God (*deus*).[100] In a line of reasoning that links up the levels of the universe, Cicero writes:

> Again, if we wish to proceed from the first rudimentary orders of being to the last and most perfect, we shall necessarily arrive in the end at deity. We notice the sustaining power of nature first in the members of the vegetable kingdom,

[94] Cicero, *De Natura Deorum (On the Nature of the Gods)*, English translation by H. Rackham, Loeb Classics (Cambridge, MA: Harvard University Press, 1951 [1933]), Book 2.

[95] Ibid., 2.XV.42. [96] Ibid. [97] Ibid., 2.IX.22.

[98] Ibid 2.IX.24: *Omne igitur quod vivit, sive animal sive terra editum, id vivit propter inclusum in eo calorem. Ex quo intellegi debet eam caloris naturam vim habere in se vitalem per omnem mundum pertinentem.*

[99] Ibid., 2.XI.30. [100] Ibid.

towards which her bounty was limited to providing for their preservation by means of the faculties of nurture and growth. Upon the animals she bestowed sensation and motion, and an appetite or impulse to approach things wholesome and retire from things harmful. For man she amplified her gift by the addition of reason, whereby the appetites might be controlled, and alternately indulged and held in check.

XIII. But the fourth and highest grade is that of beings born by nature good and wise, and endowed from the outset with the innate attributes of right reason and consistency; this must be held to be above the level of man: it is the attribute of God, that is, of the world, which must needs possess that perfect and absolute reason of which I spoke.[101]

The whole world is governed and animated by a life force that manifests as heat at one level but also as nature itself and as intellect (*nous*) and soul (*psuchē*). This force, identified with God by Cicero, holds all together in tension (*tonos*) that controls everything into what it is. Tension is the force that allows the mind (*nous*) to penetrate the stars as well as human beings and that links the *pneuma* to the *nous*.[102] In this harmonious world, each thing has its place and the world is in a state of perfection with a divine, rational quality. There is sympathy (*sumpatheia*) between all within the world because of the uniting force of the *pneuma*. God is the active principle (*archē*) in the cosmos, who acts upon the passive principle (*hule*) animating it, so creating the cosmos. Occasionally, once all the water has been used up, the universe is destroyed in a cosmic conflagration (*ekpurōsis*) in which all is reduced to fire, but then the intelligent force of God renews it again.[103] This process happens endlessly; the cosmos is destroyed in conflagration and renewed again.[104] There is a kind of vitalist ecology going on here, with the sun being sustained by vapours exuding from the oceans and in turn warming the earth. The celestial bodies move around the earth and the universe is a vast organism governed by intelligent nature, overflowing with life,[105] which for Cicero is

[101] Ibid., 2.XII.34–XIII.35: *Atque etiam si a primis inchoatisque naturis ad ultimas perfectasque volumus procedere, ad deorum naturam perveniamus necesse est. Prima enim animadvertimus a natura sustineri ea quae gignantur e terra, quibus natura nihil tribuit amplius quam ut ea alendo atque augendo tueretur. Bestiis autem sensum et motum dedit et cum quodam adpetitu accessum ad res salutares a pestiferis recessum; hoc homini amplius quod addidit rationem, qua regerentur animi adpetitus, qui tum remitterentur tum continerentur. XIII. Quartus autem gradus est et altissimus eorum qui natura boni sapientesque gignuntur, quibus a principio innascitur ratio recta constansque, quae supra hominem putanda est deoque tribuenda, id est mundo, in quo necesse est perfectam illam atque absolutam inesse rationem.*

[102] Engberg-Pedersen, *Cosmology and Self in the Apostle Paul*, p. 20.

[103] Cicero, *De Natura Deorum* 2.XLI.118. See F.G. Downing, 'Cosmic Eschatology in the First Century, Pagan, Jewish, and Christian', *L'Antiquité Classique*, vol. 64, 1995, pp. 99–109.

[104] See Richard Salles, 'Chrysippus on Conflagration and the Indestructibility of the Cosmos', in Salles (ed.), *God and Cosmos in Stoicism*, ch. 5.

[105] Cicero, *De Natura Deorum* 2.XXXIII.83.

nonetheless homo- and deo-centric in the sense that the natural order of the cosmos is for the service of gods and men.

This optimistic materialism posits nothing transcendent to the universe. We might call it an animistic materialism insofar as a force (*vitalis*) animates the hierarchy of life from the stars to the rocks. What privileges human beings for the Stoics is their intelligence (*nous*) and their ability to lead a good life, cultivated through detachment (*apatheia*, although this is not a word Cicero uses). It is into this non-dualistic materialism that Christianity falls, and that Paul has to respond to. What does Christianity have to offer that the life affirming Stoicism did not?

In his interesting study, Engberg-Pedersen shows how Paul responds not by affirming a Platonic transcendence (and therefore a Platonic dualism) but by affirming the materiality of Christianity that posits the centrality of the resurrection of the body as it occurred in Christ, who becomes a precursor of the resurrection of the dead at the end of time. Paul operates within the intellectual horizon of Stoicism and his own Christian philosophy must be understood in that context, along with Jewish apocalyptical thinking that anticipated the end of time.[106] Paul gives shape to Christianity as we know it even today and defined its basic narrative component. This narrative is that the first Adam, animated by a *psuchē*, a 'psychic Adam', fell into a state of sin and needed to be redeemed. This was achieved by the second Adam, Christ, who is animated by spirit or *pneuma*, the 'pneumatic Adam', so that 'as all die in Adam, so all will be made alive in Christ'.[107] The earthly body will die but the heavenly body made of *pneuma* will rise. At the second coming of the pneumatic Adam there will be a conflagration in which human bodies will be consumed but those that contain the *pneuma* of Christ, through having been reborn in Christ, will not be consigned to the flames but transformed into new life. Thus Paul combines Stoic cosmology with Jewish Genesis and apocalyptic tradition, moulding a narrative governed by 'a chronological sequence of the two "men" (Adam and Christ)... articulating the idea of a *change* in human beings *from* wearing the visible appearance of the earthly "clay" man *to* wearing that of the heavenly man... namely, a body that is pneumatic'.[108] The new Adam produces the life-giving *pneuma*, transforming earthly bodies (*somata epigeia*) into heavenly bodies (*somata epourania*) like the stars, sun, and moon. We are here in a Stoic cosmos that Paul has absorbed, or rather he is making sense of the Christ event within

[106] On Paul's relation to Stoic philosophy see Engberg-Pedersen, *Cosmology and Self in the Apostle Paul*, pp. 19–37; 'Stoicism and the Apostle Paul: A Philosophical Reading', in S.K. Strange and J. Zupko (eds.), *Stoicism: Traditions and Transformations* (Cambridge: Cambridge University Press, 2004), pp. 52–75.

[107] 1 Corinthians 15.22 (NRSV translation). Quoted Engberg-Pedersen, p. 31.

[108] Engberg-Pedersen, *Cosmology and Self*, p. 31.

the worldview of his time and using the categories available to him (indeed, what else could he do?).[109]

For Paul the idea of the *pneuma* is of central importance. This term, as we have seen, takes in breath, spirit, and life force within its semantic range. For the Stoics it meant the substrate through which God penetrates the cosmos and which comprises the substance of the heavenly bodies.[110] Paul uses the term in a number of senses. In what Engberg-Pedersen describes as 'a magisterial book', six are identified by Friedrich Wilhelm Horn, namely a functional sense that makes Christians act in the way that they do, a substantive sense in which the *pneuma* lives in believers, a material sense in which the *pneuma* is present in the sacraments, a sense of *pneuma* as hypostasis distinct from God, an ethical or normative sense, and an anthropological sense in which the *pneuma* is personal, belonging to me or you.[111] There are clearly nuances of meaning to the term, but what they all share throughout Paul's letters is a sense that the *pneuma* denotes new life, an energizing principle that is in contrast to earthly decay and death; *pneuma* is a force 'that transforms a people and that transforms the text'.[112] Taken from the Stoics but reframed in the context of Jewish anticipation of the Messiah and the newly emerging Christian narrative, the idea of *pneuma* captures early Christian hope and idea of potential transformation. While Stoic philosophy does contain an optimism that might be contrasted with the Pauline apocalyptic view, even this has the idea of renewed force and energy. *Pneuma* in the Stoic sense animates beings but in a Christian sense transforms them: an 'earthly' spirit of this world (*pneuma tou kosmou*) becomes a 'spirit of God' (*pneuma tou theou*). Jacob Taubes presents an image taken from a capitol in the cathedral at Vézelay that depicts Moses pouring grain into a mill from above and Paul collecting the flour in a bag at the bottom. Abbot Suger's explanation is that Paul transforms the bran of Moses into the food of angels and humans, thereby, in Taubes' reading of Paul, 'outbidding Moses'.[113] Paul, and thereby Christianity that he

[109] But see Barclay's review of Engberg-Pedersen who regards Paul's theology as 'fundamentally incompatible with Stoicism ... because his theology is configured around a narrative that is shaped, in both thought and life, around a distinctive event with its own resulting logics'. John M.G. Barclay, 'Stoic Physics and the Christian Event: A Review of Troels Enberg-Pedersen, *Cosmology and Self in the Apostle Paul*', *Journal for the Study of the New Testament*, vol. 33 (4), 2011, pp. 406–14.

[110] Jean-Baptiste Gourinat, 'The Stoics on Matter and Prime Matter', in Salles (ed.), *God and Cosmos in Stoicism*, chapter 2. See also the review by Dunn who thinks that Engberg-Pedersen begins with an important insight but pushes it too far so as to distort Paul's view. James Dunn, review, *Journal of Theological Studies*, vol. 61 (2), 2010, pp. 748–50.

[111] Friedrich Wilhem Horn, *Das Angel des Geistes: Studien zur paulischen Pneumatologie*, Forschungen zur Religion und Literatur des Alten und Neuen Testaments 154 (Göttingen: Vandenhoeck and Ruprecht, 1992), p. 60. Engberg-Pedersen, *Cosmology and Self*, p. 18.

[112] Jacob Taubes, *The Political Theology of Paul* (Palo Alto, CA: Stanford University Press, 2004), p. 45.

[113] Ibid., p. 39.

decisively shapes, takes the raw grain and transforms it: the bodily or physical *pneuma* as animating substance of life becomes with Paul a transforming force that affects the bodies of the faithful and has political consequences for the polis itself. Indeed, transformed into Geist it has great consequences for Hegel as the 'world spirit' (*Weltgeist*) that marches on the hills of Jena in the form of Napoleon.[114] Perhaps this transformed notion of *pneuma* is not so distant from Paul's emphasis on God's spirit being active in the world and having the capacity to transform the lives of those who come into contact with it.

I would not wish to paint a picture of early Christianity as a vitalist philosophy per se, but there are parallels with Stoic philosophy and it naturally draws upon its ambient discourse, although in its inception and during its formative years with Paul, the Christian communities were yet to work out a complete theology. But the coherent narrative that Paul presents, later borne witness to by the gospel accounts, forms the basis, the heart of the Christian story that life is transformed through resurrection. This is a kind of materialism in the sense that *pneuma* is a substance that animates life, and yet for Paul its source is in God and God as creator is outside of the material universe. Thus, emergent Christian theology must reject Stoic materialism, even though both offer an affirmation of life itself. The term *pneuma* when rendered as spirit has immense importance in the history of Christianity. It is a fiery substance (as for the Stoics), the cloven tongues of flame that hover over the heads of the disciples at Pentecost; it becomes the third person of the Holy Trinity; and for Hegel as Geist, it becomes the very force of history itself. When the priest prays over the bread and wine during the Christian liturgy, the anaphora, he asks the Holy Spirit to descend on the gifts and Christians are baptized in the name of the Father, Son, and Holy Spirit.[115] The Acts of the Apostles (2.1–6) present the event of Pentecost and other passages speak of the Spirit being 'poured out': 'God's love has been poured into our hearts through the Holy Spirit which has been given to us', and 'we abide in him because he has given us his Spirit'.[116] Here 'Spirit' is the English translation of the Latin *spiritus* in the Vulgate that translates *pneuma* that itself translates the Hebrew *ruach*. The single word has a broad semantic range that can cover the world soul (*animus mundi*) of the Stoics to the precise idea of the third person of the Trinity descending upon the human community. Clearly with Paul a shift of meaning has occurred, but the parallelism with Pagan philosophy of *pneuma* as the breath and life itself echoes through Paul's application and in later Christian use. By the time of Augustine, Pentecost was established

[114] Ibid., p. 43.

[115] Robert Louis Wilken, *The Spirit of Early Christian Thought* (New Haven, CT: Yale University Press, 2003), p. 101.

[116] Romans 5.5 and 1 John 4.13, cited in Wilken, *The Spirit of Early Christian Thought*, p. 104.

as a Christian festival that celebrated 'a datum of history and a fact of experience',[117] a festival that celebrated the coming of divine life into the body of the Church. Christianity took over the idea of the *pneuma* from the Greek and Roman philosophers and reformulated it in the context of its own narrative and eschatological hope. *Pneuma* as the force of life itself becomes the spirit of God manifested, 'poured out', in human affairs and in particular through the Church as the continuing witness of the continuing revelation. The place of the spirit in Christian life and doctrine has been contentious and the cause of the theological divergence between the eastern and western Churches, whether the spirit proceeds from the father *and* the son or from the father *through* the son, a split that has lasted into the modern world. The Christian conception of spirit has understood it both in transcendental and immanentist terms, on the one hand as part of trinity that is pure communication of eternal love; on the other the participation in the history of the world, which Milbank has referred to as the Catholic Transcendentalist and Protestant Hegelian positions.[118] What is important is that *pneuma*, which originally means breath and comes to mean life force, the animating principle of matter, takes centre-stage in Christian belief. The Christian view is that *pneuma* can transform the natural order into the pneumatic order, and that is a political project as much as an exegetical one, a politics that in the end marks the end of politics. It is to the later development that we now turn.

LIFE FROM GOD

Within the Roman empire Christians lived in eschatological hope of the immanent second coming, suffering sporadic persecution, until Christianity was finally, firmly established as the state religion with Constantine and in Rome the establishment of Catholic authority with the Popes attempted to ensure doctrinal conformity to complex doctrines of God and incarnation through a number of theological conferences, notably at Nicea (325 CE), where a common creed was developed in response to the Arian heresy, Constantinople (381 CE), and Chalcedon (451 CE). In these developments the category of life itself was tangential to other concerns about the nature of the incarnation, Christ as fully human and fully divine and whether Christ had two or one will, the place of the holy spirit in the ontology of God, and so on. But embedded or implicit within this theology is a doctrine of life and a positive value to materiality that comes to be developed especially by the medieval Scholastic theologians. There is certainly the Neo-Platonic idea of the vertical ascent of

[117] Wilken, *The Spirit of Early Christian Thought*, p. 103.
[118] John Milbank, *The Word Made Strange* (Oxford: Blackwell, 1997), pp. 174–86.

the soul, but there is also the value given to reproductive life and the affirmation of social values that resulted in the transformation of socio-political realm in the Reformation and the later development of what Weber called an inner-worldly asceticism, which resulted in capitalism as a dominant economic system with all the social consequences that have followed from that.

Early Christianity develops in the context of Pagan philosophies, especially Stoicism, as we have just seen, and Epicureanism, but there is also strong Middle Platonist influence. What marks Christianity out from these other systems is a denigration of the gods or demotion of them to demons, the saving power of the incarnation of God in Christ, the doctrine of God as trinity, and an affirmation of *creatio ex nihilo* in contrast to emanationism or cyclic renewal of the cosmos that we find in those other traditions. In Plato's *Timaeus*, for example, God works on pre-existing material in complete contrast to the Christian, and Judaic, idea that God creates from nothing. There is, then, a spectrum of views in the ancient world of the early Christian centuries with the strong Christian affirmation of person as a combination of soul and body in contrast to the Gnostic rejection of body and negative appraisal of the world of life as being the creation of the wicked Demiurge. Stoic monistic materialism contrasts with dualistic Gnosticism, with the Platonism of Plotinus affirming the body as the house of the soul and affirming the world of life as the emanation of the Good. Christianity has had an ambivalent attitude towards the world of life, on the one hand presenting a very positive appraisal of materiality, broadly affirmative of reproduction within the social structure of marriage, although a rejection of this value by the spiritual elite,[119] and faith in the political realization of its values while also claiming an otherworldliness, that this life in the world is but a shadow of the life to come in which the fruits of our acts and faith will be judged. In general, all of the Abrahamic religions share a positive value to life as God's creation but within which error or fault has entered, mythologically presented as the revolt of angels or the wrong choice of man who turned against God and the need to transform the psychic body into the pneumatic body, as we have seen in the theology of Paul.

Although Plato himself was not much of a direct influence on Latin Christianity as few of his works in the early centuries were translated into Latin,[120] his influence was indirect through the Neo-Platonists and the Church Fathers, especially Augustine, who had absorbed Platonic thinking and also through Dionysius the Pseudo Areopagite or the Pseudo-Denys, whose *Divine Hierarchy* absorbs the Neo-Platonic chain of being with the highest angels

[119] See Peter Brown, *The Body and Society: Men, Women, and Sexual Renunciation in Early Christianity* (New York: Columbia University Press, 1988).

[120] The *Timaeus* with Chalcedius's commentary was the first followed by translation of the *Phaedo* and *Meno* in the late twelfth century. G.R. Evans, *Philosophy and Theology in the Middle Ages* (London: Routledge, 1993), p. 26.

closest to God down to humans and animals and below them, demons. The absorption of the hierarchical cosmology of the Platonists into Christianity brings with it the theological problem of the importance of human beings and the accessibility of God. If God is immediately present in creation then he is accessible equally from all points of it, yet if there is a hierarchy of being, as the Platonists tell us, then he is closest to the top of the hierarchy and more distant from the lower realms. This is a problem with hierarchical cosmologies that we also encountered in India. Again, if we are simply mid-way in the hierarchy, well below the highest angels, where lies our significance? Furthermore, why should God become incarnate in the creation as a middle-ranking being rather than in the form of a pure angel? This is clearly an issue for Denys as indeed for all Neo-Platonic Christianity through to Bonaventure in the thirteenth century.[121] Gradations of intensity of life implied in a cosmological hierarchy cannot easily be reconciled with an equality of intensity at all levels; it is almost a paradox that the tradition lives with and reconciles through elevating human reality to a higher place above angels in their capacity for symbolizing and mediating between world and God in a way that neither angels nor animals can do.[122] But it was Aristotle rather than Plato who had a direct influence on medieval Christianity. His works were translated into Latin by the third century, especially his works on logic translated as the *Organon*, and Aristotelianism comes into the Christian world especially through Arabic translations that were themselves rendered into Latin.[123]

While many theologians are important in the development of Christian doctrine, the first being Origen who is the most important figure to introduce Platonic ideas, it is Augustine who lays the foundations for Scholastic theology. In order to approach the medieval Christian view of life itself, we need to describe briefly Augustine's contribution. Of all the Church Fathers it is Augustine who has had the strongest influence on the later tradition and whose thinking still resonates in contemporary Christianity.

Augustine was familiar with Neo-Platonic teachings, especially the teachings about the soul's inner ascent back to the One, and he tells us how Platonism translated into Latin had shown him that the word was with God in the beginning, although could not reveal to him anything of the word becoming incarnate.[124] Augustine was deeply influenced by Plotinus and Porphyry whom he cites, sometimes favourably,[125] but although he accepts

[121] Denys Turner, *The Darkness of God: Negativity and Christian Mysticism* (Cambridge: Cambridge University Press, 1995), pp. 117–20.

[122] Ibid., pp. 119–20.

[123] Evans, *Philosophy and Theology in the Middle Ages*, pp. 26–9.

[124] Augustine, *Confessions*, VII.9, translated by R.S. Pine-Coffin (London: Penguin, 1961), p. 144.

[125] E.g. Augustine, *The City of God against the Pagans*, Book X.XIV, p. 314, George McCracken (trans.), Loeb Classical Library 411 (Cambridge, MA: Harvard University Press, 1957).

the tripartite levels of being in the Platonic universe comprising the gods, demons, and humans, he must reject the Platonic account. Indeed, Augustine needs to maintain ultimate value to the unique creation from nothing. While the theology of Origen is closest to Plotinus, it is Augustine who fully Christianizes mystical ascent and offers positive value to the world of life. Plotinus' rejection of world in favour of turning the gaze inwards is almost an anti-political move in contrast to Augustine's political theology, in which eschatological hope lies in the transformation of life through Christ. The story of Augustine's turn away from Gnosticism to Christianity has been well told,[126] although the original Gnostic structure of thinking, which is close to Platonism even though Plotinus wished to distance his thinking from it, still echoes in his writing. In a way not dissimilar to Plotinus, with Augustine we have the Christian double-take on the world as a place of beauty, God's creation, and a place of sin, man's creation.

A prolific writer, Augustine was less concerned with the category 'life' and more concerned with the nature of God and the God–human relationship. Indeed, he is against the Pagan vitalism that we find in the Stoics. In his discussion of the nature of the mind in *De Trinitate*, he mentions that some philosophers regard its substance to be 'a kind of life that vivifies and animates every living body, and have attempted to prove, as each one could, that it must logically be also immortal, since life cannot be without life'.[127] He does not offer specific arguments against this view, but it is not one that he holds, although the idea of being alive is important to him. In a famous passage that foreshadows Descartes, he says that 'I know that I live' for even if a person is asleep or deceived, the one thing he knows is that he is alive; even if he is insane, he knows that he lives.[128] This is to use the idea of 'being alive' as synonymous with simply 'being', but it is not to thematize life itself in the way that we have seen in the philosophies of Plotinus and Cicero. Such an explicit thematization of life itself as an animating force does not occur in the history of Christianity until the twentieth century with the work of Teilhard de Chardin, although the category does claim some attention through Aristotle, most of whose work in Latin translation was available to medieval theologians.

I want to focus here on the way Christianity deals with the category of life in the medieval period, the period of Scholasticism, once the fundamental Christian doctrines have been settled, the boundaries of the religion are policed, and

[126] E.g. John M. Rist, *Augustine: Ancient Thought Baptized* (Cambridge: Cambridge University Press, 1994).

[127] Augustine, *On the Trinity: Books 8–15*, trans. Stephen McKenna (Cambridge: Cambridge University Press, 2002), book 10, chapter 7, p. 51. *De Trinitate* (Turnhout: Brepols, Library of Latin Texts, 2017): *quandoquidem uitam omne uiuum corpus animantem ac uiuificantem esse repererunt, consequenter et immortalem quia uita carere uita non potest ut quisque potuit, probare conati sunt.*

[128] Ibid., p. 191.

there is sufficient self-consciousness to define the tradition against Judaism and Pagan philosophies, and later against Islam. We search in vain for any vitalism of the kind that we have encountered in Pagan philosophy: there is no *anima mundi* in the medieval theologians, but there is reference to life through the theological response and absorption of Aristotle, and Christianity contains within it a philosophy of life that arguably comes to clear articulation in the theology of Duns Scotus, although it is not an explicit theology of life but rather of life understood as the unrepeatable act of being. I wish here then to take two important moments in the history of Christian theology that are in many ways antithetical to each other: the one that takes on board Aristotelian categories in a rational and systematic explication that we find in Thomas Aquinas; the other that offers a radically new view (at the time) of understanding life in terms of the univocity of being that we find in Duns Scotus. Indeed, with Scotus we find a new voice and new tradition alongside that of Platonic ascent and Aristotelian classification. Rather than attempt a general description of their theologies, in the following I will proceed in a focused manner by approaching the two theologians through their commentaries on Aristotle.

Aquinas—The Affirmation of Life

Although famous for his systematic treatment of Christian doctrine in his *Summa Theologia*, Aquinas also wrote commentaries on the works of Aristotle. Thomas is a faithful interpreter of Aristotle, trying to explain him on his own terms and without bringing in extraneous material,[129] but in his treatment we also see his Christian theology coming through and we are able to glean an understanding of a Christian philosophy of life he presents. In the *Nicomachean Ethics* Aristotle speaks about life being 'naturally good' (*vivere est naturaliter bonum*), a sentiment with which Thomas concurs, restating Aristotle's view that animal life is defined by the capacity for sensation in contrast to human life where it is defined by the capacity for perception. Thomas then makes a distinction between capacity and operation, that capacity is reduced to operation, which is therefore principal. This distinction between capacity and operation refers to Thomas' ideas about action, that an act designates an operation associated with a given capacity.[130] Life then is defined by action, by movement in the world that stems from potency, the capacity and power to act. Life 'in the full sense' refers to an act, which is also a

[129] Harry V. Jaffa, *Thomism and Aristotelianism* (Chicago, IL: Chicago University Press, 1952), pp. 6–7.

[130] Peter Weigel, *Aquinas on Simplicity: An Investigation into the Foundations of his Philosophical Theology* (Oxford: Peter Lang, 2008), pp. 90–6.

sensation, i.e. experienced through the senses, and a thought. Furthermore, Thomas says that Aristotle shows that life is naturally good and pleasant and that this is proved because life is determinate and what is determinate is good.[131]

What Thomas means here is linked to his remarks about potency, operation, and the primacy of act. That life is characterized by act means that it is determinate, i.e. specific and temporally limited, and life itself, which is act or the realized operation of potency, has within it a self-evident moral quality of goodness. To be alive is a good in itself, for Thomas as for Aristotle. It follows that the opposite is also the case, that if actuality in action is good, potentiality without act is evil. So something is evil to the extent that it is indeterminate and good to the extent that it is determinate.[132] But surely, an opponent might think, this must be incorrect because the actuality of life can be evil and if anything is designated as evil, it is certain kinds of action. Thomas responds to this kind of criticism by claiming that evil acts are indeterminate, by which he means 'lacking in proper perfection' (*debita perfectione carens*).[133] Moral evil (*malam*), or corruption, or even pain are indeterminate within the body, so on this view the more the determinacy in life, the less evil and suffering are experienced. This moral quality is therefore attractive to the virtuous person. Life, being naturally good because determinate, is good for the virtuous person and so life is pleasant for all.[134]

While all animals suffer, only humans have the capacity for evil because we are rational animals. Animals, as Aristotle says, have the capacity for perception but only humans have the additional capacity for thought and of reason. Indeed, for Thomas only as animals can human beings reason; it is a quality predicated on being animal. While the life of reason is proper to human beings, it nevertheless depends upon human life as animal life. As Denys Turner observes, 'rational beings could not do these things [thinking and ratiocinating] unless they were animals'.[135] Human life is therefore part of animal life and not distinct from it; all seek pleasure because all desire life and we are especially active concerning things we love.[136]

Thomas closely follows 'the philosopher' in his view of life and does not try to introduce a Christian anthropology into his exposition, although clearly Thomas is acting strategically in commenting on Aristotle, wishing rather that

[131] Thomas Aquinas, *Commentary on Aristotle's* Nicomachean Ethics, trans. C.J. Litzinger (Notre Dame, IN: Dumb Ox Books, 1993 [1964]), chapter 9, lecture XI, p. 575. Latin text, Stanislai Eduardi Fretté (ed.), *Sancti Thomae Aquinatis Opera Omnia*, vol. 26 (Paris: Apud Ludovicum Vivès, 1875), p. 33.

[132] Ibid., Latin text, p. 33. [133] Ibid. [134] Ibid.

[135] Denys Turner, *Faith, Reason, and the Existence of God* (Cambridge: Cambridge University Press, 2004), p. 117.

[136] Thomas, *Commentary on Aristotle's* Nicomachean Ethics, p. 606.

an Aristotelean anthropology should influence Christian theology.[137] Thomas is very close to Aristotle's conception of human life and we see this also in his treatment of Aristotle's more systematic approach to life itself, *De Anima*. As we have seen, Thacker bewails the continued translation of the book as 'On the Soul' on the grounds that *psuchē* must be understood as 'life principle' more than 'soul', the animating power that determines life. Thomas is in the stream of the theological interpretation of the text in which *psuchē* is read as 'soul',[138] but reading his commentary on *De Anima* we see that he is in sympathy with Aristotle's understanding. In the commentary on the very opening verses that the *psuchē* or *anima* is a substance or a specifying principle of a body that is potentially alive. There are three kinds of substance (*substantia*): the compound (*compositum*), matter (*material*), and form (*forma*). The living body (*corpus vitam*) is a compound but the soul (*anima*) is not, nor is it the matter of the body because the body receives life. Therefore, the soul is the form that determines a particular kind of body, namely a physical body that is potentially alive (*potentia vitam*).[139] The soul is the animating force of the body and constrains what the body is. Commenting further on this theme, Thomas writes:

> Note that he [Aristotle] does not say simply 'alive', but 'potentially alive'. For by a body actually alive is understood a living compound; and no compound as such can enter into the definition of a form. On the other hand, the matter of a living body stands to the body's life as a potency to its act; and the soul is precisely the actuality whereby the body has life. It is as though we were to say that shape is an actuality; it is not exactly the actuality of an actually shaped body—i.e. the compound of body and shape—but rather of the body as able to receive a shape, of the body as in potency to an actual shape.[140]

This makes it clear that the soul is the life of the body. A live body is a compound and a compound cannot be a form (because a form by definition is something that is uncompounded). The matter of which the body is made is

[137] Thomas' commentaries on Aristotle were among the last texts he composed and are set in the context of dispute in Paris about the status of such material. Thomas is writing implicitly to affirm the importance of classical learning at a time when ecclesiastical authority was raising questions about its heretical status.

[138] Eugene Thacker, *After Life* (Chicago, IL: Chicago University Press, 2010), p. 9.

[139] Thomas Aquinas, *Commentary on Aristotle's* De Anima *(Sentencia libri De anima)*, trans. Kenelm Foster and Sylvester Humphries (New Haven, CT: Yale University Press, 1951), 2.1.221.

[140] Ibid., 2.1.222. Latin text, Stanislai Eduardi Fretté (ed.), *Sancti Thomae Aquinatis Opera Omnia*, vol. 24 (Paris: Apud Ludovicum Vivès, 1875), p. 58: *Dixit autem 'habentis vitam potentia' et non simpliciter habentis vitam. Nam corpus habens vitam intelligitur substantia composita vivens. Compositum autem non ponitur in definitione formae. Materia autem corporis vivi est id quod comparatur ad vitam sicut potentia ad actum: et hoc est anima, actus, secundum quem corpus vivit. Sicut si dicerem quod figura est actus, non quidem corporis figurati in actu, hoc enim est compositum ex figura et corpore, sed corporis quod est subiectum figurae, quod comparatur ad figuram sicut potentia ad actum.*

that which is enlivened or transformed into life as an act is the transformation of a potency or potential. The soul is the actuality (*actus*) that gives life to the body. The term *actus* also implies the driving force, the force that can shape the body that is able and ready to receive it. So, the soul is the animating principle that enlivens the body.

This animating principle actualizes this particular body, giving it simple being (*esse simpliciter*), and they are intimately connected. The soul is a substantial form that gives being to the body and not accidental. Bare matter would be an accidental form, but the living body is animated by the substantial form of the soul. At death the body is no longer animated, but the soul still comes to take another form, for passing away always entails coming to be in a new way.[141] Thomas's doctrine of the soul closely follows Aristotle, for he sees the idea of the person as a combination of body and soul, or perhaps we can even say that the person is an animated body. A human being's having a soul, for Thomas, simply means being alive in a particular kind of way, as Turner observes. In this view he differed from his Augustinian-Platonic contemporaries and ran the risk of being accused of materialism.[142] This does leave Thomas with the problem of what happens at death if he is to maintain a Christian belief in post-mortem existence. At death the animating principle ceases to define that particular body but survives the death of the body because it is an immortal intellect. I am not my soul, for Thomas, but my soul survives me to await the resurrection at the end of time. Because of Thomas's Christian eschatology he must have some account of the soul moving elsewhere to await its reconnection with the body at the time of the eschaton, although he does not deal with this topic in his Aristotle commentary. Indeed, it seems commendable that Thomas takes the text as he has it and elucidates its meaning in its own terms, and thereby also argues for his own position regarding the nature of the person and the nature of human life.

In the classification of life, we are able to see what distinguishes the human. Thomas describes the philosopher's characterization of life as having distinct modes: the plants who are fixed in one place take nourishment, grow, and decay; some creatures are fixed but have sensation, such as shellfish; higher animals such as horses and oxen have sensation but also the ability to move from place to place; and humans in addition have mind or intellect. But it is the *anima* that is the life principle (*principium vivendi*) in them all.[143] It is the actuality (*actus*) and formal principle (*ratio*) of being that is potentially animate (*potential viventis*).[144]

The idea of the *anima* as the animating principle of the body sounds very close to Pagan vitalism that a single force animates living bodies. But Thomas is careful to distinguish his theology from this, and so to protect Aristotle from

141 Ibid., 2.1.226. 142 Turner, *Thomas Aquinas*, p. 62.
143 Thomas, *Commentary on Aristotle's* De Anima, 2.2.256. 144 Ibid., 2.2.278.

this view. Indeed, he writes rigorously against the idea of a single animating force in the universe, the emanationist cosmology of the Arab theologians. In his *Treatise on Separate Substances* (*De substantis separatis*) he describes the emanationist cosmology of Avicenna and rejects it. This is the view that all things are derived from God who is the supreme, first principle from which proceeds a second intelligence (*intelligentia secunda*), which in turn produces the soul (*anima*) of the first sphere and the first body and from thence down to the lower bodies.[145] Aquinas proceeds to reject this view on the grounds that, first, the good of the universe is stronger than that of any particular nature and, second, 'since the nature of the good and of the end is the same, if anyone withdraws the perfection of the effect from the intention of the agent, he destroys the nature of the good in the particular effects of the nature or act'.[146]

This somewhat dense refutation makes certain assumptions about the nature of the good and the end of the universe, namely that the end of the universe is itself a good and so is of the same nature as the good in itself. Thus, the effect or end of the universe is perfection (because it is good). Furthermore, we cannot separate the effect that is perfection from the intention of the agent (i.e. God). If we separate the perfection of the end from the intention of the agent, then for Thomas we erase any notion of the good in the effect. The good in the effect, the good in nature itself, cannot be generated by nature but by a source that is its cause. Thus, if we claim that perfection is a property of the effect but is distinguished from the intentionality of the cause, then we undermine the notion of the good in the effect that comes from the cause (namely, God). Thomas claims that this is also the position of Aristotle who rejected the opinion of the ancient Naturalists (*antiquorum naturalium opinionem*), who claimed that forms are generated by nature and so natural goods proceed from the necessity of matter. Thomas' criticism of Averroës is therefore that his view of emanation implies a necessity in the order of things themselves and thereby does away with the necessity of an agent or creator. But if we are to maintain that the universe is good, with perfection as its end, then we must reject the view that the good comes out of or emerges from the universe itself. Rather, this good in the universe must come from its cause and source that is external to it, namely God.

Furthermore, if God is the first cause, the unmoved mover, then his effects are in the bodies of creation. The effect of God is present in matter, the consequence of a universal agent, and so there is no need to introduce the idea of a second cause as implied in emanationism, because the first cause is

145 Thomas Aquinas, *De Substantiis Separatis, Treatise on Separate Substance*, trans. Francis J. Lescoe (West Hartford, CT: St Joseph's College, 1959), 10.53.

146 Ibid., 10.54. Brepols, Library of Latin Texts—Series A, Thomas de Aquino, *De substantis separatis*, ch. 10, no. 102, line 4 (pag Marietti 38): *Destruit autem rationem boni in particularibus effectibus naturae vel actis, si quis perfectionem effectus non attribuat intentioni agentis, cum eadem sit ratio boni et finis.*

present in the effect, which is the whole being in itself. This is a philosophy of creation that marks the Christian position clearly from the Arab and Pagan emanationists. Life is the creation of God and all substances, both material and immaterial, have their being immediately from God, thus immaterial substances, for example, have their living and intelligence (*vivant et intellectivae*) directly from God,[147] unmediated by some other principle.

Thomas thus wants to account for the living, for creation, through acknowledging Aristotle's principle of life, the *anima* or *psuchē* that animates actual living creatures, but this principle, along with the bodies it gives shape to, is created. There is no need to posit a complex hierarchy of emanation in which lower forms emerge from the higher, as God acts immediately *upon* the world but not *as* the world. And the *anima* that animates the body is not in itself to be privileged beyond its function as giving rise to person. The person, for Aquinas, is the unity of body and soul or animating principle, although in affirming this he is resisting certain Neo-Platonic tendencies in his world that would seek to see the soul as the essential person.[148] Indeed, for Thomas the human being is an animal, 'wholly animal',[149] albeit a rational one.

Aquinas represents a distinctive Christian position. Taking on board Aristotle's philosophy of the *anima* as the animating principle of living beings, Aquinas links this to a strong theology of creation in which life, as creation, is a good in itself but a good that will only find its perfection at the end of things. Aquinas' eschatological hope is affirmative of body and materiality in a way that his Neo-Platonic predecessors were not. The Christ event marks a transformation in world that will come to fruition in the future; and that future will bring the realization of the unity of body and soul in the redeemed person (at least for those saved through Christ's redeeming sacrifice). This Christian view is quite distinct, and opposed to, later Cartesian dualism because it affirms the unity of person as body or materiality and animating life principle while also affirming that death is not the end of life but a transformation of it.

Aquinas' affirmation of life without raising the idea of life itself into a metaphysical principle—his theology is not vitalist in the sense we have encountered—is partly developed and partly rejected in later years. Perhaps the second most significant theologian to emerge in the Western world in the medieval period is John Duns Scotus, who has a theology of life itself but which is articulated indirectly through the idea of the univocity of being.

[147] Ibid., 11.61.

[148] Sarah Coakley makes this point that at his time and after there is a 'trend towards a new dualism between soul and body' and this trend is seen in popular spiritual manuals. Sarah Coakley, 'Visions of the Self in Late Medieval Christianity', p. 73, in Coakley, *Powers and Submissions: Spirituality, Philosophy and Gender* (Oxford: Blackwell, 2002), chapter 4.

[149] Turner, *Thomas Aquinas*, p. 58.

Duns Scotus—Life as Unrepeatable Particular

In our inquiry into a Christian understanding of life itself we have seen how a distinct theological voice emerges, marking Christianity off from Neo-Platonic and Stoic philosophies and also from Judaism and Islam. Augustine, sympathetic to Platonic modes of thinking, lays the foundation for the future, and with Aquinas we have the establishing and thorough working out of a distinctly Christian vision, strongly informed by the philosophy of Aristotle, that broadly posits an affirmative view of life as creation. A unique voice in this history, and a voice that has much contemporary relevance, is that of Duns Scotus who, responding to Aquinas, transforms theological thinking, making theology philosophically rigorous. There is little direct talk of life itself in Scotus—we search in vain for references to the category 'life' (*vita*)—but his philosophy of being has implications for a Christian philosophy of life. Scotus is a controversial figure in the history of Christian theology, standing in many ways opposed to Thomism in his emphasis on the identity between the being of God and the being of the human,[150] but he is important for our project in that his work contains within it a philosophy of life in the particular. Although Scotus could not foresee the extent of the future influence of his thinking, he has been important for a number of significant figures and their ideas, most famously Gerard Manley Hopkins' idea of inscape and the philosophy of life of Gilles Deleuze, as we will see. Even today Scotus' work is not simply of historical interest but of contemporary philosophical importance too.[151] In broad terms, we need to approach Scotus because he emphasizes the particularity of existence and its non-repeatable instance, thereby articulating a non-negotiable feature of Christian creation, the sacrality of life itself in the present moment. The sacrality of life is seen primarily in two ideas: 'thisness' (*haecceitas*) and the univocity of being.

Thisness or *haecceitas* is a principle of individuation, that no two individuals are ever the same and each is unique unto itself. The term *haec* is simply the deixic term 'this' in Latin, and Scotus uses it to indicate the unique quality of each individual and the power of differentiation. In a sense, *haecceitas* is the force that controls events into their particularity; it makes things what they are. Each event of thisness is singular, having had no previous occurrence and

[150] For some theologians this is a wrong turn in theology that led, in time, to secular ontology with Heidegger. Indeed, Heidegger writes his thesis on Duns Scotus. See Milbank, *The Word Made Strange*, p. 45.

[151] For a survey of Scotus' research see Richard Cross, 'Duns Scotus: Some Recent Research', *Journal of the History of Philosophy*, vol. 49 (3), 2011, pp. 271–95. Cross identifies the continued relevance of Scotus in philosophy for universals and individuation, modal theory, cognitive psychology, semantics, logic, and metaethics.

without future occurrence, being unique unto itself.[152] Such thisness, unique unto itself, is the necessary condition for any general categorization, and as necessary he calls it the 'formal distinction', the objective nature of a really existing thing outside of the mind. Thisness is also linked to our own ongoing experience of personal identity and 'bare awareness'.[153] In his Oxford lectures delivered in the final year or two of the thirteenth century, Scotus raises six questions about the nature of existence. The first question concerns whether a material substance (*substantia materialis*) is of its nature a 'this'; that is, singular and unique. The background to the question is itself of interest as being about angels. On the one hand, there was a definition of person offered by Boethius as an individual substance of a rational nature, while on the other there was a debate about the nature of angels. An angel, according to Peter Lombard, whose *Sentences* is the foundational text of theological reflection, has a simple, immaterial essence, a distinct personality, a rational nature, and a free will.[154] But if this is so, how can an angel be a person, as they are immaterial, and individuation comes from matter? Thus, the two attributes of being immaterial and having distinct personalities (that is, are individuated) is a contradiction. The thirteenth century witnessed some debate about this issue and Scotus' opening Oxford lecture directly addresses it by raising the question of what it is to be individuated. What makes something what it is? In this question we have the issue of the relation between the particular and universal. That is, in what is real we find the particularity of the extramental thing and that it belongs to a community of things on which account we recognize something for what it is. In the medieval debate this was between realism (that affirmed universals) and nominalism (that affirmed particulars). Scotus is 'a moderate realist'[155] because he rejects the Platonic realism of the forms (as did Aristotle) but tries to account for the relation between general or universal definitions and particular things. That is, a universal definition is what makes something intelligible, that enables us to recognize a stone as a stone, for example, but the individuating matter, the particular stone, becomes unintelligible if divorced from any general definition. If there are no universals and individual things are simply unique unto themselves, then they become unintelligible.

To solve this problem, Scotus introduces the idea of 'thisness' (*haecceitas*), a formal principle of individuation that is universal. This preserves the unique, not duplicable quality of individuals as a 'this' or a 'that', which are differentiated from every other individual whether belonging to the same or to a

[152] See Alan B. Wolter, *John Duns Scotus, the Oxford Lecture on Individuation*, Latin text with English Translation (New York: The Franciscan Institute, 1975), pp. xi–xvi.

[153] See Wolter's interesting discussion, 'Introduction', *John Duns Scotus*, pp. xiv–xv.

[154] I am following Wolters' discussion here in his useful explication of Scotus' questions. Wolters, 'Introduction', pp. ix–x.

[155] Ibid., p. xi.

different type of thing. The thisness of a thing is its individuating nature, but we can also think of something in general terms and so thisness has a relationship with what it substantiates. The stone is a particular and contains a thisness that differentiates it into what it is, but also connects it with the general features of being a stone. The distinction between the particular stone and its stone-ness are not in the mind but in reality, and part of the way things are. Thus, prior to conceptualization, there is within things a formal distinction between the thing and what it instantiates.[156] Scotus summarizes his point:

> Just as stone is first presented to the intellect as something in its own right and not as universal or singular, neither is stone first grasped through a second intention, nor is universality a part of the meaning of the concept, but the mind understands the nature of stone for what it is in itself and not as universal or as particular or singular—so in its extramental existence stone is primarily neither one nor many numerically, yet it has its own proper unity which is less than the unity pertaining to a singular. And this is the essential being of the stone and the universal aspect according to which stone is defined, and propositions predicating this being or nature are true in the first mode of essential predication. According to its nature, then, stone is this prior state where it is not determined to be in this or that individual, has whatever is predicated of it essentially, to which as such is added the notion of being in this or in that.[157]

The stone, then, is not purely singular because, were it so, we could not apprehend it as stone, but nor is it purely universal being a particular thing. Nor is it grasped through a 'second intention' by which Scotus refers to a distinction in Avicenna, 'the commentator', between a first intention that refers to a thing and a second intention that refers to abstractions such as genus or species, that itself refers to a first intention.[158] The stone initially presents itself to the mind as something in its own right, but as stone, and so 'is neither one nor many numerically', and yet it has its own unity. The stone in its condition prior to any perception is particularized into this or that and what can be predicated of it essentially; that is, as a universal (possessing hardness or whatever). To the general notion of stone, we must add the particular features of this or that stone. It is haecceity that individuates this

[156] Scotus, *Oxford Lecture*, 6.178.

[157] Ibid., 1.32: *Unde, sicut lapis prius est aliquid intellectui secundum se, et non sub ratione universalis nec sub ratione singularis, nec intelligit secundum intentionem quando primo intelligit naturam lapidis secundum se, nec ut universalis nec ut particularis singularis,—sic in existential extra animam nec est primo una numero nec multa numero, sed habet unitatem propriam sibi, quae minor est unitate quae convenit illi singulari: et illud est 'esse' quiditativum lapidis et rei universaliter, secundum quod 'esse' datur de re definition, et verae sunt propositiones 'per se primo modo' secundum hoc 'esse'. Unde lapis in illo priore—secundum naturam suam—in quo non determinatur ut sit in hoc vel in illo, habet quidquid dicitur de eo quiditative, cui 'ut sic' accidit esse intellectum in hoc vel in illo.*

[158] Wolter, 'Introduction', p. xi.

or that stone as stone. Thus within an individual substance are, in Tweedale's terms, two individuals, the individuator and that which makes it the kind of thing it is.[159] The particular stone is what it is and yet also recognizable as stone: it contains particularizing and universalizing features but this distinction is not real in the sense that the stone is still the same stone; the distinction is only formal.

The distinctiveness and originality of this philosophy implies that a distinction between life itself and living things can be understood in this way. A living thing is simply what it is as it is, but simultaneously participates in the category of life itself: we recognize a living thing as living because it contains the formal distinction between what it is in itself while participating in the general category of life. Read in this way, the animating principle of life, Aristotle's *psuchē* or *anima*, relates to living beings in a parallel way to the category stone to particular stones. The implication of Scotus' *haecceitas* for a philosophy of life is that living entities are unique unto themselves while participating in the universal. This uniqueness, the particularity and unrepeatability of life, speaks in the same voice, the voice of life itself. Thus, what Scotus calls the univocity of being, that all being shares in the identical act of being, can be translated into the language of life, that all living shares the identical act of being alive.

If we translated Scotus' language of being into the language of life, we have a distinctive Christian philosophy of life emerging in which particular lives are both unique unto themselves and participate in an animating principle. This distinction between the living entity and life itself is a formal one in Scotus' sense. All of life is unique and unrepeatable and yet participates in something common that, in Christian terms, is *anima*. Furthermore, we can understand *anima* in terms of *pneuma*, the breath of life or spirit that is the third person of the trinity. Scotus' univocity of being means that the being of a particular entity, say my being, is identical *in act* to the being of God as being. Both God and beings, by implication, share in the act of life, so the holy spirit, *spiritus sancti*, that shares in the being of God also enlivens the animating principle, the soul, of beings.

Between Aquinas and Scotus, we have two competing visions of God: one that stresses the otherness and ineffability of God, the apophasis of God, the other that stresses what we share in the act of being, the univocity of God's being and ours. In the coming centuries this played itself out in interesting ways. We would probably call neither Thomas nor Scotus mystical theologians, but their work certainly does have implications for mystical theology: Thomas in God's unknowability and Scotus' in the mystic's sense of identity with the divine. The *hic et nunc* implication of Scotus' theology feeds into a kind of Christian vitalism that we find emerging in the twentieth century in

[159] Martin M. Tweedale, *Scotus vs. Ockham: A Medieval Dispute over Universals* [*Scotus vs. Ockham*], 2 vols (Lampeter: Edwin Mellen Press, 1999), vol. 2, pp. 402–4.

the work of Ravaisson and Teihard de Chardin, and that we also find emerging in the secular version of this kind of philosophy with Deleuze as we will see. I wish to examine more closely what Christian mysticism has to say about life itself, but before we explore that particular trajectory of Christian history, we need to take a sideways step into Islam. The Christian theologians were strongly influenced by their Arabic-language counterparts and both are the inheritors of Greek philosophy alongside Biblical and later Quranic revelation. Islam has a distinctive voice that the Christian theologians in part define themselves against, but what precisely is this voice and what does it say? By way of this sideways step I want to approach this through the work of one philosopher who is so important for the history of Christianity too, namely Averroës.

LIFE IN GOD

The history of Islamic theology has in general offered a view of life itself as created by God while yet at the same time participating within the being of God. Historical conditions were ripe for the emergence of Islam, rising between the two virtually spent empires of Byzantium and Persia. Historical knowledge of the earliest period of the history of Islam is vague as, although we have the earliest suras of the Qur'an, all the sources are late, such as Ibn Ishaq's biography of Mohammad (mid-eighth century), and the dating of the traditions (Hadith) is likewise not precise.[160] Islam spread very rapidly out of Arabia into Palestine and into what are now Iraq and Iran, where it took root among local traditions. Much of the intellectual life of Islam came to be focused on the development of law and jurisprudence (*fiq*), particularly in Iraq where there was the important influence of Judaic legalism. Islam spread throughout North Africa and into Spain. As Sunni and Shia traditions developed and Islamic polities became powerful and stable, we see the emergence of great centres of learning such as at Quam and a growing mysticism alongside legalism. The mystics or Sufis often fell under suspicion by the authorities, yet at other times Sufi thinking combined with legalism, as we find in the theology of Al Ghazzālī. The rapid political expansion of Islam was followed by an expansion of Islamic theology and an intellectual confidence in the foundations of tradition that allowed for the appropriation of external philosophical sources. The Arabic-medium scholars rendered important Greek philosophical texts, particularly of Plato and Aristotle, into Arabic from whence they

[160] Julian Baldick, 'Early Islam', pp. 315–16, in Stewart Sutherland, Leslie Houlden, Peter Clarke, and Friedhelm Hardy (eds.), *The World's Religions* (London: Routledge, 1988), pp. 313–28.

were translated into Latin. Islamic philosophy and science developed apace, enhanced by Indian influence, particularly in mathematics, and rival schools of philosophical speculation about free will and determinism came to sharply define their borders. Within Islam there is some speculation on life itself in a legal setting, but it is the more mystically orientated theologians in whom we find a discourse about the nature of life and not simply a discourse about how to live.

But we can begin with the roots of Islam in the Qur'an itself. Among the ninety-nine names of God highlighted by tradition, from the holy text itself, three are important for our theme. God is alive, the ever living (*al-hayy*) from which springs his ability to give life; he is the life-giver (*al-muhyī*), and to take life, he is the death-bringer (*al-mumīt*).[161] The nature of his kind of being is that he is alive himself and also the giver of life and destroyer of life. From the very earliest passages in the text this is clear: 'God, there is no deity except him, the ever living, the sustainer of existence'.[162] God is the creator of life and yet shares with the living the quality of being alive. But then, of course, God is outside of time as its creator and so the quality of being alive is something distinct when predicated of God to when it is predicated to creatures. God creates existence or being (*wujud*) from outside of it and yet sustains it through his infinite power. As creatures we participate in the life of God, and in Islam this is expressed in two ways. On the one hand, Islam has been prescriptive about how to live through the traditions of law (*shari'a*) for the human community in submission to God; on the other, Islam has offered reflection on the nature of life through its theology, particularly mystical theology. A number of thinkers take up this theme and it is not possible to be comprehensive here, but rather selective in illustrating the way the Muslim theologians reflected on it.

Al Ghazzālī comments on the ninety-nine names, observing that what is characteristic of living is acting and perceiving, so much so that one who does not act or perceive is dead.[163] Thus God who is the perfect and absolute living thing arranges all perceptions and activity. Nothing escapes his knowledge and activity and all things subsist in God.[164] Unlike living beings that he creates, whose life is circumscribed by their knowledge and perception, God is not circumscribed. All living creatures, says Ghazzālī, have different qualities and are ranked as angels, humans, and animals, but God is outside of this scheme

[161] Qur'an, Yusuf Ali's translation. God is the ever living (2.255, 3.2, 20.111, 25.58, 40.65), the giver of life and bringer of death (3.50, 3.156, 7.258, 57.2).

[162] Ibid., 3.2: *Allahu la ilaha illahuwa alhayyu alqayyoom*.

[163] Al Ghazzālī, *The Ninety-Nine Beautiful Names of God—al Maqṣad al-asnā fī sharḥ asmā' Allāh al-ḥusnā*, trans. David B. Burrell and Nazih Daher (Cambridge: Islamic Text Society, 1992), p. 129.

[164] Al Ghazzalī, *Love, Longing Intimacy and Contentment: Kitāb al-maḥabba wa'l-shawq wa'l-uns wa'l-riḍā, Book XXXVI of the Revival of the Religious Sciences*, trans. Eric Ormsby (Cambridge: Islamic Text Society, 2011), p. 24.

and not restricted. The quality of created, living beings is that their bodies—that is, their organs of perception—are appropriate or consonant with their worlds of experience in varying degrees of limitation. Thus, angels have pure bodies and inhabit a subtle world, with wide perception and knowledge in contrast to animals whose bodies and perception are extremely limited. Humans interact with the world through their perception and their world of experience is constrained by the kind of body they have. There is then a process of gradation regarding life. At the bottom of this hierarchy are inanimate things and those whose life is severely limited, such as plants, in contrast to those beings above them whose life is more complex yet more free. In one sense God is at the top of this hierarchy but in another sense, he is outside of it. This fundamental cosmological structure is shared by the medieval theologians, one of the most important after Al Ghazzālī being Ibn al-Arabī, whose work I wish to turn to as illustrating the Islamic, Scholastic, and mystical view of life.

Ibn al-Arabī—Imagining Life in God

Ibn al-Arabī (1165–1240) is important for his reflection on the nature of the life of God.[165] What does it mean to say that God is alive? In reflecting on these names of God,[166] he develops a theology that fills out the way creation occurs and what the nature of life on earth must be in relation to the One who is ever alive. Ibn al-Arabī was a prolific writer. Born in Spain, he travelled throughout the Muslim world, eventually passing away in Damascus, and leaving behind a legacy of manuscripts, some of which have yet to be critically edited. He is a figure that courted controversy, regarded both as a great Sufi master and as an apostate by others, and becoming famous in the West through the work of Henry Corbin, whose project was to open out 'the imaginal' or *imaginalis mundus* that gave access to direct experience of the sacred in contrast to rational thinking.[167] But while Ibn al-Arabī is certainly concerned with imagination and experiencing God as the Alive, he is rooted in the Islamic way of life and in *shari'a* along with rational argument, as Chittick's study brings out.[168]

[165] On the life of Ibn al-Arabī or in the anglicized version Ibn Arabi, the standard biography is Clause Addas, *Quest for the Red Sulphur* (London: Islamic Text Society, 1993). There is also Stephen Hirstenstein's *Unlimited Mercifier* (Oxford: Aqua Publishing, 1999). For a good general introduction see William C. Chittick, *Ibn Arabi, Heir to the Prophets* (Oxford: One World, 2007).

[166] Chittick claims that the divine names 'are the single most important concept to be found in Ibn al-'Arabī's works'. William C. Chittick, *The Sufi Path of Knowledge* (Albany, NY: SUNY Press, 1989), p. 10.

[167] Ibid., pp. ixx–xx.

[168] Ibid. See also William Chittick, 'Ibn Arabi', *The Stanford Encyclopedia of Philosophy* (Spring 2014 Edition), Edward N. Zalta (ed.), https://plato.stanford.edu/archives/spr2014/entries/ibn-arabi.

His first and most important massive work comprising 560 chapters is *al-Futūḥāt al-makkiyya*, the 'Meccan Openings', in which Ibn al-Arabī lays out his theology and prescription for a holy life.[169] An important distinction that governs his thinking in this work is between the essence or true nature of God, the He, and what is not God, the not He. Starting from the fundamental statement of faith that God is one, Ibn al-Arabī attempts to understand what this means in relation to the plurality of the experienced world. God is one and yet God is also found within his creation. Ibn al-Arabī's theology is summed up by 'the unity of being' (*waḥdat al-wujūd*), although he does not use this actual phrase,[170] which points to the oneness of God and yet, as Chittick observes, the term *wujūd* can also mean 'finding', so the phrase might be rendered as 'the unity of finding';[171] that is, finding God in the world. The relationship between Being (*wujūd*) and the existent (*mawjūd*) is one of creator to creation in which creation is a manifestation of the qualities of the divine. Every existing thing is both God, He, and not God, not He, because the world is distinct from its creator and yet the creator must pervade the world as there can be nowhere where God is not. The whole orientation of our lives for Ibn al-Arabī should be towards finding God, which is also the being of God and knowing God. This means practically to live a good Muslim life but also to pursue the supererogatory practices of the Sufi path of prayer and vision. This path of prayer and vision occurs within the imaginal world; a mediation between God and humanity.[172]

We can understand the relationship between the He and the not He as the cosmos or a cosmical hierarchy (*tartīb al-'ālam*) comprising different ontological levels (*marātib al-wujūd*) or 'the various degrees in which creatures participate in the divine presence'.[173] The metaphor of verticality is key to understanding the meaning of human life, which is to see how the human world fits into the cosmic totality. Thus, the angels, made of light, are unseen, subtle, and high, in contrast to things that are dark, visible, dense, and low. The cosmos is arranged between this polarity of beings made of light at one end and things made of clay at the other. Between the intense darkness of nothing and the intense light of being, the whole range of life is found. In Ibn al-Arabī's version of the cosmic hierarchy, the intermediate degrees of being

[169] Ibn 'Arabi, *al-Futuhat al-makkiyya*, 4 vols (Cairo, 1911). For translations see Ibn al-Arabi; selected texts of Al-Futûhât al-Makkiya, presentations and translations from the Arabic under the direction of Michel Chodkiewicz in collaboration with William C. Chittick and James W. Morris (New York: Pir Press, 2002).

[170] Chittick, *Sufi Path of Knowledge*, p. 79.

[171] Ibid., pp. 3–4.

[172] On Sufi history and practice see Nile Green, *Sufism: A Global History* (Oxford: Wiley-Blackwell, 2012); Scott Kugle (ed.), *Sufi Meditation and Contemplation: Timeless Wisdom from Mughal India* (New York: Suluk Press, 2012); Shahzad Bashir, *Sufi Bodies: Religion and Society in Medieval Islam* (New York: Columbia University Press, 2011).

[173] Ibid., p. 14. I have much relied on Chittick's explication of Ibn al-Arabī's cosmology, which has helped to guide me through the work of Ibn al-Arabī itself.

between the intense light of God and dark solidity of matter are called *barzakh* or 'isthmuses', an image that conveys the idea that each region of the cosmos is connected by a narrow strip. The term *barzakh* is also used to denote the whole realm between the corporeal and the spiritual, identified with the imagination (*khyayāl*), although by 'imagination' Ibn al-Arabī does not mean non-existent but rather more intensely real because closer to the source of light. Indeed, the barzakh, as the whole of creation, is identified with the breath of God:[174] God breathes life into creation, which is the isthmus between the living and the source. These higher realms of the cosmos are accessible to spiritual masters and it is within this realm that those on the path to God encounter visions and significant dreams. From the corporeal realm of life, the spiritual seeker must ascend to the imaginal images (*mithāl*) and through them to the divine realm of light (*nūr*). It is in the imaginal realm that we can speak of the 'spiritualization of corporeal bodies' (*tarawḥun al-ajsām*) and the 'corporealization of the spirits' (*tajjasud al-arwāḥ*).[175] This cosmological structure is described by Ibn al-Arabī as a 'horn made out of light', divided into the corporeal world, the world of imagination, and the world of the spirits, with Being as the point of the horn and nothingness as its base. All this is excellently described in Chittick's fine study that translates Ibn al-Arabī's account. The horn made of light, an image taken from the Qur'an that refers to the trumpet that sounds at the end of time, reveals manifestation. The imaginal is brought to perception through the light, which is the horn; that is, the cosmos comes into being and can be perceived as light.

This light is also the breath of life. The Alive gives life to the cosmos and the living beings within it. In this process life moves through a series of cosmological stages from the subtle to the material, and once it reaches the earth it turns inwards as the soul's interiority. The process of outer emanation is reflected in inner ascent. In this process God's grace allows the soul to behold its source. This can occur while alive, a kind of dying while living, but also at the death of the body. The Alive gives life and takes it away so that each moment is a renewal of life and the ending of what went before: life is in a process of constant becoming and transformation.[176] Ibn al-Arabī expresses this sense that the cosmos itself is alive because it reflects the aliveness of God:

[174] Chittick, *Sufi Path of Knowledge*, pp. 127–9. [175] Ibid., p. 15.

[176] William Chittick, 'Ibn al-Arabī on the Soul's Temporal Unfolding', p. 4, in A.T. Tymieniecka (ed.), *Timing and Temporality in Islamic Philosophy and Phenomenology of Life* (Dordrecht: Springer, 2007), pp. 3–10. He writes: 'In Ibn al-'Arabī's perspective, the cosmos itself is alive, because it partakes of the Living One from whom it appears. Like any living thing, it is constantly changing and transforming, but always in keeping with the principles demanded by the divine life, that is, consciousness, desire, power, generosity, wisdom, justice. To say that the cosmos is constantly changing is to say that new arrival is on-going and never-ending. One of the better known ways in which Ibn al-'Arabī expresses this notion is his doctrine of "the renewal of creation at each instant" (tajdīs al-khalq ma 'al-a-āhāt)'.

> The name of the Alive is an essential name of the real—glory be to Him! Therefore nothing can emerge from him but living things. Hence all the cosmos is alive, for indeed the non-existence of life, or the existence in the cosmos of an existent thing that is not alive, has no divine support, whereas every contingent thing must have a support. So, what you consider to be inanimate is in fact alive.[177]

This is a strong statement of the cosmos as a living entity. Even what we would usually regard as insentient, such as the stones of the earth, participate within the sentience of God and so in some sense are living because supported by or pervaded by the life of God. This is a vision of life as a constant, dynamic transformation in which each instant renews and affirms life itself in a process that articulates the overflowing abundance of God. Death is no end but a transition from one state of affairs to another. The *nafs* or soul of the person moves on. Indeed, the end of the soul's journey through the cosmos is being in the presence of God: a life before the Alive.

There is almost a vitalism in Ibn al-Arabī's vision of the cosmos. God himself is the supremely Alive one, although outside of space and time, and the cosmos that he creates is pervaded by his life in varying degrees of intensity. This is less a philosophy of reason than a wisdom or gnosis. As a young man Ibn al-Arabī met the philosopher Averroës, who was a friend of his father in Andalusia. We do not know the details of the encounter but from Ibn al-Arabī's account the meeting did not seem to go very well, and he tells us that he became aware of the distinction between knowledge through reason alone, as the pursuit of philosophy, in contrast to direct insight (*'arifun*) into the nature of life.[178] For him, the Sufi path of insight more than reason, journeying into the imaginal, provided a deeper understanding of life itself and its relation to the source of life.

In contrast to this mystical approach to understanding life and envisaging life as a journey of the soul back to God through the realms of the imaginal, Islamic philosophers were sometimes suspicious of this approach, seeking rather to understand God and creation through reason. God gives us knowledge of his will through revelation, but we need to deepen our understanding through philosophy. The Islamic philosophers were highly original in the way they sought to understand revelation through the lens of reason, with confidence in their tradition that allowed them to be influenced by ancient Greek thinking. Indeed, Plato and Aristotle were translated into Arabic (from whence into Latin). Like their Christian and Indian contemporaries, the Arabic philosophers were concerned with developing arguments for the

[177] Ibn al-Arabī, *Futūḥāt* 3.324. Chittick, trans., *In Search of the Lost Heart: Explorations in Islamic Thought* (Albany, NY: SUNY Press, 2012), p. 262.

[178] The account of the meeting is cited in R.W.J. Austin, *Ibn Al'Arabi: The Bezels of Wisdom*, The Classics of Western Spirituality (Mahwah, NJ: Paulist Press, 1980), pp. 2–3.

existence of God, developing accounts of knowledge, classifying areas of knowledge, and inquiring into the nature of reason itself, much of this work being done through commentary on the work of those who had gone before: in short, Scholasticism.

In the meeting between the older Ibn Rushd, generally known by his Latin name Averroës, and the younger Ibn al-Arabī we might say that two sides of the Islamic tradition met: the rational philosophy and empiricism of Averroës with the mystical theology of al-Arabī. I would not wish to overdraw this distinction, but these are clearly orientations within the history of Islamic theology, orientations that were the cause of dispute. Avicenna, perhaps the most important figure in the history of Islamic philosophy whose work influences the history of philosophy and of science, particularly medical science, into the early modern period,[179] draws the criticism of Al Ghazzālī for moving philosophy away, in Al Ghazzālī's view, from revelation and the injunction to act under the will of God.[180] Averroës defends philosophy against the Asharite school headed by Al Ghazzālī and the rational understanding of God and his creation: phenomena are driven by natural laws that God has decreed and we can understand these through reason.

Averroës—Life in Reason

Averroës (1126–98 AD) seems to have been unimpressed by the young Ibn al-Arabī who had not yet grown a beard, reflecting a suspicion of mystical revelation and claims to direct knowledge of God and the higher levels of creation; a suspicion of the imaginal. Averroës, it would seem, was a sophisticated, urbane thinker deeply interested in the world around him and convinced of the value of empirical knowledge, just as his Aristotelian forebears had been. Averroës is both a good scientist and a devout theologian in the sense that he believes empirical observation of the world cannot contradict revelation. Indeed, like Thomas Aquinas himself, for Averroës the truths of reason engaging with the world do not contradict the truths of revelation. If there is an apparent contradiction between philosophy and revelation, then the revealed passage in the Qur'an needs to be interpreted allegorically.[181] He wrote many works, translated into Latin, which had a direct influence on

[179] Shahid Rahman, Tony Street, and Hassan Tahiri (eds.), *The Unity of Science in the Arabic Tradition: Science, Logic, Epistemology and their Interactions* (Dordrecht, Boston, MA, and London: Springer, 2008). On his influence in the Islamic world see Yahya Michot, 'La pandémie avicennienne', *Arabica (Paris)*, vol. 40 (1993), 287–344. Also see Robert Wisnovsky, *Avicenna's Metaphysics in Context* (London: Duckworth, 2003).

[180] Ebrahim Moosa, *Ghazzali and the Poetics of Imagination* (Oxford: Oxford University Press, 2005), pp. 172–4.

[181] Ibid., pp. 206–8.

Latin Scholasticism. In contrast to his younger contemporary Ibn al-Arabī, with Averroës we find little idea of a subtle principle that animates the cosmos and that is apprehended through imaginal contemplation. We have, rather, a different approach to the category of life, an empirical approach that seeks to understand life in its actuality in the spirit of Greek science.[182] Averroës is arguably more concerned with a horizontal explanation of life than the verticality of the soul's journey back to God; indeed he disparages the idea of individual survival after death in favour of a universal intellect that we are part of. His most important work, the *Incoherence of the Incoherence* (*Tahafut al-tanafut*), is a defence of philosophy against Al Ghazzālī's the *Incoherence of the Philosophers* (*Tahafut al-falasifa*), and he wrote extensive commentaries on Aristotle, particularly on *De Anima*, the *Physics*, and the *Metaphysics*, as well as a long commentary on Avicenna's the *Canon of Medicine* (*Al-Qanun fi 't-Tibb*), with whom he disagreed on many points. These works were translated into Latin and Hebrew.

Whereas the name of God as 'the Alive' is important for Ibn al-Arabī, Averroës sees God as creator more than animator. But this idea of creation has to be reconciled with Aristotle's idea that the world has no beginning, which he tries to do by claiming that God supports creation constantly. Similarly, his doctrine of the soul that he derives from Aristotle needs to be reconciled with Quranic teachings. In his three commentaries (the short, middle, and long) he leaves no room for personal survival of death; that the soul as the essential being of a person carries on. Yet this conflicts with his religious commitments and as he does not allow for contradiction in matters of truth, he tries to resolve this problem through claiming that the soul survives as a disembodied intellect, but this intellect is universal.[183] When a person dies they remain as a single intellectual mind but not as the individual person they once were: this is Aristotle's immaterial soul. All individuating factors, namely the imagination, cease to operate at death. So, while the popular imagination might envisage life after death as a garden, this is figurative speech, for philosophy shows us that we become a disembodied intellect. It is this active intellect that gives form: the 'incorporeal mover' that infuses matter.[184] In a sense this active intellect is an animating force of living things, a view in line with the ambient emantionism from Neo-Platonic philosophy, which would have been alien to Aristotle. In his commentary on Aristotle's *Book of Animals* he writes: 'what Galen calls

[182] See Ruth Glasner, *Averroës' Physics: A Turning Point in Medieval Natural Philosophy* (Oxford: Oxford University Press, 2009).

[183] Richard C. Taylor, 'Averroës on the Ontology of the Human Soul', *The Muslim World*, vol. 102 (2–4), 2012, pp. 580–96.

[184] Herbert H. Davidson, *Alfarabi, Avicenna, and Averroës on Intellect* (New York and Oxford: Oxford University Press, 1992), pp. 232–42; Gad Freudenthal, 'The Medieval Astrologization of Aristotle's Biology: Averroës on the Role of Celestial Bodies in the Generation of Animate Beings', *Arabic Sciences and Philosophy*, vol. 12, 2002, pp. 111–37.

"formative power" [is] what Aristotle calls "the soul" (which according to him is donated by the sun and the rest of the planets) and the others call "the separate form." And it is identical to what Plato calls "the [universal] soul" and many of the Peripatetics call it "the Giver of Forms"'.[185] The active intellect is clearly an animating force of living beings. But Averroës modified his view in the course of his work to a more scientifically orientated, empirical claim that forms that impact upon matter, and so give rise to biological entities, are carried in semen or seed, and are also influenced by the heat of heavenly bodies. Davidson argues through a study of the manuscripts that Averroës changed his mind that no incorporeal mover is involved in generating living beings,[186] although the major difference between these two conceptions is the way, or *modus* in Freudenthal's terms, 'through which the forms are imprinted in matter—directly, by emanation, in one case, through the mediacy of seed or sperm and the "heats" of the stars in the second'.[187] This second, modified position is more in consonance with Averroës' increasing commitment to a scientific classification of life and move away from more vitalist tendencies within Islam. Averroës is not a philosopher of life in the sense we have described of affirming a view of life itself as verticality in which the soul ascends to a realm beyond the material world, but he is keenly interested in a more horizontal view of life in which empirical evidence allows us to develop models of the material universe. Averroës has been lauded as a precursor of Renaissance and later Enlightenment thinking, and it is clearly the case that his deep interest in furthering the work of Aristotle's naturalism impacted upon the history of other European thinking. I say 'other' because Averroës was indeed a thinker within the borders of Europe, although he died in Morocco, and the major theologians of Latin Christianity engage with his work. Averroës' medical knowledge is advanced for the time and he is interested in the natural, material forces that drive living organisms and in their scientific analysis, and it was this interest that led him to write commentaries on Aristotle and that, for a short time, led to his exile from the court of Mansur when he was accused by Jurists of adopting Pagan views about Venus being a deity.

The history of the reception of Averroës has not been without controversy, and he had more immediate impact on Christian theology than Islamic.[188] Indeed, he continued to influence the West in his rationalism and science, and

[185] Averroës, commentary on the *Book of Animals*, fol. 241b:11–13 quoted in Freudenthal, 'Medieval Astrologization of Aristotle's Biology', p. 117. Also see Ahuva Gaziel, 'Questions of Methodology in Aristotle's Zoology: A Medieval Perspective', *Journal of the History of Biology*, vol. 45, 2012, pp. 329–52. See also Peter Adamson, 'Aristotle in the Arabic Commentarial Tradition', in Christopher Shields (ed.), *The Oxford Handbook of Aristotle* (Oxford: Oxford University Press, 2012), pp. 645–89.

[186] Davidson, *Alfarabi, Avicenna, and Averroës*, pp. 232–42.

[187] Ibid., p. 136.

[188] Carol Lea Clark, 'Aristotle and Averroës: The Influences of Aristotle's Arabic Commentator upon Western European and Arabic Rhetoric', *Review of Communication*, vol. 7 (4), 2007, pp. 369–87.

even in the late seventeenth century his medical work still had impact. In 1695 a series of letters were published anonymously in London claiming to be from Averroës to a young Greek nobleman, Metrodorus, in 1149 and 1150.[189] These 'translations' have been attributed to Thomas Tryon, an interesting early advocate of vegetarianism and non-violence.[190] They are clearly not by Averroës—there are only vague details of original sources—and Averroës could not have made reference to tobacco, beer, and cider! But they are nevertheless interesting in gleaning from Averroës the spirit of empirical inquiry and an interest in health through understanding the functioning of the body, especially digestion. The letters speak of a liquid contained within the body that he calls *menstrum*, which is the basis of digestion and is even a kind of animating force in the body. About this substance he writes:

> It is generated more or less in all parts of the body, but especially in the mouth, and glandulous parts of the head, augmented by the heat and motion of the brain, and the quick and uninterrupted circulation of the blood. The more in quantity this *menstrum* arises, the sounder and healthier is the constitution, respiration is freer and less troubled, and digestion more easy.[191]

Leaving aside medical accuracy, this is interesting because it signals the continuing importance of Averroës to the age of rationalism in the West. The explanation of disease and its cure is found within the body and natural substances secreted by it. This substance that has healing properties, the menstrum, can be developed through temperance and moderate exercise in open fields, especially by running rivers 'where the air is more penetrating'. People who work and live in such regions attest the truth of this, as their health is good, being without dryness of the mouth, asthma, and other ailments. This pseudo-Averroës goes on to describe a good diet that promotes the secretion of this liquid such as rice gruel in contrast to sharp foods that inhibit its production such as 'salt meats, strong drink, and tobacco' which 'dry up this balsamic liquor'. We also need an active lifestyle against 'idleness'.[192]

Ibn Gabirol (Avicebron)—The Fountain of Life

Scholasticism as a style of philosophy and focus of concern was not restricted to Christianity and, as we have seen, is also part of Islam and Judaism during

[189] Anonymous editor (but probably Thomas Tyron), *Averroeana: Being a Transcript of Several Letters from Averroës, an Arabian Philosopher at Cordoba in Spain, to Metrodorus a Young Grecian Nobleman, Student at Athens, in the Years 1149 and 1150* (London: T. Sowls, 1695). Also contains prefatory letter in Latin with English translation by a J. Grineau of Port Royal.

[190] Craig Martin, *Subverting Aristotle: Religion, History, and Philosophy in Early Modern Science* (Baltimore, MD: Johns Hopkins University Press, 2014), p. 174. See also Nabil Matar, *Islam in Britain 1558–1685* (Cambridge: Cambridge University Press, 1998), p. 97.

[191] Ibid., p. 25. Some spelling modified to modern standards.

[192] Ibid., pp. 25–6.

this period. The most famous Jewish philosopher is Maimonides who, living in Egypt, composed his famous *Guide for the Perplexed* in Arabic. While the Jewish philosophers participated in the wider intellectual environment, especially in Muslim Spain, the concerns are often hermeneutical and focused on Torah and the commentarial tradition, although there is debate with Christian thinkers on topics such as the nature of reason in relation to faith.[193] While Jewish medieval Scholasticism does display concern for life itself in discussing creation *ex nihilo*, one work in particular stands out for its deep engagement with the way in which the universe becomes manifest, reconciling the *ex nihilo* view with an emanationism drawn from a wide range of sources from Aristotle to Neo-Platonism, to (some) Jewish throne mysticism. The *Fountain of Life* (*Fons Vitae*) by the Spanish philosopher Solomon Ibn Gabirol (1021–57), known by his Latin name Avicebron,[194] uses the Socratic dialogue as a frame within which to explicate his theory about God as the first cause and the fountain of life. Written in Arabic, of which only fragments remain, it was translated into Latin by a team of scholars in the twelfth century and it is this edition that is widely regarded as being closer to the original than the thirteenth-century Hebrew translation, because it retains the original dialogical format. Ibn Gabirol was a prolific writer famous for his poetry as well as for his irascibility.[195] In many ways the *Fons Vitae* is not representative of Jewish thinking in the sense that it does not cite Biblical passages, and its doctrines and emphasis on reason place it within Neo-Platonism, and because of this it was not recognized as a Jewish text for many years, but it is nevertheless an interesting and important book, exerting influence on Latin thinkers and further illustrating how the idea of life itself is deeply embedded within Neo-Platonic thinking. Criticized by Aquinas for misinterpreting Aristotle, Ibn Gabirol was thought to be presenting a form of Aristotelian hylomorphism, but Pessin argues that his work needs to be understood less in terms of Aristotle and more in terms of Neo-Platonism and of the influence of what has become known as the pseudo-Empedoclean tradition.[196]

The title of his book references Psalm 36.10 that presents the image of the 'fountain of life', but curiously neither the word 'life' (*vita*) nor 'fountain' occurs anywhere in the text other than in the title. But in spite of the absence of the word in the text, it is clear that for Ibn Gabirol, the source of life is God

[193] Anna Abulafia, *Christians and Jews in the Twelfth-Century Renaissance* (London: Routledge, 1995), pp. 89–93.

[194] For an excellent introduction to his life and work see Sarah Pessin,, 'Solomon Ibn Gabirol [Avicebron]', *The Stanford Encyclopedia of Philosophy* (Summer 2014 Edition), Edward N. Zalta (ed.), http://plato.stanford.edu/archives/sum2014/entries/ibn-gabirol.

[195] Ibid., section 1.

[196] Ibid., and Sarah Pessin, 'Matter, Form and the Corporeal World', in Tamar Rudavsky and Steven Nadler (eds.), *The Cambridge History of Jewish Philosophy: From Antiquity to the Seventeenth Century* (Cambridge: Cambridge University Press, 2009), pp. 269–301.

himself who is ineffable and unknowable; the source from which life flows, a doctrine in consonance with emanationist views in wider Neo-Platonism, although Ibn Gabirol also wishes to maintain that God has created the universe *ex nihilo*. So, although his general philosophy might be read theistically, that the waters of the Garden of Eden (although not referenced in his text) flow from the Lord, Gabirol presents a purely Neo-Platonic understanding about the transformation of form and substance from higher to lower levels and that the cosmos, although multiple, comprises a single substance transformed. This is the first essence (*essentia prima*, Arabic *adh-dhât al-'ûlâ*) that sustains the universe and all living things. Although he is bound to an *ex nihilo* doctrine, nevertheless he combines this with emanationism. The lower, corporeal levels of the cosmos are emanations of the spiritual levels that they recapitulate in a more solidified form: the lower reflects the higher in an imperfect form. He states bluntly that:

> The corporeal forms emanate from the spiritual forms. Now all that emanates from something is the image of the thing from which it emanates. Therefore the corporeal forms are the image of the spiritual forms.[197]

Like the earlier Neo-Platonists, Ibn Gabirol's cosmos unfolds from subtle to gross in a series of descending stages, but he presents his own understanding of this and elevates matter to a high status. So, whereas for Aristotle hylomorphism occurs at a level below the intellect, for Ibn Gabirol the conjunction of form and matter occurs from the very highest level. Pure matter and pure form are combined, permeated by the pure will of God, from which emerges the intellect (the Neo-Platonic *nous*), from whence the multiple material forms of creation in a sequence, from intellect to soul, to celestial body and thence to terrestrial body, each of these levels being a combination of both form and matter. Between the 'top' of the cosmos and God we have, for Ibn Gabirol, another membrane that—perhaps drawing on Jewish mysticism—he refers to as the 'throne', the mediation between God and creation. The lower levels in the hierarchy are denser than the higher, subtle ones and the higher forces penetrate the lower, from which they emanate. Thus, the soul pervades all lower substances, as does the intellect. He writes:

> ...and you will find that the simple substance, the substance of the universal soul, is diffused through the entire universe and that it sustains it in itself on account of its subtlety and simplicity: and you will find similarly that the substance of the universal intelligence is diffused through the entire universe and that it penetrates it. The cause of this is the subtlety of the two substances, their force and their

[197] Solomon Ibn Gabirol, *The Fountain of Life (Fons Vitae)*, trans. Harry E. Wedeck (1962), p. 63. The Latin text is found in Clemens Bäumker (ed.), *Avicebrolis (Ibn Gebirol) Fons Vitae, ex Arabico in Latinum translatus ab Johanne Hispano et Dominico Gundissalino. Ex codicibus Parisinis, Amploniano, Columbino primum* (Münster: Monasterri, 1895).

> light: and on account of this the substance of the intelligence is diffused into the interior of things and penetrates them. Therefore according to this view all the more ought the power of the holy God to penetrate all things, exist in all things, and act in all things beyond time.[198]

The power of God who is outside of creation nevertheless penetrates the cosmos through the powers of light that constitute the universal soul. In this scheme each of the layers of the cosmos comprises form and matter combined in varying degrees of subtlety, from pure force and light to solidified bodies in the material world. The soul is the cause of the lower forms, the imperfect created by the perfect, and penetrates substances that are compounded in the lower orders, as the form that subsists in the substance.[199] The lower substances envelop the light from the higher substances, and the whole cosmos envelopes the light of the first author, God.[200]

Ibn Gabirol's philosophy is a work of synthesis. It incorporates the standard Neo-Platonic model that we found in Plotinus with the Jewish theology of creation, as well as the Neo-Platonic model of the planetary spheres emanating out from the One. In his Hebrew *Kingdom's Crown* (*Keter Malkhut*), he describes how God emanates the spheres of intellect, the encompassing sphere, the fixed stars, Saturn, Jupiter, Mars, Sun, Venus, Mercury, Moon, and thence to the Earth, with the 'throne' mediating between God and cosmos beyond the sphere of the intellect.[201] He also brings Aristotle into the picture with the three kinds of soul, vegetative, animal, and rational.[202]

As with the other Neo-Platonic systems we have encountered, the purpose of life is to return; return to the source from which the soul came which is to retrace the cosmic descent back to its source: 'the path that leads to perfect happiness and that allows us to obtain true delight, that is our end'.[203] The whole complex unfolding of the cosmos as part of the nature of God in the end has as its goal the re-unification of the soul with its source in the highest layer of the cosmos, but never complete identity with the divine, as Ibn Gabirol is a committed theist. With Ibn Gabirol we have a further example of the way in which a certain kind of thinking dominated the Scholastic world. Here the Platonic hierarchical model of life reigns supreme and while in some sense we can take God to be life itself, the relationship between God and the living, which means here the manifested cosmos, must always be one of distinction. There is no univocity of being with Ibn Gabirol as God is wholly transcendent. But the innovation that he introduces is to see matter pervading the whole cosmical hierarchy, not simply emerging from form, but integrated with form in a series of levels of subtlety; light becoming denser and darker until condensed as matter.

198 Ibid., pp. 39–40.
199 Ibid., p. 58.
200 Ibid., p. 105.
201 Pessin, 'Solomon Ibn Gabirol', section 6.
202 Ibn Gabirol, *Fons Vitae*, p. 111.
203 Ibid., p. 132.

HISTORICAL TRAJECTORIES OF LIFE

We are coming to the end of this selected historical survey from ancient times through to Scholasticism of the Middle Ages, in which the category of life itself and its relation to the living has been understood in a number of ways throughout that history. In a sense the scene is set by the Judaic idea of a theistic reality who breathes life into matter that he has created *ex nihilo*, in contrast to a trajectory from Plato that sees life as animated by a transcendent source, the abstract forms, and another trajectory from Aristotle in which life itself cannot be separated from living entities, the hylomorphism that wishes to account for the living without a real transcendence. The God of Genesis, and we might add the God of the Qur'an, gives life and takes it away, becomes involved in history and for whom there is an end goal, a transformation of life in an eschatological end time. The Greek philosophers would, of course, have none of this. For them the universe has no beginning and no end, although in the Neo-Platonic version there may be intermittent relapses or contractions back to the source of life that then emanates into the cosmos once more. The Aristotelean view is again distinct in its rejection of transcendent forms that provide the blueprint of the living in favour of a view that emphasizes the material reality of living forms and for which life itself provides an explanation: a purely horizontal understanding that lays the foundations of the hard sciences and medicine. All of these trajectories find attempts at synthesis in the Abrahamic religions, and in the Scholastic period we find a conversation across traditions about the nature of life, at least in the sense that Arabic, Latin, and Hebrew theologians respond to each other's works and share the same problems, something facilitated through the translation of the Greeks into Arabic and thence into Latin and Hebrew.

In broad brushstrokes we might say that among the possibilities, first, there is a model of God transcendent to life itself and living beings, who creates life itself and animates creatures. On this view God is grace and benevolent towards the living because he is their origin as creator (the Abrahamic theist view). Second, there is a model of impersonal transcendence in which life itself is pure light as the power of the living, beheld in vertical ascent in which the spark of light, that is the soul, moves back to its origin (the Neo-Platonic view). Third, we have a model of immanence in which life itself is only apprehended within the living in which there is no transcendence of world, no outside to the universe, where the soul is the form of matter (the Aristotelian view). There is a fourth possibility, although this is hardly instantiated in the history of the West: the chthonic view of life. This is the idea that emergent into and as the living is a force of life coming from the earth. The energy of life is ongoing and overflowing, pure immanent force associated with the fecundity of nature. In modern times this autochthonous view has been associated with the idea of the Goddess, the assertion of the earth against the sky Gods of patriarchy and the assertion of female power through embodiment. On this view there is certainly no transcendence but pure immanence outflowing as the multiple forms of the world, and experienced in human life as shamanic regressions to earth energies that we neglect to our peril.[204]

With the retreat of religion from cosmology as the success of scientific models of the universe develops (the success of the GMM against the RCM), the focus on life itself diminishes. With the seventeenth century we have a deep concern with understanding the living in biology and with healing in medicine, but the philosophical interest in life itself comes to be replaced by a focus on epistemology consonant with the rise of science, a focus on the mind and body relation as mechanistic models of the material world become dominant, and a political concern about the nature of the good society and its values as massive shifts in human societies take place. But the interest in life itself does not disappear completely in modernity; rather, it seems to be articulated in different kinds of discourse, in mysticism and in poetry, and we can trace a theme from the medieval mystical theologians, such as Eckhart, to the seventeenth-century mystics that de Certeau has written about, through to the mystical poetry of Blake and into the nineteenth century through Whitman, and Rilke. We might also read the Hermetic tradition from the Renaissance—from Ficino to Bruno, to Boehme and Gitschel—as being interested in the life–living relationship. But it is some nineteenth-century philosophers who once again take up the theme of life itself, and psychoanalysis is concerned with the repression of life and the possibility of metaphorical resurrection (as we find in Norman Brown's work). It is lastly to this development that I wish to turn.

[204] The claim has been that the patriarchal religions have repressed this Goddess religion, but it can still be detected through that history in women's traditions of healing knowledge, midwifery, and witchcraft. See for example Tanya M. Luhrmann, *Persuasions of the Witches' Craft: Ritual Magic in Contemporary England* (London: Picador, 1994).

Part III

Philosophies of Life Itself

7

Philosophies of Life

I showed in Chapter 6 how medieval Scholasticism was deeply imbued with the category of life itself: from the Greek Aristotelian inheritance and from the Hebraic theology of breath, it became a focus for the living spirit in Christianity. In a sense, Christianity inherits both a substance view of life from the Greek legacy and a view of life as process from the Hebraic. Both substance and process views come to the fore at different periods, thus Aquinas is more substance orientated whereas Scotus sees life as process in the particularity of the instance. The success of the scientific explanatory paradigm, the GMM, had pushed religion back from cosmology, a retreat from empirical claims that has continued with the success of science. This has been accompanied by the retreat of Christianity from the public realm of governance to the private realm of conscience, a move precipitated by Locke. From there we have the story of the secularization of the West, along with the counter-move of the re-invigoration of the religious field in global modernity along with the development of spirituality outside of particular religious commitments.

Taking up our story from the medieval Scholastic context, I need to present an esquisse of the category of life itself in philosophical and theological modernity. On the one hand, we have the development of science from the seventeenth through the eighteenth to the nineteenth century—the foundational sciences of biology, chemistry, and physics, along with the ancillary sciences of botany, zoology, entomology, and so on—while on the other, we have a focus on life in philosophy that emerges at the dawn of the nineteenth century. Here, life is understood in terms of nature as a dynamic process linked to impulse or drive. My focus will not be on the history of the science of life, but on the philosophy of life. Partly stemming from a mystical discourse in the seventeenth century, the concern for life comes to be disseminated through the history of both Romantic poetry and Romantic philosophy. This vitalist spirit, if we might call it that, can be traced through to the twentieth century. I therefore need to show how life itself comes to be articulated through a mystical theological discourse that ends in poetry and through a philosophical discourse that ends in phenomenology. The mystical discourse comes to be redistributed through the history of poetry from Romanticism into modernity, and the philosophical discourse in a philosophy of life from Spinoza through the nineteenth century to Bergson and Deleuze.

Some philosophies of life in the nineteenth century through to the twentieth took the form of vitalism, a thesis that life is animated by a force in contradistinction to mechanical physical laws governing living beings in the life sciences. The term was only coined in the nineteenth century, but it had precursors in the animism of Georg Ernst Stahl (1660–1734) and continued to be an important concept into the twentieth century with Hans Driesch (1867–1941),[1] although it was generally rejected in twentieth-century life sciences. There are also suggestions of vitalist thinking in that broad integrator of theosophical, scientific, and philosophical thinking, Franz von Baader.[2] At a popular level from the late nineteenth century, vitalism comes to play a role in human practices orientated to health and long life in what has become known as the New Age, along with a feminist discourse that connects back to chthonic impulses and looks forward to a post-human future. Such vitalism contributed to the terrible politics of twentieth-century Germany and National Socialist ideology as well as to the positive affirmation of life in French philosophy. I will not be able to examine this history systematically, although I will make some reference to these developments in the last section, but I here wish to focus on the philosophy of life that came to coherent articulation in nineteenth-century Germany and influenced later kinds of thinking about life itself. The focus of these thinkers had a direct bearing on theological thought in the period down to the present day. But to recount this history, I need to make some remarks about its precursors in a mystical discourse of life that influences later thinking.

PRECURSORS TO THE ROMANTIC PHILOSOPHIES OF LIFE

Alongside the dogmatic theology of the Middle Ages we have encountered with Aquinas and Scotus, there developed a mystical theology, a tradition dating back to the fifth century with Denys or Dionysius the Pseudo-Areopagite and articulated through a flowering of mysticism in the medieval period. Meister Eckhart in particular stands out as an important figure, inspired by

[1] For a summary descriptive history that identifies three phases in its development see Gunnar Stollberg, 'Vitalism and Vital Force in Life Sciences: The Demise and Life of a Scientific Conception', unpublished manuscript, http://www.uni-bielefeld.de/soz/pdf/Vitalism.pdf, accessed 10 July 2017.

[2] See David Farell Krell, *The Tragic Absolute: German Idealism and the Languishing of God* (Bloomington and Indianapolis, IN: Indiana University Press, 2005), pp. 74–5.

the language of Neo-Platonism and Denys, preaching almost a Christian non-dualism.[3] By the time of the seventeenth century, mystical discourse had developed from mainstream mystical theology but also from the hermetic tradition of the Renaissance. Figures such as Ficino blurred the distinctions between magic and mysticism and sought to enhance life through manipulation of cosmological powers. This tradition continued through the sixteenth century, especially with Bruno's broad understanding of cosmology,[4] to the seventeenth where a new kind of mysticism takes off in Europe. In the mystical discourse that Michel de Certeau draws our attention to,[5] life itself finds expression in a kind of vitalism or immanentism that remains distinctively Christian while resonating with the history of life in other civilizations we have encountered.

This mysticism was not restricted to France, however, and important developments took place in Germany and Holland. A particularly important, if somewhat obscure, figure is Jacob Boehme (1575–1624), the 'Teutonic philosopher', whose mystical philosophy reconfigured knowledge, arguing that human life must be understood in relation to cosmic forces.[6] Boehme, who had a mystical experience precipitated by gazing at the light reflected in a pewter dish, was perhaps the first to understand the idea of God in evolutionary terms as a self-actualizing being through nature. Material creation is necessary for God's self-realization that comes to intense articulation in the human being. *The Aurora*, Boehme's first work, opens with the metaphor of a tree. The tree grows in full bloom in a garden that represents the world, with the soil signifying nature, the fruit human beings, with the sap as the deity: 'Now men were made out of nature the stars and elements, but God the

[3] On this history see Bernard McGinn, *The Presence of God: A History of Western Christian Mysticism*, vol. 3, *The Flowering of Mysticism* (Minneapolis, MN: Crossroad Press, 1998); Oliver Davies, *God within: The Mystical Tradition of Northern Europe* (London: Darton, Longman and Todd, 1988).

[4] The classic study of Bruno is still Frances Yates, *Giordano Bruno and the Hermetic Tradition* (London: Routledge, 1964).

[5] Michel de Certeau, *The Mystic Fable*, vol. 1, *The Sixteenth and Seventeenth Centuries*, trans. Michael B. Smith (Chicago, IL: Chicago University Press, 1992).

[6] For a recent edition of Boehme's *Aurora* and excellent introduction see Jacob Boehme, *Aurora (Morgen Röte im auffgang, 1612) and Fundamental Report (Gründlicher Bericht, Mysterium Pansophicum, 1620)*, ed. and trans. Andrew Weeks (Leiden: Brill, 2013). The original translation is by John Sparrow (London: John Streater, 1656). Also see the useful review by Lucinda Martin, *Aries: A Journal for the Study of Western Esotericism*, vol. 16 (2), 2016, pp. 241–5. For excellent studies of Boehme see David Walsh, *The Mysticism of Innerworldly Fulfillment: A Study of Jacob Boehme* (Gainesville, FL: University Presses of Florida, 1983) and Andrew Weeks, *Boehme: An Intellectual Biography of the Seventeenth-Century Philosopher and Mystic* (Albany, NY: SUNY Press, 1991). For a review of scholarship see Ariel Hessayon and Sarah L.T. Apertrei (eds.), *An Introduction to Jacob Boehme: Four Centuries of Thought and Reception* (London: Routledge, 2014).

creator reigneth in all: even as the sap doth in the whole tree'.[7] The tree is in effect the tree of life, enlivened by sap from the earth, the energy of God, and of which the fruits are human beings. Some of these fruits are corrupt, others not, but all are part of the same structure. The cosmos is a living entity expressing the nature of divinity. In the *Signature of All Things* (*de Signatura Rerum*), Boehme develops this idea in an argument that God as primordial mystery reveals himself through nature, although God is not nature.[8] In these works we see a concern with the world as a living entity enlivened by divine power; a highlighting of nature that is to come to prominence 200 years later. Boehme was influential on later philosophy, particularly Hegel who thought of him as the beginning of German philosophy, on mystical cosmology (on Johann Jacob Zimmermann),[9] on German pietism (and thence on Schelling), on English Romanticism (especially Blake and Coleridge), and even on psychoanalysis (on the work of Norman Brown).

Western esotericism from the Renaissance through the seventeenth century influenced the later Romantic philosophies of life—especially Boehme on Schelling and Hegel—but the second massively important philosopher is Baruch Spinoza (1632–77). In some ways Boehme and Spinoza are in resonance in that both regard nature as pervaded by and even identical with God, yet they are very different in the sense that Spinoza's philosophy is highly rational and critical in contrast to the cosmological piety of Boehme. For Spinoza, a brilliant, recalcitrant young man of the Amsterdam Jewish community, the personal God of tradition was mere superstition and through intellectual, rational effort we need to understand the true nature of God as the impersonal substance that upholds the universe. Rejected and vilified by the elders of the Synagogue and issued with a 'ban' (*cherem*), Spinoza spent his life grinding lenses for a living and writing philosophy, a philosophy that was to have a great impact on later thinkers. Although described by Novalis as a 'God intoxicated man',[10] Spinoza was undoubtedly a rationalist whose knowledge of God came through reason. During his lifetime Spinoza published a book on Descartes but his main work, the *Ethics*, composed in Latin, the language of Scholastic theology, was published posthumously.[11] In a rigorous if in some

[7] Boehme, *The Aurora*, Preface 7, translated by John Sparrow, *Aurora, that is, the day-spring, or dawning of the day in the Orient, or morning-rednesse in the rising of the sun, that is, the root or mother of philosophie, astrologie, & theologie from the true ground, or a description of nature* (London: John Streater, 1656), p. 2.

[8] See Pierre Deghaye, 'La théosophie de Jacob Boheme. Les trois mystères du livre *De la Signature des choses*', *Les Études philosophiques*, vol. 2, 1999, pp. 147–65.

[9] Mike A. Zuber, 'Copernican Cosmo-Theism: Johann Jacob Zimmermann and the Mystical Light', *Aries: Journal for the Study of Western Esotericism*, vol. 14, 2014, pp. 215–45.

[10] Novalis, *Schriften* (1892), vol. 3, p. 318. Referenced by Moore, *The Evolution of Modern Metaphysics*, p. 49.

[11] On Spinoza's life see Steven Nadler, *Spinoza: A Life* (Cambridge: Cambridge University Press, 2001). Also J. Israel, *Radical Enlightenment: Philosophy and the Making of Modernity*

ways simple procedure of definitions, axioms, and propositions, Spinoza demonstrates through rational argument that all is in God. He defines substance as something 'in itself', and 'conceived through itself: in other words that of which a conception can be formed independently of any other conception'.[12] Such a substance has attributes, its essence and modes, which are its modifications. Spinoza identifies God as 'absolutely infinite—that is, a substance consisting in infinite attributes, of which each expresses eternal and infinite essentiality'.[13] God is thus uncaused, necessary being. If, by definition, one substance cannot share attributes or essence with another, then this one substance must exclude other substances: 'there cannot exist in the universe two or more substances having the same nature or attribute'.[14] The one substance that is God who contains all attributes must therefore be the only substance, excluding all others: 'Whatsoever is, is in God, and without God nothing can be, or be conceived'.[15] All existent things are in God as 'modes' and cannot be discrete substances themselves, so God is in all things.

God as substance has potentially infinite attributes, but human beings perceive only two: extension and thought. Thus Descartes' dualism is for Spinoza an error in conceiving these two attributes as distinct substances, but if we regard them as attributes of a single substance, we circumvent the Cartesian mind–body problem because each is a discrete mode of operation: they do not causally interact. Thus Spinoza advocates substance monism and attribute dualism. This one substance comprises the totality of the universe and all difference can be reduced simply as its modes.

Spinoza therefore regards God as 'nature', as necessary being that is eternal and infinite. This being 'which we call God or Nature (*deus sive natura*) acts by the same necessity as that whereby it exists'.[16] This identification of God with nature proved controversial, especially in later centuries, and laid Spinoza open to the charge of pantheism (which is arguably accurate) and atheism (which is arguably false). Indeed to be a 'Spinozaist' was to be accused of atheism in the late eighteenth century. The German philosopher Jacobi famously declared that philosopher and playwright Lessing had in 1780 proclaimed himself on his deathbed to be a Spinozaist, which led Jacobi to a study of Spinoza that he published in 1785,[17] thereby precipitating what

1650–1750 (Oxford: Oxford University Press, 2001); Steven B. Smith, *Spinoza's Book of Life: Freedom and Redemption in Ethics* (New Haven, CT: Yale University Press, 2003).

[12] Spinoza, *Ethics*, I.III, trans. R.H.M. Elwes, *Benedict de Spinoza: On the Improvement of the Understanding, The Ethics, Correspondence* (New York: Dover Publications, 1955 [1883]), p. 45.

[13] Ibid., I.VI, p. 45.

[14] Ibid., I. Proposition V, p. 47.

[15] Ibid., I. Proposition XV, p. 55.

[16] Ibid., IV Preface, p. 188.

[17] See Gérard Vallée, J.B. Lawson, and C.G. Chapple (trans.), *The Spinoza Conversations between Lessing and Jacobi: Text with Excerpts from the Ensuing Controversy* (Lanham, MD, London: University Press of America, 1988).

became known as 'the pantheism controversy' (*Pantheismusstreit*). Jacobi claimed that Spinoza's philosophy was materialism in identifying God with nature as extended substance, a view that led atheism to be the consequence of enlightenment rationalism.[18] Moses Mendelssohn disagreed, arguing that pantheism and theism were virtually the same, but Jacobi's critique of Spinoza anticipated a critique of rationality that echoed in the later nineteenth century with Nietzsche, and anticipates contemporary discussion about atheism and the category of life.

Although Spinoza is a great influence on the Romantic philosophers who saw him as a philosopher of nature and also later on Deleuze, as we shall see, his philosophy is an indirect philosophy of life insofar as life itself is not thematized in his work. He certainly speaks of nature, and if we identify nature with life then this is clearly an important theme. But Spinoza is a substance monist and is less concerned with the process of becoming life. His importance for our story lies in the identification of God with nature and the pure immanence this implies, an idea that had great impact on later philosophies of life.

LIFE AS NATURE AND SPIRIT

Life itself becomes a theme in Romantic philosophy under the guise of nature. This focus finds expression through poetry and art: one thinks especially of Hölderlin in Germany and Wordsworth and Coleridge in Britain, and through philosophy that responds to Kant. The history of philosophy from Descartes became thoroughly secularized, rejecting revelation as a valid source of knowledge and advocating reason and empirical inquiry. Kant's dualism between the transcendental subject and the thing in itself that cannot be known presented a problem that later philosophers tried to overcome, rejecting the unknowable thing in itself as having any explanatory power and, by definition, being outside of understanding and knowledge. The Romantic philosophers tried to bring subjectivity and objectivity together in a concern for the reality of lived experience, and within this convergence of subjectivity and objectivity raised the issue of life itself as an explanatory paradigm. The Romantics were deeply concerned with nature, the human relationship to nature, and the expression of nature in art. Schelling in particular gave voice to these concerns.[19] And the idea of reality

[18] See Andrew Bowie, *From Romanticism to Critical Theory* (London and New York: Routledge, 1997), pp. 35–41; Wili Goetschei, *Spinoza's Modernity: Mendelssohn, Lessing, and Heine* (Madison, WI: University of Wisconsin Press, 2004), pp. 12–13.

[19] Charles Taylor, *Hegel* (Cambridge: Cambridge University Press, 1975), p. 351. Taylor remarks: 'Schelling was the philosopher who cashed in on this vogue of spiritualized physics, as it were, and supplied the philosophical vision of nature the age hungered for'.

as a single substance, that is also God, expressed by Spinoza grips the imagination of the Romantic philosophers, including or even especially Hegel.[20]

Transformations of Life: Schelling and Hegel

In 1786 in the *Metaphysical Foundations of Natural Science* (*Metaphysische Anfangsgründe der Naturwissenschaft*), Kant rejected the idea that matter was in any way animate—he warned against what he called hylozoism, the view that life is inherent within matter. Matter is dead in contrast to life, which is 'the capacity of an organism to determine itself via an inner principle to act, of a finite substance to determine itself to change, and of a material substance to determine itself for movement or rest, as a change of its state'.[21] The most distinctive form of life is the knowing subject who is irreconcilably divorced from the thing in itself, which is unknowable other than as appearance. The category 'life' does not play a great part in Kant's system, but he does occasionally refer to it. For example, in the lecture series on anthropology that he regularly presented until his retirement, he says that the strongest impulses of nature are love of life and sexual love, the former to maintain the individual, the latter the species. Although the love of life is compromised by war, some future state is possible where it will not, and a general state of happiness for the human race will be achieved.[22] But apart from this general sense, life itself is only a category set against dead matter, the most important kind of life for humans, being the animation of the thinking subject set over against the objects of cognition and experience that could only ever be encountered as appearances to consciousness. While Spinoza and Boehme, to an extent, echo the RCM of life and implicitly reject the GMM, with Kant we have the assertion of a humanism, a human-centred focus interested in freedom that is insistent on the value of the individual as an individual and not insofar as he or she participates in an impersonal cosmic power (Spinoza) or eradicates the subject of first-person predicates (the 'I') in favour of being subsumed by a direct intuition of God (Boehme). Kant's enlightenment as autonomy has no room for life as the self-assertion of impersonal, cosmic force. Indeed, life as nature is the object of consciousness for Kant, perceived by the thinking subject, a view that later Romantics would reject, echoing rather the earlier implicit vitalism of Boehme and pantheism of Spinoza.

[20] A.W. Moore, *The Evolution of Modern Metaphysics: Making Sense of Things* (Cambridge: Cambridge University Press, 2012), pp. 177–81.

[21] Kant, *Metaphysical Foundations of Natural Science*, A 120–1, cited in Andrew Bowie, *Schelling and Modern European Philosophy* (London: Routledge, 2016), p. 31.

[22] Kant, *Anthropology from a Pragmatic Point of View* (1798), 7.276, translated by Robert L. Louden, in Paul Guyer and Allen W. Wood (eds.), *Anthropology, History, and Education* (Cambridge: Cambridge University Press, 2007).

Schelling, like others of his generation, was dissatisfied with Kant's opposition between the thinking subject and the world of objects because the subject is itself part of nature. In his philosophy of nature (*Natursphilosophie*), Schelling attempts to think beyond this dualism because he argues that subjectivity itself emerges from nature and grasps nature theoretically.[23] Mind and matter are not opposed but mind springs from the causally dependent realm of nature. In contrast to Kant's dualism and Fichte's ground of being identified with pure subjectivity, the 'I', Schelling wishes to explicate an emergence of subjectivity from nature as ground. In so doing he advocates the irreducibility of human freedom and reconciles this with 'mechanistic' nature. That is, if we are to privilege and take human freedom as fundamental, as Schelling thinks we must, then we need to specify the relationship between freedom (human subjectivity) and the necessity of natural law (natural objectivity). We need therefore to overcome Kant's dualism while not falling into a priori pure subjectivism exemplified by Fichte's reduction of subjectivity and objectivity to the transcendental 'I', absolute subjectivity.

Schelling develops this philosophy in the *System of Transcendental Idealism* (*System des transzendentalen Idealismus*), a history of self-consciousness in which the reflexive mind emerges from a series of stages in a process of the absolute I's self-limitation, in which the I divides itself and objectifies itself in order to know itself.[24] There is a pre-rational identity, a force within the unmanifest absolute as 'the same' that both expands and contracts and that gives rise to manifold, objective nature as well as to subjective experience.[25] The pure subjectivity of the 'I', which is freedom, is in itself uninhibited but when it limits itself it thereby becomes inhibited, by which I think we can understand restricted. The restriction or inhibition of subjectivity is in accordance with its degree of objectification. The greater the objectification, the greater the limitation of subjectivity and the more freedom is contracted, the lesser the objectification, the lesser the limitation of subjectivity and the less freedom is contracted. The contraction or inhibition of freedom is the expansion or non-inhibition of objectivity and the contraction or inhibition of objectivity is the expansion or non-inhibition of freedom. In *On the History of Modern Philosophy* (*Zur Geschichte der neueren Philosophie*) Schelling puts this, in a passage worth quoting in full, in the following way:

> Uninhibited being is always that which does not know itself; as soon as it becomes an object to itself it is also already inhibited. Apply these remarks to the issue at

[23] F.W.J. Schelling, *First Outline of a System of Philosophy of Nature*, trans. Keith R. Peterson (Albany, NY: SUNY Press, 2004).

[24] Schelling, *System of Transcendental Idealism*, trans. Peter Heath (Charlottesville, VA: University Press of Virginia, 1978).

[25] Manfred Frank, *Ein Einführung in Schellings Philosophie* (Frankfurt: Suhrkamp, 1985), pp. 121–2, 125–7. Also see Bruce Matthews, *Schelling's Organic Form of Philosophy: Life as a Schema of Freedom* (Albany, NY: SUNY Press, 2012), pp. 6–7.

> hand the subject is, in its pure substantiality, *as* nothing—completely devoid of attributes—it is until now only Itself, and thus, as such, a complete freedom from all Being and against all Being; but it inescapably attracts itself... for it is only a subject in order that it become an object to itself, since it has been presupposed that nothing is *outside* it that could become an object for it; but *as* it attracts itself (*sich selbst anzieht*), it is no more as *nothing* but *as* something—in this self-gravitation (*Selbstanziehung*) it makes itself into something; the origin of all becoming something, or of objective, concrete being, then, lies in this self-gravitation. But the subject cannot grasp itself *as* what it is, for precisely in attracting itself (*im sich Anziehen*) it *becomes* an other, this is the basic contradiction, we can say the misfortune, in all being—for either it *leaves* itself, then it is as nothing, or it attracts itself, then it is as an other and not identical with itself.[26]

Here pure subjectivity, which in itself is freedom from being, becomes limited through objectivity. If being is unlimited or uninhibited, it cannot know itself, for only in limitation and differentiation into subject and object is self-reflection possible. Schelling is a monist: there is nothing outside of this oneness to be other to it, so otherness must come through self-limitation. The self cannot grasp itself, cannot cognize itself, in its true, non-objectified nature, in uncontracted, totally free state, so must become manifest through contraction to realize itself as something. This process Schelling calls self-gravitation in which the self either remains as nothing or attracts itself through objectification. Schelling refers here to this as a misfortune. This is fundamentally a tragic process because in its uncontracted purity, the self cannot know itself but knows itself in objectification or contraction, in which case it has become other than itself and so not truly itself. And yet this is a necessary process if subjectivity as freedom is to gain knowledge of itself and hence the tragedy (that always entails irony): the self loses itself in objectification to gain itself in freedom.

The subject, that is its own essence, wishes to become object to itself, which is contingency. This essence in itself is infinite but appears as contingency

[26] Schelling, *On the History of Modern Philosophy*, trans. Andrew Bowie (Cambridge: Cambridge University Press, 1994), p. 115. Anfred Schröter (ed.), *Schellings Werke: Nach der Originalausgabe in neuer Unordnung herausgegeben*, vol. 5, *Zur Geschichte der neueren Philosophie* (Munich: E.H. Bechund R. Didenbourg, 1928), pp. 170–1: *Unbefangene Seyn ist überall nur das, was sich selbst nicht weiß; sowie es sich selbst Gegenstand wird, ist es auch schon ein befangenes. Wenden Sie diese Bemerkungen auf das Vorliegende an, so ist das Subjekt in seiner reinen Wesentlichkeit als nichts—eine völlige Bloßheit aller Eigenschaften—es ist bis jetzt nur Es selbst, und so weit eine völlige Freiheit von allem Seyn und gegen alles Seyn; aber es ist ihm unvermeidlich, sich sich selbst anzuziehen, denn nur dazu ist es Subjekt, daß es sich selbst Objekt werde, da vorausgesetzt wird, daß nichts außer ihm sey, das ihm Objekt werden könne; indem es sich aber sich selbst anzieht, ist es nicht mehr als nichts, sondern als Etwas—in dieser Selbstanziehung macht es sich zu etwas; in der Selbstanziehung also liegt der Ursprung des Etwas-Seyns, oder des objektiven, des gegenständlichen Seyns überhaupt. Aber als das, was es Ist, kann sich das Subjekt nie habhaft werden, denn eben im sich-Anziehen wird es ein anderes, dieß ist der Grund-Widerspruch, wir können sagen, das Unglück in allem Seyn—denn entweder läßt es sich, so ist es als nichts, oder es zieht sich selbst an, so ist es ein anderes und sich selbst Ungleiches...*

where it posits itself as essence. That is, finitude—existence as we experience it—is the means through which the subject posits itself as infinite and so as freedom from being (*Freiheit vom Seyn*). Infinite freedom desires itself (and so in its desire already compromises that freedom and distorts it) and so becomes other to itself to realize this desire. Standing on the slope, it cannot hold itself back, says Schelling: 'For either it remains still (remains *as* it is, thus pure subject), then there is no life and it is itself as nothing, or it *wants* itself, then it becomes an other, something not the same as itself (*sich selbst Anziehens*), *sui dissimile*'.[27] But—and here we see the tragedy again—the subject cannot experience itself in an immediate way but only mediated as being, but then this 'is not what it really wants'.[28] The self-objectification of the subject progresses through a series of stages in which the subject as infinite, that Schelling calls 'A', is identical with itself (A=A) and then transforms into the finite ('B') such that A=B. The pure subject or self-identical essence (A) makes itself other to itself (as B) in a process of self-gravitation (*Selbstanziehung*), although it cannot be B, cannot be objectified, without at the same time retaining its essence as infinite. 'Therefore it cannot be B without being as A *uno eodemque actu* [in one and the same act], not as far as it is B but in another form (*Gestalt*) of its essence'.[29] This is a process of objectification in which the subject as freedom becomes limited as object and contingency, and yet in which that appearance reveals more of the subject than its unmanifest state because through manifestation it realizes its identity at a higher level, as it were. A=B is not merely a disguised form or A=A for Schelling, but a new kind of cognition in which subject knows itself as subject through object. Earlier philosophies—Schelling mentions Descartes and Spinoza—have missed this understanding through seeing extension as devoid of thought.[30]

Schelling's philosophy of life must be seen in the context of this self-limitation that is also a self-revelation. The first moments of the infinite self-positing subject is *life*, and 'the first moments of this *life* are moments of *nature*'.[31] The subject is clearly animated and alive, as are objects of experience. Even matter, contra Kant, is animated insofar as it emerges out of a power that gives rise to both subject and object. But this animation is not vitalism for Schelling. Indeed, he is keen to distinguish his position from vitalism on the grounds that the idea of a vital force is contradictory. A force, he says, is something finite but no force is finite unless opposed by

[27] Ibid., p. 116. *Schellings Werke*, p. 171: *Denn entweder bleibt es stehen (bleibt, wie es ist, also reines Subjekt), so ist kein Leben, und es selbst ist als nichts, oder es will sich selbst, so wird es ein anderes, sich selbst Ungleiches, sui dissimile.*

[28] Ibid.: *ob es gleich nicht das ist, was es eigentlich will.*

[29] Ibid., p. 117. Ibid.: *Es kann also nicht B seyn, ohne* uno eodemque actu *als A zu seyn, nicht sofern es B ist, wohl aber in einer andern Gestalt seines Wesens.*

[30] Ibid., p.119. Ibid., p. 174:... *die ersten Momente dieses Lebens Momente der Natur sind.*

[31] Ibid., p. 120.

another force, so when we consider matter as a force, we must necessarily think of an opposing force. Thus, life understood in terms of force is limited in the sense that force is necessarily an oppositional concept: to be a force implies resistance from another force. There could be no unresisting force animating beings. As a river flows and creates eddies due to resistance, so opposing forces restrict the flow of life: forces in nature 'must eternally flee each other in order to eternally seek each other'.[32] In a text cited by Bowie, *On the World Soul*, Schelling distinguishes between a formative force (*Bildungskraft*) and a formative drive (*Bildungstrieb*), the former being inherent within matter itself, the latter being the drive within organisms. Rather than Kant's sharp distinction between the living and non-living matter, Schelling wishes to maintain the unity of subject and object through an account of life in which matter itself contains a formative force that controls its form. The essence of life is a free play of forces of nature and the principle of life is simply the cause of a certain form of being.[33] The principle of life must be understood not as a single animating force, but as a plurality of forces out of which living beings emerge and that can be identified only through its instantiations. Schelling thus attempts to avoid a vitalism that posits a life force that itself is inexplicable. In Schelling's philosophy of nature, nature is enlivened by a dynamic interplay of unconscious forces within it from which conscious subjectivity arises. Thus, nature is invisible mind (*Geist*) and mind is visible nature, so the system of nature is actually also the system of our mind.[34] Freedom that is characteristic of human subjectivity is inscribed within nature itself. Freedom producing action based on subjective impetus (desire or will) reflects the unconscious production of forms in nature itself. The productivity of nature is the way in which forms become manifest through forces that counter each other: the productivity of nature is the ground from which forms of nature appear.

The human subject is part of nature for Schelling. This is no Christian Platonism but almost a Christian animism in the positive affirmation of the living and Schelling's thematization of nature, with subjectivity emerging from it and reflexively considering it. Productivity is the reality that pours forth both subject and object as nature. Although it cannot now be taken as a serious account of organic life, Schelling does present a description drawing on the natural science available to him and he himself contributed to the conceptualization of the life sciences.[35] The three forces of magnetism, electricity, and chemistry form life as objective differentiated matter. Through a

[32] Schelling, *On the World Soul*, p. 325, quoted in Bowie, *Schelling and Modern European Philosophy*, p. 37.

[33] Ibid. [34] Ibid., p. 39.

[35] Robert J. Richards, *The Romantic Conception of Life: Science and Philosophy in the Age of Goethe* (Chicago, IL: Chicago University Press, 2002), pp. 125–9.

series of stages organic substance develops from plant life to animal life, in which the differentiation of an animal's organs indicates 'the steps which the whole organic process has climbed',[36] a process that foreshadows Darwin's discovery to some extent in recognizing the developmental sequence of species. The spirit of life (*der Geist des Lebens*), the spirit of organic nature (*der Geist der organischen Natur*), works with the potentials of living beings to make them what they are. A living being is less a transformation of matter than a form that gives expression to it: 'Life depends upon the form of the substance, or the form has become that which is most essential'.[37] While matter is necessary for living beings, it is not enough to account for their life: we need the spirit of life as a complex of forces within them.

He does not wish to mystify nature as incomprehensible, seeing it in terms of mystical forces, but nor does he wish to see nature in purely mechanistic terms; rather he offers, in Löw's phrase, a 'hermeneutics of nature'[38] in which a continuum of causation from living beings to inorganic matter is controlled by forces that emerge from the ground of subject/object differentiation. The philosophy of nature that contains a science of nature as analysis of material forces that control it is complemented by a philosophy of identity. This identity is ultimately the sameness of subjectivity and objectivity transformed to a higher level of realization.

For Schelling this process is also a theology of sorts because it explains God. Not the result of an objective process but a subjective one, this God is the transformed subject outside of which there is nothing. God is at the end as he was at the beginning and posits himself as Becoming.[39] The subject going through the whole process of objectification as nature, as life, is already God, but for Schelling God must be self-realized in this self-positing. The philosophy of nature is therefore a philosophy of religion in the sense that subjectivity as infinity pervades nature as limitation, as light pervades matter, to use Schelling's analogy. In his later philosophy Schelling identifies God with absolute subjectivity, which is freedom. Not that freedom is an attribute of the absolute subject; eternal freedom *is* the absolute subject and this freedom is God. The one subject goes through everything and remains in nothing,[40] from not knowing itself in its beginning, it moves into knowledge of itself and

[36] Schelling, *On the History of Modern Philosophy*, p. 123. *Schellings Werke*, p. 176: *über welche der gesammte organische Naturproceß emporgestiegen ist.*

[37] Schelling, *On the History of Modern Philosophy*, p. 122. *Schellings Werke*, p. 175: *Das Leben hängt an der Form der Substanz, oder für das Leben ist die Form das Wesentliche geworden.*

[38] Reinhold Löw, *Philosophie des Lebendigen. Der Begriff des Organischen bei Kant, sein Grund und seine Aktualitat* (Munich: Suhrkamp Verlag, 1980), pp. 227–9.

[39] Ibid., p. 133.

[40] Schelling, *On the Nature of Philosophy as a Science*, discussed and quoted by Andrew Bowie, *Schelling and Modern European Philosophy*, pp. 130–40. I have relied heavily on Bowie's explication of this material for the first time in English, especially his placing it in relation to Hegel, indeed as Schelling's critique of Hegel.

so beyond the beginning, then back to its beginning again: 'the restored beginning is the end of all knowledge'.[41] As Bowie observes, this is close to Hegel but differs crucially from Hegel. In the move from the immediacy of being at the start to its self-reflection, it is not clear for Hegel how, in Bowie's words, 'what is mediated can know itself to be identical with what is immediate without simply presupposing this identity'.[42] Schelling gets around this problem, Bowie argues, by claiming that the real process of development is *repeated* in thought but is not the truth of the process itself. Thus the absolute subject cannot be understood through thinking, and indeed reflection itself is the result of a process that cannot itself be grasped by thinking. If I might rephrase this as follows: the absolute subject, that is freedom, becomes objectified life (both living entities and matter) and this process is reflected in thought, but thought cannot grasp it as object, being itself the result of the process: thinking the absolute depends on something that is itself not thought. The process of the absolute subject becoming life is prior to consciousness that reflexively can only grasp this in ecstasy (*Extase*).[43] Reflection on life itself cannot really become knowledge as object, because the identity of thought and being, which become separated in the process of productivity, cannot be proved but must be presupposed. Contra Hegel, this process cannot be articulated within reflection.[44] But Schelling's idea is close to Hegel in so many ways and Hegel articulated his own philosophy after Schelling's initial publication on transcendental philosophy. Lastly in this section, I wish to turn to Hegel as a second exemplar of a philosophy of life that has deep implications for religion and has given us a legacy that has come through the centuries.

Schelling and Hegel are not philosophers of life in the sense of vitalism—indeed Schelling distances himself from this—but they address the question of life directly and relate it to their systems focused on the different categories of being, subjectivity, and objectivity. Hegel, Schelling's student roommate, overlaps with Schelling in many ways but forges a way ahead that understands life as the result of a dialectical process resulting from the negation of 'being' in 'nothing' and their unity in the process of 'becoming'. While presenting a philosophy of nature initially close to, and inspired by, Schelling, Hegel wishes to distance himself from what he regarded as the Romantic fantasy of nature based on intuition.[45] In a nutshell, Hegel's system of philosophy claims that from pure abstract being-in-itself, the concept (*Begriff*) or spirit (*Geist*) becomes alienated from itself in a process of being-for-itself and moves through, or as, history, through the stages of art, religion, and philosophy to its final self-realization in being-in-and-for-itself. There is therefore the telos towards which spirit is moving in a dynamic process that should not be

[41] Ibid., p. 132. [42] Ibid. [43] Ibid. [44] Ibid., p. 136.
[45] Taylor, *Hegel*, p. 352, note 1.

regarded as simply an end state but as including the process that led up to it. Moore uses the analogy of a musical performance: 'It is the *telos* of the process in something like the way in which the *telos* of a musical performance is the entire performance, not just the playing of the last note'.[46]

This movement and transformation of spirit occurs through a dialectical process in which higher stages are transformations of lower. The spirit 'is the unmoved solid *ground* and *starting point* for the action of all, and it is their purpose and goal, the in-itself of every self-consciousness expressed in thought'.[47] The movement towards this goal of absolute idea Hegel describes in the *Encyclopaedia* as comprising three sides: the understanding or abstract side, the dialectical or negative side, and the speculative or positive reason.[48] In the dialectical stage things pass into their opposite and are transformed in a progression culminating in the endpoint of realization. This movement is, says Hegel, the power of God and nothing is outside of this, as all things in the finite world are subject to the dialectical destruction that transforms them.[49]

Hegel sometimes uses the category life to denote an initial object of desire for the subject who is yet alienated from it, while at the same time desiring it. This life is really spirit and spirit resolves contradiction and conflict. Spirit, as we have seen, embraces the idea of life itself within its semantic range, so with Hegel we have the explicit idea of life itself becoming matter and living beings in a process of self-actualization and realization.[50] But the category 'life' does not play a prominent role in Hegel's conceptual universe, although it does appear as a category in the *Logic* where it functions to describe the immediacy of soul realized as body, the individual living thing.[51] He also identifies 'imperishable life' or life itself with the 'absolute idea', the concept in its fully developed form, which is in fact the totality of the system.[52] The absolute idea is the unity of subjective and objective idea, the goal to be realized through an evolutionary process: it is the unity of the idea of life with the idea of cognition. The absolute idea contains within it both life itself and the cognition of life itself. That is, the absolute idea has an existential dimension realized in immediacy: the idea of life. Life itself and reflection on life itself constitutes the absolute idea identified with being-in-itself and being-for-itself.[53]

[46] Moore, *The Evolution of Modern Metaphysics*, p. 171.

[47] G.W.F. Hegel, *Phenomenology of Spirit*, trans. A.V. Millar (Oxford: Oxford University Press, 1977), VI. 439, p. 264. *Phänomenologie des Geistes* (Stuttgart: Reclam, 2009 [1807]), p. 311: [der Geist] . . . *ist der unverrückte und unaufgelöste* Grund *und* Ausgangspunkt *des Tuns Aller,—und ihr* Zweck *und* Ziel, *alsdas gedachte An-sich aller Selbst-bewusstsein.*

[48] Hegel, *Logic*, §79, p. 113. [49] Ibid., §80, p. 118.

[50] I have found the best overview still to be Charles Taylor, *Hegel*, although Inwood's *Hegel* (London and New York: Routledge, 1983) is a remarkable achievement and model of clarity.

[51] G.W.F. Hegel, *Logic*, trans. William Wallace (Oxford: Clarendon Press, 1975 [1873]), §216, p. 280.

[52] Ibid., §237, p. 293. [53] Ibid., §236, p. 292.

Life itself as articulated in the living is therefore teleological and transformative in the sense that the spirit, through the world, undergoes transformations into higher stages. Thus nature, in the sense that it excludes the sense of 'I', is being 'there and then', attaining its realization or truth in mind, which in turn absorbs nature within itself.[54] That is, the natural world cannot be understood independently of subjectivity because the very concepts we use to understand and categorize it are from the mind. The transformations of life within nature are understood by the mind and are themselves transformations within the mind. If existence is the unity of 'reflection-into-itself' (*Reflexion in sich*) and 'into another' (*das andere*),[55] then nature is penetrated by the mind. The term Hegel uses for the transformation of one state to another is *Aufhebung*, the archetypal Hegelian word that means both 'annulment' and 'preservation', indicating the transformation of a lower stage into a higher stage such that the lower is superseded by the higher but is a necessary condition for it.[56] Thus there is a transition from the lowest level in a hierarchy of being, the level of matter and spatio-temporal extension, to higher levels of life to spirit itself and the development of human history.[57] Nature is the outermost and external extension of spirit that must undergo transformation, and so a philosophy of nature is a philosophy of the way the absolute idea moves into life as a movement of its own freedom. Hegel ends the first part of the *Encyclopaedia of the Philosophical Sciences* by saying that absolute liberty, enjoyed by the Idea, goes forth freely as nature.[58] Nature is extended through the stages of nature as mass, the mechanical sphere that realizes itself as the solar system, the stage of nature as the unity of form and matter in which differences of form (the forms of beings, for example) become differentiated only within matter, and the stage of the for-itself totality of spirit, in which individual entities are differentiated[59] and so most alienated from spirit in its pure in-itself phase. Life as nature is the externalization of spirit from itself in order to return in a transformed way.

But it is in the *Lectures in the Philosophy of Religion* that we find a more sustained consideration of the category. Here in the context of a defence of the teleological argument, Hegel offers an account of life itself in relation to the living through a reflection of the soul as life force, the idea of a purpose, the

[54] Ibid., §96, p. 141. [55] Ibid., §123, p. 179.

[56] Ibid., §96, pp. 141–2. On the verbal form *aufheben* Hegel remarks: 'We mean by it (1) to clear away, or annul: thus, we say, a law or a regulation is set aside; (2) to keep or preserve: in which sense we use it when we say: Something is well put by. This double usage of language, which gives to the same word a positive and negative meaning, is not an accident, and gives no ground for reproaching language as a cause of confusion. We should rather recognize in it the speculative spirit of our language rising above the mere 'either-or' of understanding'. Also see Moore, *The Evolution of Modern Metaphysics*, p. 163.

[57] Taylor, *Hegel*, pp. 122–3. [58] Hegel, *Logic*, §244, p. 296.

[59] Taylor, *Hegel*, p. 355.

relation of the organic to the inorganic, and the emergent third thing that gives coherence. All living creatures contain a purposive determination that Hegel identifies with the soul that is also the subject. He writes:

> Life as subject is the soul, and the soul is purpose, i.e. it posits itself, it brings itself to fruition. So the product is the same as the productive [activity]. But what is living is an organism, and organs are means. The living soul implies a body. And only with that body does it constitute a whole, something actual. The organs are the means of life, and these means—the organs—are also that in which life comes to fruition and preserves itself; they are also material. This is what self-preservation is; what is alive preserves itself, it is the beginning and the end; the product is also what starts [the activity].[60]

Echoing Aristotle, the soul is the life force of living beings and the aspect of life that is subjectivity. This life force is self-positing through the living, through the organic bodies of living creatures and so—in the spirit of Aquinas as well as Aristotle—the soul is nothing without body for through the body the soul becomes actual. Implicit here is the idea that the soul experiences a world through body and through the senses, the organs of the body are 'the means of life' (*die Mittel des Lebens*) through which a life realizes itself, which means attains the goals of its purposive activity. The body is a coherent system in which its organs support each other and through maintaining each other maintain themselves, although in fact this is the preservation of a single being. The life principle pervades the organism and the organs of the body serve to anchor the living being in the world to enable it to interact with an environment and to promote its purpose of self-preservation. What is alive preserves itself and, Hegel claims, produces itself, containing within itself 'the material for its own emergence'.[61] That is, organisms are self-sustaining in one view, but they also need to absorb material from their environment in order to live. Life forms are self-directed, and intentional in the sense of movement towards an end, but they also need to absorb nutrients from the environment to sustain themselves.

In being teleological the living are self-sustaining. Indeed, it is not sufficient to understand life purely in terms of mechanistic and chemical processes for

[60] Hegel, *Lectures on the Philosophy of Religion*, vol. 2, *Determinate Religion*, ed. Peter C. Hodgson (Oxford: Oxford University Press, 2007), p. 710. Walter Jaeschke (ed.), *Vorelsungen über die Philosophie der Religion* Teil 2, *Die bestimmte Religion* (Hamburg: Felix Meiner, 1985), *Der Teleologische Beweis nach der Vorlesung von 1831* (*Sekundäre Überlieferung*), pp. 599, lines 231–40: *Das Leben als Subjekt ist die Seele, diese ist Zweck, d.i. sie setzt sich,vollbringt sich selbt; also das Produkt ist dasselbe als das Produzierende; das Lebendige ist aber ein Organismus, die Organe sind die Mittel; die lebendige Seele hat einen Körper an ihr selbst, mit diesem macht sie erst ein Ganzes, Wirkliches aus; die Organe sind die Mittel des Lebens, und dieselben Mittel, die Organe sind auch das, in dem sich das Leben vollbringt, erhält; sie sind auch Material. Dies ist die Selbsterhaltung; das Lebendige erhält sich selbst, ist Anfang und Ende; das Produkt ist auch das Anfangende.*

[61] Ibid.: *das Material des Hervovbringens*...

such processes cannot render an account of life itself. Rather we need a *tertium quid*, a third thing, to explain the relation of the organic to the inorganic. This third thing, beyond both, brings them together in this higher unity, for chemical processes (such as circulation of the blood) and mechanical processes (such as the decomposition of foodstuffs) 'cannot render an exhaustive account of what life itself is, so that a *tertium quid* must be assumed which has established these processes'.[62] There is a harmony or 'higher unity' to the processes of life as we see in the structure of the organs—Hegel gives the example of the circulatory system, the nervous system, the intestines, stomach, liver, and so on—but this unity, while being within the subject, must come from outside of it because the processes themselves are purely mechanical and chemical. This harmony, furthermore, this *tertium quid*, is the subject. He goes on: 'But in fact this unity, this harmony of the organism, is precisely the subject; this unity, however, also involves the active relation of the living subject to external nature, which has only a contingent, indifferent being vis-à-vis the subject'.[63]

We might read this in terms of purpose and meaning, that external nature is indifferent, mechanical process but given meaning by the subject, or rather the relation of the subject to external nature is the meaning. Indeed, Hegel argues that the correspondence of 'the concept', which he identifies with the organic, to 'reality', which he identifies with the inorganic, is 'nothing but the meaning of life itself'.[64]

This is a strong claim that Hegel grounds in ancient philosophy, specifically in Plato's idea of the *nous*, the higher mind that designates or orders the universe into a harmonic whole, 'an organic life that is determined according to purposes'.[65] Hegel goes on to identify this *nous* with reason or *logos*, that itself is identified with 'the world soul' (*Weltseele*) that posits the vitality or life principle, the soul, of the living. Thus, he identifies the *nous* with the *logos*, with the life principle that animates the organic and renders meaningful the relation between the organic and the inorganic, by which he means renders life teleological. Plato furthermore identifies God with the immortal life principle (*zōon*), the one principle and living system of the universe, the single subject of an organism whose organs are the planets and the sun: a coherent system based on the life principle.[66]

[62] Ibid., p. 711 (German text p. 600): ... *durchwelche Verläufe aber nicht erschöpft warden kann, was das Leben selbst ist; dabei müsste ein Drittes angenommen warden, welches diese Verläufe gesetzt hätte.*

[63] Ibid.: *In der Tat aber ist diese Einheit, diese Harmonie des Organismus eben das Subjekt; doch bei dieser Einheit ist auch das Verhalten des lebendigen Subjekts zur äusserlichen Natur, welche nur als gleichgültig und zufällig gegen dieses ist.*

[64] Ibid., p. 715 (German text p. 604): ... *nur die Bedeutung des Lebens selbst.*

[65] Ibid.

[66] Ibid., p. 716.

The telos of living beings is diverse; the purposes of animals and indeed of human beings is varied and finite, finite because all these purposes perish and each form of life or each purpose is built upon the death of other forms of life. In a hierarchy of purposes from animal survival to human goals of ethical life and civilization, every purpose comes to the same end: it perishes as we see in the earth covered with ruins. Both petty purposes of life and the essential 'come to grief' and 'we must regard all these purposes as finite and subordinate, and ascribe the destruction that has befallen them to their finitude'.[67] Even the highest human purpose in the end comes to nothing in the face of decay and destruction.

We therefore need to present an account of life beyond the life principle of Greek philosophy. The idea of the *zōon* gives us coherence to the universe that is not merely a collection of accidents, and this is clearly important, but in itself it still does not render an adequate account because all goals within the realm of the living are finite and perishable. This is insufficient because it does not involve the category of spirit and the idea of the highest good. On top of the Platonic life force we must introduce the idea of the spirit as the highest or infinite purpose beyond the finite purposes of living beings. Indeed, the highest purpose (*der höchste Zweck*) cannot be found in experience because it is without limit and can be described as 'the good, the universal final goal of the world'.[68] While we can rejoice in their unity, the life of living beings does not itself constitute the higher goal of life itself, which is the good. This goal is grounded in the category of reason that is itself the limit of spirit. Thus, the purpose of the good stands over against both the goals of physical nature and limited human purposes that contain so much evil. But the good does not actualize itself in experience so remains as an abstract 'ought to be' and is, in effect, powerless to realize itself. We therefore need, says Hegel, the *tertium quid* to actualize the final goal of the world, namely God, although God's determinate being can only ever be an ought within human subjectivity.[69]

In sum, for Hegel, although he does not much use the language of life, life itself is the animating principle of the living in relation to the inorganic from which the organic sustains nourishment. The life principle is the animating force of universal history from the inorganic to organic animal forms, to human life. But the relation of the organic to the inorganic is given meaning by a third thing that is the subject, even identified with God and so intimately related to the highest purpose of life, which is the good as pure potential. The movement of the life force through history is itself insufficient to account for

[67] Ibid., p. 717 (German text p. 605): *Alle jene Zwecke, so sehr sic uns interessieren, müssen wir als endliche, untergeordnete ansehen und ihrer Endlichkeit die Zerstörung zuschreiben . . .*

[68] Ibid. (German text p. 606): *. . . das Gute, der allgemeine Endzweck der Welt . . .*

[69] Ibid., p. 718. Indeed, for Hegel the morally good can only subsist in its battle against evil. It is a relational term.

the highest purpose of life, always yet to be realized for Hegel, which culminates in the spirit's self-realization as the in- and for-itself.

Indeed, before leaving Hegel we need to offer a further account of life in relation to his idea of history. Although, as I have noted, the theme of life is not so explicit in his work, the Hegelian view of historical progress can be understood as a particular form or expression of life itself: it is possible to read the dialectic as a philosophy of life working through history. We see this especially in Hegel's depiction of force, the power of spirit moving through time and the dialectic in which the force of negation goes against affirmation. Kojève's reading is particularly helpful here; he offers a guide to reading Hegel in terms of historical development and Hegel's famous distinction between the master and slave or bondsman.[70] In this account, at the beginning of history the future master and future slave are within the realm of nature, determined by the natural world, caught in the pure immediacy of life and so are not yet historical beings and so truly human. The master raises himself above his animal nature, thereby creating himself as a historical being, and forces the slave to work, who thereby also becomes raised above nature through the impact his work has upon the natural world. It is the slave who changes the world through his action, through his work, and so changes himself in contrast to the master who is changed only through the slave. As Kojève says: 'To be sure, without the Master, there would have been no History; but only because without him there would have been no Slave and hence no Work'.[71] But through work the slave can cease to be a slave because work makes the natural world human; it cultivates the human and produces culture (*Bildung*) such that through such education the slave comes to question, challenge, and overthrow the master, as happened in the French Revolution and as the phenomenon of Napoleon shows. But before this point of realizing freedom, the slave imagines a series of ideologies: first, Stoicism in which the slave believes himself to be free by simply knowing it; second, Solipsism in which all reality other than the 'I' is denied and scepticism is born, which gives rise to Nihilism, but this cannot last because the consistent nihilist would commit suicide ('Only the nihilist who remains alive is interesting').[72] The nihilist must see the contradiction in his existence and to be aware of this is to go beyond it and wish to remove it. The slave then tries to justify his life by a new ideology that resolves the contradiction between the ideal of freedom and the reality of slavery. This ideology is Christianity in which the slave accepts the contradiction in life but imagines another world beyond the senses. In this world the master is as much in bondage as the slave but true freedom occurs in the life beyond this one, so there is no need to struggle against the master.

[70] Alexandre Kojève, *Introduction to the Reading of Hegel* (Ithaca, NY and London: Cornell University Press, 1969), pp. 52–70.

[71] Ibid., p. 52. [72] Ibid., p. 54.

In that world, the slave and master are equal but under the supreme master, who is God. But to become free of even that absolute master, the slave must throw off Christian ideology and realize freedom in the world 'as a human being, autonomous and free'.[73] This is realized in the French Revolution that completes the history of the Christian world and inaugurates the stage of philosophy (culminating in Hegel) and in the modern state (culminating in Napoleon).[74]

Now while this may read as a 'just so' story from the twenty-first-century perspective, it is important for our historical narrative of life itself in relation to the living because it introduces a historical and political dimension to this relationship for the first time in the history of modern philosophy. Hegel's developmental sequence comes to be questioned and transformed, above all by Marx, but the master–slave metaphor reveals a structure about human life in relation to life itself that arguably echoes later in a political philosophy of life, linked to human environmental concerns. At the heart of Hegel's metaphor is a concern for the relation between life and death, of affirmation and negation that highlights an arguably universal structure of any account of life itself. Hegel comments that consciousness depends upon life as its natural setting and death is the negation of life and so of consciousness. In their struggle, the master risks life against nature and the slave too through work and the transformation of the world of life; the transformation of nature into culture. The life of the natural world, characterized by unconscious drive, is transformed into the life of the cultural world, transformed into the struggle of history. In a passage that depicts this structure, Hegel writes:

> In this experience [of consciousness transformed beyond nature], self-consciousness learns that life is as essential to it as pure self-consciousness. In immediate self-consciousness the simple 'I' is absolute mediation, and has as its essential moment lasting independence. The dissolution of that simple unity is the result of the first experience; through this there is posited a pure self-consciousness, and a consciousness which is not purely for itself but for another, i.e. is a merely immediate consciousness, or consciousness in form of thinghood. Both moments are essential. Since to begin with they are unequal and opposed, and their reflection into a unity has not yet been achieved, they exist as two opposed shapes of consciousness; one is the independent consciousness whose essential nature is to be for itself, the other is the dependent consciousness whose nature is simply to live or to be for another. The former is lord, the other is bondsman.[75]

[73] Ibid., p. 57. [74] Ibid., p. 70.

[75] Hegel, *Phenomenology of Spirit*, §189, p. 115. *Phänomenologie des Geistes* (Stuttgart: Reclam, 1987), pp. 144–5: *In dieser Erfahrung wird es dem Selbstbewusstsein, dass ihm das Leben so wesentlich als das reine Selbstbewusstsein ist. Im unmittelbaren Selbstbewusstsein ist das einfache Ich der absolute Gegenstand, welcher aber für uns oder an sich die absolute Vermittlung ist, und die bestehende Selbstständigkeit zum wesentlichen Momente hat. Die Auflösung jener einfachen Einheit ist das Resultat der ersten Erfahrung; es ist durch sie ein reines Selbstbewusstsein, und ein*

In this characteristically dense formulation Hegel posits a structure about the relation of human historical consciousness to life and death. First, life is essential to self-consciousness, for without life it would not exist. Second, self-consciousness as the subject of first-person predicates, the 'I', is absolute mediation between the for-itself and for-another, the immediacy between present and future, that undergoes transformation through its dissolution or negation. The 'I' posits itself as living or the expression of life that can be negated as subjectivity in objectivity of thinghood. Self-consciousness becomes other-consciousness, both a transformation and a negation. The master is self-consciousness that has risen above nature itself and the other consciousness, thinghood, is the slave who is posited as other to the master but who will in due course supersede the master through historical action, through work. Consciousness depends upon life itself but consciousness affirms itself as self-consciousness through awareness of its possible negation in death and posits its other, thinghood, which in turn will become self-conscious and overcome the master through the dialectic of history. Life itself becomes historical life through the master–slave dialectic.

We see here that life itself enters history, and the narrative of civilizations and religions that Hegel presents is an unfolding of life into a political history and a political philosophy of life. Thus, there is a progressive hierarchy as spirit that we can read as life itself moving through nature religions, to the religion of God incarnate in Christianity, and from thence into philosophy. In all of the different religions life is present, depicted as God, the object in religion, and as subject, the human being 'that comports itself toward that object, religious sensibility, intuition, etc'.[76] Life itself on this reading is the animating force of the living, impelled through history by force that is a synonym for God, striving towards its final end.

There are striking parallels between Hegel and Schelling in their accounts of life in relation to nature. God for Hegel is the activity of production and likewise for Schelling, productivity becomes the heart of the animating force of life. Life itself pours forth as blind abundance from which self-consciousness emerges as reflection. But Hegel develops a historical awareness of life articulated through a process of reason and takes temporality as being central to understanding the relationship between life and the living in a way that Schelling did not. For Hegel, the absolute idea comes to expression through

Bewusstsein gesetz, welches nicht rein für sich, sondern für ein Anderes, das heisst, als seiendes Bewusstsein oder Bewusstsein in der Gestalt der Dingheit ist. Beide Momente sing wesentlich; da sie zunächst ungleich und entgegengesetzt sind, und ihre Reflexion in die Einheit sich noch nicht ergeben hat, so sind sie als zwei entgegengesetzte Gestalten des Bewussteins; die eine; die eine das selbständige, welchem das Für-sich-sein, die andere das unselbständige, dem das Leben oder das Sein für ein Anderes das Wesen ist; jenes ist der Herr, dies der Knecht.

[76] Hegel, *Lectures on the Philosophy of Religion*, vol. 1, *Introduction and the Concept of Religion*, ed. Peter C. Hodgson (Oxford: Oxford University Press, 2007), p. 186.

the historical dialectic and the process of *Aufhebung*, the annulment of nature and the preservation of its central features, into self-conscious reflection through various stages, including the representation of the absolute in art. On the other hand, Schelling, foreshadowing later philosophical critique, regarded Hegel's as a closed system that privileges logic over ontology and because it is a system enclosed within itself, it cannot be wrong. But philosophy cannot end here and needs to be open to the future.[77] We see openness to endless possibility in art. Schelling regarded art as the immediate articulation of absolute subjectivity, but not as a stage on the way to philosophy; rather art is immediate and direct in a way that the philosophical expression of the absolute could never be.[78] Art, and in this we include poetry alongside the plastic arts, could approximate to a pure expression of the animating force of the living in a way that some poets—Coleridge, Hölderlin, and Goethe in particular—attempted to theorize. I need to make some comment on this important development: the Romantic theorization of art as expression of life itself.

Life in Art

For Schelling, the absolute as total freedom cannot manifest itself as itself because limiting itself in manifestation contradicts that total freedom.[79] But the absolute can come into view indirectly through art. Art is revelatory of the absolute in a way that science is too, but art embodies infinity whereas science endlessly pursues it.[80] The characteristic of art is imagination (*Einbildungskraft*) that gives expression to the absolute that can only be disclosed through art, only through which it can come into view.[81] The painter John Constable, although no philosopher, wrote in a letter that art was the union of nature with imagination.[82] This neat formulation is a good expression of what Romantic artists were doing. Constable's remarkable, massive landscapes contained nature within the boundary of the canvas and gave expression to nature contained within a human frame; his landscapes depict not simply nature in the raw but within a rural, human setting in contrast to England's newly industrialized cities in the North. Courbet would be a parallel example in France. Romantic longing—that the theologian Schleiermacher characterized

[77] See Bowie, *Schelling and Modern European Philosophy*, pp. 143–4.

[78] For a discussion see Krell, *The Tragic Absolute*, pp. 35–6.

[79] For a discussion of this issue see Bowie, *Schelling and Modern European Philosophy*, pp. 139–40.

[80] Bowie, 'Introduction', in Schelling, *On the History of Modern Philosophy*, pp. 13–14.

[81] F.W. Schelling, *System of Transcendental Idealism*, introduction by Michael Vater, trans. Peter Heath (Charlottesville, VA: University of Virginia Press, 1993), part 6.1.

[82] C.R. Leslie, *Memoirs of John Constable Composed Chiefly of his Letters* (London: Phaidon Press, 1951 [1845]), p. 177.

as longing for the infinite[83]—was both for nature as vitality within a human frame and for the past as a bygone age through the depiction of ruins.

The force of life itself was seen in nature that human beings' urban lives were becoming more distanced from. This was accompanied by a parallel concern with feeling as a more authentic indicator of life itself, as opposed to a Hegelian emphasis on thinking or a Kantian emphasis on will. In the Preface to his *Lyrical Ballads* (1802), Wordsworth writes that he chose incidents from 'common life' that he describes in realistic language commonly used, and that his emphasis on 'low and rustic life' reflected 'elementary feelings' that exist in a greater degree of simplicity.[84] Furthermore, such 'passions' are assimilated into a state of nature. Although a short preface, Wordsworth's remarks display the idea that there is an affinity between an unsophisticated, rural life that expresses simplicity of feeling and the state of nature. And this state of nature is itself a permeating force that penetrates the living, as we see in Wordsworth's 'Lines Written a Few Miles above Tintern Abbey', where the sentiment is vitalistic and pantheist.[85]

Paralleled in poetry, Romantic poets were concerned with nature—the archetypal image being Coleridge and Wordsworth's depictions of the Lake District—and Coleridge theorized this concern in a discussion of Wordsworth's *Lyrical Ballads*, partly inspired by his engagement with German philosophy, especially that of Schelling, but also inspired by Spinoza. In his text Coleridge articulates a vision of the poetic imagination in relation to an absolute ground of being that he derives from German Romantic philosophy. In this interesting short book, Coleridge traces a history of philosophy from Renaissance Neo-Platonic thinkers such as Ficino, through Bruno to the mystic Jacob Boehme (Behem) and thence to Kant, Fichte, and Schelling, whom he regards as the founder of a philosophy of nature begun by Bruno. This dynamic philosophy even constituted 'a revolution in philosophy'.[86] Coleridge says that his own philosophy, arrived at independently, is very close to Schelling, although this is understandable given that they both learn from Kant, Bruno, Boehme, and Richard Saumarez.[87] This philosophy shows that there are two forces at work in the philosophy of nature: one a force

[83] Friedrich Schleiermacher, *On Religion: Speeches to its Cultured Despisers*, trans. Richard Crouter (Cambridge: Cambridge University Press, 1988), p. 6.

[84] William Wordsworth and S.T. Coleridge, *Lyrical Ballads 1798 and 1802* (Oxford: Oxford University Press, 2013), pp. 95–116.

[85] 'Lines Written a Few Miles above Tintern Abbey', in *Lyrical Ballads*, pp. 193–6.

[86] S.T. Coleridge, *Bibliographia Literaria*, ed. James Engell and W. Jackson Bate (Princeton, NJ: Princeton University Press, 1983), pp. 161–3.

[87] Ibid. On Bruno's influence see Miklos Vetö, 'Jacob Boehme et l'idéalisme postkantien', *Les Études philosophiques*, vol. 2, 1999, pp. 167–80. Jean-Louis Vieillard-Baron, 'Schelling et Jacob Boehme: Les Recherches de 1809 et la lecture "la lettre pastorale"', *Les Études philosophiques*, vol. 2, 1999, pp. 223–42.

of infinite expansion and the other a force that finds itself in this expansion.[88] We might read this as the force of life moving out in infinite expansion accompanied by reflective understanding, reflection upon the process itself that comes to articulation in Romantic philosophy but above all in art, especially poetry, where imagination, divided into primary and secondary, becomes a reality quite distinct from 'fancy'. Primary imagination is the 'living power and prime agent of all human perception and as a repetition in the finite mind of the eternal act of creation in the infinite I am'. Secondary imagination is an echo of the former 'coexisting with the conscious will, yet still identical with the primary in the kind of its agency... It is essentially *vital*, even as all objects (*as* objects) are essentially fixed and dead'.[89] Imagination then is the human power of reflecting the force of nature and the wellspring of life itself expressed through art, in contradistinction to mere 'fancy' that is simply a mode of memory 'emancipated from time and space'.[90]

In a parallel way to what we have seen in philosophy, the Romantic poets' emphasis on imagination was a way of articulating a sense of vital life that could be approached more intensively as feeling and so expressed more authentically through imagination. Art becomes the vehicle for an implicit philosophy of life that privileges feeling over thinking, identifies imagination as a faculty through which life comes to articulation in the immediate medium of art, and begins to express an irrationalism that comes to cultural expression later in Modernist movements, such as expressionism and abstract expressionism in painting, and which can even be heard in Stravinsky's *Rite of Spring*. We might even say that the philosophies of life in modernity since the seventeenth century emphasize life itself as a process, whereas the implicit philosophy of life in art emphasizes life itself as a drive or force. But I now need to take up the philosophical trajectory in relation to religion into the modern period.

[88] Ibid., p. 196. [89] Ibid., p. 202.

[90] Ibid. See Douglas Hedley, *Coleridge, Philosophy and Religion: Aids to Reflection and the Mirror of the Spirit* (Cambridge: Cambridge University Press, 2000).

8

The Philosophy of Life as the Field of Immanence

In Chapters 1–7 I have described what we know of the biological foundations of life and its articulation through the history of three civilizations. In this macro-history we see that religions are systems that have expressed life itself through the ordering of space and time, through systems of salvation, and in their ontological commitments, especially in their orientation towards vertical ascent. But a system is nothing without a person and while a person is constrained by the system, and perhaps we can even say a person is a person because of the system, there is an irreducible nature to personal and interpersonal experience; there is the frame of the irreducible datum of human freedom. This frame, the horizon of a person's world, their lifeworld, is constrained by systems through language and institutions such as law, economies, and the ordering of space and time, and collective experience in turn constrains and influences the system. The radical experience of a Buddha or Jesus or Mohammad can impact upon and transform the systems they were born into. This two-way reciprocity between system and person that might be expressed as a reciprocity between the circle of life and the lifeworld has been theorized in a number of ways, and we can trace a trajectory in the history of philosophy that places emphasis either on system or on person and freedom. While Hegel and Kierkegaard might be the two exemplars of these orientations that come to mind, some thinkers have come to a view of the universe less in terms of a system and more in terms of a cosmos animated by a metaphysical principle.

In very general terms, for philosophies within this orientation, freedom is not incompatible with the circle of life. This trajectory that includes kinds of Romantic pantheism, pan-psychism, and vitalism has been an important current in the history of thought from the Pre-Socratics through Platonism to Christian mystical sources (Eckhart, for example) that are themselves rooted in Greek thinking, through to the rationalist Spinoza whose impact on the nineteenth century we have seen. From Spinoza we can trace a theme to German idealism and thence to Nietzsche, Bergson, Deleuze, Latour, and even Butler, or, in a slightly different trajectory, through Nietzsche to Heidegger and to Jonas.

Nor must we forget the life-sociology of Simmel. This kind of thinking is parallel to the biological semiotics that we find in Uexküll and the anthropology of Ingold. There is also an important theological train of thought here from Bergson to Rousselot, to Teihard de Chardin and de Lubac, that itself draws on deeper Catholic, mystical roots, particularly the mysticism of the body.

The circle of life as a system precedes us and exceeds us as individual persons. In the language of Chapter 2, it is the niche that supports and nurtures individuals and in the specificity of religion, it is the integrated system spoken of by Luhmann, whose work I will address in Chapter 11, that offers the hope of salvation or redemption, orders space and time, and provides the frame for individual and community ontological commitments. The system that provides the imaginaire for the circle of life constrains persons into the particularity of who they are in space/time. As we have seen, religions as systems provide frameworks and the ordering of the world within which people live. Persons in turn inform the circle of life and their collective action produces the niche within which people can flourish. This is institutional through law, governance, and education, but also physical in the dwellings we build, the space we manage, and the cities we create to live in.

If the philosophers and theologians give us an account of the circle of life as articulated in human life, we also need a more thickly textured account of human life as experience or lifeworld that we find in personalism and in existential phenomenologies of life, especially in a Francophone tradition from Merleau-Ponty to Michel Henry. The philosophy of life comes to be expressed as a system, on the one hand, and experience on the other, and we need to understand both to gain some leverage on life itself and human reflection upon it, although this very distinction between person and system has been questioned within network or web theories of life in which freedom or agency is seen to be distributed through the network. To speak of a person does not entail a self distinct from the world of life, but to draw attention to, and to thematize, a specificity within the field of immanence: the specificity denoted by the deixic pronouns 'my' and 'your' as they function within the network. On the one hand, we have philosophers of systems or circles of life; on the other, philosophers of the experience of life. This relationship between system and person, between constraint and freedom, between circle of life and lifeworld, goes to the heart of the civilizations I have described here in their understandings and articulations of life itself. It is not simply that system and experience are in propinquity, but each constrains the other or, in a different kind of language, flows into the other: to speak of the person entails the system, the niche or environment within which the person exists, and to speak of system entails the person, without which it is empty abstraction. The system embodies the power to facilitate freedom in person and the person embodies the ability to articulate the system; the macro-world of system and the micro-world of person are recapitulated in each other.

We now need to begin to express in more general terms the particularity of the categories of life we have so far encountered and develop a more general thesis about the system or circle of life, taking the languages of traditions seriously but not being restricted by them. In the belief that the religions we have examined have something relevant to say about life itself, we need to articulate a more current language through which to abstract and generalize the particularity of theological language in both realms of system and person. In this chapter I wish to bring into sharper philosophical focus the idea of life itself that I think has theological implications across religions. Articulating a view of life in terms of immanence, we can see how this comes to be theologically expressed in religions and also how traditional theologies diverge.

A number of thinkers have articulated the dynamic nature of the circle of life and the place of the person within it: Ingold in anthropology is a fine exemplar, as well as Latour's actor network theory, Deleuze's plane of immanence, and even Haraway's cyborg feminism. All these thinkers have emphasized the relationality of human life and the network, or meshwork to use Ingold's term, we inhabit. This way of seeing life itself in terms of a dynamic network has precursors in philosophy from Spinoza to Bergson that leads into Deleuze, a view of life that also has parallels in biological semiotics, some political materialism, and contemporary non-dualist anthropologies. I wish here to bring into sharper relief a view of life as immanence, drawing on this trajectory. What we might call the immanence view of life makes claims that are especially important: that immanence encapsulates the whole story of life, there is no outside or something beyond, no transcendence, and that the circle of immanence which is life itself needs to be conceptualized in terms of a mesh or a network, and individuals as such are no more than integrated forces within the mesh. Seeing the person purely in terms of forces within an integrated web is problematic, as we shall see, but we need to travel some way down this path to understand the force of this position and its contemporary importance for understanding life itself. This is another way of articulating a problem that Luhmann dealt with concerning the relation of system to person. To bring the issue into sharp relief, I will present an esquisse of the philosophy of life from Bergson through Deleuze, who is one of the most important thinkers to articulate a view of life that disrupts our usual categories and forces us to think afresh about what we mean by life and its relation to the person.

THE VITAL PRINCIPLE

Henri Bergson (1859–1941) is a fine example of a philosopher who responded to the most recent biological science of his day and addressed the philosophical questions raised by that science. Although he does address standard philosophical issues about philosophy of mind (especially the importance of memory) and epistemology, his most significant focus was on evolution and its philosophical implications.[1] In *Creative Evolution* (*L'évolution créatrice*) Bergson wishes to account for the continuity of life or rather living beings and for their multiplicity, not in terms of evolutionary adaptation—that answers the question 'how'—but in terms of a deeper philosophical or metaphysical reason that answers the question 'why?' He addresses this question through developing his theory of a vital principle accounting for diversity of forms, time and duration, and framing this in terms of instinct and intelligence. From a study of the contemporary biology available to him Bergson arrived at a theory of life itself that he named the vital principle or, more literally, 'momentum' (*élan vital*), which could be rendered as 'life impulse', an animating force that we can trace through evolution and that is indeed necessary in order to make sense of evolution. It is necessary to posit such a force in order to stave off either a mechanistic view of life or a teleological view, both of which are deterministic and so compromise human freedom and possible futures.

Life moves from one state to another, as an invisible current that passes through a series of bodies from generation to generation that becomes 'divided among species and distributed among individuals without losing anything of its force, rather intensifying in proportion to its advance'.[2] This current contains a continuous progress upon which individual lives 'ride'; a power that animates and gives rise to all organisms. Although it is the same impetus, the force of life follows different lines of evolution in a process of unceasing creation. It is not as though there were a single line of progress, but rather complementary trajectories of development have evolved, in particular plant life, non-human animal life, and human life, characterized by 'vegetative torpor, instinct, and intelligence' respectively.[3] These are not developmental, as Aristotle thought, successive stages of the one impulse, but rather three separate directions of evolution although connected and animated by the same principle. We might even see animal and human life as transformations of the energy from the sun absorbed by plant life (a view that echoes Bataille described in Chapter 4). We see how different forms of life are animated by the same impulse in the development of vision in which a simple function, seeing,

[1] For a lucid description of Bergson's philosophy, even though in the end seeing no value in it, see Bertrand Russell, 'The Philosophy of Bergson', *The Monist*, vol. 22, 1912, pp. 321–47. The first part is good exposition and the last part astute critique in which Russell accuses Bergson of replacing argument with metaphor and florid language. For a fine introduction to his philosophy see Keith Ansell Pearson, *Henri Bergson: An Introduction* (London: Routledge, 2011); Moore, *The Evolution of Modern Metaphysics* (Cambridge: Cambridge University Press, 2012), ch. 16; Vladimir Jankélévitch, *Henri Bergson*, trans. Nils V. Schott (Durham, NC and London: Duke University Press, 2015 [1959]).

[2] Henri Bergson, *Creative Evolution*, trans. Arthur Mitchell (London: Dover, 1998 [1911]), p. 26.

[3] Ibid., p. 135.

is enabled by the eye, an organ of great complexity that evolved across the evolutionary spectrum. The eye reveals the existence of a 'virtual' unity to life, characteristic of the *élan vital*, through being a structure shared across the spectrum of life forms: a solution produced by light to a problem of an organism's finding its way in the world.[4] We also see the implicit diversity in the development of embryos; the embryo of a bird hardly differs from a reptile, which diversify as they develop. Up to a certain point, the individual life recapitulates the species (ontogeny recapitulates phylogeny), although Bergson does not think in terms of progressive development but a multiplicity of forms along distinct trajectories. What Bergson calls higher forms in the stream of life are more diversified in their split from the preceding.

This is a dynamic view of life. It is not that there is a deterministic biological law, but rather life throws up new forms endlessly in new directions, with each new form organically related to earlier forms and to the first impulses of life inscribed upon matter. Indeed, species contain within them the trace of the earlier life form and evolution entails the persistence of the past into the present. Time and therefore duration is a feature inscribed 'somewhere' within living organisms in a constant upsurging of new forms that inevitably come to fruition and fade away: duration 'gnaws on things and leaves on them the mark of its tooth'.[5] The present effect contains the germ of the past within it, now in space/time. There is then a participation of organisms in the totality of the universe, in its duration in which everything participates. This is, says Bergson, 'participation in spirituality'.[6] But along with duration, change is fundamental to the nature of organisms: from the law of duration there is, in Jankélévitch's phrase, 'the gushing forth of an always complete and always new existence'.[7] To exist is to change, but not according to some predetermined order; the future is open, and this is guaranteed by the vital principle. Thus, Bergson argues forcefully against a mechanistic determinism and also against what he calls radical finalism, a view that the end is predetermined at the beginning. Both entail that we could in principle calculate the future and the past from the present, that 'all is given',[8] but even a super intelligence could never offer such a prediction because of the infinity of possibilities lying before the organism. The future cannot be read from the present. Unpredictable new forms ceaselessly arise, become present, and immediately fall back into the past where the new reality becomes old and 'falls under the glance of the

[4] For a discussion see Alia Al-Saji, 'Life as Vision: Bergson and the Future of Seeing Differently', pp. 149–51, in Michael Kelly (ed.), *Bergson and Phenomenology* (London: Palgrave Macmillan, 2009), pp. 148–73.

[5] Ibid., p. 46.

[6] Bergson, *Introduction to Metaphysics: The Creative Mind*, trans. Mabelle L. Andison (Totowa, NJ: Littlefield Adams, 1965), p. 33. See the discussion in Moore, *The Evolution of Modern Metaphysics*, pp. 412–13.

[7] Jankélévitch, *Henri Bergson*, p. 182.

[8] Bergson, *Creative Evolution*, pp. 37, 45.

intellect whose eyes are ever turned to the rear'.[9] Thus human consciousness is always after the fact and knowledge is always exceeded by the constant overflow of life impelled from behind by the vital momentum. This momentum is '*vis a tergo*: it is given at the start as an impulsion, not placed at the end as an attraction'.[10] The force of life impelling temporal development is not leading towards some end, a telos, but rather multiplies in myriad forms and develops along a particular line. What distinguishes the notion of cause from the *élan* is that the former remains distinct from its effect whereas with the latter the force coincides with the process itself.[11] The future is open, and Bergson calls this the virtual that can come into actuality if realized in the present.

One contemporary critic of Bergson, Claude Bernard, claimed that it would be possible to avoid the problematic postulation of vitalism without falling into mechanism through the notion of an organizing idea (*l'idée organisatrice*). This is not a force but an idea that can be attributed to the living, comprising organization, generation, nutrition, evolution, and decay (*caducité*).[12] But such a principle would not be strong enough for Bergson, who needs an account of life itself that is not an abstraction in a realm of ideas, but a principle that can be argued as explaining the experience of time and human life as it moves from virtuality to actuality. This virtual/actual distinction is a way of accounting for duration and so is crucial to the mechanism whereby the vital principle proceeds through the generations. In a sense, the virtual constrains the possibilities of the actual; what actually happens in the world is because of the possibilities contained in the virtual, but these possibilities are not deterministic. The virtual is not a telos or a mechanistic process of unfolding what is in a sense predetermined, but it is a constraint upon the possibilities of the future. Nature, from this view, is in a condition of constant self-creation in which the virtual moves into the present as duration.[13]

Bergson is aware of his thinking as a kind of vitalism. Indeed, he is conversant with the vitalist theories of his day[14] and he needs to be seen in that context, although he is critical of Neo-Darwinism, which he saw as mechanistic and so inadequate in its account of the development of life. Bergson identified adaptation as the main concern of Neo-Darwinism, the study of the 'elimination of the unadapted',[15] which he contrasts with his own concern with what life survives and how it does so. One of the most important Neo-Darwinists is August Weismann, whom Bergson cites on a few occasions in his *Creative Evolution*. Indeed, it is true that Weismann saw evolution as

[9] Ibid., p. 47. [10] Ibid., p. 106. [11] Jankélévitch, *Henri Bergson*, p. 111.

[12] Marie Cariou, *Lectures bergsoniennes* (Paris: Presses universitaires de France, 1990), p. 88.

[13] Moore, *The Evolution of Modern Metaphysics*, pp. 416–19.

[14] Bergson, *Creative Evolution*, p. 42. Following up the sources that Bergson cites gives a picture of the general state of evolutionary theory in France at the end of the nineteenth and beginning of the twentieth century. The foundations of modern biology are laid here.

[15] Ibid., p. 56.

'a blind machine'[16] that reproduces itself through what he called the germ plasm, perhaps a theoretical forerunner of DNA, that controls the pattern of a particular life. Bergson cites Weismann's famous claim based on experimental evidence that sex cells divide differently from other somatic cells that make up the rest of the body: 'they pass on their properties directly to the sexual element of the organism engendered'.[17] Bergson, accepting this general principle, reinterprets it to mean that there is a 'continuity of genetic energy' passing through the germ plasm. From this he concludes that 'life is like a current passing from germ to germ through the medium of a developed organism'.[18] This is a dynamic view of life that went against the mechanistic current of Neo-Darwinism while citing the same sources of experimental evidence. If we treat Bergson's claim in terms of philosophical argument rather than as an empirical claim about the vital principle, then it can be credited with plausibility in that the *élan vital* is akin to the notion of life itself, which can be abductively posited as that which gives rise to the myriad particularities of the living. Furthermore, this impulse is the guarantor of freedom because it is not determined and is not compelled to a particular resolution. There are echoes of Schelling here, for whom freedom is the ability to abstain.[19] And although the idea of a vital impulse has been subjected to criticism on a number of grounds, such as C.S. Lewis' dismissing of the idea that it does not go beyond death,[20] Bergson's idea has been taken very seriously and has been influential in the fields of psychology and philosophy, in particular on Gilles Deleuze.

THE FIELD OF IMMANENCE

Drawing on Bergson, Spinoza, and Nietzsche, with Deleuze we have an argument for life itself in a unique way, in terms of immanence. While Deleuze is influenced by bio-philosophy, especially Bergson, he should not be read simply in terms of that, as Ansell Pearce has well argued.[21] Deleuze's philosophy draws inspiration from Spinoza and Nietzsche. In a good summary of Deleuze's position, A.N. Moore writes:

> For at the heart of what they [Spinoza and Nietzsche] most fundamentally share is a celebration of activity, an affirmation of life, in all its diversity. Deleuze, like

[16] Keith Ansell-Pearson, *Germinal Life: The Difference and Repetition of Deleuze* (London: Routledge, 1999), p. 5.

[17] Ibid., p. 26. [18] Ibid., p. 27. [19] Jankélévitch, *Henri Bergson*, p. 183.

[20] C.S. Lewis, *The Weight of Glory* (London: Harper Collins, 1980 [1941]), p. 32.

[21] Ansell-Pearson, *Germinal Life*: 'It would be inadequate to restrict Deleuze's project to the merely biological and to claim it solely and exclusively for a novel "philosophical biology"' (p. 4).

> both of them, rejects the idea that life needs somehow to be justified, whether by some *telos* towards which everything is striving or by some transcendent structure in terms of which everything makes sense. Nature has no grand design. Nor is there anything transcendent to it. The celebration of activity and the affirmation of life are the celebration and the affirmation of immanence. And they reside in an ethic of empowerment, a concern with how things can be not in a morality of obligation, a concern with how things ought to be.[22]

With Deleuze, as for his precursors, we have a rejection of the idea of transcendence. Life must be understood in its own terms and not in terms of something outside of it, and indeed, we can account for the totality of our experience wholly within this field of immanence. Deleuze's thinking about immanence brings into sharper focus the issue of the circle of life in relation to the person because Deleuze, in emphasizing the immanence of *a* life, moves away from any individualism or personalism. Let us seek to understand Deleuze's complex thought by focusing on three themes: first, through his account of immanence; second, in his distinction between the real and the virtual; and third, in his reflection on the face.

In the last paper he published, 'L'immanence: une vie…',[23] Deleuze presents a pithy account of some his major themes. This condensed exposition is perhaps best remembered for its evocation of an image from Dickens. In *Our Mutual Friend*, a universally scorned rogue is dying and brought into the house to be cared for. The carers show 'ardent devotion and respect, an affection for the slightest signs of life', but when the dying man comes towards consciousness and his malevolence returns, the carers contract and grow cold. So long as he was not conscious, the carers displayed care for this particular life: 'The life of the individual has given way to a life that is impersonal but singular nevertheless'.[24] He uses another literary image in *A Thousand Plateaus* (written with Felix Guattari) from Virginia Woolf's *Mrs Dalloway*. Mrs Dalloway taking a walk is *haecceitas*, the unique unrepeatability as she will never again say 'I am this, I am that, he is this, he is that',[25] an idea also echoed in Borges' line 'there is a door I've closed to the end of the world'.[26] This particularity without individuality is the plane of immanence; bare life,

[22] Moore, *The Evolution of Modern Metaphysics*, pp. 547–8.

[23] Gilles Deleuze, 'Immanence: une vie…' *Philosophie*, vol. 47, Sept. 1995, pp. 3–7. Translated by Nick Millet, 'Immanence: A Life…', *Theory, Culture, Society*, vol. 14 (2), 1997, pp. 3–7.

[24] Ibid., p. 6.

[25] Gilles Deleuze and Felix Guattari, *A Thousand Plateaus: Capitalism and Schizophrenia*, trans. Brian Massumi (Minneapolis, MN and London: University of Minnesota Press, 1987), p. 263.

[26] Jorge Luis Borges, 'Limits (or Goodbyes)', *Selected Poems 1923–1967*, ed. Norman Thomas di Giovanni (London: Allen Lane, the Penguin Press, 1972), pp. 256–7. 'There is a line of Verlaine that I'm not going to remember again. / will not be able to remember. / There's a nearby street that is forbidden to my footsteps, / There's a mirror that has seen me for the last time. / There's a door I've closed until the end of the world. / Among the books of my library

we might say, as a 'transcendental field', 'a pure a-subjective current of consciousness', which is pre-reflexive and devoid of self.[27] Of course, the term 'immanence' is relational and only makes sense in contrast to 'transcendence', but for Deleuze this relationality is still within life and does not refer to anything outside. The old idea of transcendence is rejected[28] for there is nothing that the immanent is immanent to, not to God, not to the self, not even to being, but only to itself.[29] This is close to Spinoza for whom non-dual substance has two attributes that we can know, extension and time, and to Bergson, both of whom Deleuze invokes. Deleuze also refers to the late Fichte for whom the transcendental field is a life that does not depend on the notion of Being but is 'an absolute immediate consciousness whose very activity no longer refers back to a being but ceaselessly posits itself in a life'.[30] The field of immanence is life itself characterized by duration and extension but lacking any reference point outside of itself.

This is not a monism that reduces all particularity to sameness or to a single substance, but rather a philosophy that emphasizes particularity and distinction. Indeed, Deleuze is trying to think pure difference outside of sameness, and to support this draws on Duns Scotus' idea of *haecceitas*, 'thisness', and univocity.[31] For Scotus the univocity of being claims that 'there is nothing whose being has to be understood differently from the being enjoyed by us',[32] and so life itself is pure immanence that does not need explication by something outside of it. It is not that there is only one being, beings are distinct and multiple, but it does mean that Being is articulated in a single voice (hence it is univocal). Deleuze writes:

> The univocity of being signified that Being is Voice that it is said, and that it is said in one and the same 'sense' of everything about which it is said. That of which it is said is not at all the same, but Being is the same for everything about which it is said. It occurs, therefore, as a unique event for everything that happens to the most diverse things, *Eventum tantum* for all events, the ultimate form for all of the forms which remain disjointed within it, but which bring about the resonance and the ramification of their disjunction.[33]

(I'm looking at them) / There are some I'll never open again. / This summer I'll be fifty years old. / Death invades me constantly' (trans. Alan Dugan).

[27] Deleuze and Guattari, *A Thousand Plateaus*, p. 3.

[28] The origin of the transcendence-immanence distinction is quite late, first coined in the mid-nineteenth century. See Johannes Zachhuber, 'Transcendence and Immanence', in Daniel Whistler (ed.), *The Edinburgh Critical History of Nineteenth Century Christian Theology* (Edinburgh: Edinburgh University Press, 2017), chapter 9.

[29] Moore, *The Evolution of Metaphysics*, p. 579.

[30] Deleuze, 'Immanence: une vie…', p. 4.

[31] Deleuze, *Difference and Repetition*, trans. Paul Patton (London: The Athlone Press, 1994), pp. 35–40.

[32] Moore, *The Evolution of Metaphysics*, pp. 548–9.

[33] Deleuze, *The Logic of Sense*, trans. Mark Lester (London: Bloomsbury, 2004 [1990]), p. 205.

The field of immanence is univocal in that all forms of existence from the inanimate to the human share the single voice of their Being; the pure immanence of what they are. Univocity is an idea that, according to Deleuze, finds expression thereafter in Spinoza, Nietzsche, and thence to his own univocal ontology.[34] What occurs and what is said are both features of univocity, although these are not distinct, forming just a single event: the 'one Being and only, for all forms and all times, a single instance for all that exists, a single phantom for all the living, a single voice for every hum of voices and every drop of water in the sea'.[35] The being of extended bodies is the same and yet this is not the affirmation of sameness for Deleuze but rather a radical articulation of difference. This is no monochrome identity that might be subject to Hegel's criticism of being the night in which all cows are black. The univocity of being articulates the never-ending difference of its instances. Difference is at the heart of being articulated in a single voice in each instant. This is the true meaning of Nietzsche's eternal return; not that events repeat themselves, but rather what repeats itself is the univocity that is pure immanence expressed in the endless originality of becoming.

At first blush we might well think Deleuze's position to be inconsistent insofar as, on the one hand, he wishes to maintain the univocity of being, that being is a single event, and yet on the other he wishes to affirm difference in repudiation of sameness. He articulates this issue in an image in *Difference and Repetition* in which we have 'a black nothingness' contrasted with 'a white nothingness'. The black nothingness is that in which there is no difference, sameness, 'in which everything is dissolved', in contrast to the white nothingness in which differences are so unique that they are not connected,[36] that in Moore's formulation 'do not make sense'.[37] The problem for Deleuze, in Moore's summary, is how to avoid the dissolution of difference in the black nothingness while yet affirming meaning: 'he wants to make sense of how differences become connected'.[38] Without losing distinction Deleuze affirms difference as a distinct event, as the affirmation of singularities in which what is virtual becomes actual.

This is close to a position that emphasizes the *hic et nunc*, the only truth is here and now, the pure event. Deleuze's non-dualism, if we might call it that, is not a non-dualism of substance (as for Spinoza) but rather that Being speaks in one voice, the univocity of Being entails a certain equality between the being of the stone and the being of the living person, an equality at any rate in terms of being itself. But this does not deny difference and distinction. Indeed, it applauds it as the very heart of life itself. Being is particularized every moment

[34] Daniel W. Smith, 'The Doctrine of Univocity: Deleuze's Ontology of Immanence', pp. 168–70, in Mary Bryden (ed.), *Deleuze and Religion* (London: Routledge, 2001), pp. 167–83.

[35] Deleuze, *Logic*, p. 205. [36] Deleuze, *Difference and Repetition*, p. 28.

[37] Moore, *The Evolution of Metaphysics*, p. 553. [38] Ibid., p. 554.

in infinitely distinct kinds of duration. But as we have seen, the insistence on particularity, on *haecceitas*, is not an emphasis on individuated life, on my life in a personal sense, but rather the unrepeatability of this life, of bare life stripped of individuality but nevertheless unique and particular. 'There is no abstract universal beyond the individual or beyond the particular and the general: it is singularity itself which is "pre-individual"'.[39] This is life itself comprising a series of singular points and is not dissimilar to the Śaiva non-dualism we encountered in Chapter 5, where Abhinavagupta does not wish to deny the reality of the plural world in favour of a substance monism but rather that the singularity of consciousness is the same in all its instances, driven by pure power (*śakti*). This might help us appreciate Deleuze's understanding of the field of immanence as pure potentiality. The power of life is not individual but nevertheless unique as the realization of the actual emerging from the virtual. The virtual is 'fully real' and the necessary counterpart to the actual.[40] Through differentiation the virtual becomes actual endlessly and the actual falls away into the past and so returns to the virtual. This needs to be distinguished from the possible. The possible is not real and the difference between them is that the virtual is a quality of ideas that can realize existence, whereas what is merely possible cannot. The possible is constituted only after the fact, as it were, 'retroactively fabricated in the image of what resembles it'.[41] For Deleuze this is more than a matter of definition as it distinguishes what is real from what is not. The virtual is as real as the actual, but not yet actualized, in contrast to the possible that is formed only after we know what is real. In a sense the possible is impossible to actualize for Deleuze whereas the virtual is pure force or energy; the potential that feeds the actual.

Where, then, does this discussion lead us in our account of life itself? It shows us that Deleuze understands the world as a philosophical realist and that this realism is life itself as a field of immanence. There is no transcendence, no elevation to some higher state of being, no verticality. Thus, for Deleuze (although he does not discuss this), we might say that religion is possible but not actual or even potential. The ladders of ascent that religions speak of to climb outside of the universe cannot be real for Deleuze because there is no outside to go to. The field of immanence comprises the totality of world and the totality of human experience within it. We have the endlessly repeated formation of the virtual in the actual and the actual in turn dissolving into the virtual in a never-ending process. This affirmation of life is the rejection of those philosophies and religions in which the denial of life leads to virtue, 'life against life' in which life repudiates itself for some higher order.[42]

[39] Deleuze, *Logic*, p. 176. [40] Ibid., p. 208. [41] Ibid., p. 212.

[42] Daniela Angelucci, *Deleuze and the Concepts of Cinema*, trans. Sarin Marchet, *Deleuze Studies*, vol. 8, no. 3 (Edinburgh: Edinburgh University Press, 2014), p. 368. On Deleuze's

This never-ending process, the constant renewal of life itself in the present, the coming to actuality of the virtual and its dissolving back into the virtual, to use Bergson's terminology, is a horizontal philosophy, although Deleuze himself does use the metaphor of verticality to present a typology of philosophy. On the one hand, we have an emphasis on height, that we see in Plato for example, the height where Apollo lives, in contrast to which we have the depth of Dionysus and the pre-Socratic philosophers of fire and water. The Platonic ideal of height is linked to asceticism in that the ascetic rises up, thereby becoming purified in a process that Deleuze refers to as 'ascensional psychism'[43] and that we find innumerable examples of in the history of religions. In contrast, Empedocles jumps into Mount Etna and the volcano spews out his lead sandal. Empedocles is a philosopher of the depth for Deleuze, a philosopher of the earth, not the height:

> To the wings of the Platonic soul the sandal of Empedocles is opposed, proving that he was of the earth, under the earth, and autochthonous. To the beating of the Platonic wings there corresponds a pre-Socratic subversion. The encased depths strike Nietzsche as the real orientation of philosophy, the pre-Socratic discovery that must be revived in a philosophy of the future, with all the forces of a life which is also a thought, and of a language which is also a body.[44]

But the contrast between Plato's wings and Empedocles' sandal is somewhat of an illusion for Deleuze because in truth there is no height or depth. What we have is surface and the philosophers of surface par excellence are the Stoics and the Cynics who wrap themselves up in 'the surface, the carpet and the mantle'.[45]

I am reminded here of the 'blue clad' sect (*nīlāmbara*) of tantric Śaivism who performed sexual congress in public wrapped only in a blue cloth, much to the consternation of the wider public,[46] as a performance of going beyond social norms in the confidence of the non-dual reality of life. And indeed Dan Ingalls has brought our attention to the similarities between the Cynics and the Śaiva ascetic group, the Pāśupatas.[47] Deleuze locates himself in this middle ground between Platonism's verticality or conversion to height and the autochthonous depth of subversion with the Pre-Socratics: 'perhaps we can call it "perversion"'.[48] This middle realm of surface is the realm of Hercules rather than Apollo or Dionysus: not height above the earth, or depth below it,

relation to biology and Darwin in particular see Nathan Eckstrand, 'Deleuze, Darwin and the Categorisation of Life', *Deleuze Studies*, vol. 8 (4) (2014), pp. 415–44.

[43] Deleuze, *Logic*, p. 145. [44] Ibid., p. 146. [45] Ibid., p. 150.

[46] Jayanta Bhaṭṭa, *Nyāyamañjari*, trans. V.N. Jha (Delhi: Sri Satguru Publications, 1995), p. 562.

[47] Daniel H.H. Ingalls, 'Cynics and Pāśupatas: Seeking Dishonor', *Harvard Theological Review*, vol. 55 (4), 1962, pp. 281–98.

[48] Deleuze, *Logic*, p. 151.

but a middle ground that asserts the field of immanence; a philosophical operation that we can understand as the assertion of life itself that reduces all verticality—and indeed depth—to surface in an affirmation of sensual life, an affirmation of bare life, and an affirmation of pure potentiality. The middle-ground philosophers, the philosophers of the earth's surface, are disdainful of elevation in the sense of pure transcendence of the world and keen to affirm the particularity of being here and now.

There is an ethical dimension to Deleuze's thought and an insistence on social justice throughout his work. On the one hand, we have univocal being that is particular yet impersonal; on the other, we have the insistence on social justice and a critique of impersonal power. The negative side of the univocity of Being is the political oppression of multi-vocality and the attempt to reduce all otherness to sameness. We see this in the way that Deleuze and his colleague Guattari deal with the face. One would have thought, perhaps, that Deleuze would find assertion for the univocity of Being in the face, but his thinking is ambivalent here. Faciality is distinct from body: fundamentally a white wall with black holes, 'a white wall-black hole machine' that is an index of power in that the primary face has been that of Christ or 'your average ordinary White Man'.[49] Rather than the face representing or articulating the univocal voice of Being, it is something we grow into and indicates historical power, particularly with regard to racism: 'racism operates by the determination of degrees of deviance in relation to the White Man face'[50] against which all other faces are judged. The face or faciality is a politics for Deleuze that comes out of a repressive social order, despotism, and induces a semiotic of subjectification of those who do not conform.[51] The machine of faciality is the product of power and itself seeks to be a centre of control. The abstract machine produces the face and the face expresses 'the assemblages of power that require that social production'.[52]

This is perhaps a rather sceptical and bleak view of the face. The face is the face of power and only that, so is in a sense the negative side of the univocity of Being. The voice of Being that we hear in each instant certainly articulates the power of life itself, but this power has produced systems of oppression and inequality. But another reading of the face is that the face is an index of bonding and cooperation as the science of social cognition teaches us and, as we have seen, lies at the root of fundamental human communication and at the root of religion. Certainly, the face of Christ has been important in the history of the West—especially the history of art—although of course this is not the earliest face, as such representations go back to the Neolithic death masks in Jericho. While Deleuze emphasizes the non-individuality of the face,

[49] Gilles Deleuze and Félix Guattari, *A Thousand Plateaus: Capitalism and Schizophrenia*, trans. Brian Massumi (Minneapolis, MN: University of Minnesota Press, 1987), p. 178.

[50] Ibid. [51] Ibid., p. 181. [52] Ibid.

its abstract nature, we might understand this differently that faciality asserts pure interrelationality without individuality: the communicative act itself as articulation of the field of immanence.

On this view, religions are a fundamental articulation of the field of immanence that comes to expression through them. In the language of Chapter 2, the bio-sociology of life itself is expressed in religions that reflexively attempt to control the face-to-face interaction, and we can now see that this is the expression of the field of immanence even though theistic religions would resist being re-described in this way. There are a number of overlapping ideas: the field of immanence is more or less synonymous with the circle of life that we find in Luhmann's religion as system. Deleuze's immanence that entails a rejection of transcendence also signals for him the triumph of philosophy over religion, although that is arguably premature as religions are not, of course, intellectual positions but ways of life that function to bond communities, as we have seen. The philosophy of life as a field of immanence has alerted us to the way in which life itself can be articulated at the level of thought, but the question remains not so much about the particularity of voice but the individuality of voice, and the power and significance of the deixic pronoun in '*my* voice'. That is, to what extent does the philosophy of immanence entail the rejection of the significance of experience and the interpersonal? Where does this leave person? Does the impersonal particularity of Deleuze's non-dualism (and Abhinavagupta's, for that matter) leave us with a denuded sense of experience and human person? I don't think the answer to this is clear-cut. On the one hand, impersonal being as pure immanence asserts itself through its univocity, yet on the other the one voice is also my voice, the particularity of my being.

NETWORK AND MESHWORK

One way of addressing this dilemma has been in a series of related positions in which agency, or freedom in an older language, needs to be understood as part of a broader network or frame. The human person characterized as the one who acts needs to be understood as part of a wider system. The field of immanence that asserts itself through voice asserts itself through particular voices. One form of the immanence view with a particular understanding of agency is actor network theory (ANT) or, in an important variant, meshwork.

In a nutshell, ANT, which develops from a science and technology context, claims that actors are part of a network and that within such a network the technical and the social cannot be clearly differentiated: the social and technological are fused together. This is a materialist semiotics in which not only persons, but objects can function as actors within the network. This is not the place to describe a history of this approach to the social,[53] but it generally maintains that the social is a principle of connections, a network, that cannot be separated from biological organisms and that sociology is 'a kind of inter-psychology'.[54] It raises the question about agency, about who else is acting when we act, and how an actor is made to act by many others. That is, we have to understand the social world and reality more broadly in terms of an intersection of lines or trajectories of organisms in which agents are only agents because of the broader context in which they function. It is not that the particularity of act is the result of social system but rather that the collectivity of particular acts produces system.

Tim Ingold has developed this idea in his concept of 'meshwork', developing the term from Henri Lefebvre,[55] that we need to understand the world in terms of the flow of intersecting lines and fluid nature of the life process. The body does not stop at the skin, so to speak, nor does the mind, but we are all interrelated in a web of significance. According to ANT, we live in the midst of flows and counter-flows of materials in which agency 'is distributed throughout the network',[56] but even more radically, viewing life in terms of a web or mesh we find that 'action is not the result of an agency that is distributed around the network, but rather emerges from the interplay of forces that are conducted along the lines of the meshwork'.[57] Agency is less a matter of individual will than the consequence of a set of relationships and so always functions within a web of constraints that are themselves beyond the control of the agent. This is not to deny agency but to complexify and specify how it arises.

Meshwork as a description of life is a different way of describing what Odling-Smee and his colleagues called niche construction as a way of accounting for organisms' adaptation to their environment. The organism sits within a niche that is certainly the result of evolution, but also the result of cultural innovation and the two lines of development, natural and cultural evolution, feed into each other. What this range of thinking suggests is that what I have called the system of life comprises a network or webwork of signification in which agency is given rise to and which in turn can act upon the web. Deleuze's unrepeatability of Being, the univocity of Being, occurs within a mesh or web of signification, the intersecting lines of which produce the agent. The field of immanence—articulated through the metaphor of the web or, in

[53] For a survey of the literature see John Law, 'The Actor Network Resource', http://www.lancaster.ac.uk/fass/centres/css/ant/ant.htm, accessed 27 August 2016.

[54] Bruno Latour, *Reassembling the Social: An Introduction to Actor Network Theory* (Oxford: Oxford University Press, 2005), p. 13.

[55] Tim Ingold, *Being Alive: Essays on Movement, Knowledge and Description* (London and New York: Routledge, 2011), p. 84.

[56] Ibid., p. 90. [57] Ibid., p. 92.

Deleuze's case, the network of mycelium—is the context within which social systems and religious systems have developed.

This is closely related to the non-dualist anthropology in which non-dualism refers to an ontological orientation, in which anthropologists share a consensus that an emergent ontology in the discipline 'puts relations before entities', in the words of Michael Scott, one of its major proponents, and that this approach to ethnography is counter to the Cartesian-Kantian ontology of clear distinctions between subject and object.[58] The web finds articulation of itself within the system. Here the person is the agent as a function within the system; a view that we need to interrogate in Chapter 9, but for now we are in a position to return to religion and to the importance of life itself in the explanation of religious life.

RELIGION AND THE FIELD OF IMMANENCE

The field of immanence is important for us insofar as it articulates a view of life that we can see coming to expression in the history of religions and insofar as it highlights specific features of the understanding of life in the traditions we have studied. It is not that the immanence view as the latest philosophy of life corresponds more closely to the truth and so we can judge the past in terms of it, but rather that the immanence view has current intellectual value and so as a sign of where we stand is useful for us as a lens through which to view the past. I do think that the immanence view articulates well with contemporary evolutionary and cognitivist accounts of human reality, but in the end metaphysical positions cannot be proved or disproved through empirical inquiry, and we are still left with an impasse between contemporary immanentism and traditional theologies. Yet there are some analogues of the immanence view in the theologies of the traditions we have studied. Śaiva non-dualism and Duns Scotus' *haecceitas* that Deleuze develops are good examples of immanentist thinking. In this concluding section I wish to perform a comparison through the lens of the field of immanence that shows analogues to this way of thinking in the past across cultures and histories, thereby adding weight to the speculation that life itself comes to articulation through religions and their self-reflection in theology.

[58] Michael Scott, 'Steps to a Methodological Nondualism', in Venkatesan Soumhya et al. (eds.), *The Group for Debates in Anthropological Theory (GDAT), The University of Manchester: The 2011 Annual Debate—Nondualism Is Philosophy Not Ethnography*, ed. Soumhya Venkatesan et al., pp. 303–8, 356, in *Critique of Anthropology*, vol. 33 (3), 2011, pp. 300–60. Also see Joseph S. Alter, 'Gattungswesen: The Ecology of Species-Being: Alienation, Biosemiotics, and Social Theory', *Anthropos*, vol. 110, 2015, pp. 515–31.

In aligning himself with the Stoics, Deleuze implicitly aligns himself with tantric Śaivas and he shares much with the metaphysics of Abhinavagupta. For both philosophers the field of immanence is the truth of life as power identified with consciousness, which is not so much the denial of the pluralism of world as the affirmation of a univocity. All beings in the universe speak in a single voice that for Abhinavagupta is the power of speech (*vāc*), the supreme power (*parā śakti*). The abundance of the world asserts itself in 'appearance that arises as an object of knowledge ever new'.[59] And, perhaps like Deleuze, Abhinavagupta draws his illustrations from the realm of the senses: like a picture of a woman on a wall 'with deep navel and elevated breasts', so the diversity of cognitions are connected with the one wall of the universal light of consciousness.[60] Had Abhinavagupta seen the cinema he would, no doubt, have drawn his illustrations from there, as does Deleuze. I would not wish to exaggerate conceptual connection between these two philosophers—Abhinavagupta is committed at one level to a kind of theism, albeit one in which the 'theos' (*īśvara*) is essentially light (*prakāśaikātmā*)—but there is an affinity and parallelism in that Deleuze identifies with the Stoic ethos, Abhinavagupta with the ascetic Kāpālika ethos. Both are transgressive and both philosophers see the abundance of life itself overflowing in their thought. While Abhinavagupa maintains that his tradition is superior to others because it contains both immanence (*viśvātmā*) and transcendence (*viśvottīrṇa*), we could take this to mean that the pair immanence/transcendence that entail each other are still within the field of immanence as the totality of all that is, as Deleuze has done. Indeed, the monism of Abhinavagpta would have to deny the distinction, as it must, in the end, refute all spacial analogy of height and depth.

With Duns Scotus we have a different theological world, although a world not without analogues with Abhinavagupta. As we have seen, Scotus develops the univocity of Being that is given expression in each particular, singular act of existence itself. As a realist, Scotus was convinced that all of creation shared this unique property of *haecceitas*, thisness, articulated through time and extension. God is the transcendent source of creation, standing outside of it, but even God cannot interfere with the unrepeatable singularity of the act of Being. While this is not vitalism, it shares much with the kind of vitalism expressed in Deleuze in which Being is a force of constant becoming where the virtual becomes actual and returns to virtuality. For Scotus each unrepeatable act of Being has potential (*potentia*) for the generation of the next event, although in truth there is a seamless flow of existence. Although he does not privilege the category of life (*bios, zoe*), Scotus's concept of Being is close

[59] Abhinavagupta, *Īśvarapratyabhijñāvimarśṇī*, Kriyādhikāra, āhnika 3, 16: *ābhāso navanavaprameyaunmukhyāt*...K.C. Pandey and K.A.I.S. Iyar, *Bhāskarī* (Delhi: MLBD reprint 1986 [1938], vol. 1.

[60] Ibid., āhnika 3, 1–2, p. 73.

to Deleuze, which draws on Bergson's *élan vital* and Spinoza's substance monism. Scotus's theology is not Spinozism—Scotus could not be accused of atheism—yet the unrepeatable act of Being in the innumerable instances of its occurrence, all of which share a singularity of voice, draws Scotus close to Deleuze, close to Avicenna on whom he draws for his understanding of Aristotle, and even philosophically close to Abinavagupta for whom the world is the infinite instantiation of the power of consciousness. And all shared a worldview in which personalized, invisible forces play a role and impact upon human reality; angels for Scotus and Avicenna, yoginīs for Abhinavagupta.

Yet in spite of these parallels, we are still left with the impasse between the moderns and the theologians of the past. For Deleuze there is a primacy to human reality in that the gods exist only as projections and could be real (that is, have agency) only in the sense that the cinema is real, and its characters have agency. Although we have addressed the issue that we have to be methodologically agnostic, the implicit worldview of modernity is generally horizontal and atheistic, at least in the North Atlantic, where to believe in God in the twenty-first century, as Charles Taylor says, is unusual in contrast to 1500 where it would be the norm and almost impossible not to.[61] While accepting that the world of life is a projection on a screen, Abhinavagupta nevertheless would have maintained that the projection of human life was but one type in a spectrum of categories (*jāti*) that included not only domestic and wild animals but a myriad of supernatural beings as well, from tree spirits to great Gods such as Śiva and his emanations. Duns Scotus lived in a world in which angels and demons were common encounters, as did Avicenna. Modernity implicitly claims that the worldviews of Duns Scotus, Abhinavagupta, and Avicenna are purely veridiction; even when we bracket out the ontology of the non-material beings of religions, the implication is that this is only methodological, that we moderns know better and that the objective reality of the world is measurable and potentially knowable through science. Latour writes eloquently on this point:

> Up to now, the Moderns thought they had to refer to 'fictional beings', 'gods', 'idols', 'passions', 'imaginings', as *real things* only *out of charity*—critical rather than Christian charity. It was understood that with such things it was only a matter of 'representations' 'taken for beings' by people whose 'convictions' had to be respected, to be sure, but whose 'phantasms' had to be 'feared' at the same time, and one had to protect oneself above all against 'an always possible return of the irrational and of archaism'. The real nature of these beings devoid of existence 'quite obviously' came from elsewhere, since it could not lie in 'material' things.[62]

[61] Charles Taylor, *The Secular Age* (Cambridge, MA: Harvard University Press, 2007), p. 27.

[62] Bruno Latour, *Modes of Existence: An Anthropology of the Moderns* (Cambridge, MA: Harvard University Press, 2013), p. 376.

This is a difficult issue. The view that 'we moderns know better' is a kind of hubris and arrogance, and yet there is an overwhelming sense of the power of science along with its accompanying technology that has furnished knowledge of the world unsurpassed in history. We have seen the fruits of some of this labour from evolutionary anthropology and social cognition in Chapters 2 and 3. But science also teaches us humility, that there are no closed doors to knowledge, and that a genuine methodological agnosticism needs to be just that and therefore open to the possibility of wider ontologies. In the long view of things, it may turn out that the medieval cosmologies contained a truth about the world that modernity has closed off; that there is indeed a superabundance of life that, up to now, we have only represented and seldom, although not never, encountered. Again, to return to Latour, to dismiss an emic account (although he does not use that term) of the reason why somebody should go on a pilgrimage to a monastery ('because the Virgin Mary called me to go') is problematic, because that is still part of the agency in the world.[63] The social field as the site of multiple agents, in which human agency is facilitated by a range of other agents, makes the specification of human agency complex in the sense that human agency is but one within a range of social actors who mutually affect one another. According to Jane Bennett, these actors might include non-animate bodies in a vitality 'intrinsic to matter itself'.[64] That is, agency is distributive across a network and comprises 'assemblages', in Deleuze and Guattari's terminology, of 'vibrant matter'. These assemblages are affective bodies that are not controlled by a 'head', but the network of interaction serves to de-stabilize the human/non-human divide and pressurizes us to develop non-human-centred theories of action.[65] Bennett's argument is that we have to account for the effect on events of non-human forces, including 'inert' matter that actually serves as agential force, as electricity did, or lack of it, during a large-scale blackout in the United States. Although there is a substrate of 'brute matter' that we call nature, there is also matter as generative and purposive, on the one hand 'nature naturing', *natura naturans*, on the other 'nature natured', *natura naturata*, to use Jaspers' terms for Spinoza's distinction, 'the uncaused causality that ceaselessly generates new forms'.[66] Bennett's argument widens the range of agency but always within a materialist paradigm. It is the natural world that acts upon us rather than any supernatural world.

But for historians or phenomenologists of religion, suspending the question of the being behind religious appearances is important, yet we also need to

[63] Bruno Latour, *Reassembling the Social: An Introduction to Actor Network Theory* (Oxford: Oxford University Press, 2005), p. 48.

[64] Jane Bennett, *Vibrant Matter: A Political Ecology of Things* (Durham, NC: Duke University Press, 2010), p. xiii.

[65] Ibid., p. 24.

[66] Ibid., p. 117.

embrace those appearances within the specification of constraint that controls an event into its outcome, to use Bowkeresque language. Latour's example of the Virgin Mary calling the pilgrim to the monastery is within the network of agency that we need to specify as part of the explanation for that particular outcome, the pilgrim going to the monastery. That is, the Virgin Mary needs to be specified within the network of constraints as a force upon the actor, although this does not entail the acceptance of the pilgrim's explanation by the social scientist. In a sense there is equality between natural and supernatural affect if we suspend the natural attitude, although arguably material causation has precedence—even phenomenology would need to agree to this at the level of genetic phenomenology, the source of the appearance we encounter.[67] In a sense, even for Deleuze's plane of immanence, this does not really matter because whatever the kind of event, whatever the kind of being or agent, all Being speaks in one voice and is reducible to the unique event. While we might dispute the kind and range of agents we can encounter, the act of Being is singular.

So, the specification of constraint is important. For Bennett this might include not only neurons in the human brain but a complexity of preceding temporal events: 'If one extends the time frame of the action beyond that of even an instant, billiard ball causality falters. Alongside and inside singular human agents there exists a heterogenous series of actants with partial, overlapping, and conflicting degrees of power and effectivity'.[68]

The problem is, where are we to draw the line? To account for human action, we need the specification of constraints, and such specificity will be for particular reasons or based on particular assumptions. In one sense, the whole history of the universe stands behind any action, whether significant such as Caesar crossing the Rubicon or inconsequential such as Swann walking a different way home, but the level of constraint that we specify is crucial to the explanation and analysis given. While Bennett's identification of a wide range of constraints, indeed the whole of life, as contributing to a particular action is commendable, a responsible account of human action needs to specify a narrower range of constraint. Both Latour's and Bennett's work is an appeal to account for a much wider range of controls and agencies. Indeed, the circle of life itself within which all action occurs is the boundary limit of our acts. The actor is part of a network or meshwork, certainly, and agency can be understood to be distributed across a network, but still the nature of human agency, especially the reasons we give for our actions, needs to be taken into account. The plane of immanence or the circle of life must be related to the person (as we will see in Chapter 9).

[67] See Donn Welton, *The Other Husserl: The Horizons of Transcendental Phenomenology* (Bloomington and Indianapolis, IN: Indiana University Press, 2000), pp. 221–58.

[68] Bennett, *Vibrant Matter*, p. 33.

IMPLICATIONS FOR COMPARATIVE RELIGION

So far in this chapter we have presented a view of the circle of life in terms of a plane of immanence that we can trace as a trajectory through the history of modern philosophy. Implicit here has been the possibility of comparing this view with the traditions' understanding of life itself that we have encountered in the history of religions described in Part II of this book. There are two kinds of claim here: first, a strong philosophical claim that the immanence view expresses a truth about life itself, supported by recent science, against which the history of religions can be measured; second, a weak philosophical claim that modern articulations of life itself are no more adequate than those of tradition, but the modern view is simply another approximation in expressing the field of immanence. Deleuze, Ingold, and Bennett represent versions of the strong claim about life itself. They are not interested in the history of religions, but implicitly this view asserts itself against other worldviews. It asserts itself against historicism in maintaining that the field of immanence is a truth about the nature of life that we can now claim. There is progress in philosophy. The weak view implies that there is no progress in philosophy, but modern immanence philosophies of life are simply different instances or versions of a series of related views; simply the most recent articulation of ways of thinking instantiated throughout history.

This is a complex issue. We can argue that the strong view has support from recent empirical evidence. The extended evolutionary synthesis and niche construction show coevolution between nature and culture and that life itself comes to articulation in this way. There is no outside to this process that implies a philosophical non-dualism of the kind we have encountered in Deleuze: life asserts itself in the flow of its constant repetitions. The problem here is that while science progresses, the empirical claims it makes cannot be translated into metaphysical claims because empirical claims can be measured, while metaphysical claims cannot. And yet there is a sense in which philosophy has always related to the contemporary science available to it and this is no exception. Life itself coming to articulation in the philosophy of Deleuze might have empirical support in what science shows us of the evolution of life. The weaker claim is more distanced from empirical inquiry, and while a modern philosophy of life might have more appeal to 'us moderns', its claims to truth cannot be judged on empirical or scientific grounds. The weak view stands outside of itself in maintaining itself as a version of the field of immanence that has found different articulation in different times and places. Thus, on this view, Deleuze, Scotus, and Abhinavagupta are articulating a truth about life itself, but this is only ever seen in its instantiations. It is not as though we can compare Deleuze and Abhinavagupta to a truth that can be objectively articulated in a language other than that of these two philosophers, but those languages are the consequence of modes of Being outside of them.

On balance, then, I think we must argue not for the strong claim but for the weak claim. The weak claim allows us to observe the proximity of philosophies of life that we find throughout history, while not wishing to privilege one language as its true articulation. The weak claim is genuinely plural while at the same time wishing to maintain its truth that it approximates to, or points to, life itself. It is not that there is progress in philosophy, but on this view philosophies of life that are non-dualist in their insistence that there is no outside to the world are closer to the truth of the world than older theologies of transcendence. This is not, however, to deny verticality. Indeed, while Deleuze might wish to raise depth and reduce height to surface, the history of civilizations is replete with the experience of verticality and what Sloterdjik calls 'height psychology'[69] and 'vertical attraction'. But such transcendence can only ever be within the world, albeit a world which, in the end, is unfathomable, ungraspable, and strange. The weak version of the field of immanence also has implications for comparative theology in that it might promote comparison from within a horizon of faith, that the field of immanence is in fact a theological claim, albeit one in which theos is not outside of universe.

We began this chapter with a contrast between the circle of life understood as a system and the person or agent, and raising the question as to whether an emphasis on system was at the cost of the person, an issue that we need to return to. The philosophies of life that emphasize life as a plane of immanence in which there is no outside and no transcendence beyond the world have expressed a modern kind of non-dualism that is compatible with contemporary developments in neuroscience, social cognition, and what we know of evolution. Deleuze's plane of immanence is a life, the unrepeatable act of Being that constantly renews and reasserts itself, emerging from virtuality to actuality. We saw how this philosophy is a kind of vitalism that owes much to Spinoza and Bergson, and even offers a structure of interpretation through which to view the theologies of the past. Unlike those earlier theologies, however, Deleuze functions with a materialist paradigm of material causation; there is little room for a transcendent theos in his work other than as cultural trope. The implications of this view for agency or freedom are that we must understand agency or freedom as distributive, that there are many kinds of event that function as agents mutually affecting each other. Individualism needs to be understood within a network of agency, a network of freedom, rather than in terms of the isolated individual actant. We saw how this plane of immanence view articulates a philosophy of life that we have seen in other times and places, in distant civilizations, but we must now address this final problem of how life itself relates to persons as such, and whether the plane of immanence supports individual actors or simply particulars, an important

[69] Peter Sloterdjik, *You Must Change Your Life*, trans. Weiland Hoban (Cambridge: Polity Press, 2013), pp. 111–30.

idea that goes beyond the analysis of religion. To repeat a point, as I will develop in Chapter 11, there is a symbiosis between religion as a system, as a higher-level communication process that functions to bond communities, and religion as personal experience. It is this relationship between person and system, between experience and plane of immanence, that we need to examine further, in terms of theological response, in Chapter 9, and finally in terms of human bio-sociology.

9

The Phenomenology of Life

In this chapter I wish to take up more explicitly the problem I drew attention to in Chapter 8 concerning the relationship between life understood as a system or process and life in relation to the person. This is to re-tell the story, as it were, of the relation of life to the living from a different perspective of twentieth-century philosophy. Deleuze's plane of immanence I have read as a kind of vitalism in which agency is distributive across a network of actants. There is a tension between a living system rooted in a materialist metaphysics and the lived reality of persons, between system and lifeworld. I believe the best way to approach this issue, as with previous chapters, is to present a historical narrative that traces the development of the phenomenology of life focused on human experience, juxtaposing this with a new materialism that emphasizes the objective constancy of life itself as material reality beyond the human. The picture is somewhat complex as thinkers such as Deleuze clearly draw on the phenomenological tradition, and phenomenologists draw on broader materialist philosophies from Marxism and critical theory to what might be called eliminativist materialism. Part of the task of the present chapter is to delineate the contours of the discussion and to show how this debate is relevant to contemporary religious understandings of the person.

Phenomenology takes up the theme of life, especially Heidegger who brings the idea of life itself into the framework of phenomenology whose origins are in Husserl, but who follows a different course. Drawing on the mainstream phenomenological tradition, we also have a deep concern for the lifeworld, the social being of human reality that sees a fusion of phenomenology with sociology. But all these philosophies take human experience as central in contrast with some new kinds of materialism that de-centre the human, a philosophical project that also ties in with the philosophy of ecology and the idea of the post-human. New defences of religious subjectivity need to be seen in this context too, although here I must restrict my comments to the matter at hand, the way they articulate with a philosophy of life or more particularly how these philosophies deal with religion as a system, as a higher-level communication process, and religion as personal experience. It is appropriate to begin with Heidegger, that giant of twentieth-century philosophy, because with his early thinking he explicitly thematizes life and understands it within the context of the history of philosophy.

HEIDEGGER AND THE PHILOSOPHY OF LIFE

Martin Heidegger is a philosopher of life in the sense that he places human life, lived life, at the centre of his philosophy. Although he replaced a concern for life itself with the fundamental concern for Being (*Sein*) in relation to beings (*Seiende*), he remained focused on human being, being there or here (*Dasein*), and our ways of being in the world. Indeed, Heidegger's concern about the relationship between Being and beings, with what he calls the ontological difference, parallels the concern of this book about the relationship between life and the living. Clearly Being is a wider category than 'life', through which Heidegger raises the most fundamental question about the nature of our life and all of metaphysics: why is there something rather than nothing? This is a question that might deeply concern us or 'strike but once like a muffled bell that rings into our life and gradually dies away', or 'it may merely pass through our lives like a brief gust of wind'.[1] This 'first of all questions' (*die erste aller Fragen*) eludes propositional answer because the very notion of Being cannot adequately come to articulation but remains strange and mysterious. Being does not confront us directly and stand before us, for this would be to confuse Being with being, with particular existent. Thus, life itself is related to the category Being and to the fundamental question. Why is there life rather than no life? This would be a parallel but different-order question that could evoke a purely scientific response, which in some ways denudes the question's power, reducing it to mere biologism. The philosophical approach to the question respects its power and does not lose a sense of its existential immediacy. But Heidegger is not a philosopher of life in the sense of, say, Bergson, who articulates the idea of a force in life, and nor is he a pure existentialist concerned with 'brute life' as Campbell says, but he is certainly concerned with life as dynamic occurrence.[2]

[1] Martin Heidegger, *An Introduction to Metaphysics*, trans. Ralph Mannheim (New Haven, CT and London: Yale University Press, 1959), pp. 1–2; *Einfühlung in die Metaphysik*, *Gesamtausgabe*, vol. 40 (Frankfurt: Vittorio Klostermann, 1983), p. 2: *Vielleicht nur einmal angeschlagen wie ein dumpfer Glockenschlag, der in das Dasein herintönt und mählich wieder verklingt... nur durch unser Dasein ziehen wie ein flüchtiger Windstoss*. The question was first raised by Leibniz.

[2] Scott M. Campbell, *The Early Heidegger's Philosophy of Life: Facticity, Being, Language* (New York: Fordham University Press, 2012). Campbell writes: 'He is not a life philosopher because, for him, the flow of life is not simply a flux of sense impressions...Nor is he an existentialist,

Heidegger's early philosophy begins with the question of life, with life lived in human experience and the primacy of this experience. In his early lectures on Aristotle, Heidegger looks at the word 'life' in terms of a transitive and intransitive verb—in an intransitive sense 'to be alive, to really live (= to live intensely), to live recklessly, dissolutely...' (*am Leben sein, jemand lebt* (*im Sinne: er lebt intensiv*); or in a transitive sense '"to live life," "to live one's mission in life"' (*das Leben leben, seiner Aufgabe leben*)—and in terms of the noun, 'life' (*das Leben*).[3] These are not simply grammatical considerations but reflect fundamental meanings of the word as enacted in life itself. Indeed, the verb 'to live' indicates care about one's own life and so is a 'formal indication' of the life that is lived. The intransitive sense of the word indicates the idea of something to live 'in', namely a world. The term 'world' names what is lived and the noun 'life' is therefore a relational term concerned with living in the world. Life experience is within a shared world, a world towards which life has a directionality and this directionality is caring, which is also simultaneously meaning.[4] In his early lectures on Aristotle, Heidegger already expresses the idea of caring as the basic orientation of life. Care (*Zorge*) is about the relationality of life and Heidegger analyses it in three structures of inclination, distance, and sequestration, all of which indicate movement as a feature of life. By inclination life is directed towards its world, is pulled in a certain direction and has a particular weight. There is a pull from life itself towards a maturation, which can also be described as a proclivity. This proclivity is towards the world where life is experienced as a world:

> In proclivity, life itself is experienced essentially as world; i.e. life itself, in facticity, exists always in the form of its world, its surrounding world, its shared world, its own world; every life is in the form of my, your, his, her, their world; our life, our world.[5]

because it is not simply brute life or sheer human existence that he investigating but rather that rich mysterious "something" that motivates and dynamizes the happening of life' (pp. 23–4). Campbell has written an excellent book showing the importance of the philosophy of life for the early Heidegger that has been most useful in orientating my thinking with regard to Heidegger's early works. Other important guides have been Theodore Kisiel, *The Genesis of Heidegger's* Being and Time (Berkeley, CA: University of California Press, 1993) and, with reference to religion, Benjamin Crowe, *Heidegger's Phenomenology of Religion: Realism and Cultural Criticism* (Bloomington, IN: Indiana University Press, 2008).

[3] Martin Heidegger, *Phenomenological Interpretations of Aristotle*, trans. Richard Rojcewicz (Bloomington and Indianapolis, IN: Indiana University Press, 2009), p. 63; *Phänomenologische Interpretationen zu Aristotles: Einführung in die Phänomenologische Forschung, Gesamtausgabe*, vol. 61 (Frankfurt: Vittorio Klostermann, 1985), p. 82.

[4] Ibid., pp. 76–7.

[5] Ibid., p. 76: *Phänomenologische Interpretationen zu Aristotles*, p. 101: *In der Geneigtheit zu seiner 'Welt' hat wesentlich als Welt erharen, d.h. in der Faktizität ist selbst Leben immer in der Gestalt seiner Welt, Umwelt, Mitwelt, Selbstwelt; jedes eigene Lebens as meiner, deine, seinige, ihre Welt; unser Leben, unser Welt.* In the first sentence, the German text has 'Welt' although Rojcewic translates it as 'life,' which makes sense here.

Life is always experienced as expressed as a world and characterized by inclination or directionality towards the world that also 'carries along' distance. This second feature of caring in fact means the overcoming of distance. With care there is 'upsurge toward', the abolition of 'the before': in proclivity life does not maintain distance and lives in dispersion.[6] This is not easy to grasp at first, but what I take Heidegger to mean is that the fundamental characteristic of life itself is care, which is characterized by proclivity or directionality, impulse towards the world, which is its own realization and fulfilment. This is simultaneously accompanied by eradication of distance so that the forward movement of impulse or proclivity negates distance, automatically, as it were. The affirmation of care in proclivity is simultaneously the negation of distance as antithetical to care and antithetical to life itself. The third feature of the caring orientation is sequestration. This means that because of proclivity and the erosion of distance, the possibility of the 'before' (*vor*) becomes deferred and so in the actualization of care in this way, 'a possibility of life has become lost' (*eine lebensmässige Möglichkeit in Verlust greaten*).[7] But even though other possibilities are no longer possible because a particular situation has been actualized through the characters of proclivity and distance, the care (*Zorge*) or relationality that life is now 'becomes factically visible' (*er wird faktisch sichtbar*).[8]

Translating this structure of care which is the central relationality of life into a different kind of language, we might say that life itself comes into its particularity through the constraints of proclivity or directional impulse, that which drives movement, of the collapse of distance such that there is no 'before', and of sequestration (*Abriegelung*) in which other possibilities become lost or are excluded by the actualization of the possibility that, in fact, comes to be. Thus, the relationship between life itself and the living is here characterized by care, where care is the fundamental feature of life and the force—analysed into the above three features—that controls the event of life into its particularity. The more future possibilities are occluded, the more certain life becomes of itself in its present realization. Heidegger writes:

> The more life increases its worldly concern and the 'before' is lost in the increased proclivity and expulsion of distance, all the more certainly does life then have to do with itself.[9]

The event of life, the particularity of its occurrence, excludes other possibilities, and as that event increases in intensity of its self-realization, the more other possibilities, 'the before', are excluded. Through the structures of care

[6] Ibid. [7] Ibid., p. 79, p. 106. [8] Ibid.

[9] Ibid., p. 80, p. 107: *Je mehr das Leben seine welthafte Besorgnis steigert, d.h. in der gesteigerten Geneigtheit, Abstandsverdrängung das 'vor' umkommen lässt, umso sicherer hat es dabei mit sich selbst immer zu tun.*

and the exclusion of other possibilities we have now arrived at factical life, life lived in the world. This is where philosophy needs to begin, in the middle of experience.

We are always already in the midst of things, and so making sense of life is always making sense from within it. As Moore says, summarizing Heidegger's general position: 'We who make sense of things are planted firmly in the midst of the things of which we make sense'.[10] Heidegger's concern with understanding our lives, understanding Dasein, already assumes a pre-understanding; we already have 'a fore-having, fore-sight, and fore-conception'[11] that determines the structure of what is meaningful for us. It is not that we project meaning or bring meaning along with us, but rather that life is already charged with meaning that we encounter through the structures of pre-understanding. The flow of life is not indifferent but meaning (*Sinn*) is inherent within it.[12] Life in one sense cannot be other than meaningful because the flow of life is directional. There is an end goal, being towards death, of course, but in another sense a goal of life realized authentically. In his analysis of Greek terms, Heidegger tells us that the fundamental characteristic of the living thing, the existent, is 'the end', *to telos*, which means not so much an ending as a fulfilment (*Vollendung*).[13] In this sense we can understand Aristotle's term for being, *entelecheia*, holding or preserving itself in the limit (*Sich-in-die Grenze-herstellen*).[14] Thus, the feature of a living thing, an existent, for the Greeks is that it places itself in its limit, it completes itself or has form, which Heidegger remarks is a long way from the degeneration of Greek thought that the term 'entelechy' was to have in biology, as merely designating something alive. A living thing, something that has form, stands within itself and so within its limit, within its end. The meaning of a living entity is inseparable from its form and from its end; that is, to be what it is. This idea of particularity and unrepeatability is reminiscent of the particularity of being and its univocity that we have encountered in Duns Scotus, Deleuze, and Hopkins. The particularity of a life, standing at the limit of what it is, is furthermore the meaning of that life: everyday facticity reveals a being's meaningful (teleological in the above sense of standing within itself) structures. Facticity is the specific way Dasein is in the world, the way the self participates in life. It is the way something is as it stands in its limit and the way it is with other beings. Heidegger writes: 'The concept of facticity implies that an entity 'within-the-world' has Being-in-the-world in such a way that it

[10] A.W. Moore, *The Evolution of Modern Metaphysics: Making Sense of Things* (Cambridge: Cambridge University Press, 2012), p. 463.

[11] Martin Heidegger, *Being and Time*, trans. John Macquarrie and Edward Robinson (Oxford: Blackwell, 1962), p. 193.

[12] Campbell, *The Early Heidegger's Philosophy of Life*, pp. 23–4.

[13] Heidegger, *An Introduction to Metaphysics*, p. 60; *Einfühlung in die Metaphysik*, p. 65.

[14] Ibid.

can understand itself as bound up in its 'destiny' with the Being of those entities which it encounters within its own world'.[15] In a sense we have no power over the 'when and where' of our being; the life in which we find ourselves is our 'thrownness' (*Geworfenheit*) into the world.

Dasein is human life characterized by being-in-the-world, a kind of life that could not be otherwise in the sense of our fundamental comportment to world. This human comportment is distinct for us as rational animals, the living entity (*zōon*) that speaks (*logos*), because the human is a creature whose existence is an issue for it. It is not that 'being' is a property of being alive, but rather a living human being has a comportment towards or way of being in the world that defines who and what it is, from which it cannot be otherwise. It is not that Dasein is something alongside the world; this is far from the case because Dasein is in the world as a living entity from which it cannot be abstracted or set aside. A living being's 'environment' (*Umwelt*) is not something it has choice in possessing but is a necessary condition for being-in-the-world in the way it is in the world. Even biology must presuppose the idea of an environment, which, Heidegger says, it can never define because environment is a priori. Environment is part of the ontological structure of Dasein, integral to the kind of life that Dasein is. Heidegger writes:

> Only in terms of an orientation towards the ontological structure thus conceived can 'life' as a state of Being be defined *a priori* and this must be done in a privative manner. Ontically as well as ontologically, the priority belongs to Being-in-the-world as concern.[16]

Life itself is a state of Being but to formulate such a notion is to abstract it from the living world of Dasein. The 'privative manner' (*auf dem Wege Privation*) means that life as a category is depleted in its abstraction from the lived reality of being here, of the living person; it becomes a privation.[17] Rather than life itself, the priority of analysis must be Being-in-the-world, which is Dasein, and we can understand Dasein ontically—that is, in the lived reality of its everydayness—and ontologically, in its Being.

Dasein as a living entity, as a human being, encounters other beings in the world, both animate and inanimate, and these beings reveal themselves in two modes, either as what Heidegger calls 'ready-to-hand' (*zuhanden*) or

[15] Heidegger, *Being and Time*, p. 82; *Sein und Zeit*, *Gesamtausgabe*, vol. 2 (Frankfurt: Vittorio Klostermann, 1977), p. 75: *Der Begriff der Faktizität beschliesst in sich: das In-der-Welt-sein eines 'innerweltlichen' Seienden, so zwar, dass sich dieses Seiende verstehen kann als in seinem 'Geschick' verhaftet mit dem Sein des Seienden, das ihm innerhalb seiner eigenen Welt begegnet.*

[16] Ibid., pp. 84–5; ibid., p. 78: *Aus der Orientierung an der so begriffenen ontologischen Struktur kann erst aug dem Wege der Privation die Seinsverfassung von 'Leben' apriorisch umgrenzt werden. Ontisch sowohl wie ontologisch hat das In-der-Welt-sein als Besorgen den Vorang.*

[17] Ibid.; see Macquarrie and Robinson's note 1, p. 85.

'present-at-hand' (*vorhanden*). This distinction is important for his analysis of being humanly alive in that entities that are ready-to-hand find a place in use within the human world, within the world of experience, in contrast to their being simply present-at-hand as objects of potential theoretical consideration. A human life encounters the world in terms of readiness-to-hand in its use of those things it encounters. Heidegger's famous example of the hammer illustrates this. The meaning of the hammer reveals itself through being used, it is ready to hand, and through hammering, the hammer's manipulability is uncovered.[18] The hammer contemplated theoretically, as merely present-at-hand, does not reveal its fullness to us; it has to be used for its meaning to become apparent, and once broken it falls back into mere presence. We discover the nature of things through their use and although we can approach the world theoretically as present-at-hand, this is to miss something fundamental about living, for only through readiness-to-hand do we appreciate the fullness of experience. Nature, for example, seen theoretically, does not stir us: 'The botanist's plants are not the flowers of the hedgerow; the "source" which the geographer establishes for a river is not "the springhead in the dale"'.[19] This is the same issue of the structural connection between life and world that Romano has highlighted and the task of phenomenology being the elucidation of the lifeworld.[20] The abstract life of beings is present-at-hand and only as ready-to-hand enters a fullness for human being. Heidegger thus places the experience of human living in the world as the centre of philosophical reflection, the analytic of Dasein. Through this analysis we see how Dasein is characterized by concern and care towards the world and its own being, reflected in its fundamental states of anxiety and as a being towards death. The meaning of a life is characterized by its comportment towards its end—towards its death, certainly, but also towards its way of being in the world.

For Heidegger, then, the idea of life itself is a form of Being but we have to privilege the living, and in particular the human life as Dasein, in any analysis. Indeed, this is foundational to all the sciences, and even biology, 'the science of life', is founded upon it. On the relation between Dasein and the category 'life', Heidegger writes:

> Life in its own right is a kind of Being; but essentially it is accessible only in Dasein. The ontology of life is accomplished by way of a privative interpretation; it determines what must be the case if there can be anything like mere-aliveness [*Nur-noch-leben*]. Life is not a mere Being present-at-hand, nor is it Dasein.

[18] Ibid., p. 98.

[19] Ibid., p. 100; p. 94: *Die Plflanzen des Botanikers sind nicht Blumen am Rain, das geographisch fixierte 'Entspringen' eines Flusses ist nich die 'Quelle im Grund'.*

[20] Claude Romano, *The Heart of Reason*, trans. Michael B. Smith and Claude Romano (Evanston, IL: Northwestern University Press, 2015), p. 526.

> In turn, Dasein is never to be defined ontologically by regarding it as life (in an ontologically indefinite manner) plus something else.[21]

Dasein is not life plus some extra entity that, for example, inhabits the body but is bound up with the fabric of world. All other realms of human experience must be understood with this, with Dasein, as the primary reality and the focus of philosophical analysis. This philosophical analysis is phenomenology, the fundamental 'science of life' that is presupposed in all the sciences and disciplines. We are in the world from which we cannot step out of in some notion of pure objectivity; we find ourselves here already with pre-understanding. Thus, phenomenology is not a worldview in Jaspers' sense.

The distinction between phenomenology and worldview is important because it privileges ontology as the inquiry into life itself over mere historical contingency. It is not that phenomenology is not concerned with the historical, far from it, because we can only understand Dasein as a historical entity constrained by time, but the analytic of Dasein is not simply worldview, which is pre-critical. In fact, philosophy develops out of worldview. Thus, we have, says Heidegger, workers and peasants who have their worldviews, political parties that have worldviews, and philosophy that aims at achieving a universal understanding of human life. The separation of philosophy from worldview is no mean task because it demands a confrontation with the truth of historical reality that it itself is. He begins his critique with an analysis of the word 'Weltanschauung' (worldview), a purely German word that first occurs in Kant's *Critique of Judgement*, says Heidegger, where it denotes contemplation of the world given to the senses or the apprehension of nature.[22] The term is used by Schelling who links it not to sense experience but to intelligence and the possibility of developing a schema of different worldviews; Hegel speaks of a moral worldview; and Jaspers' book *Psychologie der Weltanschaung* (*The Psychology of Worldview*) speaks of worldview as ideas, the shape of the world, and what is ultimate to human beings.[23] A worldview is thus a view of life and reflection on Dasein at a particular time. Philosophy, by contrast, is fundamentally ontology; not concerned with views of the world at a certain time and place but with the basic issue of being, of human life and vitality. As such philosophy is ontological in contrast to worldview that is ontical, concerned with particular beings or groups of beings to the neglect of the fundamental

[21] Heidegger, *Being and Time*, p. 75; *Sein und Zeit*, p. 67: *Leben ist eine eigene Seinart, aber wesenhaft nur zugänlisch im Dasein. Die Ontologie des Lebens vollzieht such auf dem Wege eigner privativen Interpretation; si bestimmt das, was sein muss, dass so etwas wie Nur-noch-leben sein kann. Lenen is weder pures Vorhandensein, noch aber auch Dasein. Das Dasein wiederum ist ontologisch nie so zu bestimmen, dass man es ansetzt als Leben—(ontologisch unbestimmt) und alos überdies noch etwas anderes.*

[22] Martin Heidegger, *The Basic Problems of Phenomenology*, trans. Albert Hofstadter (Bloomington and Indianapolis, IN: Indiana University Press, 1988), pp. 4–5.

[23] Ibid., pp. 5–6.

question of Being itself.[24] This is prior to understanding the experience of beings themselves; understanding beings presupposes a pre-understanding of Being that Heidegger elsewhere refers to as fore-having, fore-sight, and fore-conception.

Philosophy as the science of life is not a description of worldviews but an analysis of human reality, of Dasein. This deeper inquiry goes beyond worldview in raising questions about the truth of human life and its comportment towards the world, in its analysis of factical life. Thus, when we approach religious life we are already given over to a certain orientation and understanding. Heidegger emphasizes that we are in the midst of things and so Dasein is already an inevitable comportment towards the world, and so our pre-understanding gives us a key or link to religious life as human life. Factical life experience is the point of departure for understanding any realm of human life, including the religious. There is a 'formal indication' (*die formale Anzeige*) that allows us to penetrate religion because religion is always fundamentally historical. To understand religion is to understand the historical.[25] Philosophy can approach religion as the historical but from the perspective of the analytic of Dasein, and so Heidegger understands Christianity as factical life experience that is historical.[26] I will need to return to this methodological orientation of phenomenology to religion, but Heidegger's categories based on factical life arguably provide a way of explicating religious life. The structures of care—proclivity, distance, and sequestration—allow us to see religions as orientated towards the world such that they contain an impulse towards the world, a collapse of distance in the immediacy of the present, and the losing of other possibilities of life; the ascetic orientation to life excludes hedonism, the desire for transcendence entails the discipline of practice, and so on.

In sum, Heidegger does offer a philosophy of life, not as a vitalism but as an analysis of everyday human experience. The concern with life itself in relation to the living is paralleled by his concern with Being in relation to beings, of the ontological in relation to the ontical. This emphasis on human life as formally indicative of others' lives and history means that phenomenology is a kind of philosophy of life that explicates and describes factical everydayness and uncovers the forgetfulness of Being entailed in living. The focus on human experience revealed in phenomenological analysis—or we should probably say phenomenological-hermeneutical analysis—is not to privilege the harder sciences and to disclaim pure objectivity. This has been contested, of course, but before we discuss that contestation, I wish to illustrate more complexity in the phenomenological philosophy of life with a reflection on Heidegger's student Hans Jonas.

[24] Ibid., pp. 11–12.

[25] Martin Heidegger, *The Phenomenology of the Religious Life*, trans. Matthias Fritsch and Jennifer Anna Gosetti-Ferencei (Bloomington and Indianapolis, IN: Indiana University Press, 1995).

[26] Ibid., p. 57.

HANS JONAS AND THE PHENOMENON OF LIFE

Heidegger had emphasized life itself as a mode of Being, an ontological category in the service of illuminating, bringing into the clearing, a deeper understanding of life in relation to human living, to Dasein. But his work was not without controversy and his students (such as Hannah Arendt and Hans Georg Gadamer) developed phenomenology in different directions. Hans Jonas was one of these who raised critical questions about his teacher's philosophy, and while he clearly owes a great debt to Heidegger's thinking, he took phenomenology in an ethical and indeed theistic direction. In particular he was critical of Heidegger's brief involvement with National Socialism; for how, as the translator put it, could a philosophy be so blind as to consider this to be 'Germany's authentic destiny?'[27] It is not simply an idiosyncrasy of Heidegger, but something embedded within his philosophy that allows it to move in such a direction, thinks Jonas. That is, the abstract, impersonal Being overwhelms or denudes the personal being, the individual, such that the personal voice is lost. The call of one person to another or of human being to things is 'drowned in the voice of being to which one cannot say, No'.[28] The overwhelming imperative of Being for Jonas must be wrong on existential, ontological, and theological grounds.

Jonas begins with a historical claim about early humans imbuing nature with life in a panvitalism in which death rather than life was the anomaly, the 'disturbing mystery' that counteracted the dominant primacy of life. Death is the mystery that needs to be assimilated to life, and so early thought considered death to be but a transformation of life, a transition to another condition, as we see borne witness to in early tombs.[29] It is only with modernity since the Renaissance that the situation becomes reversed and death is thought to be the natural thing with life as the anomaly, and panmechanism replaces panvitalism as the comprehensive hypothesis: 'Vitalistic monism is replaced by mechanical monism, in whose rules of evidence the standard of life is exchanged for that of death'.[30] As panvitalism begins to question death, one solution develops to resolve the contradiction between life and death, which is dualism, that the body is a tomb of the soul as the animating principle (*soma-sema* in the Orphic formulation). The soul is the living visitor in the body and only at death does the body 'return to its original truth, the soul to hers'.[31] With Gnosticism, of which Jonas made an important study, dualism comes to be applied to the entire universe such that the totality of the cosmos is seen as a tomb, a prison house of the soul. With modernity dualism fades and we are left with the idea of mechanistic matter in which life is an improbable accident: 'All modern theories

[27] Lawrence Vogel, 'Foreword', p. xii, Hans Jonas, *The Phenomenon of Life: Toward a Philosophical Biology*, trans. L. Vogel (Evanston, IL: Northwestern University Press, 2001).
[28] Ibid., p. 258. [29] Ibid., pp. 7–9. [30] Ibid., p. 11. [31] Ibid., p. 13.

of life are to be understood against this backdrop of an ontology of death, from which each single life must coax or bully its lease, only to be swallowed up by it in the end'.[32] A further response to dualism was the appreciation that as matter can be without spirit, spirit could exist without matter, and so the two possibilities develop of materialism, on the one hand, and idealism, on the other.

Jonas' history of life thus moves from a panvitalism or monism in which the cosmos is filled with life, to a dualism in which spirit as animating force is considered distinct from essentially dead matter, to a monism that privileges spirit in idealism or a monism that privileges matter in materialism, to a mechanistic materialism in which life is the alien presence and accident that needs to be explained. Human beings move from an animistic monism, to dualism, to the materialistic monism. While we can take the importance of the philosophical point that Jonas is making, that modernity offers a materialism in which life becomes a problem to be explained, the comparative history of civilizations shows that the picture is more complex than Jonas' scheme. In India, for example, as we have seen, the dualism of the Saṃkhya system existed alongside the monism of Advaita, and materialism was already present in an early age. Similarly, Buddhism does not fit at all well with the schema, although we can take Jonas' broader point shared by dualisms that the material universe is considered to be a prison in which the soul or the process of the living entity is trapped.

Natural selection replaces a teleological explanation as the directing principle of life that Jonas accounts for. But now materialism, of which evolution theory is a form, needs to account for the emergence of the mind (a problem that dualism did not have because the mind was regarded as from a different realm). The mechanistic view cannot account for life or mind and neither can the dualistic view that perceives matter as essentially passive, inanimate substance because both see life and mind as essentially alien to matter. What Jonas proposes is a philosophy of biology in which mind and life emerge along with matter from Being. Being has the potentiality of life within it and organic life is a natural emergence. Furthermore, the mind naturally arises from organic life and is prefigured in the lifeworld and still further, ethics itself is inherent within life. It is not that ethics is only restricted to the human but is part of the structure of life, part of the ontology of being: 'a philosophy of mind comprises ethics—and through the continuity of mind with organism and of organism with nature, ethics becomes part of the philosophy of nature'.[33] Ethics may be grounded in the very structure of Being, and although Jonas admits this to be speculative, the main point is that we need to acknowledge ethics beyond the human within the universe itself.

Jonas' vision of the phenomenon of life reacts against the religion and philosophy that he had deeply studied. Gnosticism is the religion of the exiled soul, trapped in the world not of its own making, through a cosmological

[32] Ibid., p. 15. [33] Ibid., p. 282.

catastrophe (a doctrine that we have seen both in India and Europe, although not indigenous to China). The material world on this view is evaluated purely negatively as ultimately a kind of nihilism. This legacy remained in modernity and echoes, according to Jonas, in existentialism, especially as articulated by Heidegger, in which we are thrown into the world not of our making and in a state of anxiety with regard to it. The world itself has no meaning. Thus nihilism—Nietzsche's strange guest at the door—is shared by both Gnosticism and existentialism.[34] But the theistic vision Jonas presents sees the world not as empty of meaning but as imbued with the potential for complete meaning by God who created the world but left it to its own unfolding, with human reality as the highpoint of its evolution. God needs humanity to help fulfil the potential of creation and humans have the freedom to fulfil that potential or go against it and destroy the human experiment. Jonas' God is an absent God but a God to whom we owe allegiance and the fulfilment of creation.[35]

Although Heidegger's analysis of Dasein is in a different league to the philosophy of his student Jonas, he does raise the critically important point about how Heidegger's philosophy of life could support the National Socialism of his day, and he raises the issue that life needs to be seen in a broad spectrum beyond the human. In this he is a precursor of more recent philosophy of life that lays emphasis on the wider planet as ecosystem beyond the human. In broad brushstrokes, Jonas paints a picture of a history that moves through demarcated stages to a modern materialism that itself needs to be overcome through a philosophy of life that sees humans as the guardians of nature in the service of fulfilling the potential of creation by a *deus absconditus*. In addressing life itself directly and in relation to the biological science of his day, Jonas can be seen as presenting a philosophy that moves towards a demotion of the human and imbuing life with an ethical structure that comes to articulation through evolution and the resulting human community. It is, in the end, a story about the repair of God and the long-term seeking for a fullness of human life lived in harmony with nature; a very Jewish theological vision.

§

Phenomenology has privileged human experience in its analysis and explanation of the world; the world becomes the lifeworld, the realm of human dwelling. This has been the case from Husserl to Heidegger to contemporary phenomenology.[36] Although Jonas lays emphasis on what is beyond the human and the ethical structure of nature itself, in the end he still falls back

[34] Ibid., 'Gnosticism, Existentialism, and Nihilism', pp. 211–34.

[35] Ibid., 'Immortality and the Modern Temper', pp. 262–81.

[36] Even Claude Romano is focused primarily on human perception and understanding prelinguistic experience. See Claude Romano, *At the Heart of Reason*, trans. Michael B. Smith (Evanston, IL: Northwestern University Press, 2015).

on the primacy of human reality as caretaker of nature. Life, for Jonas, is in a cosmos sanctioned by an ontotheological reality, an ontotheology that has itself come under critique even before Jonas with Heidegger. Our story has now to be brought up to recent times and has reached a point where life itself, independent of human life, has become an important theme in philosophy, especially in the face of environmental issues, debates about bio-technology, and social responsibility to the planet. In some ways the system has taken precedence over the person, with the turn to a new realism and materialism that critically responds to the phenomenology of life. The new realism moves away from a human-centred focus to an attempt to broaden the philosophical horizon and reach a pure objectivity of life itself, an objectivity that exceeds a purely human reality; a philosophy that goes by the name of speculative materialism. There are emergent philosophies of life with various points of origin and philosophical perspectives, but which share a reaction against both phenomenology and postmodern philosophies that they regard as unjustifiably privileging the human over the world of other beings and objects. This new realism, which itself covers a wide range of philosophical perspectives, in the Anglophone world draws on a range of philosophical resources from the process of philosophy of Whitehead to the realism of Hilary Putnam, while in the Francophone world it draws on the philosophy of Badiou and other materialisms. The new realism, then, is both a critique of androcentrism and a positive political philosophy.

The turn to the real, to the extra-human planet, is a general orientation that covers a spectrum of philosophical positions. At one end we have a response to phenomenology in a new realism that calls itself speculative realism or speculative materialism, while on the other we have a new vitalism that I will call vital materialism, characterized by an emphasis on life beyond the human. Both of these types of philosophy are atheistic in orientation in the sense of not only rejecting any onto-theology, but also rejecting any mystical panpsychism. Alongside these developments there is a religious vitalism associated, on the one hand, with Protestant fundamentalism and, on the other, with a theological trajectory that comes out of secular process philosophy.

THE NEW REALISM AND SPECULATIVE MATERIALISM

According to Maurizio Ferraris, the phrase 'new realism' was coined in a discussion at a specific time and place, on 23 June 2011 at 13.30 in a restaurant in Naples.[37] This movement is characterized by weariness with the general

[37] Maurizio Ferraris, *Introduction to New Realism*, trans. Sarah De Sanctis (London: Bloomsbury, 2015), p. 2.

postmodern view that reality is constructed through language, conceptual schemas, and the media. Gadamer's famous phrase, 'the being that can be understood is language', is surely mistaken, thinks Ferraris, for so much of our human experience is extra-linguistic. Certainly, one can see the sense of this phrase regarding social reality—institutions, conferences, social class—but it does not apply to realities beyond human social formations, such as mountains and seas, to nature. Philosophy has taken a wrong turn on this view by confusing ontology, an account of what there is, with epistemology, an account of what we know about what there is.[38] Postmodern relativism and the focus on the construction of reality is but 'a terminal stage', thinks Ferraris, of 'a crisis in German Idealism. All reality belongs to science and philosophy is left with nothing much'.[39] This critique has a long pedigree, from Aristotle's criticism of Plato to Schelling's philosophy of nature, and is characterized by the claims that reality does not depend upon human perception and ideas, and that reality can have a causal impact on the human world. In terms of the category of life, we therefore have a philosophy that life itself stands within the living as objective reality, testifying to a trans-human space that exceeds the human.

In terms of the philosophy of life, one of the most interesting articulations of the new realism is in speculative materialism, a term coined by the doyen of the movement, Quentin Meillassoux. Meillassoux's *After Finitude* (*Après la finitude*) is a critique of what he calls correlationism. Correlationism is the phenomenological doctrine that there is a correlation between thought and reality, between thinking and being, and that the only access to reality we have is through thinking. Meillassoux's book is a sustained attack on this idea in the service of highlighting the reality independent of human experience. Meillassoux wants to reach 'the great outdoors' and to go beyond correlationalist error that has held sway in the history of Western philosophy and particularly in the phenomenology of Husserl and Heidegger.

Meillassoux's argument takes the following steps. First, he wishes to bring back the old distinction between primary and secondary qualities that has largely been abandoned in philosophy. Primary qualities are those inseparable from the object: the properties of a thing without my perception, independent of me, such as geometrical proof, length, depth, and so on, aspects of an object that can be formulated in mathematic terms and so 'meaningfully conceived as properties of the object itself'.[40] Secondary qualities are those within perception and so, in a sense, both in the object itself and in my perception of it. The redness of the book is in my perception but also a quality of the book itself. So, first, Meillassoux wishes to question the rejection of the distinction between primary and secondary qualities on the grounds that certain

[38] Ibid., p. 6. [39] Ibid., p. 20.

[40] Quentin Meillassoux, *After Finitude: An Essay on the Necessity of Contingency*, trans. Roy Brassier (London and New York: Bloomsbury, 2009), p. 3.

properties of objects, the primary qualities or those that can be defined mathematically, stand before any human perception of the object. Second, Meillassoux argues for 'ancestrality', which refers to any reality that existed prior to the emergence of the human species, or indeed any life form. Related to this, the 'arche-fossil' is a trace that not only indicates prior life on earth but any material event or reality that provides evidence for ancestral phenomena, such as 'an isotope whose rate of radioactive decay we know, or the luminous emission of a star that informs us to the date of its formation'.[41] How can correlationism interpret statements about the arche-fossil? Of course it cannot, he says, because the correlationist has no account of ancestrality. While the scientist might say, 'event Y occurred x number of years ago', the correlationist will always add 'for humans' or 'for the human scientist'. This, says Meillassoux, is 'the codicil of modernity'.[42] Rather than being a property of an object or event that preceded human life, scientific statements about ancestrality are taken to mean that such statements are intersubjectively agreed by the community of scientists, and this guarantees its truth, not the objectivity of the event itself. Kant claimed that knowledge is centred on human knowing and that we cannot know the thing in itself, thereby putting limits on knowledge, making it finite. Meillassoux's task is to refute this limitation, not to go back to a naïve pre-Kantian metaphysics, but to move to a position after finitude in which statements about ancestrality and the arche-fossil are meaningful independently of human perception.

There is a spectrum of views from dogmatic or naïve realism (that things in themselves exist and we can know them), to weak correlationism (that things in themselves may not be knowable but they are thinkable), to strong and very strong correlationism (that things in themselves are unknowable and unthinkable, but they may exist), and finally to absolute idealism (things in themselves are unknowable, unthinkable, and meaningless so therefore impossible).[43] Meillassoux illustrates this with an aptly religious theme. Imagine, he says, two dogmatists, an atheist and a Christian, arguing about life after death. The Christian claims that there is post-mortem life with God that can be shown by finite reason to be incomprehensible to finite reason. The atheist, of course, claims that death annihilates us completely. But then a correlationist comes along defending an agnosticism. As I cannot know the in-itself without converting it into a 'for-me', so I cannot know which of the two theories about post-mortem existence is the correct one. It is contradictory to know what happens when one is no longer in the world because knowledge entails being in the world. But a newcomer arrives on the scene, the subjective idealist who decries the claims of the agnostic on the grounds of being just as

[41] Ibid., p. 10. [42] Ibid., p. 13.

[43] These positions are clearly summarized by Graham Harman, *Quentin Meillassoux: Philosophy in the Making* (Edinburgh: Edinburgh University Press, 2011), p. 23.

inconsistent as the two realists, because all think that there could be an in-itself, either God or nothing, radically different to the world. But these are outside of knowledge, as is my post-mortem existence. I cannot think of myself as non-existing because this entails a contradiction. I can only think of myself as existing and always existing, therefore I, or my mind, am immortal. Radical transcendence, of God or nothing, is annulled by the idealist because an in-itself other than the 'for-us' cannot be thought. The speculative realist then comes on the scene claiming that all these positions are in error because what is called absolute is simply the capacity to be other. This is the capacity beyond reason to be other and marks a transition state, and furthermore this is a kind of knowledge 'of the very real possibility of all these eventualities'.[44] The speculative philosopher thinks that we can know that the possibility of any of these is true, and this without reason. These possibilities actually lie outside of the correlate and are indeed possibilities. It is here that the speculative materialist gains leverage over correlationism and the opens the door to the wide horizon of existence beyond the correlation.

Thus, Meillassoux positions himself with the strong or very strong correlationists but transformed into speculative materialism. The very strong correlationist position moves up to the limit of thought, as held by Heidegger and also Wittgenstein, says Meillassoux, in maintaining that the strong correlation between thinking and being does not completely reject the notion of something existing outside of the correlation but that it would be meaningless. There are two stages here. First, Meillassoux argues that the correlationist position entails that we do not have access to the in-itself (as the Kantian position) and so rather than leave it at that, that the in-itself is unknowable, the correlationist argues that 'the correlation is the only veritable in-itself'.[45] Here the ignorance of the thing in-itself is transformed into the certainty of the correlation. Thus 'it is not the correlation but the facticity of the correlation that constitutes the absolute'.[46] Second, Meillassoux argues that this position that entails turning facticity into an absolute is absurd because the correlation is supposed to express thought's inability to reveal the reason why the thing in itself exists as it is. This limit to thought that comes up against the unreason of the thing in itself must be turned into a property of the thing in itself: the absence of reason becomes the ultimate property of the entity. Meillassoux writes:

> We must convert facticity into the real property whereby everything and every world is without reason, and is thereby *capable of actually becoming otherwise without reason*. We must grasp how the ultimate absence of reason, which we will refer to as 'unreason', is an absolute ontological property, and not the mark of our finitude of our knowledge . . . for the truth is that there is no reason for anything to

[44] Ibid., p. 56. [45] Meillassoux, *After Finitude*, p. 52. [46] Ibid.

be or to remain thus and so rather than otherwise, and this applies as much to the laws that govern the world as to the things of the world. Everything could actually collapse: from trees to stars, from stars to laws, from physical laws to logical laws; and this not by virtue of some superior law whereby everything is destined to perish, but by virtue of the absence of any superior law capable of preserving anything, no matter what, from perishing.[47]

We see here partly a return to Hume in the stress on contingency and the idea that there is no rational justification for causation but simply constant conjunction of events. If I have understood Meillassoux correctly, thought comes up against its limit, that which cannot be thought, and this that cannot be thought is therefore outside of reason and can become anything. Facticity is the knowledge of the absence of reason, and this knowledge is itself sufficient to escape from correlationism; this knowledge of facticity is the path to 'the great outdoors' beyond the thought–event correlation. But facticity maintains that the thought–world correlation is a simple fact so cannot really be factical because it admits the possibility of the world being other than it is, of the world exceeding the correlation. This is the 'non-facticity of facticity' for which Meillassoux coins a new word, *factualité*, translated as 'factiality'.[48] Factiality is the existence of world deduced from the correlation, from the factical, and is the way in which we gain access to reality beyond thought. It opens up the possibility of mathematical and scientific knowledge and opens up knowledge of ancestrality. Through factiality the arche-fossil becomes not simply 'for me' or 'for the scientist' but a truth of the world beyond thought, beyond the human in the great outdoors.

But we are faced with a difficulty. On the one hand, we have the phenomenological claim (in both Husserl and Heidegger) that phenomenology is primary science of the human because it begins with Dasein, with the human lifeworld as the necessary perspective through which to view the regional sciences (physics, biology, chemistry, and so on), while on the other, we have the argument against correlation, against the 'for-me' view, in which pure objectivity takes precedence. In the former, life itself as articulated in human life, in Dasein, is the way human beings must begin to make sense of the world, while in the latter we can only make sense of the world in terms of the transhuman objectivity of the in-itself. On this view we have access to the world, the world of objects, through geometrical proof, length, breadth, depth, size, and so on, all of which can be formulated mathematically. This reorientation to the objectivity of life is important in the narration because it indicates a future direction towards the objective sciences, towards a scepticism not about the world but about subjectivity. It comes to be a philosophical articulation of a broader cultural orientation towards the harder sciences and

[47] Ibid., p. 53. [48] Ibid., p. 79.

implicitly the power of technology to affect change. This apparent impasse between competing accounts of human life, between first- and third-person accounts, might only be apparent insofar as they might be seen as two ways of framing human life; life seen in terms of evolution and the bio-sociology of our hominin inheritance, and life seen in terms of the lifeworld and the I/you of intersubjectivity. But one further response needs to be taken into consideration: one that I have already introduced which, following Bennett, I will call vital materialism.

VITAL MATERIALISM

Unlike speculative realism, vital materialism, whose roots are in the philosophy of Deleuze and Guattari, and so before them in Bergson and Spinoza, is concerned with establishing a political philosophy of materiality in which human agency is placed within a spectrum of agential forces. In particular, Bennett's work offers a broad view of life that incorporates what we would usually think of as the inanimate. While Meillassoux is interested in the arche-fossil, Bennett is keen to articulate a voice from the present material reality of things. Although she does not use the term 'correlation', her position is inherently anti-correlational, like Meillassoux's. Agency is distributed across a network and this network includes the inanimate. On this view vitality includes objects—metals, edibles, storms—that act like agents in their capacity to affect human life and in having their own trajectories through time/space. There is material agency to non-human things. Bennett tries to illustrate this through a series of examples and descriptions of earlier vitalist positions, to conclude with 'a litany... for would-be vital materialists':

> I believe in one matter-energy, the maker of things seen and unseen. I believe that this pluriverse is traversed by heterogeneities that are continually doing things. I believe it is wrong to deny vitality to nonhuman bodies, forces, and forms, and that a careful course of anthropomorphization can help reveal that vitality, even though it resists full translation and exceeds my comprehensive grasp. I believe that encounters with lively matter can chasten my fantasies of human mastery, highlight the common materiality of all that is, expose a wider distribution of agency; and reshape the self and its interests.[49]

This is all very well, but the problem is that Bennett's vital materialism mistakes a web of causation or rather the web of constraints with agency. Clearly an electricity blackout, the eruption of a volcano, or even the properties of metal have effects on the human world, but to refer to the source of these

[49] Ibid., p. 122.

effects as agency is to stretch the semantic range of the term so wide that it loses the force of the fundamental distinction between the living and the non-living. This is, indeed, Bennett's intention, but the claim to distributing agency across a field takes away the particularity and distinction of human agency or freedom being linked to decision and intentionality. That is, agency implies purposive behaviour linked to human meaning, a kind of intentionality that is located within the brain, within the neo-cortex where most language ability is located.

The issue, then, is that on the one hand, we have the phenomenology of the human lifeworld in which we make decisions and impact upon the world, while on the other, we have a world that exists independently of human reality—as shown through ancestrality and the arche-fossil—and has impact upon the human. What is at issue is important for the explication of how we are as we are in the world.

But we do not necessarily have to choose between the 'in-itself' and the 'for-me'; between human being and the great outdoors. The problem of Kantian dualism between the appearance and the thing in itself still applies, it seems to me, to the speculative materialist position, and its overcoming is not through a reorientation to objectivity but through a re-narration of the subject/object dichotomy in terms of non-dualism partly derived from science itself that proclaims the discovery of the in-itself with human interiority. Put briefly, the human speaks from within the in-itself. The speculative realist still accepts the distinction but if there is a non-dualism at a deep physical level, the quantum brain, then the human subject cannot posit itself outside of the world. In this sense phenomenology is correct in claiming that as humans we are already in the midst of things and so leverage on life itself is always from within life; it can be from nowhere else. The sophistication of the speculative materialist argumentation in the end comes down to accepting the Kantian presupposition, but new forms of thinking about the world in terms of science challenge this in ways that have implications for explaining religion. The Meillassoux position is a 'difficult atheism',[50] but the new science points to ways of integrating human life and world that are inclusive of religious approaches to reality. In a sense both correlationism and anti-correlationism are mistaken ways of viewing the problem, because the human is always within the world and the world penetrates and permeates human reality through a network that locates our brain/body within matter.

This debate about the location of agency and the emphasis on the world versus an emphasis on human life has important cultural and political implications, certainly on issues such as climate change, but also more fundamental

[50] Christopher Watkin, *Difficult Atheism: Post-Theological Thinking in Alain Badiou, Jean-Luc Nancy and Quentin Meillassoux* (Edinburgh: Edinburgh University Press, 2011), p. 135: here is an illustration of some of the difficulty in paradoxical statements such as the following: 'philosophy believes in God because God does not exist'.

philosophical concerns about the place of agency and how the human species relates to the world. The importance of life itself and its political ramifications comes into view over the issue of what life itself is, where it begins, and how we delineate it from non-life. In particular this issue comes up in relation to technologies of life and molecular biology that presents strong challenges to some traditional religions.

Helga Nowotny argues that in the modern age, understanding life means changing life. The technology that enables us to observe the building blocks of life, the cells, can be used to change those cells; knowledge can be expressed as power and as a result, the molecular level of the living becomes an area of contestation when the question is raised as to whether science should intervene in the molecular, even though it can. Nowotny presents the hypothesis that the more we know about our biology, the less can we relate this knowledge to a coherent whole. This means that new knowledge gained at the molecular level, that is never complete, is essentialized in the sense that it is taken out of its embodied social contexts in contrast to earlier days when reproductive knowledge, although technically and technologically inferior, was set within a complex of social and cosmological relationships: 'There was a continuum that provided scope for gods and wet nurses, for immaculate conception and multiple fatherhood, whereas today the genetic view of things prevails exclusively'.[51] Following from this, a second hypothesis Nowotny presents is that assisted reproductive technology is perceived to be a threat because it strips reproduction of social context: 'A "social bond" in the truest sense of the term, which was clothed for centuries in social conventions and in love and power relationships, now presents itself in its stark genetic nakedness'.[52] Modern technology essentializes the molecular and takes it out of the social, whereas in earlier times reproductive practices were social and relational. That is, the development of molecular science has privileged that level of biological constitution over any social formation of the person, thereby stripping the human of social context.

§

We have, then, roughly three modern philosophical trajectories that I have examined in this and Chapters 7 and 8, that express a view about life itself in relation to the living, which we might call the immanentist, the phenomenological, and the new realist philosophies. First, the philosophers of immanence, particularly Deleuze, inherit a vitalist tradition that emerged in the Romantic philosophers and comes to expression in Bergson, a tradition that also informs the vital materialism of Bennett. This tradition recognizes the necessity of

[51] Helga Nowotny and Giuseppe Testa, *Naked Genes: Reinventing the Human in the Molecular Age* (Cambridge, MA and London: MIT Press, 2010), p. 12.

[52] Ibid., p. 13.

non-dualism, that the philosophical subject is always part of the world, and some of these philosophers such as Schelling try to link their philosophy with the contemporary science of the time. The phenomenological tradition began with the distinction between subject and object and the suspension of the question of the being behind appearances (with Husserl), but has moved beyond this in recognizing the human reality of already being in the world (with Heidegger). Here, life is identified with a mode of Being to which human life gives expression—the most important kind of life for we to whom it belongs. Second, a new materialism or realism has responded to the anthropocentrism of phenomenology in its attempt to claim objectivity and transcend the collapse of objectivity, the thing in itself, into subjectivity with idealism. On this account, life is objective existence beyond the human that can be measured and analysed through mathematics. In actual fact this view ties in with a wider conception of science that eradicates the human and sees life in terms of objective process. As science develops, especially in molecular biology, the possibilities of manipulation and technological intervention in natural processes increase manifold. There is in this orientation a tendency to abstract life processes from the living to the neglect of ethical issues raised through such molecular intervention, as Nowotny has drawn our attention to.

It seems to me that undoubtedly the central tension is between the first-person accounts of life in phenomenology and the third-person accounts of life in the new materialism. Yet both perspectives seem to contain truth and both are important for understanding not only the histories of civilizations but also contemporary, global civilization, especially as the rapid development of science and its inseparable connection to technology pushes the human future to new ways of inhabiting the world. Science and technology advance, yet people remain the same anatomically as homo erectus, with the same desire for life and the same need for self-repair. Science and technology have allowed the construction of a sophisticated, urban-based human niche that has become more important in our development than biological evolution. Within this complex of rapid development the issues that religions have traditionally been concerned with—how to live a good life, what is the purpose of life, how do communities interact, and what is the best way to be orientated towards the animate and inanimate other—come into sharp focus because traditional ways of living are inevitably changing, and this raises new challenges and necessitates new conceptualizations of the world and the human person. Among these emergent conceptualizations, religions potentially play an important role because, as we have seen, they not only articulate conceptions of life and the living but have developed practices that seek to repair the human by orientating people towards the world such that they live more fully in the world. Although religions can bring destruction and can be anti-life and anti-civilization, they can also promote human fullness of life, and human flourishing in which the natural proclivity of life towards meaning through its temporal structure becomes open to new narratives of holiness. Life understood through the fullness of tradition, on the one hand, and scientific explication, on the other, is potentially of central importance to the emergence of a trans-global, cosmopolitan civilization. Such a civilization has not closed its doors to a discussion about and with religion and, lastly, I need to delineate the boundaries of an emergent global discourse that sees the power of life as a liberating force, as transformative healing. We might read this in terms of political theology and a reconfiguration of religious discourse, especially in Asia, that critically engages with the post-colonial legacy and asserts a philosophy of life that seeks to both repair and transcend secular philosophies.

10

Bare Life and the Resurrection of the Body

I here want to take up a theme from Chapter 6 and following from Chapter 9, namely the question of political theology in relation to the philosophy of life.[1] As we have seen, Taubes characterized Paul's theology as transforming the bran from Moses into the food of angels and humans, thereby instituting a 'catholic', that is universal, view of subjectivity. The physical *pneuma* as the animating force of life becomes transformed into a theology relevant to both the bodies of the faithful and the wider polis. Paul's vision is both eschatological and political, a vision that shapes Christianity into a project inseparably political in its orientation, in spite of an incipient Gnostic narrative within it and in spite of the separation of church and state from an early period. The philosophies of life that we have examined, and the life force as a theme within the history of religions, have tended to present life itself as transcendent to political concerns. But throughout the history of Christianity the life force as spirit (*pneuma*) or Geist has impacted upon formations of the political, especially in Hegel, as we have seen, although it has been played down in the subsequent history of Hegelianism in which Marx's materialist reading feeds into the Frankfurt School with little regard for the idea of life itself.

In this penultimate chapter I wish to bring the narrative into modern times and unpack the somewhat complex relationship between the philosophy of life and Christian theology, with an emphasis on political theology in particular. On the one hand, in a secular context, the philosophy of life becomes, as we saw in Chapter 9, a vital materialism that imbues life with positive value and interfaces with environmentalism. But there is another kind of vitalism in which the political colonizes life in a way that brings into question the value of life itself and brings life into proximity with nihilism. We might call this a dark vitalism, which we see emerging in the European body politic in the twentieth century with National Socialism, so astutely analysed by Agamben, but whose

[1] This chapter is derived in part from an article published in *Religion*, 'The Political Sacred and the Holiness of Life Itself', *Religion*, vol. 47 (4), 2018, pp. 688–703.

roots are much older. While this stream of thought, like all vitalisms, can be read as an attempt to heal the past through creating a utopian and messianic future, and so is a form of human self-repair, it nevertheless negates the values of life and undermines its healing project because fundamentally locked into a form of nihilism, thereby negating life-affirming values. In contrast to this, spiritual philosophies of life (and I would include the environmentalist imperative here) have a view of material life imbued with meaning and higher-order value and offer a counter-narrative to the dark vitalism that has held such a grip on nations in the last hundred years. This dark vitalism is the embodiment of sovereign power and antithetical to the spirit of vitalism that we found in Paul. That impulse, by contrast, comes to be articulated in the twentieth century by a number of theologians including de Lubac and Rahner in Catholicism and Barth and Milbank in Protestant theology, and we see it coming to articulation in the doctrine of the resurrection of the body and Christ's ascension to heaven; even Teilhard de Chardin's theology, which is positively vitalist but not overtly political, offers implicit critique of dark vitalism. Agamben's analysis of bare life is relevant here in arguing for what he regards as the inextricably political nature of life itself and the twentieth-century response to bare life in 'the camp'.

DARK VITALISM

Agamben begins with the Greek distinction between *zoē* and *bios*; both words translate 'life', the former being a purely physical entity, a quality common to all living beings, the latter being 'the form or way of living proper to an individual or a group'. Thus, Aristotle could speak of the life of the philosopher (*bios theōrētikos*), the life of pleasure (*bios apolaustikos*), and the political life (*bios politikos*).[2] *Zoē*, when it enters into the polis, is the politicization of bare life as such that according to Agamben 'constitutes the decisive event of modernity'.[3] Natural life comes to be the focus of sovereign power; the management of biological life and its colonization by the political. In itself *zoē* is prior to or outside language in contrast to *bios*, which is in language. Bare life (*nuda vita*) is the politicization of biological life, of the non-linguistic *zoē*. Agamben then links this with law and sovereign power, following Carl Schmitt, as that which defines the state of the exception (Schmitt's *Ausnahmezustand*). On this model, law speaks through sovereign power that can define what is and is not an exception to it. Furthermore, in Roman law, this is

[2] Giorgio Agamben, *Homo Sacer: Sovereign Power and Bare Life*, trans. David Heller-Roazen (Stanford, CA: Stanford University Press, 1998), p. 1.

[3] Ibid., p. 4.

associated with the 'sacred man' (*homo sacer*) as he who can be killed but not sacrificed. This is not only to see sacredness as something set aside, as Durkheim would have it, but as constituted within a legal system that retains its deep ambivalence as both wholesome and polluting. The man who can be killed without committing murder, but cannot be sacrificed, is a liminal figure, an outcast in law who is the exception. As such the sovereign himself is that figure. This absorption of the sacred into the judicial order is simultaneously the absorption of bare life or the politicization of biological life. Agamben writes:

> If our hypothesis is correct, sacredness is...the originary form of the inclusion of bare life in the juridical order, and the syntagm *homo sacer* names something like the originary 'political' relation, which is to say, bare life in so far as it operates in an inclusive exclusion as the referent of sovereign decision. Life is sacred only in so far as it is taken into the sovereign exception, and to have exchanged a juridico-political phenomenon (*homo sacer*'s capacity to be killed but not sacrificed) for a genuinely religious phenomenon is the root of all equivocations that have marked studies both of the sacred and of sovereignty in our time.[4]

Here the characteristic feature of Western discourse has been the politicization of life. The sacred is thus not an exclusively religious sphere but is crucial to the body politic as a realm transposed into the legal system. The sacred is the state of the exception, and biological life understood in these terms becomes bare life and identified with the exception, and the exception is the *homo sacer*. Sovereignty as the embodiment of state power, whether invested from below in a Hobbesian manner or derived from above in a Divine Right of Kings way, defines and controls the sacred, which means that sovereign power defines the state of the exception and so bare life. We might say that bare life is the politicization of life itself under the sign of sovereign power.

In the Hobbesian tradition of political philosophy sovereign power lies in the sovereign in whom the people invest their trust. In exchange for loyalty and obedience, the sovereign protects the people and creates a functioning order for the maximal prosperity of citizens. Conformity to sovereign power is the price paid to ensure that society does not revert to the state of nature, a state of war, in which life, to use his famous phrase, is 'solitary, poor, nasty, brutish and short'.[5] This is the Hobbesian social contract *avant la lettre*. The body politic functions through investment in sovereignty, whether this be dictator, king, or democracy, so long as citizens obey and conform to sovereign dictate. Sovereign power is articulated through law that, in Esposito's terms, ensures the survival of the community and functions like an immune

[4] Ibid., p. 85.

[5] Thomas Hobbes, *Leviathan*, edited with notes by Edwin Curley (Indianapolis, IN and Cambridge: Hackett, 1994), p. 76.

system in the body.[6] Hobbes has an essentially pessimistic (some would say realistic) view of human nature that the human natural state is at war, driven by mimetic desire (of course, he did not use that Girardian phrase but describes the same thing).[7] The health of the social body is ensured by sovereign power that rules both secular state and church, ensuring obedience through law.

Thus, for Hobbes, humanity's natural condition is brutal unless controlled and transformed by politics. People's desire for peace can only be assured by the sovereign's iron fist. Hobbes' is no overt philosophy of life, but the body politic is nevertheless an organic unity in which each part plays a role and contributes to the overall health of the body. Indeed, Esposito reasonably suggests that with Hobbes the question of life embeds itself within political theory and practice, even that with modernity 'life brings into being or "invents" modernity as the complex of categories capable of answering the question of the preservation of life'.[8] But there is no Christian spirit animating the body politic and indeed Hobbes rejects the idea of a top-down transference of power from God to sovereign; rather, power is bestowed upwards to the monarch from the citizens. This is the birth of the modern conception of the state, one in which the vision is not so much repair as construction. Hobbes is not concerned with the animating principle of the body politic but only with its function, namely to prevent a reversal or regression to the state of nature, which is the state of war, and thereby institutes a biopolitics to control life.

This model was to prove fundamental in centuries following Hobbes, both in the political actualizations of sovereign power in the coming centuries and in political theorizing, a fundamentally Hobbesian worldview coming again to the fore in the politics of Carl Schmitt in the twentieth century. Schmitt follows Hobbes in his assessment of human nature as conflictual; indeed, conflict and potential violence are inherent in the human condition and so politics itself cannot be stripped of warlike elements. Politics as the conflictual realm of human interaction reflects the conflictual nature of life, a view that draws on the Christian, and particularly Catholic, view of original sin: humanity is fallen to a state of nature characterized by violence.[9] In this pessimism life itself is characterized as energy, certainly, but a violent energy that seeks domination over the other. The secular politics of power is a

[6] Roberto Esposito, *Immunitas: The Protection and Negation of Life*, trans. Zakiya Hanafi (Cambridge: Polity Press, 2011), p. 21. See also J. Sørensen, 'Religion, Evolution, and an Immunology of Cultural Systems', *Evolution and Cognition*, vol. 10 (1), 2004, pp. 61–73.

[7] Hobbes, *Leviathan*, p. 75: 'if any two men desire the same thing, which nevertheless they cannot both enjoy, they become enemies'.

[8] Roberto Esposito, Rhiannon Welch, and Vanes Lemm, *Terms of the Political: Community, Immunity, Biopolitics* (New York: Fordham University Press, 2012), p. 70.

[9] Carl Schmitt, *Political Theology: Four Chapters on the Concept of Sovereignty* (Chicago, IL: University of Chicago Press, 2005).

transformation of a theological ontology of sin. Realistic politics, thinks Schmitt, is not irenic but assertive of human proclivity to violence that needs to be controlled through the state, through the mechanism of dictatorship. Even democracy contains elements of dictatorship defined as the capacity for the sovereign to decide the state of the exception.

Thus, the creation of biopolitics, the politicization of life, is a feature of modernity whose roots are in Hobbes, but that comes to articulation especially in Nietzsche for whom, as Esposito remarks, life is the sole subject and object of politics as the will to power.[10] Foucault develops the analysis of the history of biopolitics, describing how natural life comes to be included within state power at the beginnings of modernity and through that control of the biological health of a nation, it becomes possible for the state to both 'protect life and to order a holocaust'.[11] Who is defined in terms of bare life is a politico-legal decision that has played out so disastrously in the history of the twentieth century. The Jews and others were relegated to the category of bare life with terrible consequences of the state privileging death as the primary mode of the category, and so bringing bare life close to nihilism. The dark, autochthonous vitalism of National Socialism partly derived from Nietzsche's affirmation of life becomes in the end an affirmation only of death and a triumph of machinic efficacy: the final triumph of death and nothingness over life and in Heideggerian terms, ironically as he initially supported the National Socialist vision, the triumph of *technē* over organism. The philosophy of life turns into the practice of death. In Esposito's words, Nazism was 'the realisation of biology' as the apex of 'a thanatopolitical drift'.[12]

Relevant to this analysis is a volume that stands behind Foucault, and is cited by him, namely *The Productive Body* by Guéry and Deleule. These authors present an analysis of the way in which capitalism has appropriated the body—and we might read this as life itself—into its mode of production. They argue that there are three bodies: the biological, the social, and the productive. Capitalism harnesses the labour-power inherent within the biological body through incorporating it into a social body by means of a productive body. In this process of appropriation, the social body is in fact diminished in the sense that capitalism replaces socialization with a

[10] Esposito et al., *Terms of the Political*, p. 72. See also Roberto Esposito, *Bios: Biopolitics and Philosophy*, trans. Timothy C. Campbell (Minneapolis, MN: University of Minnesota Press, 2008), on 'thanatopolitics' and eugentics (p. 115).

[11] Foucault, *Dits et écrits* (Paris: Gallimard, 1994), vol. 3, p. 719, quoted by Agamben, *Homo Sacer*, p. 3.

[12] Esposito et al., *Terms of the Political*, pp. 72–3. Nikolas Rose also makes a point that contemporary emphasis on human corporeality is 'to free ourselves from an overly intellectualist and rationalist account of contemporary politics, economics and culture', in which biology is 'translated into ontology, ontology is transmuted into politics', a move which 'should give us pause'. Nikolas Rose, 'The Human Sciences in a Biological Age', *Theory, Culture and Society*, vol. 30 (1), 2013, pp. 3–34.

privatization of social functions.[13] Capitalism creates a productive body by eliminating the social nature of work and the social nature of the body, thereby creating an individualized biological body. This reduction of a tripartite scheme that was historically the case, into a binary opposition of productive and biological bodies, disconnects individuals from a sense of shared identity that might resist exploitation by capitalism. Indeed the term 'capitalism' is itself derived from the Latin *capitulum* or *caput*, 'head', thereby indicating the head as the seat of knowledge held by managers to control the body of the workers from whom collective knowledge has been taken away.[14] Thus the biological body is 'produced as an autonomous body trapped in the workings of the productive body in its machinified representation'.[15] On this account, one that Foucault also takes up, the natural body becomes codified with machine-like qualities in a system geared up for maximal economic production (and so maximal profits). The individual, biological body is overcoded with a value system that sees it in terms of productivity rather than any intrinsic worth and in which the non-productive body must, inevitably, be set aside. The Guild, in Marxist terms, is the earliest manifestation of this system in which each biological body is 'machinifed',[16] a process that continues to the full flowering of capitalism where the biological body is a cog in the machine comprising productive bodies.[17]

Foucault was to describe the absorption of the biological body into the productive body, and its consequent machinization in capitalism, as the ordering of the body through regimes of power that are also systems of knowledge, in which politics becomes biopolitics. With the increased power of the state there is an increased concern of sovereign power for the health of the population and the political control of the biological body,[18] which has positive consequences for the productive body of capitalism and the maximizing of profit.

Now clearly the inscribing of the biological body by sovereign power to produce the productive body is not the development of dark vitalism per se, but arguably is the precondition for it. The way regimes of health were promoted in National Socialism for the appropriate body can be contrasted with the denigration of the non-Aryan body and the denial of any life force flowing within it. But the denigration and finally destruction of the non-Aryan body in regimes of machinic efficiency exhibit a laudation of death that contradicts and undermines the affirming of any life principle by the state. In Agamben's terms, sovereign power's relegating the non-Aryan body to the state of the exception, the one who can be killed without committing homicide, is the triumph of nihilism over life. The state is the harbinger of death and there are no higher

[13] François Guéry and Didier Deleule, *The Productive Body*, trans. Philip Barnard and Stephen Shapiro (Winchester and Washington, DC: Zero Books, 2014), p. 51.

[14] Ibid., pp. 92–5. [15] Ibid., p. 53. [16] Ibid., p. 66. [17] Ibid., p. 126.

[18] Foucault, *Dits et écrits*, vol. 3, p. 719

values here than pure force. The body politic has become not the organic whole of Hobbes but a machine-like mechanism of oppression in which technology is violence against life.

HOMO SACER

We can see Agamben's bare life as a negative consequence of a Nietzschean drive for life. Nietzsche's celebratory Dionysian impulse strips down human life to that bare impulse and thereby opens the *apolitical* drive for life itself to becoming absorbed or colonized by a *political* drive to bring all within its remit: in Agamben's terms, the politicization of *zoē* through *bios*. This is the production of bare life that stands outside the law while at the same time being included in the law as the state of the exception. Furthermore, this is the production of sacred life, the *homo sacer*, who as the state of the exception can be killed without homicide being committed. Agamben traces this in Roman law to a statement by a Pompeius Festus who says that 'it is not permitted to sacrifice this man, and yet he who kills him will not be condemned for homicide'.[19]

While the production of the state of the exception is understandable, what then of the prohibition on sacrifice? Why should the one who is in the state of exception, vulnerable to death without murder, be excluded from the sacrifice? This has been the topic of some debate. On the one hand, some scholars, such as Mommsen and Bennett, see this idea of sacrality as a leftover from an earlier time when capital punishment was regarded as a sacrifice to the gods: penal and religious law were not yet distinguished. On the other, there are those such as Kerényi who argue that sacred man cannot be sacrificed by virtue of already belonging to the gods; the sacred man cannot be consecrated because he is already in the divine realm. Agamben points out that the first group of scholars can account for the idea of *impune occidi*, being killed without the culpability of homicide, but this cannot account for the prohibition on sacrifice, while the second group of scholars cannot account for why he can be killed with impunity.[20] Agamben's account tries to resolve the dilemma through arguing for the political nature of *homo sacer* before any distinction between sacred and profane, in which *homo sacer* already belongs to God and so is not sacrificeable, yet is part of the community in being able to be killed. This is the 'sovereign sphere' in which the sovereign has the power to create the *homo sacer*, the state of exception (and so the *homo sacer* cannot be sacrificed by virtue of already belonging to the divine law, *ius divinum*), and

[19] Cited by Agamben, *Homo Sacer*, p. 71. [20] Ibid., pp. 72–3.

in which all men can act as sovereigns in their capacity to kill the *homo sacer* without culpability (and so he belongs to human law, *ius humanum*).[21]

On this view, the sacredness of life is an idea formed in an originary political context of sovereign power that decides the state of the exception. Sacrality is not inherent in life itself but is constructed within the political order as the demonstration of sovereign power. The sacrality of life seen in the figure of the *homo sacer* is constructed and constituted within the context of political power and the early formation of rules of human conduct. This political understanding of sacrality is a critique of apolitical views such as Rudolf Otto for whom sacrality was confined to the psychological realm of emotion,[22] and is also distinguished from anthropological accounts that see sacrality in terms of prohibition or taboo. Clearly the anthropological notion of taboo is operative in the case of the *homo sacer* who is imbued with good and bad sacred power in the sense of being both divine and polluting, but it is the political origin in the state of the exception that is most important here, not simply as a historical analysis but because it has had modern political impact in the notion of 'the camp'. The camp is the 'biopoliticization of life',[23] a phrase Agamben takes from Karl Löwith who observed how totalitarian states in the twentieth century—Marxist Russia, Fascist Italy, National Socialist Germany—politicize 'even the life that had until then been private'.[24]

But even bourgeois, liberal democracies have a concern with biopolitics, and the ease with which democracies turned into totalitarian states and totalitarian governments into democracies is explained, says Agamben, by the focus on bare life. Indeed, he traces a history from the Magna Carta (1215), which says that 'no free man (*homo liber*)' may be placed outside the law, to the 1679 writ of *habeas corpus* in which 'free man' has become replaced by 'body' (*corpus*). That is, 'body X' by whatever name must appear before the court, so 'you will have a body to show' (*habeas corpus ad subjiciendum*) becomes the central legal foundation of democratic law.[25] This, claims Agamben, is the way that even modern democracies take bare life and attribute it to each individual and the body becomes the central metaphor of political community from Hobbes' *Leviathan*, bearing both liberty and rights as well as being subjected to sovereign power, being subjected to the capacity to be killed.[26]

Here lies the potential for decisions about life worthy of being lived or otherwise, exposing the underlying logic that allows democracies to become totalitarian states that determine which life can be lived and which life is to be eliminated. Thus, National Socialism's relegation of particular bodies or types

[21] Ibid., pp. 82–4.

[22] Agamben somewhat caustically remarks: 'That the religious belongs entirely to the sphere of psychological emotion, that it essentially has to do with shivers and goose bumps—this is the triviality that the neologism "numinous" had to dress up as science' (p. 78).

[23] Ibid., pp. 119–25. [24] Quoted ibid., p. 121. [25] Ibid., pp. 123–4.

[26] Ibid., p. 125.

of body to bare life is simply the converse of the relegation of particular bodies to the ideology of fullness and health. If originally life and politics were distinct, linked together by the state of the exception, by bare life, then the history of totalitarianism—in Stalin's Russia, Hitler's Germany, and, we might add, Mao's China—brings them together such that, in Agamben's words, 'all life becomes sacred and all politics becomes the exception'.[27] Within the philosophy of National Socialism, life, or rather 'good' life as determined and controlled by the state, defines its opposite as the 'bad' life that can be exterminated with impunity. The vitalism that this ideology draws on is thus a dark vitalism and brings the affirmation of life in the affirmation of death. In this dark vitalism—whose impulse is indeed traceable to Nietzsche, although he is simply an articulate exemplum of forces deep within the juridico-political structure of the West—there is no value outside of life itself, there is no transcendent value in either a Gnostic or a theistic narrative. In contemporary terms, bare life—the simple fact of being born—comes to be the arena of biopolitics in which genetic control becomes the site of political contestation, as we have seen. The philosophy of life becomes a dark philosophy of life because the vital force is seen to be exclusive with degrees of intensity linked to degrees of value for the state: thus life developed from the purity of blood and earth comes to be accorded high value, defined in contradistinction to the low value of the 'degenerate' other defined both racially ('Jewish blood') and politically (Socialist ideology).

That the biological body can be colonized by the state and relegated to bare life and the state of exception so easily is an indication for Agamben of the proximity of democracy to totalitarianism. The sacrality of life is political gesture created with modernity and the move towards secularization. The person is overdetermined as biological life and politically formed either as citizen or as bare life. The creation of bare life or the *homo sacer* is ironically the de-sacralization of the person and the complete objectification of the body as owned by the state: I am reminded of Kafka's terrible story of the machine that executes criminals by inscribing their crime on their body.[28] The Agamben view charts a dark vitalism and a de-humanization. It is because of this stripping a person down to bare life that the *homo sacer* cannot be sacrificed because sacrifice entails, on the contrary, that the victim take on the sins of the sacrifice or community and so must proximate to the human in some way. Hence in Bataille's Aztecs the human sacrificial victim has to be brought within the realm of intimacy, within the realm of human life to take on the properties necessary for cathartic expulsion through sacrificial violence. Similarly, with the Christ figure, which Agamben does not address, we have the Roman state

[27] Ibid., p. 148.

[28] Franz Kafka, 'In the Penal Colony', trans. Stanley Corngold, in *Kafka's Selected Stories*, Norton Critical Edition (New York: Norton, 2007), pp. 35–59.

creating Jesus as the state of exception and stripping him to bare life. But here the body condemned through Roman judicial order is overcoded with the Christian vision and so set up as the counterpart of bare life. The *homo sacer* is not outside of the sacrifice in this case, but is regarded as the sacrifice of all sacrifice, in fact to redeem the biological body and to question the very secular power that had attempted to de-humanize the body in the judicial process. There is then an alternative vision.

§

So much for the analysis of the conditions under which the politicization of life itself can turn into a dark vitalism, but what are the discourses that counter this? A purely negative, political sacrality as the dark side of a secular age does not go unchallenged. On the one hand, we have a still secular discourse of an emergent vital materialism linked to environmental concerns that we examined in Chapter 9. On the other hand, we have a theological discourse that is both critical of dark vitalism and affirmative of life through its grounding in a Christian metaphysics and the highlighting of the person in relation to life: persona as the consequence of life itself and Christ as *homo sacer* redeems through disrupting the category in being understood as the one who can and must be sacrificed. The redemption of bare life and its transformation through death and the proposed resurrection of Jesus provides a counter-narrative to a purely secular history.

This is to see sacredness not as bare life or the politicization of biological life, but as integral to the nature of the person as a participant in a trans-political order. The trans-political is not non-political, as Christian theological participation in an order of sacrality that goes beyond the contingencies of human power relationships includes those relationships within it. Carl Schmitt's political theological project, whose analysis of sovereign power integrates so well with Agamben's, is not representative of mainstream theological thinking on the matter of life. Rather, the Christian project has been participative in a sacred order beyond the human—in pre-modernity it has been fundamentally cosmological—and yet has been imbued and implicated within human political life. Taubes is surely right in his characterization of Paul and in identifying the fundamentally political nature of Paul's project, a project that embodies the idea of a transformed state with the realization of Christian eschatology along with an evangelizing imperative inspired by the life of the spirit. The Christian theological narrative is affirmative of life itself and both implicitly and explicitly critical of state power that through the centuries has attempted to control it. The biopolitics analysed by Foucault and Agamben, and even Arendt, is a site of resistance in Christian theology where the sacredness of life is not defined by the secular state, as Agamben's *homo sacer*, but is rather defined by transcendence beyond it: by the God who creates *ex nihilo*. The Christian affirmation of life is thus set against secularization and de-humanization of human life and also against the Gnostic narrative that negatively evaluates the world.

GRACE AND NATURE

Within modern Christian theology the ordering of the political realm has taken the form of argument against secular politics and humanist atheism, and political theology has also set itself against the dark vitalism and relegation of the body to bare life that Agamben has so astutely analysed. De Lubac, living through the dark vitalism of National Socialism in France, perceived that Christian theology needed to go beyond rationalism or 'extrincicism'[29] and return to the sources of the Catholic tradition, a *ressourcement* to the Fathers of the Church and to medieval exegesis. De Lubac presents a Christian vision in which secular bare life could not occur because of the sacrality of life pervaded by the supernatural order and by grace, and offers an analysis of how a purely secular narrative developed (that could give rise to bare life). Moving to a machinic understanding of life, in terms discussed above, to a transformation of the biological body into the productive body, secularism misses the sacrality of life understood as participation.

De Lubac thought that Catholic theology had become divorced from everyday concerns and irrelevant to those concerns because it was focused on Neo-Scholasticism, simply a commentary on Thomas Aquinas, which, in de Lubac's view, had widened the gap between the sacred and the everyday, between grace and nature. One of de Lubac's key, if controversial, texts is *Surnaturel*.[30] In this book published in 1946 he argued that the distinction between the supernatural and the natural, that maps on to the distinction between the realms of grace and nature, had meant that theology had in a sense given up on the natural order through focusing on the supernatural. Such relinquishing of a realm of discourse in favour of a kind of transcendence from the seventeenth century had meant that theology developed an 'extrinsic' approach and allowed the gap thereby created to be filled by secular

[29] Henri de Lubac, *A Brief Catechesis on Nature and Grace*, trans. Richard Arnandez (San Francisco, CA: Ignatius, 1984), pp. 37–41.

[30] For a good overview of de Lubac's writings on the supernatural see Georges Chantraine, 'Surnaturel et destinée humaine dans la pensée occidentale selon Henri de Lubac', *Revue des sciences philosophiques et théologiques*, vol. 85 (2), 2001, pp. 299–312. For a good discussion of the issue of contextualizing de Lubac in contemporary theology see Bryan C. Hollon, *Everything Is Sacred: Spiritual Exegesis in the Political Theology of Henry de Lubac* (Eugene, OR: Cascade Books, 2009), pp. 75–94.

philosophy and secular political philosophy, such as that of Hobbes and Locke.[31] On the contrary, as creatures we are inseparably participant in creation as essential to our nature; creation is not simply an extrinsic fact, but our essence. De Lubac writes:

> Forgetting, or at least not fully realizing our situation as creatures, we reason more or less as if creation were only a fact, the pure extrinsic condition of our origin, and not our essence.[32]

Theology thus has a healing task of making us aware of how we are integrated within creation and the inseparability of grace and nature; that life is nature and the supernatural interpenetrated.

The relinquishing of the secular realm by Catholic theology had meant its becoming increasingly irrelevant to mainstream political and philosophical discourse in the secular world. On this view, the church is there to save souls and as a consequence had neglected the concerns of everyday life. This, following the Catholic philosopher Maurice Blondel, is what is meant by extrinsicism,[33] a theology extrinsic to the needs of everyday human concerns. An alternative could be presented in which the supernatural pervades the natural: life itself comes to express its divine source. This *nouvelle théologie*, an ironic title in that it wished to return to the Christian sources in the Church Fathers and to medieval exegesis of scripture, presented the affirmation of the everyday through claiming that participation in the divine realm shows in the inseparability of nature and grace. The separation of nature and grace in Neo-Scholasticism is a dualistic ontology for de Lubac that distinguishes too rigidly the natural ends of human life through humanity's own efforts from the supernatural end of life formed through grace.[34]

Sin has harmed nature, but this can be healed through supernatural beatitude and participation for de Lubac, which is a return to a view prior to the sixteenth century in which grace, the supernatural realm, penetrated into nature. Theology after the sixteenth century wished to maintain that human nature had its own natural end in contrast to the beatific vision that was a

[31] See William Cavanaugh, *Torture and Eucharist: Theology, Politics, and the Body of Christ* (Oxford: Blackwell, 1998), pp. 19–20.

[32] Henri de Lubac, *Surnaturel: Études Historiques* (Paris: Aubier, 1946), p. 485: *Oubliant, ou du moins ne réalisant pas à fond notre situation de créatures, nous raissons plus ou moins comme si la création n'était qu'un fait, pure condition extrinsèque de notre origine, et non jusq'à notre essence.* My translation.

[33] Maurice Blondel, *Action: Essay on a Critique of Life and a Science of Practice*, trans. Oliva Blanchette (Notre Dame, IN: Notre Dame University Press, 2004 (1893)). De Lubac derives his distinction from Blondel. He writes that Blondel 'overcame the opposition between an extrinsicism which ruined Christian thought and an immanentism which ruined the objective mystery which nourishes this thought'. *Nature and Grace*, p. 38.

[34] Henri de Lubac, *Le surnaturel* (Paris: Vrin, 1946), p. 437. See Brian Daley, 'The Nouvelle Théologie and the Patristic Revival: Sources, Symbols and the Science of Theology', *International Journal of Systematic Theology*, vol. 7, 2005, pp. 362–82.

supernatural end, freely given through grace. There is a 'pure nature' quite distinct from the supernatural realm of grace. If the two realms were conflated, this meant a kind of compulsion on grace to grant the beatific vision. By contrast, de Lubac rejects this view on the grounds that the dignity of the human needs to be protected against nihilism and atheist humanism, showing through his scholarship that the idea of pure nature only develops long after Aquinas in the Thomistic tradition.[35] Thus following Blondel, secular historicism, that there is no foundation, is also to be combatted as much as extrincism.

For de Lubac, Christian anthropology entails the idea of the supernatural in which the human is made in the image of God.[36] This supernature is fundamental to Christian thinking and articulates the idea of the Spirit that in Pauline terms is called *pneumatikos*,[37] thereby linking the supernatural to the idea of life itself as understood in the early Christian sources. The supernatural, as being intrinsic to life itself because it pervades nature, lays stress on a more tactile or material understanding of grace that de Lubac's friend and fellow Jesuit Teilhard de Chardin called 'physicism'.[38] The importance here lies in the idea that the supernatural does not simply elevate nature, but transforms it.[39] On this view, the central Christian message is the transformation of creation from a state of fallenness, to use Christian mythological language, and such transformation must inevitably be political because it is concerned with human affairs and the 'natural' transactions of people.

Thus, in contrast to the bare life that is produced through a secular political ideology in which life itself is colonized by a *bios* that actually strips it of any human features, de Lubac's supernatural imbues life with divine power and functions against the danger of relegating the political purely to nature and thereby divorcing it from any moral vision. Moral vision is not really the language of de Lubac, but rather, in the more robust language of Christian salvation, he writes that the 'darksome' and 'sinful drama' of humanity can be healed through the supernatural intervention of Calvary, which is actually a return of humanity to life.[40] The formation of bare life and its consequent horror can be counteracted according to this Christian paradigm by the transformation of life itself in the salvific act of the sacrifice of Christ, translated into the political arena that is part of nature. Through colonizing nature, supernature means that transcendence is always present in human life and moves against a de-humanizing secularization in which human life can become merely bare life in what de Lubac would see as a corruption of life itself.

The intrinsic desire for a supernatural end is a characteristic feature of the human, a longing to see God that is the necessary condition for civilization

[35] De Lubac, *Surnaturel*, p. 487. [36] De Lubac, *Nature and Grace*, pp. 17–18.
[37] Ibid., pp. 26–7. [38] Ibid., p. 50. [39] Ibid., p. 81. [40] Ibid., p. 135.

resulting from it; the necessary impetus to transform human life both in the personal realm and in political reality. Human beings have a supernatural finality imprinted on their natures, an orientation towards a final end,[41] which can be read as the sanctification of life itself. Thus, for de Lubac the crucial redemptive structure for human hope is the church that expresses God's grace and performs through history the resurrection.[42] It is this structure and institution that can prevail against the de-humanization of secularism and its anti-foundationalism, a process analysed by Agamben that results in the transformation of democracy into totalitarianism, and it must do this through privileging the idea of the person.

CHRISTIAN ANTHROPOLOGY

In Christian, and especially Catholic, anthropology the centrality of the institution of the church is complemented by the centrality of the sanctity of the person. With de Lubac's non-dualism between nature and grace, the supernatural pervading the natural, the person participates in God who is the source of life. The centrality of person militates against de-humanization and locates human beings within an environment, within an *Umwelt*, that is orientated towards their support. This Christian anthropocentrism contrasts with a vitalism that sanctifies all of nature—as we see in the affirmation of nature in work by thinkers such as Donna Haraway—but that has the potential to become the dark vitalism so devastatingly well articulated in the twentieth century. The Christian vision of the de Lubacian kind places the person redeemed through Christ at the centre of a cosmology characterized by love.[43] This is a covert political theology that affirms a Christian philosophy of life. Privileging the centrality of the person is to promote a stance against any state totalitarianism that seeks to demote the person in the interests of sovereign power. The affirmation of the person against its totalitarian eradication becomes an overt political act, as we have seen with so many examples from Etty Hilsum to Dietrich Bonhoeffer or, perhaps less dramatically but nevertheless poignantly, in the affirmation of the person against bureaucratization as depicted, for example, in the films of Ken Loach, especially *I, Daniel Blake*.

[41] De Lubac, *Surnaturel*, p. 487. [42] De Lubac, *Nature and Grace*, pp. 109–15.

[43] See Oliver Davies, *Theology of Compassion: Metaphysics of Difference and the Renewal of Tradition* (London: SCM, 2010), p. 77: 'love represents an attempt to re-enact on the temporal plane the immutability of the divine realm: it is an attempt to accomplish on earth the divine *ousia*'; also on the theology of love, Werner Jeanrond, *A Theology of Love* (London: T and T Clark, 2010), pp. 25–44.

In the de Lubacian view, the person comes to the fullness of life through a participative ontology in which my being as person has its fulfilment in the being of the Christ, specifically in the resurrection, articulated through the church in the Eucharistic transformation. Person, on this view, is less individualistic and more subjective in the sense that participation in Christ is an intensification of life itself and the development of a Christian-specific inwardness that is intensely personal while simultaneously transcending personality or ego in a mystical ascent.[44] But even without the language of inwardness, Christian anthropology presents a view focused on the experience of life lived in the world—as Spaemann says, who develops this line of thinking, 'what it means to be alive is something we know from experience'.[45] With the person at the centre of a Christian ontology, life itself becomes personalized in a way that always resists reduction to bare life. For Spaemann, 'we experience what it is to live' when we experience life as our being, and 'to have life is what it is to be a person'.[46] Experiencing life, a person has a different sense of sacredness to the grim legalism of the *homo sacer*. In a pre-modern cosmic Christianity, the person participates in a cosmos imbued with the life of Christ and it is this sense of participative sacrality that is lost with modernity and the retreat of religion from cosmology that is replaced by mechanistic science and mechanistic body. But with thinkers such as de Lubac, and even more so with his friend Teilhard de Chardin, Christianity offers a participative ontology in which the person is given an intensity of life through grace that pervades nature. But this participation does not mean the eradication of the person, and one reading of Christianity wishes to maintain a strong personalism in which each is unique with a single name, 'a name which only one person bears and God alone knows'.[47] To see this structure of grace is to see history in terms of a narrative of fall and redemption, the central Christian narrative, that for Christianity is a 'true myth' in which the violence of history as we see articulated in the *homo sacer* is challenged ironically by the violence of the cross that is actually an irenic gesture 'to repair the things that have been broken', to paraphrase Walter Benjamin.[48] The secular violence of the *homo sacer* when read from the Christian theological perspective is the negation of life itself as gift. De Lubac's emphasis on participation and Spaemann's focus on the person are within the sphere of a political theology that challenges the secular narrative that is in

[44] For a more sustained argument see my *The Truth Within: A History of Inwardness in Christianity, Hinduism, and Buddhism* (Oxford: Oxford University Press, 2013), pp. 69–101. See also Paul S. Fiddes, *Participating in God: A Pastoral Doctrine of the Trinity* (London: Darton, Longman and Todd, 2000).

[45] Walter Spaemann, *Persons: The Difference between 'Someone' and 'Something'*, trans. Oliver O'Donovan (Oxford: Oxford University Press, 2012), p. 157.

[46] Ibid. [47] Ibid., p. 150.

[48] Walter Benjamin, 'Theses on the Philosophy of History', in *Illuminations*, trans. Harry Zorn (London: Pimlico, 1999), p. 249.

danger of falling into nihilism, as we see with the *homo sacer*. As Milbank has highlighted, Christianity confronts a purely secular politics with a vision of life that is transformative, offering an alternative to a politics resting purely on power, the power of the strong over the weak.[49] In the Christian vision, life is inherently ethical, as Ward points out, intrinsic to the nature of creation and not simply an add-on to bare life.[50] In the Christian story, a political theology offers critique of secular politics in which power can create the *homo sacer* and that looks to a denouement at the end of history.

A Christian anthropology is therefore a philosophy of life that is also necessarily a political theology. In a fundamentally Christian structure, three terms are important in their interrelationship: life (related to *pneuma*, spirit), being (related to cosmos), and gift (of Christ to the world). Underlying this structure is the idea of God as the perfection of goodness and the supremacy of 'good' over 'right', a view that entails a participation in a cosmic order, a view that Milbank following de Lubac has developed.[51] Here participation means a community of individuals focused on the transformation of life that the church sees as enabled through Christ's resurrection, the central idea of all Christian churches. That is, participation is the true spirit of Christian faith rather than the dry extrinsicism of believing a set of propositions, a view that is decidedly modernist and counter to the traditional understanding. Both extrinsicism and historicism in the de Lubacian view are against the tradition and do not embrace the full implication of the resurrection.

THE RESURRECTION OF THE BODY

The argument against extrinsicism and for participation can be seen in particular with regard to the central Christian dogma of the resurrection. The literature on this is vast, but Enlightenment and post-Enlightenment modernity has inevitably struggled with the doctrine, and its claim to Christian uniqueness in the face of a plurality of religions has been an ongoing issue. But the death on the cross and resurrection of Christ is central to the logic of Chrystianity: as the Orthodox Easter homily of John Chrysostom declares: 'Christ is risen from the Dead, by death hath he trampled down death, and on

[49] As Milbank says, 'The Cross was a political event and the "apolitical" character of the New Testament signals the ultimate replacement of the coercive *polis* and *imperium*, the structures of ancient society, by the persuasive Church, rather than any withdrawing from a realm of self-sufficient political life'. John Milbank, *The Word Made Strange: Theology, Language, Culture* (Oxford: Blackwell, 1997), p. 251.

[50] Graham Ward, *How the Light Gets in: Ethical Life* I (Oxford: Oxford University Press, 2016), pp. 290–1.

[51] Ibid., pp. 24–32.

those in the graves hath bestowed life'.[52] Stanley Spencer's *The Resurrection, Cookham* shows the dead climbing out of the graves in an English country churchyard, an eschaton enabled in Christian belief through the death and bodily resurrection of Christ. But as Karl Rahner asks, what do we really mean by the resurrection of the body?[53]

There is a range of views on this, the central mystery of Christianity, all of which implicate human history in a history of salvation. For Rahner, the resurrection means the 'the termination and perfection of the whole man before God, which gives him "eternal life"'.[54] This termination of history is not really conceivable other than in vague terms as 'the perfection and total achievement of saving history',[55] but it is the endpoint towards which creation is moving and the fulfilment of life that has been enabled by the resurrection of Christ. Christ, as it were, paves the way for universal redemption in which life overcomes death. In the New Testament account, shortly after the resurrection, Christ ascends to heaven, iconographically depicted as Christ emanating a glorious light that blinds the disciples. In the words of de Bérulle quoted by de Lubac, Christ received from the Father a body 'far more glorious than the sun' that contains 'within its immense grandeur both earth and sun, all the stars and all the expanse of the heavens, a body that rules all bodies and all heavenly spirits', an image that resonates with Teilhard's cosmic vision of Christ as the omega point.[56] In this ascension into heaven in de Bérulle's terms, Christ becomes light and expands to the limit of the universe, thereby transforming the world such that humanity can follow in Christ's wake: we too will become light in a new creation.

The resurrection of the body is the central Christian metaphor and believed spiritual truth that stands as a bulwark against a purely secular understanding of history and the denigration of life, made possible through the dark vitalism that is its consequence. The resurrection signifies the transformation of the world and of the person, offering a deep eschatological hope for those within the paradigm. Echoing Rahner, Oliver Davies asks, 'where is Jesus Christ now?'[57] Davies answers this by returning to the distinction between the resurrection and the ascension, the latter doctrine having been somewhat neglected in Christian

[52] *A Prayer Book for Orthodox Christians* (Boston, MA: Holy Transfiguration Monastery, 1987), p. 163.

[53] Karl Rahner, *Theological Investigations*, vol. 2: *Man in the Church*, trans. Karl H. Kruger (London: Darton, Longman and Todd, 1963), p. 210.

[54] Ibid., p. 211. [55] Ibid., p. 213.

[56] Henri de Lubac, *Teilhard de Chardin, the Man and his Meaning*, trans. Réne Hague (New York: Hawthorn Press, 1964), p. 57.

[57] Oliver Davies, *Theology of Transformation: Faith, Freedom, and Christian Act* (Oxford: Oxford University Press, 2013), p. 5. For a discussion of the same theme also see Douglas Farrow, *Ascension Theology* (New York: T and T Clark, 2011); Anthony J. Kelly, '"The Body of Christ Amen!" The Expanding Incarnation', *Theological Studies*, vol. 71, 2010, pp. 792–816.

theology since the sixteenth century.[58] Christ ascended reaches a boundary of cosmic possibility and we might say in this theology, the ascended Christ becomes coterminous with life itself. For Thomas Aquinas the ascended Christ was at the highest point of heaven and for us, now that we know so much more about the physical universe, that highest point is at the boundary of the known,[59] the boundary that is total light.

This cosmological retrieval of Christ as transformed into the universe contains an implicit political theology in which Christ ascended represents the transformation of the world and the negation of negation: the affirmation of life transformed is the negation of bare life and the assertion of a different kind of sacrality in which the person could never be stripped to bare life. The state machine implied by bare life and the reduction of the biological body to the productive body is inveighed against in the Christian narrative that seeks to speak truth to power. While we must acknowledge how the Christian voice has so often been used in the interests of state power to oppress and even to render others into a state of slavery, the heart of the political theology articulated by Paul is one of affirmation of spirit that is affirmation of life in the hope of resurrection, a message renewed through its history in which tradition can itself be read as subversion.[60]

The psychoanalyst Norman Brown observes that the resurrected body is the transfigured body, a body reconciled with death.[61] Brown comments that the specialty of Christian theology is its rejection of the Platonic view of the body—a rejection of what I have called the Gnostic narrative—in favour of the affirmation of the body in an eternal life. Eternal life can only be in the body according to the Christian narrative, and it is Jacob Boehme who takes up the theme, seeing death not as nothing 'but as positive force either in dialectical conflict with life (in fallen man), or dialectically unified with life (in God's perfection)'.[62] In Brown's reading of Boehme, he sees the affirmation of life itself as the affirmation of play and a resolution of anxiety and neurosis because of the necessity to accept life as life in the body, a view that later Protestantism represses (along with Boehme's writings). This acceptance—one might add joyful acceptance for Boehme—of the life of the body is simultaneously an acceptance of death. This tradition of bodily affirmation according to Brown runs from Luther to Boehme and thence to the poets Blake and Novalis, Goethe, and even Hopkins and Rilke, and one should add

[58] Ibid., p. 8: 'Even such a conservative doctrinal theologian as Karl Barth tends to conflate the ascended and resurrected Christ'.

[59] Ibid., pp. 34–57.

[60] James Hanvey, 'Tradition as Subversion', *International Journal of Systematic Theology*, vol. 6 (1), 2004, pp. 50–68.

[61] Norman Brown, *Life against Death: The Psychoanalytic Meaning of History* (London: Routledge and Kegan Paul, 1959), p. 309.

[62] Ibid., p. 310.

the philosophers Hegel, Berdyaev, and even Freud.[63] For Brown, a psychoanalytic reading of the Christian theme means that the resurrection of the body as an idea is the affirmation of life, of Eros, but an affirmation that can only occur with the acceptance of death, of Thanatos. In a sense Brown's version of the resurrection is an attempt to offer a secular view of resurrection as a symbol that nevertheless has the power to move against the political colonization of life itself to produce bare life. The joy and affirmation of the resurrected body is the affirmation of a civilization that has reduced collective anxiety and neurosis in the affirmation of a secular sacred against nihilism.

This is certainly a weaker version of the transformed body, the resurrected body, than the mainstream Christian one, relegating it to the realm of metaphor and imagination, albeit a real imagination that has effects in material reality. But clearly Brown understands the image—or we might say the icon—of the resurrected body as a necessary force against bare life, a counter to the secular sacred of *homo sacer*, and a secular alternative to the theistic vision. Brown recognizes the need in modernity to counteract the nihilistic desire for death with the positivistic force of life, but in contrast to the Christian strong ontology of the resurrection can only offer a weak ontology, but perhaps an ontology that because of its weakness has wider appeal in a secular age. The Christian solution to *homo sacer* that meets bare life with the resurrection of the body is a political theology that may have wider appeal in many areas of the globe, but in the North Atlantic meets with scepticism in the secular, democratic faith that democratic institutions through law, along with a culture of human rights, will be robust enough to resist any future attempts for politics to revert to a politics of bare life. When Freud left Vienna in 1939—writing ironically, 'I can most highly recommend the Gestapo to anyone' in a document he was made to sign[64]—he could reflect that the analysis of repression had not been enough to prevent the mass tide of de-humanization and scapegoating. But then neither had the Christian resurrection of the body been enough, and it could be argued that civilization needed to enshrine human self-repair in institutions legally regulated. The United Nations legislation concerning the global implementation of human rights might be seen as a gesture towards this.

§

The resurrection of the body is a powerful idea and image, translated as political theology that stands against any secular de-humanization of life. Of course, there is no guarantee that political theology does not produce de-humanization—the history of Christianity is replete with such examples—but it can function in the contemporary world as a counterfoil to totalitarian

[63] Ibid., pp. 311–13.

[64] Peter Gay, *Freud: A Life for our Times* (London and Melbourne: J.N. Dent and Sons, 1988), p. 628.

de-humanization and the reduction of life to bare life that we have seen so abundantly on a mass scale in the twentieth century. But the Christian vision, while claiming universality, cannot function as theology in a global, mostly secular political context. The affirmation of life itself that we see in the strong theology of the resurrection comes to be replaced by covert political theologies, especially in Asia where religion functions as a force for political change. Thus in the complexity of a contemporary global history of ideas, we have life itself coming to articulation in a spiritual-material environmentalism, a popular retrenchment against this in a return to fundamentalist Christianity, and a covert Asian vitalism as critique of sovereign power manifested in a number of local religious movements such as Daoist possession cults in South-East Asia.[65] Furthermore, Islam as a global political force stands at an angle to these developments, taking the form of anti-democratic values that it sees as counter to the sacred order of life. How all this will play out on the world stage has yet to be seen.

Political theology is a category that has developed within the Abrahamic religions. Arguably Islam is inherently political, and we cannot separate the 'spiritual' and 'political' in a worldview that has such concern with the way we live our lives and the imperative of correct governance in consonance with God's perceived law. Judaism likewise has a concern with the political and governance under God's law, although here the concern is with the in-group and is non-universalizing; indeed, it is universalism that ends up with the evil of the holocaust, and life affirmation in Germanic vitalism ironically ends in the affirmation of death through stripping life to bare life. With Christianity, theologies of life have intersected with mainstream theology as we have seen, from Paul to Aquinas, but here there has always been the 'Mary' option of following a path of life focused purely on spiritual development to the neglect of the 'Martha' option of engaging with secular politics. Life force in Asia is a different story, with Daoist ideas influencing the Confucian court in China, and in India the vitalism of yoga becoming a political force in post-independence Indian polity.

In one of his last books, *The Nomos of the Earth*, Carl Schmitt presented a history of the global order rooted in Europe and the discovery of the New World. This is an optimistic work that argues for establishing a world order based on international law that has come out of a European context of Occidental rationalism that produced the sovereign state, an important European achievement.[66] Leaving aside the historical question about whether Europe is unique in coming up with the sovereign state—it could be argued that China

[65] Kenneth Dean, 'Underworld Rising: The Fragmented Syncretic Ritual Field of Singapore', unpublished paper presented at 'Political Theologies and Development in Asia', Asia Research Institute, NUS, 21 February 2017.

[66] Carl Schmitt, *The Nomos of the Earth in the International Law of the Jus Publicum Europaeum*, trans. G.L. Ulman (New York: Telos Press, 2006).

established itself as a united polity long before—the major point of Schmitt's text is to argue that the USA is the main force for global order, a view that is resonant sixty years or more after its publication.

Political theology that taps into the life energy of civilization—as Stravinsky's *Rite of Spring* might be said to have done on the eve of the First World War—has come to be relevant to discourse once again, but this time not only a Christian political theology but implicit political theologies that operate at local levels, especially in Asia, that come to be highly relevant in shaping global politics. I have here attempted to articulate a problem that lies at the heart of Western views of life, namely the secular creation of sacrality in the *homo sacer* identified by Agamben, and I have shown how Christianity offered a counter-narrative through the affirmation of life itself in the resurrection of the body; an anticipation of fullness and the healing of the entire cosmos through a participation in a trans-human order. But this is not enough for a contemporary pluralist world in which we see the need for political institutions to address the affirmation of life and desire for life itself. The history of civilizations expresses a history that sustains and reinforces the human proclivity to assert life, and for institutions to articulate the impulses of life and the need to control those impulses through law and narrative. It is to this transformational structure, rooted in the kind of beings that we are, to which we finally need to turn.

11

Religion and the Bio-Sociology of Transformation

Civilizations address the need for human self-repair and the desire for life through practices that seek to bring people more fully into life and philosophies of life that seek to explain human location in a cosmos. In the story I have told of the ways in which this happens, there is a remarkable convergence of ideals of what a holy life is and the cultivation of human practices conducive to it. Even when challenged, as we saw with some tantric practices, there is a sense in which holiness is normative across civilizations. I have narrated the ways in which this has occurred. Now it remains to more explicitly bring out an underlying explanation of this pattern of holiness borne witness to in the history I have narrated. I want finally to return to the theme of social cognition I began with, to link this to systems theory of religion and thereby gain leverage on the underlying pattern we can identify of human self-repair.

A crucial question that faces contemporary humanities that is highly relevant to wider social, cultural, and political issues is: how do we account for the persistence of religion on the global stage? In spite of the development of technologies that have transformed communication, in spite of massive secularization in the Western or North Atlantic world, religions persist and are growing in some parts of the world through population increase but also through missionary activity. Even in what Taylor identifies as the secular age, religions continue to be central to identity politics, to nationalism, and to the formation of human meaning in the many kinds of spirituality that inform Western democracies. Even China is a burgeoning religious field where religion may well be central to social transformation in the future. Although some commentators disparage religion as a cause or explanation of political conflict, many agents in those conflicts identify religion as the major driving force of their activity. This is clearly the case with Islamic State, with Christian/Hindu and Christian/Muslim conflict in South Asia and Africa, and has been true of conflict in the Balkans and in Northern Ireland. Religion is a major factor on the political world stage with massive impact, both positive and negative, on the lives of millions of people.[1]

In this study I have focused not on the destructive power of religion in the history of civilizations but on its constructive power as a force to heal lives and bring hope through eschatologies that link people to life. The task of understanding and explaining religion is therefore of major importance, not only as an area of academic inquiry, but because it has impact on governance and economics throughout the world. The explanation of why Muslim youth in the UK identify with fundamentalist Islam in Syria needs to inform policy decision and formation, and therefore law.

To give an account of the complexity of religion and how it functions in the complexity of the world, society, and politics, we need to draw on complex disciplines that offer comprehensive, coherent, and corroborated explanations of human reality. Science holds such a position of privilege in contemporary society and has done since its rise to dominance in the seventeenth century. Science has offered comprehensive accounts of the world that are open to empirical investigation and rational inquiry and has undergone several revolutions or paradigm shifts in its development.[2] While acknowledging that it is difficult to speak of science in general, we might say that the contemporary sciences, particularly biological and neurological sciences, offer understandings of who we are as mind and body in the material world. These scientific accounts intersect with cultural understandings and there is an imperative to explain religion in such terms. Throughout the current work I have developed an integrated approach to human reality and here wish to bring out more explicitly the scientific side of that integration. Science has developed technologies that affect embodiment and offers powerful explanations of how humans act and interact; science has something to offer us in any account of human desire for life and the search for healing and completion. Comparative religious studies has an opportunity to absorb the findings of science in relation to religion[3] and through recognizing the explanatory

[1] John Bowker, *Religion Hurts: Do Religions Cause More Harm than Good?* (London: SPCK, 2018).

[2] On the social history of science see Steven Shapin, *A Social History of Truth: Civility and Science in Seventeenth-Century England* (Chicago, IL: Chicago University Press, 1995).

[3] A development in this area that I do not explicitly address is the Cognitive Science of Religion (CSR). One view within this field is that 'religious predilections are understood as by-products of our natural cognitive capacities' (Robert N. McCauley, *Why Religion Is Natural and Science Is Not* (Oxford: Oxford University Press, 2011), p. 154). There is a very extensive literature here that is not possible to review, from the origins of the cognitivist approach in, among others, Dan Sperber's influential book *Rethinking Symbolism* (Cambridge: Cambridge University Press, 1975), to Robert N. McCauley and E.T. Lawson, *Bringing Ritual to Mind: Psychological Foundations of Cultural Forms* (Cambridge: Cambridge University Press, 2002). Also see Armin W. Geertz, 'Cognitive Approaches to the Study of Religion', in P. Antes, A.W. Geertz, and R.R. Warne (eds.), *New Approaches to the Study of Religion*, Volume 2: *Textual, Comparative, Sociological, and Cognitive Approaches* (Berlin: Walter de Gruyter,

power of the harder sciences can transform those into culturally relevant meanings, and indeed explanations of religion that have impact and political relevance on the world stage.

In the spirit of such inquiry this final chapter seeks, first, to account for some recent explanations of human interaction in terms of cognition and evolutionary neuroscience, and second to show how social cognition is transformed through religion at a cultural level as a system in relation to the environment; religion is a form of what we might call bio-sociology.[4] This transformation is echoed in the history of homo sapiens as a move from sign to symbol, suggesting, third, an abductive philosophical claim that life itself comes to articulation through religion; religions are the transformation of bio-energy expressed at an interpersonal level in human face-to-face encounter that is re-articulated at structurally higher levels of religious systems comprising practice, doctrine, narrative, and law. This transformation of human bio-sociology into religion is the way in which civilization seeks to repair the human and to bring us more acutely into life through the integration of higher linguistic consciousness with deeper, pre-linguistic forms of life. Religion as bio-sociology is the transformation of bio-energy at the level of culture, serving to reflexively control the bio-energic source in social cognition, which is the reintegration of higher, alienating brain function with a deeper affirmation of life.

2004), pp. 347–99; C.S. Alcorta and R. Sosis, 'Ritual, Emotion, and Sacred Symbols: The Evolution of Religion as an Adaptive Complex', *Human Nature*, vol. 16 (4), 2005, pp. 323–59; D.J. Slone and J. Van Slyke, *The Attraction of Religion* (London: Bloomsbury Academic Press, 2015); I. Pyysiäinen, *How Religion Works: Towards a New Cognitive Science of Religion* (Leiden: Brill, 2001); Pascal Boyer, *Religion Explained: The Evolutionary Origins of Religious Thought* (New York: Basic Books, 2001); Scott Atran, *In Gods We Trust: The Evolutionary Landscape of Religion* (Oxford: Oxford University Press, 2002); J. Bulbulia et al., *The Evolution of Religion: Studies, Theories, and Critiques* (Santa Margarita, CA: Collins Foundation Press, 2008); H. de Cruz and R. Nichols, *Advances in Religion, Cognitive Science, and Experimental Philosophy* (London: Bloomsbury, 2016); U. Frey (ed.), *The Nature of God: Evolution and Religion* (Marburg: Kubitza, Heinz-Werner, Tectum Verlag, 2016); J.H. Turner, A. Maryanski, A. K. Petersen, and A.W. Geertz, *The Emergence and Evolution of Religion: By Means of Natural Selection* (London: Routledge, 2017). For a good overview of the cognitivist debate see James W. Jones, *Can Science Explain Religion? The Cognitive Science Debate* (Oxford: Oxford University Press, 2016); F. Watts and L.P. Turner (eds.), *Evolution, Religion, and Cognitive Science: Critical and Constructive Essays* (Oxford: Oxford University Press, 2014). On the application of the CSR approach see, for example, Glen A. Hayes, 'Possible Selves, Body Schemas, and Sādhana: Using Cognitive Science and Neuroscience in the Study of Medieval Vaiṣṇava Sahajiyā Hindu Tantric Texts', *Religions*, vol. 5, 2014, pp. 684–99, DOI: 10.3390/rel5030684. Also Timalsena Sthaneshwar, *Tantric Visual Culture: A Cognitive Approach* (London: Routledge, 2015). My own view is that cognition is too restricted as an explanatory framework and we need a wider evolutionary perspective of the kind I have presented here.

[4] A related term, 'bio-sociality', for overcoming the nature/culture divide is probably coined for the first time by Paul Rabinow, 'Artificiality and Enlightenment: From Sociolobiology to Biosociality', in Paul Rabinow (ed.), *Essays on the Anthropology of Reason* (Princeton, NJ: Princeton University Press, 1996), pp. 91–111.

THE FACE-TO-FACE ENCOUNTER

New work in cognitive neuroscience has argued that the human face-to-face encounter is fundamental to the development of the evolution of the human species. A recent development, focused on what has become known as second-person neuroscience from a research team in Cologne, has attempted to understand and explain human social transformation in terms of social neuroscience, which 'has begun to illuminate the complex biological bases of human social cognitive abilities'.[5] What is happening at the neurological level has explanatory power for human behaviour, in particular human social practices and the development of language. Social cognition is the broad field in which brain and social behaviour are linked, and in particular work on the face-to-face encounter has shown how complex processes of interaction between people are at work even below conscious awareness. In particular, there is a mirroring process at a cellular level that occurs in face-to-face cognition reflected at higher levels in imitation of gesture, eye movement, and facial expression. These very subtle signals have evolved in our species to maximize sociality and thereby ensure optimal survival potential. From these signs we directly perceive, rather than infer, information about gender, age, status, and other factors important for decision-making in social situations. From the complexity of information given in the face-to-face encounter, humans have evolved abilities to transform that fundamental process into higher-level forms of cognition and social engagement. The successful building of coalitions 'rests in part on the ability to extract and recall the identities of others' and people who attend to faces can establish 'social control and affiliation'.[6] This ability to recognize and track emotional responses in the face is established from a very early age in the human infant in whom the ability to 'read' faces seems to be an innate ability; babies as young as thirty minutes old can track faces and privilege the mother's face.[7] Eye movements and the gaze in particular are important indicators of group affiliation and the direction of gaze has been an important factor in establishing a joint focus

[5] Leonard Schilbach, Bert Timmermans, Vasudevi Reddy, Alan Costall, Gary Bente, Tobias Schlicht, and Kai Vogeley, 'Towards a Second Person Neuroscience', *Behavioural and Brain Sciences*, vol. 36 (4), 2013, pp. 393–414. Also see related issues in Chris D. Frith, *Making up the Mind: How the Brain Creates our Mental World* (Oxford: Blackwell, 2007) and Ethan Kross and Kevin K. Ochsner, 'Integrating Research on Self-Control across Multiple Levels of Analysis: Insights from Social Cognitive and Affective Neuroscience', in Ran R. Hassin, Kevin N. Ochsner, and Yaacov Trope (eds.), *Self Control in Society, Mind, and Brain* (Oxford: Oxford University Press, 2010), pp. 76–92.

[6] Kurt Hugenberg and John Paul Wilson, 'Faces Are Central to Social Cognition', p. 168, in Donal E. Carlston (ed.), *The Oxford Handbook of Social Cognition* (Oxford: Oxford University Press, 2013), pp. 167–93.

[7] Ibid., p. 169; B. De Gelder and R. Rouw, 'Beyond Localization: A Dynamic Dual Route Account of Face Recognition', in *Acta Psychologica*, vol. 107, 2001, pp. 183–207.

of attention and in developing trust and social cohesiveness.[8] There is a priority to our face-to-face interactions through which we infer the mental states of others and anticipate their future actions that might be crucial to the success of the group.[9]

Work on facial recognition has been going for a long time. Paul Ekman did preliminary research on the cross-cultural expression of emotions, showing that six emotional expressions are commonly recognized, namely happiness, sadness, anger, disgust, fear, and surprise.[10] While fear and surprise are sometimes confused, happiness is not. Faces in Ekman's study are a means of social communication. Happy faces signal affiliation to the group, sad and fearful expressions signal submissiveness, whereas anger signals dominance.[11] That is, the face becomes a sign, bearing social messages and according to the second-person account is always in response to the other within a particular context. The face-to-face encounter is therefore crucial both phylogenetically in the evolution of culture that stems from group interaction and the development of what Tomasello calls shared intentionality,[12] and ontogenetically in the infant's ability to become part of the group.[13] The philosopher Peter Sloterdijk refers to the human face-to-face as 'the species wide interfacial

[8] Hugenberg and Wilson, 'Faces Are Central to Social Cognition', p. 176.

[9] Ibid., p. 176; J. Hyönä Nummenmaa and J.K. Hietanen, 'I'll Walk This Way: Eyes Reveal the Direction of Locomotion and Make Passersby Look and Go the Other Way', *Psychological Science*, vol. 20, 2009, pp. 1454–8; M.L. Smith, F. Cottrell, F. Gosselin, and P.G. Schyns, 'Transmitting and Decoding Facial Expressions', *Psychological Science*, vol. 16, 2005, pp. 184–9.

[10] Paul Ekman, 'Universals and Cultural Differences in Facial Expressions of Emotion', in J. Cole (ed.), *Nebraska Symposium on Motivation 1971*, vol. 19 (Lincoln, NE: University of Nebraska Press, 1972), pp. 207–83; Paul Ekman and W.V. Friesen, 'Constants across Cultures in the Face and Emotions', *Journal of Personality and Social Psychology*, vol. 17, 1971, pp. 124–9; Paul Ekman, W.V. Friesen, M. O'Sullivan, A. Chan et al., 'Universals and Cultural Differences in the Judgments of Facial Expression of Emotion', *Journal of Personality and Social Psychology*, vol. 53, 1989, pp. 112–17; A.J. Fridlund, *Human Facial Expression: An Evolutionary View* (San Diego, CA: Academic Press, 1994); N.H. Fridja and A. Tcherkassof, 'Facial Expressions as Modes of Action Readiness', in J.A. Russell and J.A. Fernandez-Dols (eds.), *The Psychology of Facial Expression* (New York: Cambridge University Press, 1997), pp. 78–102. However, the universality of Ekman's study has been brought into question. See R.E. Jack, Oliver G.B. Garrod, Hui Yu, Roberto Caldera, and Philippe G. Schyns, 'Facial Expressions of Emotion Are Not Culturally Universal', *Proceedings of the National Academy of Sciences of the United States of America*, vol. 109 (19), 2012, pp. 7241–4.

[11] B. Knutson, 'Facial Expressions of Emotion Influence Interpersonal Trait Inferences', *Journal of Non-Verbal Behaviour*, vol. 20, 1996, pp. 165–82; B. Parkinson, 'Do Facial Movements Express Emotions or Communicate Motives?', *Personality and Social Psychology Review*, vol. 9, 2005, pp. 278–311.

[12] Michael Tomasello, *The History of Human Thinking* (London and Cambridge, MA: Harvard University Press, 2014), pp. 80–123. See also David Sloane Wilson, *Darwin's Cathedral: Evolution, Religion, and the Nature of Society* (Chicago, IL: Chicago University Press, 2003).

[13] James E. Swain and S. Shaun Ho, 'Baby Smile Response Circuits of the Parental Brain', *Behavioral and Brain Sciences*, vol. 33 (6), 2010, pp. 460–1.

greenhouse effect'.[14] When we deal with the face-to-face, we are dealing with a fundamental structure of the human.

The scientific study of the human face-to-face interaction is being developed in second-person neuroscience that through experimental data builds models of how we interact.[15] Schilbach and his colleagues have shown that social cognition entails two networks in the brain, the mirror neuron system that allows us to identify someone else as an 'I' and the mentalizing system that allows us to understand someone else as 'she' or 'he'. But models of cognition based on these two systems that give us a grasp of first- and third-person accounts are limited because they presuppose the very theoretical framework that they intend to test. The experiments that demonstrate both the mirror neuron and mentalizing systems are themselves based on an 'isolationist paradigm' in which there is no interaction with those who are being observed. But once interaction between researcher and researched takes place, then this produces different results in our neural networks. The significance of this is that social cognition occurs between people and the very interaction produces differences in brain processing. Schilbach et al. observe that '(a)fter more than a decade of research, the neural mechanisms underlying social interaction have remained elusive and could—paradoxically—be seen as representing the "dark matter" of social neuroscience';[16] the unconscious processes in the brain, seen through neuro-imaging, which remain outside of conscious awareness. This 'dark matter' is the arena of second-person neuroscience in which the interaction of people in the face-to-face encounter affects the neurology between them.

This new way of understanding human cognition through the face-to-face encounter means that the older model of the 'theory of mind' (ToM), whereby we impute mental states in others analogically due to our own, needs to be modified because it is based on a questionable mind/body dualism. ToM entails an *inference* of other minds from behaviour. But rather than a theory of mind, we 'do not need to impute consciousness to others if we directly perceive the qualities of consciousness in the qualities of action'.[17] The second-person approach shows that the understanding of other minds is not inferred but rather a direct response, embodied in social interaction in which brains themselves are mutually changed through the interpersonal encounter. The individual brain is part of a dynamic process and the neuroscientists Di Paulo and de Jaegher have proposed that an Interactive Brain Hypothesis (IBH) is

[14] Peter Sloterdjik, *Bubbles*, trans. Wieland Hoban (Los Angeles, CA: *Semiotext*, 2011), p. 169.

[15] Schilbach et al., 'Second Person Neuroscience', p. 395; Jonas Chatel-Goldman, Jean-Luc Schwartz, Christian Jutten, and Marco Congedo, 'Non-Local Mind from the Perspective of Social Cognition', *Frontiers in Human Neuroscience*, April, vol. 7, Article 107, 1–7; 2013.

[16] Schilbach et al., 'Towards a Second Person Neuroscience', p. 395.

[17] Asch 1952, p. 158, cited in Schilbach, p. 395.

more adequate to the data than the 'mindreading' of ToM.[18] The brain alone is not enough to explain social cognition; we need to take into account bodies in interaction that enact a world through the face-to-face encounter that has developed in human social evolution. That is, the body plays a role in the formation of brain mechanisms that are shaped by persons in interaction rather than simply passively receiving inputs and processing social interaction. Such an enactment model provides empirical support for philosophical positions long since held about the embodied nature of human experience, from Francisco Varella[19] to Merleau-Ponty's work on the body earlier in the last century as an instrument or field of perception giving access to and creating the world.[20]

Second-person neuroscience develops three modes of social interaction. First, that any situation of interaction involves an initiator and a respondent; second, that within the interactive context new shared intentions arise; and third, interaction always entails that social situations have to be understood in relation to a past.[21] Human interactions entail communication, minimally a sender and receiver, and such interaction entails the development of joint intentionality but also beyond this a shared or collective intentionality. Tomasello has argued through experimentation with pre-linguistic children and higher primates that human children are born with an innate tendency towards altruism or, to be more specific, what he calls 'mutualism' in which we benefit from cooperation: the tendency to help others without any immediate reward for the self.[22] This mutualism is conducive to collective intentionality or the movement of the group towards a common goal. Such collective intentionality that developed in our early hominin ancestors is future orientated but also entails reference to a past or to a cultural memory of some kind. We make sense of social situations with reference to a past and a future goal that collective intentionality entails. The common neurological processes generated by the face-to-face interaction are transformed through mutualism into a higher-order collective intentionality in which a group can pursue a common goal (such as locating food or shelter): clearly a strong evolutionary advantage over other primates. Crucial here has been the development of language in the neo-cortex and the central role of language in the success of our species. While the origins of language are still being debated, what is clear

[18] Ezequiel Di Paulo and Hanne de Jaegher, 'The Interactive Brain Hypothesis', *Frontiers in Human Neuroscience*, vol. 6, June 2012, article 163, p. 1.

[19] Francisco J. Varela, Eleanor Rosch, and Evan Thompson, *The Embodied Mind: Cognitive Science and Human Experience* (Cambridge, MA: MIT Press, 1992).

[20] Maurice Merleau-Ponty, *The Phenomenology of Perception* (London and New York: Routledge, 1958), p. 169.

[21] Schilbach, 'Second Person Neuroscience', p. 397.

[22] Michael Tomasello, *Why We Cooperate* (Cambridge, MA: MIT Press, 2009), pp. 46–7.

is that to facilitate this mutualism we needed to develop a mode of communication beyond the here and now, as we saw in Chapter 2.

Second-person neuroscience shows us the ways in which we interact in the face-to-face encounter that in a group situation becomes intensely complex, particularly through eye contact and other expressions, which can be seen in neuroimaging. Our faces imitate each other, as do our brains through neural coupling through which we are able to make judgements about the trustworthiness of members of the group. Making judgements in complexity has been crucial for homo sapiens as a species and judgements occur initially below consciousness. In interactivity the brains of all participants are affected, and this process constitutes a material reality beyond individual participants. Put simply, the social cognition system in the brain (more specifically comprising the mirror neuron and mentalizing systems) is located below the neo-cortex where language is stored. Our social cognition brains, as it were, interact directly with each other beyond linguistic, neo-cortex interaction.

How, then, is this relevant to religion? In recent times there has been much work on connecting religion to cognitive processes in the brain, particularly through to brain imaging technology,[23] and good arguments have been made that cognitive science and evolution do not in and of themselves 'challenge any particular worldview'.[24] Social or narrative rationality, in contrast to colder, scientific reasoning, has been associated with religious belief and also empathy and compassion.[25] But a different mode of research is orientated towards a more interactive approach. To answer this question, we need to introduce new terminology that links the bio-sociology of the face-to-face encounter with higher-level cognitive abilities, namely language in the neo-cortex, and broader historical processes. We might claim that Sloterdijk's 'species wide greenhouse effect' of the face-to-face encounter is the articulation of what might be called bio-energy; the power of life itself coming to articulation through social cognition that has material or physical effects on the brains and behaviours of social actors. The bio-energy of the face-to-face is

[23] See for example the important work of Harvey Whitehouse and Robert N. McCauley, *Mind and Religion: Psychological and Cognitive Foundations of Religiosity* (Walnut Creek, CA and Oxford: Altamira Press, 2005); Harvey Whitehouse and James Laidlaw, *Religion, Anthropology, and Cognitive Science* (Durham, NC: Carolina Academic Press, 2007).

[24] Justin Barrett, 'Cognitive Science, Religion, and Theology', p. 99, in Jeffrey Schloss and Michael J. Murray (eds.), *The Believing Primate: Scientific, Philosophical, and Theological Reflections on the Origins of Religion* (Oxford: Oxford University Press, 2009), pp. 76–99.

[25] Anthony I. Jack, Jared Parker Friedman, Richard Eleftherios Boyatzis, and Scott Nolan Taylor, 'Why Do You Believe in God? Relationships between Religious Belief, Analytic Thinking, Mentalizing and Moral Concern', *Plos One*, vol. 11 (3), 2016, DOI: 10.1371/journal.pone.0149989. However, the two kinds of thinking are mutually exclusive. Anthony I. Jack, Abigail J. Dawson, Katelyn L. Begany, Regina L. Leckie, Kevin P. Barry, Angela C. Ciccia, and Abraham Z. Snyder, 'fMRI Reveals Reciprocal Inhibition between Social and Physical Cognitive Domains', *Neuroimage*, vol. 66, 2013, pp. 385–401.

transformed or recapitulated in religion as a higher-level system that then feeds back to the face-to-face encounter, controlling social interactions, particularly through law. While the face-to-face is clearly a system of communication, what marks out homo sapiens' communication is language. It is all very well focusing on social cognition as face-to-face interaction, but it is human communication through language where distinctively human capacities come into play and in which we see the transformation of the face-to-face into a higher-order sphere of collective intentionality. The social cognition of the pre-linguistic face-to-face encounter is arguably the root of language that is the precondition for religion; indeed, we might say that the face-to-face encounter *as index* becomes articulated *as sign* in language, which then undergoes further transformation *as symbol* in religion.

The transformations of social cognition from index to symbol occur through the plasticity of the brain.[26] The human brain responds rapidly to changing events and grows or diminishes in relation to particular situations. This plasticity is linked to memory that is intrinsic to close relationships and intrinsic to symbolic (distantiated) communication. Memory comes to play an important role in human face-to-face encounters and is intrinsic to such relationships. Transposed to the cultural level, we see the plasticity of the brain operating in personal identity, the sameness of who we are through time or *idem* identity, and in what Ricoeur has called narrative identity or *ipse*, the cultural formation of the person in relation to others.[27]

These two kinds of identity, *idem* and *ipse*, are important to the formation of human persons in providing a sense of sameness that links to the past and a sense of narrative and directionality that links to the future. Both are crucial in the formation of human meaning; as humans, we need *being*, the personal identity or sense of unchanging identity, and *becoming*, the potential for change and the selfhood that entails narrative.

The transformation of the micro-level of the face-to-face at the macro-cultural level occurs through personal and narrative identity both of self and of community. A person's narrative identity is partly formed by the communities to which he or she belongs and the sense of being part of a story bigger than the self. Religions have traditionally located the person in a community, which has traditionally extended beyond the human to cosmological entities, to gods, angels, and demons. The forms of *idem* and *ipse* identity are therefore generated at the cultural level in ritual that transforms or establishes *idem* identity and through *ipse* that is formed in the narratives of tradition and in law where its implementation occurs through enactment, through the body; the

[26] Catherine Malabou, *The Ontology of the Accident: An Essay on Destructive Plasticity* (Cambridge: Polity, 2012).

[27] Paul Ricoeur, *Oneself as Another*, trans. K. Blamey (Chicago, IL: Chicago University Press, 1992), p. 18.

somatic exploration that constitutes the practices of religion[28] and experiences of transcendence[29] that have served to facilitate human evolutionary development.[30]

FROM INDEX TO AUTOPOIESIS

To account for this shift, I need to explicate and clarify a number of terms and offer some reflection on the change from indexicality to structurally higher forms of communication through language and symbol. My claim is that one of the most important constraints on religion is the transformation of the bio-energy that finds expression in the face-to-face encounter. This is a claim within the horizon of evolutionary theory: that religions have developed as human niche construction by building on the human potentialities for survival and intra-species communication. It is also a philosophical claim about meaning and the articulation of life as human self-repair in the niche that is civilization. We can understand religions as higher-level transformations of bio-energy that are meaningful; that is, religions bring the bio-energy of the face-to-face into the realm of meaning and therefore of language. The social cognition in the deeper brain is brought to resonance with the linguistic brain of the neo-cortex. In that transformation we see a shift from systems of communication that are *indexical* to systems of communication that are *signal* through signs and symbols that are not ostensive in any singular way. Indeed, we might see religions as particular kinds of symbol system, as Geertz claims,[31] that function as autopoietic systems that are transformations of social cognition. But to arrive at this point we need to unpack the conceptual development from indexicality, to sign, to symbol and to examine the ways in which the index, the sign, and the symbol signify meaning. These abstract notions must be related to human practices, for if they are anything, religions are above all systems of practice that seek repair and healing.

My first question concerns the meaningfulness of the face-to-face encounter. The answer to this question partly depends upon what kind of communication event the face-to-face encounter is. On the one hand, if we take meaning as arising within non-linguistic contexts, then the face-to-face

[28] John Bowker, *The Sacred Neuron: Extraordinary Discoveries Linking Science and Religion* (London: Tauris Press, 2005), pp. 49, 111, 165.

[29] C. Urgesi, S.M. Aggioti, M. Skrap, and F. Fabbro, 'The Spiritual Brain: Selective Cortical Lesions Modulate Human Self-Transcendence', *Neuron*, vol. 65, 2010, pp. 509–19.

[30] I. Pyysiainen and M. Hauser, 'The Origins of Religion: Evolved Adaptation or By-Product?', *Trends in Cognitive Science*, vol. 14 (3), 2010, pp. 104–9.

[31] Clifford Geertz, 'Religion as a Cultural System', in *The Interpretation of Cultures* (New York: Basic Books, 1973), pp. 87–125.

encounter that is primarily non-linguistic could be meaningful (and animal communication could therefore be meaningful). If, however, we restrict meaning to linguistic contexts, to utterance, then the face-to-face encounter is a different kind of communication. One approach to this problem might be to begin with the old structuralist definitions of index, sign, and symbol that are still helpful. In a somewhat overlooked book, Edmund Leach defines an index as something that indicates something else in a non-arbitrary relationship, A indicates B as smoke indicates fire, whereas with a signum that comprises two subsets of sign and symbol, A stands for B as a result of arbitrary human choice.[32] On this account the difference between a sign and a symbol is one of metonymy and metaphor: on the one hand, a sign bears a metonymic relationship to what it signifies, as the crown is a sign of royalty; on the other, a symbol bears a metaphoric relationship to what it signifies, as a snake is a purely metaphorical (i.e. arbitrary) symbol of evil. Leaving aside the question of whether Leach's scheme is credible, it is useful for our purposes in sharpening the nature of face-to-face communication in relation to language. Clearly a face can be all of these, but the first question we need to address is whether the face-to-face encounter is an index or a signum.

It could be argued that the primary relationship of the face-to-face encounter is as an index. That is, it indicates communicativity between two or more persons, a communication field that can be filled with different contents but primarily filled with the six modes of emotional expression identified by Ekman. The face-to-face in its pre-linguistic form is not a signum because it points to nothing other than itself; to nothing other than the encounter of two or more persons even when there could be a shared intentionality arising within that encounter. The shared intentionality is within or part of social cognition. In its bare, primate form, the face-to-face is a pre-linguistic mode of communication, an index of group belonging and trustworthiness as we have seen, and so stands outside of meaning if we wish to restrict meaning to language, and therefore to signum.

If human behaviour, as Leach argues, can be divided into natural activities of the body such as breathing and the metabolic system, technical actions that alter the physical world, such as digging a hole, and expressive actions that wish to say something about the world, then language functions in both technical and expressive senses.[33] It is in the expressive sense that language is symbolic and the primacy of the pre-linguistic face-to-face is transformed through language both into communication that ensures the practical, more effective realization of shared goals and into communication that transforms language into a symbolic realm. But before we examine this transformation

[32] Edmund Leach, *Culture and Communication: The Logic by Which Symbols Are Connected* (Cambridge: Cambridge University Press, 1976), pp. 23–4.

[33] Leach, *Culture and Communication*, p. 9.

more closely, we need more carefully to examine the relationship of language as sign to the face-to-face encounter as index.

There is a remarkable sociality facilitated through language that is unparalleled in all other species. Even though the face-to-face encounter is fundamental to human sociality, language has transformed the force of this bio-energy such that language is shared and yet it points to me, the possessor of language, at the same time. It recognizes in its very structure relationality and the inevitable sociality of the human alongside the subjectivity that claims language, that claims the indexical 'I' as pointing to me, and that claims the deixic 'here' to indicate my present location and time. As Luhmann observes, language can be used psychically as well as communicatively.[34] In the dialectic between system and person, language plays a crucial role: language in general (*langue*) is part of system yet language in particular belongs to me as speech (*parole*). If the indexicality of the face-to-face forms collectivities through the shared plasticity of the brain that changes in the interactive moment, then indexicality in language itself, pragmatic indexicality, serves to individuate language, to give it a particular voice, my voice. We might say that language transforms index (pointing with finger or eye) into sign where 'pointing' becomes substituted by a linguistic marker indicating particular persons, places, and times.[35] The sign becomes a defining feature of language beyond non-linguistic index in which the sign 'designates that which it means in the context of the use of a particular language'.[36] That is, the sign comprises a signifier and a signified (as de Saussure posited), but more than this, it has a communicative capacity for the person or group that receives it. The structure of signal relation, between signifier and signified, is not dyadic (as for de Saussure) but rather triadic (as for Pierce) in that the signifier signifies a signified for someone or for some community. In one sense language serves to distance us from the face-to-face encounter through the ability to abstract and project into past and future, although in another sense it enhances social cognition through articulating thought about action: what should we do now? What direction should we go in? Where will we find shelter? As Ricoeur has remarked, on the one hand, language as identifying reference tells us about world; on the other, language as speech act changes world.[37] Indeed, for Ricoeur language is a link between agent and action and therefore world. Ricoeur is helpful here in clarifying the issues at stake.

[34] Niklas Luhmann, *Introduction to Systems Theory*, trans. Peter Gilgen (Cambridge: Polity Press, 2013), p. 203.

[35] On deixis see Stephen Levinson, *Pragmatics* (Cambridge: Cambridge University Press), pp. 54–96; Gérard Genette, *Métalepse: De la figure à la fiction* (Paris: Seuil, 2004).

[36] Luhmann, *Introduction to Systems Theory*, p. 207.

[37] Paul Ricoeur, *Oneself as Another*, trans. Kathleen Blamey (Chicago, IL: Chicago University Press, 1992), p. 34.

In order to understand the way in which language as sign is distinct from social cognition as index, we need to understand something of the way in which language signifies. One of the features of language (comprising a series of signs) is that it identifies something for someone or group. Language has identifying reference whereby, in Ricoeur's explication, it makes something apparent to others 'amid a range of particular things of the same type, of which one we intend to speak'.[38] The way in which language signifies, the way signs work, is through mechanisms or procedures, individualization operators, namely definite descriptions such as 'the first man on the moon' (Ricoeur's example), proper names (Felix the cat), and indexicals such as I, you, here, there, now. Definite descriptions convey meaning even when not true, proper names particularize individuals (and so language is differential), while indexicals are always ostensive or pragmatic, operating in particular situations.[39] That is, once we have language a whole range of communicative potential is opened up beyond the face-to-face, and communication in the sense of collective intentionality becomes subtler and also more specific.

We therefore might wish to restrict meaning to the way in which signs signify for someone or some group in this way, through the three individualizing operators. This goes beyond the immediacy of social cognition in allowing communication between social actors about distant events outside of the range of ostensive act, and distant temporal events in the past and the future. With language we have the possibility of metaphor and analogy, the presentation of sequences of events and so the opening up of narrative, the ability to articulate structural relationships within the society, and the ability to articulate reasons for action. This formation of relationships beyond the immediacy of social cognition transcends individual actors and the complexity thereby created takes on a life of its own, so to speak, in which the structure or system becomes self-producing. Using the term *autopoiesis* coined by the biologists Maturana and Varella, Luhmann applies it to social systems that he understands as self-regulating in relation to the environment in which they occur. Maturana saw autopoiesis as a defining feature of living systems that accounted for the way in which they are closed but also self-generating and in some relationship with their environment. He writes:

> A closed network of molecular productions that recursively produces the same network of molecular productions that produced it and specifies its boundaries while remaining open to the flow of matter through it, is an autopoietic system, and a molecular autopoietic system is a living system.[40]

[38] Ibid., p. 27. [39] Ibid., pp. 28–9.

[40] Humberto Maturana Romesin, 'Autopoiesis, Structural Coupling and Cognition: A History of These and Other Notions in the Biology of Cognition', p. 7, *Cybernetic and Human Knowing*, vol. 9 (3–4), 2002, pp. 5–34.

This is a specific, physical process that defines living entities in terms of molecular structure and interaction, and Maturana claims that autopoiesis is only this and no more: it refers to a biological, molecular process.[41] In identifying autopoiesis as the basic mode of cellular organisms to succeed or not, Maturana seems to be identifying a structure of life itself, the way in which bio-energy transforms itself either remaining in a condition of stasis or moving to a different condition constrained by its environment. The structural coupling between organism and environment is necessary for the articulation of the organism's bio-energy and controlling its outcomes.

But while Maturana might disclaim the extension of his discovery to other spheres of human behaviour, particularly to the human institutions and structures we inhabit, the extension of the idea by Luhmann to social systems is more than analogical or metaphorical in that social systems can be understood in terms of self-contained structures that seem to be self-generating in relation to an environment. In a different language (not used by Maturana or Luhmann) we might claim that the bio-energy of life itself, expressed through the autopoiesis of the molecular level, is transformed into a higher-level autopoiesis as social system, which is a kind of niche construction. The same mechanism is at work in the macro-level structure as in the micro-level organism.

A social system is an 'operationally closed' system of communication or niche construction that is self-replicating, adapting to its environment to which it is 'structurally coupled'. For Luhmann this 'environment' is the person, the 'psychic system' who stands outside of the social system but who is related to it through structural coupling. Persons can therefore influence the social system (because they stand outside of it as environment) and in turn the social system can influence persons. Indeed, it is language that is the operative force of structural coupling in linking person to systems defined in terms of communication. The defining feature of a social system is that it is communicative; a social system replicates itself through the generations in a process of communication that involves, says Luhmann following Wil Martens, information, utterance, and understanding. Sociality is a consequence of the fusion of these three components.[42] Persons, Luhmann's 'psychic systems', relate to social systems through a process in which there is a 'structural coupling' between them: the person as self-generating system exists within the social system that feeds back to the person in a process of reinforcing status and norms. Society as a process of communication stands outside of persons as its environment and yet is changed by persons who live within it.

[41] Maurana, 'Autopoiesis', p. 7. Yet Luhmann seems to agree with this when he writes that autopoiesis 'is a principle that can be realized only in living cells and only as life' (*Systems Theory*, p. 192), while yet making autopoiesis central to his theory of society and communication.

[42] Luhmann, *Systems Theory*, p. 192.

Language is the medium of structural coupling in that it links person to society through allowing a person to reflect on what s/he is going to say, uttering, and simultaneously being understood. The function of communication within the social system is the reduction of complexity and so meaning, the heart of communication, is in fact about the reduction of complexity[43] that allows for the structure to operate and replicate itself. Meaning on this account is produced through the systems of signs that is language and this functions to reduce complexity that has arguably evolved to maximize species survival; a selective process that blocks out overwhelming and unnecessary information coming from the world. Although meaning is highly complex, on this view it has evolved to reduce complexity in the interests of system survival.

With the emergence of social systems facilitated through the communication that language allows, we have transcended the immediacy of social cognition with the generation of social systems as autopoietic. The indexicality of the face-to-face encounter is transformed with the development of sign into language that creates the metaphorical space for the creation of meanings that reach back into the past and project a future. With the extended temporality that language allows, the range of the way language signifies through definite descriptions, proper names, and indexicals—Ricoeur's individualization operators—we have the ability to create socialities rooted in social cognition but transformations of it. In particular, two modes of language develop that transform the sign into symbol, namely narrative and law, both of which are intimately connected to action and the formation of practices. Narratives about the group give social actors identity and allow human beings to locate themselves in relation to others and in relation to the wider environment or cosmos, while law ensures group cohesion and conformity to group norms that is conducive to the flourishing of the majority. These linguistic modes are linked to practices in that stories and frameworks for behaviour that we might call law are enacted or performed. It is this combination of narrative and law through language, enacted though embodiment in the ritual act and the moral act, to which we take the term 'religion' to refer. In particular, law is a sanction that limits what is permitted and in this sense is sacred.

On this view, the origin of religion lies in a system of autopoiesis, a system of communication that simultaneously expresses the system's orientation and articulates its boundary with the environment. Religion develops from the language facility of the neo-cortex expressing the social cognition of the deeper brain. Thus, with religion as an autopoietic system, the boundary is determined by practices that are themselves suffused by a worldview and these practices and philosophies of life are structurally coupled with the environment. By environment I mean a world governed by material causation and

[43] Niklas Luhmann, *Theory of Society*, vol. 1, trans. Rhodes Barrett (Stanford, CA: Stanford University Press, 2012), p. 23.

also, in Luhmann's terms, a person who practises the religion. That is, religion is a realm of meaning distinct from the ambient environment of the world that it simplifies. It creates a boundary between itself and the infinitely complex world and also, one might add, the infinitely complex person. Structural coupling between system and environment allows the control of environment through religious law that controls the environment of persons and world through regulating the modes of its technological manipulation: civilizations tell us how to dispose of the dead, how to grow food, and what technologies are acceptable.

Luhmann himself has written a systems theory of religion. His question in this book is 'what lets us recognize religion?'[44] His answer is that we recognize religion as a form of communication, as a code for dealing with complexity that functions as a differentiated social system in providing meaning (and so the reduction of complexity).

This in many ways is a compelling account in that religion must be distinguished from environment, functioning as a realm of meaning to reduce complexity. But there are problems with it. One of Luhmann's theoretical goals is to replace an account of religion in terms of humanity with an account of it in terms of communication, thereby replacing a theory of religion centred on anthropology with one centred on society.[45] Indeed, we might link autopoiesis with the bio-energy of the face-to-face relation insofar as such an autonomous system nevertheless derives its power from the biological interaction; religion on this account is a bio-sociology.

While the logic of this is coherent within Luhmann's systemic account, it is nevertheless problematic in demoting the importance of the human. Part of this is the claim that sociology is the discipline par excellence that can explain religion and through privileging communication over humanity, Luhmann highlights the importance of social systems over individual persons or interpersonal relationships. This, of course, is the classic debate about what takes precedence, the individual or society, but in placing emphasis on religion as a system, Luhmann misses a fundamental function of religion, namely to provide meaning to persons and to offer structured wonder. That is, while a social system emphasizes the important role of abstract meaning in a communication system, it is persons who nevertheless communicate and religions open a world of experience that is beyond the common sense, beyond mundane everyday activity. This might be simply saying a prayer in a field over the crops or a more ecstatic adventure of a pilgrimage. We therefore need phenomenology that draws on the biological sciences, not only sociology, to account for religion. The explanation of religion grounded by the biological

[44] Niklas Luhmann, *A Systems Theory of Religion*, trans. David Brenner (Stanford, CA: Stanford University Press, 2000), p. 3.

[45] Ibid., pp. 5–7.

sciences of social cognition needs to take account not simply of social system but of persons, their intentions, forms of inwardness, and practices. That is, we need to account for the ways in which people internalize and transform the triangle of narrative, law, and practice that comprises religions within collective life. The bio-sociology of religion needs to be related to person and purely autopoietic account brought back to the realm of human encounter and meaning.

Nevertheless, if we bring meaning back from social systems, and so from the realm of sociology, to the person or interpersonal communication, and so to the realm of hermeneutical phenomenology, we are still dealing with language at the heart of both system and person. If we take from Luhmann that religions understood as autopoietic systems are concerned with meaning, and so with language, we can concede the important point that meaning is the reduction of complexity that functions to support evolutionary development. But this is paradoxical to some extent. On the one hand, human communities are faced with the overwhelming complexity of the world, with information coming into the brain through the senses, and so there is a pressure to control the flow of that information and arguably language and meaning function to control this pressure. Yet on the other hand, language creates the opportunity for the development of a further complexity. We might say then that while meaning might reduce the complexity of the world, it increases the complexity of imagination. Through the temporal extension that language allows, religion creates a metaphorical space for the development of meanings that would otherwise not become evident. Thus, through the power of language religions can formulate the meanings of death, formulate accounts of the origin of the world and the purpose of life, and formulate accounts of transcendence. All this is achieved through the capacity of language to signify both in terms of sign, where the arbitrary relationship between the signified and signifier is articulated through the modes of individualization operators, and in terms of symbol, where the relationship between signifier and signified is semantically dense and multivalent. The narrative power of language allows communities to develop a more complex shared intentionality beyond the present or immediate future and so allows the development of identities and the articulation of formalized relationships through identifiable roles (hunter, mother, child, dominant male, healer, and so on), which the system can define and control. But these roles are of persons who use and control language in an act of radical freedom that can affect the social system. Indeed, every utterance of persons is the demonstration of radical freedom to choose particular words, to say particular sentences, and to express particular sentiments. A social system as such does none of these things, which can be brought to light through phenomenology, although utterances are always historical and situational and so hermeneutical as well.

We began with persons in interaction as imaged in social cognition where indexicality rules over signification. If indexicality is a pure communication event, then signification is a communication event with meaning. Once we have the signifying system of language, we have the development of symbolic systems and of enacted narrative and law. Thus, while Luhmann's account of meaning as the reduction of complexity has credence, it is the reduction of a certain kind of complexity, the complexity of world experienced as environment, but that opens to a new kind of complexity; the complexity of imagination and the power to image and represent what is not immediately present. And it is persons rather than systems that enact this imagination.

To recap our findings so far: in human evolution the primary social encounter has been face-to-face social cognition that we experience in the present but that reaches deep into our hominin past and is located in the deep brain, the importance of which we can appreciate ontogenetically in the development of children's social capacities, even in newly born infants. This capacity for sociality expressed through indices of bonding (such as facial expression and pointing) becomes transformed with the development of language into a system of signification that can recapitulate the past and project a future. With the system of signs, i.e. the arising of language, we have the further development of more complex signification through symbol; the ability of language to represent complexity by creating group narrative and by creating a system of acceptable behaviour conducive to the common good (or at least to the ruling elites) that we might call law. All this is developed from the neo-cortex where language is located. Moreover, such systems of signification are never purely abstract but encoded in forms of action; stories are told and laws are enforced; above all, self-enforced.

We might say that religion is the mode of group sociality that transforms the bio-energy of the face-to-face into higher-level signifying systems of narrative and law that reflexively control the face-to-face encounter, that channel the desire for life, and that seek to address alienation from life. With religion we have emphasis on identity and group belonging that is reinforced through strict laws of human interaction; religions are deeply concerned with reproduction, who we can and cannot have sexual relationships with, and with alimentation, what we can and cannot eat. While some religions are more concerned with these things than others—we think of Judaism and Islam on kosher and halal food, Hinduism, Jainism, and Sikhism on vegetarianism, caste systems, and codes of sexual morality in Buddhism and Christianity—all religions seek to control these fundamental modes of human interaction and survival. The control of food and sex through religious law has been crucial to the modification of human sociality and the development of civilization.

Religions have been extraordinarily successful as transformations of human bio-energy into culture, more successful than the smaller social unit of the family that is necessarily restricted to an immediate environment, more successful than social units such as a tribe or clan, and more successful than more recent nationalisms that are restricted to a particular geographical location. Secular ideologies can and have functioned as transformations of social cognition and so the remit of comparative religious studies is not simply restricted to what we have traditionally understood as religions, but extends to other patterns of social and political transformation whose forms are complex in complex contemporary societies. Religions and their analogues provide us with an elaborate social apparatus through ritual and ethical codes expressed as law that sustain the vulnerability of the face-to-face while at the same time transforming this into a corporate identity (that often defines itself in relation to other corporate identities).

PHILOSOPHICAL IMPLICATIONS

The general thesis that religions are the transformation of the bio-energy of face-to-face social cognition is, as we have seen, an empirical/historical claim about bio-sociology, but also a philosophical claim insofar as we can begin to articulate its implications for the way we understand the human; there is an implicit philosophy of humanity here which is also a philosophy of life. The data of social cognition and its transformation in religion implies a realist ontology that we can understand in terms of a non-dualist vitalism. The idea that religions function to provide social bonding is, of course, close to Durkheim, but differs from Durkheim in drawing the claim from recent scientific knowledge of social cognition and wishing to emphasize not only sociology and the group function of religion but also phenomenology and the individual experience of religion. Similarly, the position is close to Luhmann insofar as religions can be seen as autopoietic communication systems that are self-regulating and rooted in human biology, but again wishing to bring religion back from a purely systems theory approach to an emphasis on the person (Luhmann's 'psychic systems') and the second-person relationality thereby entailed. We must necessarily bring explanation into the realm of phenomenology, but a hermeneutical phenomenology that recognizes the particularity of location in space and time and the particularity of the interpersonal.

There is then a symbiosis between religion as system, as a higher-level communication-generating and bonding process, and religion as personal experience, although always historical, that enacts or performs the practice of religion, transforming bio-energy into action. Because of the necessity of historical location, the need to begin the study of religion with the historical as Heidegger insightfully observed,[46] experience is always necessarily interpreted.

[46] Martin Heidegger, *The Phenomenology of the Religious Life*, trans. Matthias Fritsch and Jennifer Anna Gosetti-Ferencei (Bloomington and Indianapolis, IN: Indiana University Press, 2004), pp. 38–9.

If the face-to-face encounter is the primary human experience biologically rooted in mother–child connection, then this is always mediated through language (as Ricoeur teaches us) and through the mechanism of the higher-order system (such as religion). There are a number of philosophical implications here, perhaps first the phenomenological and post-phenomenological understanding that we are embodied, both mind and body in world (as many philosophers and cultural theorists have now emphasized, particularly Merleau-Ponty, Streets-Johnson, Varela, and Csordas). Second, in an embodied understanding of religion, language is central as the medium of communication, particularly the mechanism of identification between person and system. Third, religion as the transformation of bio-energy entails a realism and a materialism that articulates with contemporary science but that maintains continuities with the cultural philosophies. Religion as social cognition is in sympathy with new realist philosophies now emerging and re-establishes seeing religion in terms of process, in terms of life itself, and in terms of human transformation. I need to say something about them here.

(1) That mind and body are integrated within a single system or person is widely, although not universally, accepted in philosophy and neurobiology.[47] The structure of human neurobiology is reflected in consciousness and consciousness reflected in our neurobiology. This is hardly surprising but important insofar as the integration of mind and body in relation to religion points to the importance of practice; that action affects mind or rather action and cognition form an inseparable unity. In his later thoughts Husserl recognized the importance of the lifeworld that consciousness was part of, but that could reflexively stand outside and inquire into its origins. The possibility of such true distanciation was questioned by later phenomenologists, particularly Heidegger and Merleau-Ponty, although even their thinking has come under critical scrutiny for remaining theoretical and not pragmatic enough in their understanding because reflection, for Varela and his colleagues, is itself experience.[48] This view is in consonance with Deleuze's field of immanence previously discussed, where each particularity, being integrated into a system, speaks with a single voice, a univocity. Within this cluster of philosophical views—integrationist and immanentist philosophies of life—there is generally a strong

[47] Anthony Chemero, *Radical Embodied Cognitive Science* (Cambridge, MA: MIT Press, 2011). Alfred Mele, *Effective Intentions: The Power of Conscious Will* (New York: Oxford University Press, 2009); Alva Noe, *Action in Perception* (Cambridge, MA: MIT Press, 2006); Maxine Sheets-Johnstone, *The Corporeal Turn: An Interdisciplinary Reader* (Charlottesville, VA: Imprint-Academic-Dotcom, 2009); Varela et al., *The Embodied Mind*; Adam Zeman and Oliver Davies, 'A Radical New Way to Understand the Brain', *Standpoint*, Sept., 53–5, 2013.

[48] Varela et al., *The Embodied Mind*, pp. 18–19, 27.

non-dualist claim about the integration of consciousness and neurobiology as embodied act. This is important for understanding religion as primarily focused on practice that is considered transformative of person and community. Through liturgical movement, through prayer, fasting, and meditation, through pilgrimage, and through social action people enact religion and integrate themselves into religion as system.

(2) Second, in the integration of embodied person into system—another way of putting this would be the appropriation of system by person—language forms a crucial dimension. Through language the integrated mind–body system establishes a deep connection to tradition. Language is shared by everyone: it is an objective system (*langue*) that is also my own (*parole*), as the Structuralists long ago observed. One of the important ways to understand this is through pragmatics, the inquiry of the ways in which language functions in the world. Of particular importance is indexicality and the way the first-person pronoun brings with it the identification of person with tradition, which is crucial to identity formation. Urban has convincingly argued for the importance of the 'indexical-I', the first-person pronoun used by all human beings, that in different social circumstances comes to be identified with the 'I of discourse', the 'I' implicit within text or tradition. In Urban's account there is a spectrum of use of the first person from ordinary indexicality indicating, for example, desire ('I want that') to possession by a god in which indexicality is overwhelmed by another presence, by another 'I'.[49] Between the indexical-I and possession are various degrees of identification of the first person with the I of discourse, thus we have theatre in which the indexical-I is partially subsumed by the I of discourse (the actor takes on the role) or poetry in which the indexical-I identifies with the poem. In one sense we are seldom ourselves, much of the time being identified with an I of discourse. This is important because it shows the mechanism whereby identities are formed. Strong identification of the I with tradition can lead to a person, perhaps, undertaking roles that they would not in other circumstances. The overwhelming of the indexical-I by a narrative of martyrdom might lead someone to become a suicide bomber, for example.

(3) The orientation presented here thirdly implies a realism and philosophical materialism that is not positivism but nevertheless takes seriously materiality and responds to cultural philosophies of postmodernity of a previous generation. The realism of understanding religion in terms of

[49] Greg Urban, 'The I of Discourse', in Benjamin Lee and Greg Urban (eds.), *Semiotics, Self and Society* (Berlin and New York: Mouton de Gruyter, 1989), pp. 27–51. I have written on this elsewhere. See Gavin Flood, *The Ascetic Self: Subjectivity, Memory and Tradition* (Cambridge: Cambridge University Press, 2004), pp. 218–22.

social cognition is not a crude materialism. Rather, it implies that the bio-energy of life itself is articulated through human social cognition. This is a position that seeks to meet religion and explain religion on its own level through maintaining that this understanding cannot be reduced to biology, but neither can it be reduced to consciousness. The true understanding of religion lies in this realm of embodied cognition that exceeds both genetics and consciousness because it operates at a higher level of system. The bio-energy of life is a force that cannot be explained purely biologically, but nor can it be accounted for in consciousness because it clearly exceeds it. Religion as the transformation of bio-energy entails a new mode of explanation that draws on both the natural and human sciences. Indeed, there could even be theologies of embodied action that assume the transformation of bio-energy that we are discussing. We might say religion is a way that life thinks itself, a reflection not restricted to consciousness but articulated through action and practice in a process that is open because enacted through time. Terry Pinkard makes a point from Hegel, that religions are forms of self-reflection through symbols and stories, through which, for Hegel, the spirit knows itself as spirit. The form that religious reflection takes, its representation (*Vorstellung*), which includes religious practice, is a way of moving 'individual agents out of their natural, egoistic state (their merely personal point of view) into a communal state in which they accept reasons, obligations and proscriptions that for them place objective limits on belief and action'.[50] In modernity this, for Hegel, is a move from the 'unhappy consciousness' of the pre-modern world that has lost faith in itself to the modern secularity of a religion of morality, an insight that reflects the shift to a regnant disenchantment in the modern North Atlantic, along with the possibility of a transformed sociality in the future. If religions contain closed eschatologies (and arguably they do), then understanding religion as the articulation of life itself entails a challenge to traditional religions in implying an open eschatological explanation of religion. An infinitely open universe, as contemporary physics claims,[51] requires an infinitely open notion of life that can account for the eschatological structures that have transformed that bio-energy of life and that have been necessary structures in the development of the human experiment.

[50] Terry Pinkard, *Hegel's Phenomenology: The Sociality of Reason* (Cambridge: Cambridge University Press, 2014), p. 223.

[51] Geraint F. Lewis and Luke A. Barnes, *A Fortunate Universe: Life in a Finely Tuned Cosmos* (Cambridge: Cambridge University Press, 2016), 'the universe will never run out of inflating space' (p. 298).

In sum, to return to our opening question of how we can account for the persistence of religion, we can say that religions are still forms of life that give meaning to human communities and provide ways of living and dying.[52] They form and influence life in the unrepeatable nature of the ethical act—they constrain people's choices—and in the non-identical repetition of the ritual act—they lay claim to the social space enacted in repeated acts of worship and meditation. That the churches are emptying in much of northern Europe does not indicate the demise of religion; indeed analogous forms of culture will inevitably develop that lay claim to the face-to-face and attempt to control social cognition. Of course, other societal patterns can do this—even secular practices of socialization such as sport or the shared experience of plastic art or music—but only religions are cultural forms with historical depth and semantic density that to date have successfully negotiated and developed the primacy of the human encounter. It seems to be the case that other forms of cultural life are developing in which people choose different aspects of religion to engage with and people might participate in more than one religion in the modern world.[53] If religions are central to human communities because they stem from life itself and reflexively control social cognition, then religious studies as a critical discipline is crucial for understanding these cultural forms, both in terms of cognition and the processes in the brain that accompany religious activity, and in terms of cultural production and the representation of communities and people.

[52] Gavin Flood, *The Importance of Religion: Meaning and Action in our Strange World* (Oxford: Blackwell, 2013), pp. 53–75.

[53] Sondra Hausner and David Gellner, 'Category and Practice as Two Aspects of Religion: The Case of Nepalis in Britain', *Journal of the American Academy of Religion*, vol. 80 (4), 2012, pp. 971–97.

Epilogue

Modernity and the Life of Holiness

I have told a long story about the category of life itself, of the human desire for life, of human longing, of the need for repair, and of how civilizations have sought to address this need through the religions that have driven them and the philosophies that have guided them. The endpoint we have arrived at, although not the end of the story, is an ending in which life itself articulated through civilizations can be understood in terms of repair that can be envisaged as a kind of holiness of life, a bringing of human reality into an intensity of life, and a repairing of shattered communication. This intensity is human integration of life itself into modes of culture at the level of linguistic consciousness. We might say that civilization and the religion that drives it have bridged the evolutionary gap between a pre-linguistic mode of being human and the linguistic one facilitated through the neo-cortex articulated in the structures of civilization. Reflecting this distinction, I have attempted to integrate two modern accounts of human life into a coherence: on the one hand, a tradition of humanist, particularly philological, scholarship on traditions within the broad parameters of Indic, Chinese, and European/Middle Eastern civilizations, and on the other, a tradition of scientific, particularly evolutionary discourse about human life. It seems to me that we need both humanism and science to offer descriptions adequate to the complexity of life in human history and the ways in which civilizations have attempted to repair the human condition. These are distinct modes of description within which to frame both the constraints on human reality along with its freedom. Throughout this story we have seen how the human desire for life has been commonly recognized as a deep human trait and how this desire is transposed as a longing for completion, fulfilment, or even redemption. The religions directly address this desire that is related to a desire for meaning in life. But what are we to do once religions are no more? What cultural forms address the desire for life and the need for repair? What is the human future without religions?

It seems to be certain that in what Charles Taylor has called the North Atlantic region, traditional religions, particularly Christianity, are in rapid decline and the churches are emptying.[1] This is not the case, however, outside of this world in Asia, South America, and Africa where Christianity and Islam in particular are growing. Nor must we forget the individualistic spirituality in the North Atlantic world developing in the wake of Christianity's decline.[2] I have argued in this book that civilizations have sought repair of the human through the great edifices of religion and have claimed that an important explanation of this is rooted in our evolutionary history and human niche construction. Civilizations have developed as a kind of niche for the protection not so much of the genotype as the phenotype as the driver of evolutionary change. The linguistic creatures that we are, allowed by the development of the language-bearing neo-cortex, has been at odds with the deeper and older social cognition that takes us more immediately into life. The repair that civilizations seek is also to address the desire for life itself and to bring linguistic consciousness into harmony with social cognition. Once religions cease, there is nevertheless still the need for repair and human fulfilment, the human longing for meaning and place. In a thought-provoking book Marcel Gauchet sees religion in terms of a negation of an established order that has allowed freedom to critique the political and has developed a strong notion of the individual. But even for him, 'the religious after religion' will still have a place within human subjectivity where it functions to provide meaning in the face of the discontinuity of religion's social function. This will be 'a major source for cultural innovation'.[3] But it is not clear how the post-religion world will address deeper questions of human meaning and belonging; the human desire for life in the face of death that erases any trace of our presence. The life of holiness that all religions have conceptualized in some form as something to be cultivated will remain, although traditional lives of holiness will be lived through individual choice in the secularized world rather than as an institutionalized goal.

The question of life itself in relation to the living along with the human desire for life has shifted in modernity from the imaginal realm of mediation between human and divine fostered in religions, in the Religious Cosmic Model (RCM), to a contraction of imagination or rather a re-configuring of the imaginal through technology, and a dominance of 'the screen' as a primary mode of human interaction with the world and fellow humans (in the Galilean Mathematical Model (GMM)). The contemporary, disenchanted citizen of the globe is a buffered self that has largely replaced the cosmological, porous self of

[1] Charles Taylor, *The Secular Age* (Cambridge, MA: Harvard University Press, 2007), p. 1; Steve Bruce, *God Is Dead: Secularization in the West* (Oxford: Blackwell, 2002).

[2] E.g. Paul Heelas and Linda Woodhead, *The Spiritual Revolution: Why Religion Is Giving Way to Spirituality* (Oxford: Blackwell, 2005).

[3] Marcel Gauchet, *The Disenchantment of the World*, trans. Oscar Burge (Princeton, NJ: Princeton University Press, 1997), p. 200.

pre-modernity.[4] Certainly rapid developments in technological innovation have led some thinkers to envisage a post-human future encapsulated by one of the founders of this ideal, Donna Haraway's concept of the human as cyborg in which the organic and the mechanistic are fused,[5] and in a condition almost in which questions of meaning are redundant. This vision of the future is itself a proposal for human self-repair. More immediately and more practically, developments in science raise questions about human meaning through new possibilities in biomedicine. Helga Nowotny raises important questions about reinventing the human in the molecular age and about human value in relation to bioethical questions. In contemporary research, life has a 'new visibility' through molecular biology that allows interventions into life processes, such as cloning or the transfer of cell nuclei, which would not have been possible, or thinkable, in the past. In the new world of the molecular, old distinctions between knowledge and application, science and technology, are outdated: 'Under the hegemony of the molecular glance, knowledge has become action. Today the fact is that *understanding life means changing life*'.[6]

It is possible, thinks Nowotny, that we stand on the brink 'of an epochal rupture in the history of humankind'[7] in that the rapid developments in assisted reproductive technologies mean that for the first time humans have control over their future genetic inheritance, and drugs are developed to increase happiness and achievement, from cognitive enhancers such as Modafinil and Ritalin to modifiers of somatic states such as Viagra and Prozac.[8] These developments that offer a biochemical route to a better life are parallel in many ways to the goal of human self-repair. Perhaps with the demise of traditional religions in the secularized North Atlantic, new technologies (other than prayer, fasting, and liturgy) seek to enhance life and to bring us into deeper contact with life. The distinction between vitalism and mechanism breaks down in the molecular age. These new technologies challenge traditional religions' solutions to the need for self-repair through re-conceptualizing what a person is and through new imaginings of the universe at a molecular level. The universe that science took over from the cosmos of religion, beginning in the seventeenth century, has now reached a point of modelling reality and changing reality in radically new ways that undermine old certainties and dualities

[4] Taylor, *The Secular Age*, pp. 37–42.

[5] Donna Haraway, 'The Cyborg Manifesto', in *Simians, Cyborgs and Women: The Reinvention of Nature* (London: Free Association Books, 1991), pp. 149–81. On the worldview implicit in Haraway's work see Richard Roberts, *Religion, Theology and the Human Sciences* (Cambridge: Cambridge University Press, 2002), pp. 269–91. For the continuity of nature and artificial life and its hope for the future see Craig Venter, *Life at the Speed of Light: From the Double Helix to the Dawn of Digital Life* (London: Little, Brown, 2013).

[6] Helga Nowotny and Giuseppe Testa, *Naked Genes: Reinventing the Human in the Molecular Age*, trans. Mitch Cohen (Cambridge, MA, London: MIT Press, 2010), p. 5.

[7] Ibid., p. 6. [8] Ibid., p. 19.

(such as culture vs. nature, male vs. female, human vs. animal). Bioethical discourse has become a political process that 'undermines the cultural plausibility of the code of meaning that religions can offer'.[9] The emergent new meanings generated by new research inform a public that in a globalizing world people can become citizens who imagine life in new ways and whose desire for life is transformed into specific wants for life enhancement. New technology allows us to be 'somatic individuals'[10] in a world in which traditional eschatologies have been abandoned by the educated elites and freedom of choice—and choices that are not trivial but potentially significant in life enhancement—is privileged over conformity to normative, religious regulation.

Nikolas Rose has called the contemporary potential for transformation of the human condition 'ethopolitics', 'the self-techniques by which human beings should judge and act upon themselves to make themselves better than they are'.[11] That is, ethopolitics aspires to human self-repair perhaps as the latest development in civilization to address this need. Although traditional religions may pass away—and religion itself could be a purely historical phenomenon with a beginning and an end, as Gauchet thinks[12]—then there still remains the need to address the question of satisfying the desire for life or healing the broken consciousness of human lives. Ethopolitics has developed to address these questions of the meaning of life, and coalesces around a new vitalism that sees value in life itself. Rose expresses the issue quite well when he writes:

> While ethopolitical concerns range from those of life-style to community, they coalesce around a kind of vitalism, disputes over the value accorded to life itself: 'quality of life,' 'the right to life,' or 'the right to choose,' euthanasia, gene therapy, human cloning and the like. This biological ethopolitics—the politics of how we should conduct ourselves appropriately in relation to ourselves, and in our responsibilities for the future—forms the milieu within which novel forms of authority are taking shape.[13]

These novel forms of authority are not the traditional authority of religions but experts in various scientific technologies and a range of 'somatic experts' keen to offer advice on living life to the full. Developments in expertise that has been emerging over the last half-century in the West are sometimes linked to new forms of religion and spirituality, such as what Paul Heelas has called

[9] Ibid., p. 61, citing Danielle Hervieu-Léger, 'The Role of Religion in Establishing Social Cohesion', in M. Michalski (ed.), *Religion in the New Europe* (New York: Central European University Press, 2006), pp. 45–63.

[10] Nikolas Rose, *The Politics of Life Itself: Biomedicine, Power, and Subjectivity in the Twenty-First Century* (Princeton, NJ, Oxford: Princeton University Press, 2007), p. 26.

[11] Ibid., p. 27. [12] Gauchet, *The Disenchantment of the World*, pp. 21–2.

[13] Rose, *The Politics of Life Itself*, p. 27.

'the self-religions' and the New Age focus on self-actualization,[14] in which the orientation is to the individual over the collective, to ease of life over ascetic striving, to experience over conceptualization.

Such new forms of authority, de-centred and individualistic, assume a new kind of political order and perhaps an emergent global community or global civilization that is probably moving forward inexorably, in spite of counter-globalization forces and a politics of retrenchment that is a worldwide phenomenon at the time of writing. If we can analyse political orders into ideal types, then we might argue that there are two axes or lines of demarcation along which political systems can be plotted. A horizontal axis of political power or political system that denotes the degree of the centralization of power such that there could be strong or weak centralization, and a vertical axis of the law or legislative system that denotes the degree and strength of legislation. There is sometimes conflict or tension between the two axes of power. Thus, according to such a grid, where there is a system of strong centralization and strong legislature we might call an autocracy—or a theocracy—in contrast to a system of weak or no centralization and weak legislature that is really an anarchy. Peter Kropotkin's optimistic view of human nature saw this eradication of the state and of law to be the ideal utopian community, whereas in reality we have dysfunctional states where the law of the strongest rules. Strong centralization and weak legislature likewise results in a somewhat dysfunctional state as centralized power, as in a tyrant or dictator, is not regulated or controlled by law. The opposite here would be a strong legislature combined with weak centralization in which power is distributed across a field of local networks. This is democracy in varying degrees.[15]

It seems to me that weak centralization along with a fairly robust legislature is conducive to pluralism and the kind of de-centralized authority that Rose implies. The emergent global civilization would need strong legislation to regulate excesses of free market capitalism, but would also need to be de-centralized to allow power for local communities to make decisions about cultural values and ways of living. Such a system might augur contraction of the importance of the nation state and a re-affirmation of the importance of the city, as Aristotle thought. In terms of the theme of life itself and the repair of human life, it could be that weak centralization and strong legislation allow for human enhancement within a legal framework that protects human

[14] Paul Heelas, *The New Age Movement* (Oxford: Blackwell, 1996), p. 31. See also Heelas' stimulating account of the link between contemporary spirituality and life: Paul Heelas, *Spiritualities of Life: New Age Romanticism and Consumptive Capitalism* (Oxford: Wiley-Blackwell, 2009).

[15] The literature on religion and politics is of course vast, but a good survey article is John Madeley, 'Religion and the State', in Jeffrey Haynes (ed.), *Religion and Politics* (London: Routledge, 2009), pp. 174–91.

rights. Problems emerge, of course, over who decides the legislation within a particular state and what controls the legislation. In the realm of biotechnology, religious constraints are clearly an important issue and religions are important stakeholders in the formation of legislation that affects people's lives, from legislature about stem cell research to abortion laws.

Living a life of holiness to facilitate human repair within this context is a challenge for those committed to particular religious views, but the broader idea that human beings seek self-repair through the structures that have developed to nurture their communities is salient. Religions are systems that unify, yet they are exclusive; but the social cognition system that religions access is universally inclusive. Indeed, I have made a case for seeing civilizations as human niche constructions for human self-repair that have evolved from our biology, but that exceed that biology. The affirmation of life as more than simply survival, or even as more than simply pleasure, has been a constant struggle through history, and we have seen how this struggle has played out in the histories of three civilizations and has also been articulated in secular philosophies of life. Religions as the drivers of civilizations in earlier eras have offered philosophies of life to articulate and even explain human alienation and prescribe practices of life to offer repair and cultivate holiness. The holiness of life that religions have developed, while being specific to each, shares a sense of compassion and openness to future possibility. While religions, as we have seen, are not all the same, they do share ways of addressing the desire for life and bringing people into the fullness of life which, I have argued, is partly explained through the integration of our linguistic apparatus with social cognition. The Dalai Lama and the Pope are revered by their followers as holy men, and exemplify modes of existence whereby the universality of their philosophies of life comes to articulation in the specificity of their action. Both have completely different understandings of the cosmos, but both share a common vision of humanity as having a potential fullness of life that militates against destruction. It is not an issue of particular persons—whether the Pope or the Dalai Lama really are exemplars of true holiness—but rather a structural claim that religions at their best function to intensify life. At their worst they destroy life and contract human reality to a shadow of what it could be. The holiness of life is, I think, what Derrida meant by the politics of love that I referred to earlier, as a necessity for modern civilization and a necessary condition for any far distant future in which self-repair is achieved as 'A condition of complete simplicity', where 'all shall be well and / All manner of things shall be well', 'with the drawing of the Love and the voice of this calling'.[16]

[16] T.S. Eliot, 'Little Gidding', p. 43, *The Four Quartets* (London: Faber and Faber, 1995).

Bibliography

Primary Sources and Translations

Indic Traditions

Āgamaḍambara by Jayanta Bhaṭṭa, trans. Csaba Dezso, *Much Ado about Religion* (New York: New York University Press, 2005).

Bhagavad-gītā. Gavin Flood and Charles Martin, *The Bhagavad Gita* (New York: Norton, 2015).

Bṛhadāraṇyaka-upaniṣad. Patrick Olivelle, *The Early Upaniṣads* (Oxford: Oxford University Press, 1998).

Chāndogya-upaniṣad. Patrick Olivelle, *The Early Upaniṣads* (Oxford: Oxford University Press, 1998).

Īśvarapratyabhijñāvimarśṇī by Abhinavagupta, trans. and ed. K.C. Pandey and K.A.I.S. Iyar, *Bhāskarī* 3 vols (Delhi: MLBD reprint 1986 [1938]).

Nyāyamañjari by Jayanta Bhaṭṭa, trans. V.N. Jha (Delhi: Sri Satguru Publications, 1995).

Patañjali Vyākaraṇa Mahābhāṣya, ed. S.D. Joshi (Poona: University of Poona, 1968).

Ṛg-veda. Stephanie Jamieson and Joel Brereton, *The Rig Veda*, 3 vols (Oxford: Oxford University Press, 2015).

Ṣaḍdarśanasamuccaya, with a Commentary Called Laghuvṛtti by Maṇibhasra, by Haribhadra, ed. Damodara Lal Goswami (Benares: Chowkhamba Sanskrit Book Depot, 1905).

Śatapatha Brahmaṇa. Julius Eggeling, *Śatapatha Brahmaṇa according to the Text of the Madhyandina School*, Sacred Books of the East (Oxford: The Clarendon Press, 1900).

Somaśambhupaddhati. Hélène Brunner, *Somaśambhupaddhati*, 4 vols (Pondicherry: Institut Français d'Indologie, 1963, 1968, 1977, 2000).

Śvetāśvatara-upaniṣad. Patrick Olivelle, *The Early Upaniṣads* (Oxford: Oxford University Press, 1998).

Tantrasāra by Abhinavagupta, ed. M.S. Kaul (Śrīnagāra: Kashmir Series of Texts and Studies, 1918).

Tantrāloka by Abhinavagupta with the *viveka* by Jayaratha, vol. 1 ed. M.S. Śāstrī, vols 2–12 ed. M.S. Kaul (Śrīnagāra: Kashmir Series of Texts and Studies, 1918–39); chapters 1–5, French translation by André Padoux, *La Lumière sur les tantras* (Paris: CNES, 1998).

Tattvārthasūtra by Umāsvāti, trans. with introduction by Nathmal Tatia, *That Which Is* (London: Harper Collins, 1994).

Vaiśeṣika-sūtra with Śaṅkaramiśra commentary. Nandalal Sinha, *Vaiśeṣika-Sūtra of Kaṇāda (with the Commentary of Shanaka-Miśra and Extracts from the Gloss of Jayanārāyaṇa)*, Sacred Books of the Hindus (Allahabad: The Panini Office Bhuvaneswari Arama, 2nd ed. 1923 [1911]).

Vaiśeṣika-sūtras. Candrānanda's *Vṛtti* (Baroda: Gaekwad's Oriental Series, 1961).

Vedāntasāra of Rāmānuja, trans. and ed. V. Krishnamacharya and M.B. Narasimha Ayyangar (Adyar: Adyar Library, 1953).

Yogaśāstra by Hemacandra, trans. Olle Quarnström, *The Yogaśāstra of Hemacandra: A Twelfth Century Handbook on Śvetambara Jainism* (Cambridge, MA and London: Harvard University Press, 2002).

Chinese and Japanese Traditions

Chuangzi. Burton Watson, *The Complete Works of Chuang Tzu* (New York: Columbia University Press, 1968).

Daodejing. Hans-Georg Moeller, trans., *Daodejing: A Complete Translation and Commentary* (Chicago and La Salle, IL: Open Court, 2007).

Huainanzi. H. Roth, trans., *The Huainanzi: A Guide to the Theory and Practice of Government in Early China* (New York: Columbia University Press, 2010).

Hui-neng. Red Pine, translation and commentary, *The Platform Sutra: The Zen Teaching of Hui-neng* (Berkeley, CA: Counterpoint, 2006).

Lunyu. Arthur Waley, *The Analects of Confucius* (New York: Vintage Books, 1989 [1938]).

Mengzi. James Legge, *The Chinese Classics: A Translation, Critical and Exegetical Notes, Prolegomena, and Copious Indexes vol. II, The Works of Mencius* (London: Trübner and Co., 1861).

Mozi. Ian Johnson, *The Mozi: A Complete Translation* (New York: Columbia University Press, 2010).

Shobo Genzo. Kazuaki Tanahashi, *Treasury of the True Dharma Eye: Zen Master Dogen's Shobo Genzo* (Boston, MA and London: Shambhala, 2012).

Classical and Abrahamic Traditions

Al Ghazzālī, *The Ninety-Nine Beautiful Names of God: al Maqṣad al-asnā fī sharḥ asmā' Allāh al-ḥusnā*, trans. David B. Burrell and Nazih Daher (Cambridge: Islamic Text Society, 1992).

Al Ghazzalī, *Love, Longing, Intimacy and Contentment: Kitāb al-maḥabba wa'l-shawq wa'l-uns wa'l-riḍā, Book XXXVI of the Revival of the Religious Sciences*, trans. Eric Ormsby (Cambridge: Islamic Text Society, 2011).

Aquinas, Thomas, *Commentary on Aristotle's* De Anima *(Sentencia libri De anima)*, trans. Kenelm Foster and Sylvester Humphries (New Haven, CT: Yale University Press, 1951).

Aquinas, Thomas, *De Substantis Separatis, Treatise on Separate Substance*, trans. Francis J. Lescoe (West Hartford, CT: St Joseph's College, 1959).

Aquinas, Thomas, *Commentary on Aristotle's* Nicomachean Ethics, trans. C.J. Litzinger (Notre Dame, IN: Dumb Ox Books, 1993 [1964]).

Aristotle, *Nicomachean Ethics*, trans. H. Rackham, *Aristotle IXX*, Loeb Classics 73 (Cambridge, MA: Harvard University Press, 1932).

Aristotle, *Politics*, trans. H. Rackham, *Aristotle XXI*, Loeb Classics 264 (Cambridge, MA: Harvard University Press, 1932).

Aristotle, *De Anima*, trans. W.S. Hett, *On the Soul; Parva naturalia; On Breath*, Loeb Classics 288 (Cambridge, MA: Harvard University Press, 2014 [1957]).

Augustine, *The City of God against the Pagans*, trans. George McCracken, Loeb Classical Library 411 (Cambridge, MA: Harvard University Press, 1957).

Augustine, *Confessions*, trans. R.S. Pine-Coffin (London: Penguin, 1961).

Augustine, *On the Trinity*, trans. Stephen McKenna (Cambridge: Cambridge University Press, 2002).

Averroeana: Being a Transcript of Several Letters from Averroes, an Arabian Philosopher at Cordoba in Spain, to Metrodorus a Young Grecian Nobleman, Student at Athens, in the Years 1149 and 1150, anonymous editor but probably Thomas Tyron (London: T. Sowls, 1695).

Cicero, *De Natura Deorum (On the Nature of the Gods)*, English translation by H. Rackham, Loeb Classics (Cambridge, MA: Harvard University Press, 1951 [1933]).

Duns Scotus, *John Duns Scotus, the Oxford Lecture on Individuation*, Latin text with English translation, Alan B. Wolter (New York: The Franciscan Institute, 1975).

Homer, *Odyssey*. *Odyssey* vol. 1, trans. A.T. Murray, Loeb Classical Library 104 (Cambridge, MA: Harvard University Press, 1919).

Ibn Arabi, *Ibn Al'Arabi: The Bezels of Wisdom*, trans. R.W.J. Austin, The Classics of Western Spirituality (Mahwah, NJ: Paulist Press, 1980).

Ibn Gabirol, Solomon, *The Fountain of Life (Fons Vitae)*, by Solomon Ibn Gabirol; specially abridged edition; translated from the Latin by H.E. Wedeck, introduction by Theodore E. James (London: Peter Owen, 1963). The Latin text is found in Clemens Bäumker (ed.), *Avicebrolis (Ibn Gebirol) Fons Vitae, ex Arabico in Latinum translatus ab Johanne Hispano et Dominico Gundissalino. Ex codicibus Parisinis, Amploniano, Columbino primum* (Münster: Monasterri, 1895).

Lucian, 'On Sacrifices', in A.M. Harmon, trans., Lucian volume 3, Loeb Classics 130 (Cambridge, MA: Harvard University Press, 2014), pp. 155–71.

Pausanias, *Description of Greece*, in W.H.S. Jones, Loeb Classical Library 93 (Cambridge, MA: Harvard University Press, 1918).

Philo, 'Special Laws', in Philo vol. VII *On the Decalogue, On the Special Laws Books 1–3*, with trans. by F.H. Colson, Loeb Classics (Cambridge, MA: Harvard University Press, 1937).

Plotinus, *Enneads*. A.H. Armstrong (trans.), *Plotinus: Porphyry on Plotinus*, Loeb Classical Library 440 (Cambridge, MA: Harvard University Press, 1966).

Qur'an, trans. Yusuf Ali (London: Islamic Foundation, 1975).

Modern and Secondary Sources

Abramova, Zoya A., *L'art paléolithique d'Europe orientale et de Sibérie* (Grenoble: Jérome Millon, 1995).

Abulafia, Anna, *Christians and Jews in the Twelfth-Century Renaissance* (London: Routledge, 1995).

Adamson, Peter, 'Aristotle in the Arabic Commentarial Tradition', in Christopher Shields (ed.), *The Oxford Handbook of Aristotle* (Oxford: Oxford University Press, 2012), pp. 645–89.

Addas, Clause, *Quest for the Red Sulphur* (London: Islamic Text Society, 1993).

Adler, Joseph A., *Reconstructing the Confucian Dao: Zhu Xi's Appropriation of Zhou Dunyi* (Albany, NY: SUNY Press, 2014).

Agamben, Giorgio, *Homo Sacer: Sovereign Power and Bare Life*, trans. David Heller-Roazen (Stanford, CA: Stanford University Press, 1998).

Aklujkar, Ashok, 'Candrānanda's Date', *Journal of the Oriental Institute Baroda*, vol. 19, 1969/70, pp. 340–1.

Al-Saji, Alia, 'Life as Vision: Bergson and the Future of Seeing Differently', in Michael Kelly (ed.), *Bergson and Phenomenology* (London: Palgrave Macmillan, 2009), pp. 148–73.

Alcorta, C.S. and R. Sosis, 'Ritual, Emotion, and Sacred Symbols: The Evolution of Religion as an Adaptive Complex', *Human Nature*, vol. 16 (4), 2005, pp. 323–59.

Allen, Nick, 'Tetradic Theory and the Origin of Human Kinship', in Nick Allen et al. (eds.), *Early Human Kinship: From Sex to Social Reproduction* (Oxford: Blackwell, 2008), pp. 96–112.

Allison, Henry E., *Kant's Theory of Freedom* (Cambridge: Cambridge University Press, 1990).

Alter, Joe, *Yoga in Modern India: The Body between Science and Philosophy* (Princeton, NJ: Princeton University Press, 2004).

Alter, Joseph S., 'Gattungswesen—the Ecology of Species-Being: Alienation, Biosemiotics, and Social Theory', *Anthropos*, vol. 110, 2015, pp. 515–31.

Ames, Roger T., 'The Meaning of the Body in Classical Chinese Philosophy', in Thomas Kasulis, Roger Ames, and Wimal Dissanayake (eds.), *Self as Body in Asian Theory and Practice* (Albany, NY: SUNY Press, 1993), pp. 157–77.

Ames, Roger T. and David L. Hall, *Focussing on the Familiar: A Translation and Philosophical Interpretation of the Zhongyong* (Honolulu, HI: University of Hawaii Press, 2001).

Ames, Roger T. and Henry Rosemont Jr., *The Analects of Confucius: A Philosophical Translation* (New York: Ballantine Books, 1998).

Angelucci, Daniela, 'Deleuze and the Concepts of Cinema', trans. Sarin Marchet, *Deleuze Studies*, vol. 8 (3), 2014, pp. 311–414.

Angle, Stephen C., *Contemporary Confucian Political Philosophy* (Cambridge: Polity Press, 2012).

Angle, Stephen C. and Marina Svensson (eds.), *The Chinese Human Rights Reader: Documents and Commentary 1900–2000* (Abingdon and New York: Routledge, 2001).

Armstrong, Karen, 'The Golden Rule across Faiths', Resource on Faith, Ethics and Public Life, 17 November 2008, Georgetown University Center for Religion, Peace, and World Affairs, https://berkleycenter.georgetown.edu/quotes/karen-armstrong-on-the-golden-rule-across-faiths.

Asouti, Eleni and Dorian Q. Fuller, 'From Foraging to Farming in the Southern Levant: The Development of Epipalaeolithic and Pre-Pottery Neolithic Plant Management Strategies', *Vegetation History and Archaeobotany*, vol. 21, 2012, pp. 149–62.

Assmann, Jan, *Moses the Egyptian: The Memory of Egypt in Western Monotheism* (Cambridge, MA: Harvard University Press, 1998).

Atkinson, Michael, *Plotinus, Ennead V.1: On the Three Principal Hypotheses: A Commentary with Translation* (Oxford: Oxford University Press, 1983).

Atran, Scott, *In Gods We Trust: The Evolutionary Landscape of Religion* (Oxford: Oxford University Press, 2002).

Baldick, Julian, 'Early Islam', in Stewart Sutherland, Leslie Houlden, Peter Clarke, and Friedhelm Hardy (eds.), *The World's Religions* (London: Routledge, 1988), pp. 313–28.

Barrett, Justin, 'Cognitive Science, Religion, and Theology', in Jeffrey Schloss and Michael J. Murray (eds.), *The Believing Primate: Scientific, Philosophical, and Theological Reflections on the Origins of Religion* (Oxford: Oxford University Press, 2009), pp. 76–99.

Bar-Yosef, O. 'The Natufian Culture in the Levant: Threshold to the Origins of Agriculture', *Evolutionary Anthropology*, vol. 31, 1998, pp. 159–77.

Bar-Yosef, O. and R.H. Meadow, 'The Origins of Agriculture in the Near East', in T.D. Price and A.B. Gebauer (eds.), *Last Hunters—First Farmers: New Perspectives on the Prehistoric Transition to Agriculture* (Santa Fe, NM: School of American Research, 1995), pp. 39–94.

Barclay, John M.G., 'Stoic Physics and the Christian Event: A Review of Troels Enberg-Pedersen, Cosmology and Self in the Apostle Paul', *Journal for the Study of the New Testament*, vol. 33 (4), 2011, pp. 406–14.

Barnard, Alan, *Genesis of Symbolic Thought* (Cambridge: Cambridge University Press, 2011).

Bartley, Christopher, *The Theology of Rāmānuja: Realism and Religion* (London: Curzon, 2002).

Bary, Wm Theodore de, *Neo-Confucian Orthodoxy and the Learning of the Mind-and-Heart* (New York: Columbia University Press, 1981).

Bary, Wm Theodore de, *The Message of the Mind in Neo-Confucianism* (New York: Columbia University Press, 1989).

Bataille, Georges, *La Part maudite* (Paris: Les Editions de Minuit, 1967).

Bataille, Georges, *Théorie de la religion* (Paris: Gallimard, 1973).

Bechert, Heinz and Richard Gombrich (eds.), *The World of Buddhism* (London: Thames and Hudson, 1984).

Bedau, Mark A. 'The Nature of Life', in Steven Luper (ed.), *The Cambridge Companion to Life and Death* (Cambridge: Cambridge University Press, 2014), pp. 13–29.

Bell, Catherine, *Ritual Theory, Ritual Practice* (Oxford: Oxford University Press, 1992).

Bellah, Robert N., *Religion in Human Evolution: From the Paleolithic to the Axial Age* (Cambridge, MA and London: The Belknap Press of Harvard University Press, 2011).

Bellah, Robert N. and Hans Joas (eds.), *The Axial Age and its Consequences* (Cambridge, MA: Harvard University Press, 2012).

Bénatouil, Thomas, 'How Industrious Can Zeus Be?', in Richard Salles (ed.), *God and Cosmos in Stoicism* (Oxford: Oxford University Press, 2009), pp. 23–45.

Bendall, D.S. (ed.), *Evolution from Molecules to Men* (Cambridge: Cambridge University Press, 1983).

Benjamin, Walter, 'Theses on the Philosophy of History', in *Illuminations*, trans. Harry Zorn (London: Pimlico, 1999).

Bennett, Jane, *Vibrant Matter: A Political Ecology of Things* (Durham, NC: Duke University Press, 2010).

Benveniste, Émile, *Vocabulaire des institutions indo-européennes* (Paris: Editions de Minuit, 1969).

Bergson, Henri, *Creative Evolution*, trans. Arthur Mitchell (London: Dover, 1998 [1911]).

Bergson, Henri, *The Creative Mind: Introduction to Metaphysics*, trans. Mabelle L. Andison (New York: Dover, 2007 [1949]).

Berthrong John, 'Transmitting the Dao: Chinese Confucianism', in Wonsuk Chan and Leah Kalmanson (eds.), *Confucianism in Context* (Albany, NY: SUNY Press, 2010), pp. 9–13.

Biardeau, Madeleine, *Théorie de la connaissance et philosophie de la parole dans le brahmanisme classique* (Paris: Mouton, 1964).

Biardeau, Madeline, *Histoire de Poteaux. Variations védiques autour de la déese hindou* (Paris: Publications de l'Ecole Française d'Extrême Orient, 1989).

Biardeau, M. and C. Malamoud, *Le Sacrifice dans l'Inde ancienne* (Louvain and Paris: Peeters, 1996).

Bickerton, Derek, *Adam's Tongue: How Humans Made Language, How Language Made Humans* (New York: Hill and Wang, 2007).

Blondel, Maurice, *Action: Essay on a Critique of Life and a Science of Practice*, trans. Oliva Blanchette (Notre Dame, IN: Notre Dame University Press, 2004 [1893]).

Bodewitz, H.W., 'Prāṇa, Apāna and Other Prāṇas in Vedic Literature', *Adyar Library Bulletin*, vol. 50, 1986, pp. 326–48.

Boehme, Jacob, *Aurora*, trans. John Sparrow (London: John Streater, 1656).

Boehme, Jacob, *Aurora (Morgen Röte im auffgang, 1612) and Fundamental Report (Gründlicher Bericht, Mysterium Pansophicum, 1620)*, ed. and trans. Andrew Weeks (Leiden: Brill, 2013).

Bol, Peter K., *Neo-Confucianism in History* (Cambridge, MA: Harvard University Press, 2008).

Boland, Mechtilde, *Die Wind-Atem Lehre in der alteren Upaniṣaden* (Münster: Ugarit-Verlag, 1997).

Borges, Jorge Luis, *Selected Poems 1923–1967*, ed. Norman Thomas di Giovanni (London: Allen Lane, the Penguin Press, 1972).

Bowie, Andrew, *From Romanticism to Critical Theory* (London and New York: Routledge, 1997).

Bowie, Andrew, *Schelling and Modern European Philosophy* (London: Routledge, 2016).

Bowker, John, 'On Being Religiously Human', *Zygon*, vol. 16, 1981, pp. 365–82.

Bowker, John, *The Meanings of Death* (Cambridge: Cambridge University Press, 1991).

Bowker, John, *Is God a Virus? Genes, Culture, and Religion* (London: SPCK, 1995).

Bowker, John, *The Sacred Neuron: Extraordinary Discoveries Linking Science and Religion* (London: Tauris Press, 2005).

Bowker, John, *Why Religions Matter* (Cambridge: Cambridge University Press, 2015).

Bowker, John, *Religion Hurts: Do Religions Cause More Harm than Good?* (London: SPCK, 2018).

Boye, Katherine et al. (eds.), *Rethinking the Human Revolution: New Behavioural and Biological Perspectives on the Origin and Dispersal of Modern Humans* (Cambridge: McDonald Institute for Archaeological Research, 2007).

Boyer, Pascal, *Religion Explained: The Evolutionary Origins of Religious Thought* (New York: Basic Books, 2001).

Bragué, Remi, *The Legend of the Middle Ages: Philosophical Explorations of Medieval Christianity, Judaism, and Islam*, trans. Lydia G. Cochrane (Chicago, IL: Chicago University Press, 2009).

Bronkhorst, Johannes, *Greater Magadha: Studies in the Culture of Early India* (Leiden: Brill, 2007).

Bronkhorst, Johannes, *Buddhism in the Shadow of Brahmanism* (Leiden: Brill, 2011).

Bronkhorst, Johannes, 'Can Religion Be Explained? The Role of Absorption in Various Religious Phenomena', *Method and Theory in the Study of Religion*, vol. 29 (1), 2016, pp. 1–30.

Brown, Norman O., *Life against Death: The Psychoanalytical Meaning of History* (London: Routledge and Kegan Paul, 1959).

Brown, Peter, *The Body and Society: Men, Women, and Sexual Renunciation in Early Christianity* (New York: Columbia University Press, 1988).

Bruce, Steve, *God Is Dead: Secularization in the West* (Oxford: Blackwell, 2002).

Brunn, Stanley D. (ed.), *The Changing World Religion Map: Sacred Places, Identities, Practices, and Politics* (New York: Springer, 2015).

Brunner, Hélène, 'Le Sādhaka, personnage oublié dans le Śivaisme du Sud', *Journal Asiatique*, vol. 263, 1975, pp. 411–16.

Bulbulia, J. et al., *The Evolution of Religion: Studies, Theories, and Critiques* (Santa Margarita, CA: Collins Foundation Press, 2008).

Burkert, Walter, *Homo Necans: The Anthropology of Ancient Greek Sacrificial Ritual and Myth*, trans. P. Bing (Berkeley, CA: University of California Press, 1983).

Burl, H.A.W., 'Henges: Internal Features and Regional Groups', *The Archaeological Journal*, vol. 126 (1), 1969, pp. 1–28.

Campbell, Scott M., *The Early Heidegger's Philosophy of Life: Facticity, Being, Language* (New York: Fordham University Press, 2012).

Caplan, Arthur L. (ed.), *The Sociobiology Debate: Readings on Ethical and Scientific Issues* (New York: Harper and Row, 1978).

Cariou, Marie, *Lectures bergsoniennes* (Paris: Presses universitaires de France, 1990).

Carrette, Jeremy, *Religion and Culture by Michel Foucault* (Manchester: Manchester University Press, 1999).

Carrithers, Michael, Steven Lukes, and Steven Collins (eds.), *The Category of the Person* (Cambridge: Cambridge University Press, 1985).

Cavanaugh, William, *Torture and Eucharist: Theology, Politics, and the Body of Christ* (Oxford: Blackwell, 1998).

Cave, Stephen, *Immortality: The Quest to Live Forever and How It Drives Civilization* (London: Biteback, 2013).

Certeau, Michel de, *The Mystic Fable, vol. 1: The Sixteenth and Seventeenth Centuries*, trans. Michael B. Smith (Chicago, IL: Chicago University Press, 1992).

Chan, Albert, *The Glory and Fall of the Ming Dynasty* (Norman, OK: University of Oklahoma Press, 1982).

Chan, Wing-tsit, *A Sourcebook in Chinese Philosophy* (Princeton, NJ: Princeton University Press, 1963).

Chan, Wing-tsit, *Chu Hsi, Life and Thought* (Beijing: Chinese University Press, 1987).

Chang, Carson et al., *A Manifesto for a Reappraisal of Sinology and Reconstruction of Chinese Culture in Carson Chang, The Development of Neo-Confucian Thought*, vol. 2 (New York: Bookman Associates, 1962).

Chang, Hsun, 'Incense-Offering and Obtaining the Magical Power of Qi: The Matzu (Heavenly Mother) Pilgrimage in Taiwan', PhD (Berkeley, CA: University of California, 1993).

Chantraine, Georges, 'Surnaturel et destinée humaine dans la pensée occidentale selon Henri de Lubac', *Revue des sciences philosophiques et théologiques*, vol. 85 (2), 2001, pp. 299–312.

Chatel-Goldman, Jonas, Jean-Luc Schwartz, Christian Jutten, and Marco Congedo, 'Non-Local Mind from the Perspective of Social Cognition', *Frontiers in Human Neuroscience*, vol. 7 (107), 2013, pp. 1–7.

Chemero, Anthony, *Radical Embodied Cognitive Science* (Cambridge, MA: MIT Press, 2011).

Chemparathy, George 'The Testimony of the Yuktidīpikā concerning the Īśvara Doctrine of the Pāśupatas and Vaiśeṣikas', in *Wiener Zeitschrift für die Kunde Sudasiens*, vol. 9, 1965, pp. 119–46.

Ch'en, Kenneth Kuan Sheng, *Chinese Transformation of Buddhism* (Princeton, NJ: Princeton University Press, 1973).

Cheng, Chung-ying, 'Li and Ch'i in the I Ching: A Reconsideration of Being and Non-Being in Chinese Philosophy', *Journal of Chinese Philosophy*, vol. 14, 1987, pp. 1–38.

Chérif, Mustapha, *Islam and the West, a Conversation with Jacques Derrida*, trans. Theresa Lavender Fagan (Chicago, IL: Chicago University Press, 2008).

Chidester, David, *Savage Systems: Colonialism and Comparative Religion in Southern Africa* (Charlottesville, VA: University of Virginia Press, 1996).

Childe, Gordon W., *Man Makes Himself* (London: Watts, 1936).

Chittick, William C., *The Sufi Path of Knowledge* (Albany, NY: SUNY Press, 1989).

Chittick, William C., *Ibn Arabi, Heir to the Prophets* (Oxford: One World, 2007).

Chittick, William C., *In Search of the Lost Heart: Explorations in Islamic Thought* (Albany, NY: SUNY Press, 2012).

Chittick, William, 'Ibn Arabi', *The Stanford Encyclopedia of Philosophy* (Spring 2014 Edition), Edward N. Zalta (ed.), https://plato.stanford.edu/archives/spr2014/entries/ibn-arabi.

Chong, Kim-Chong, 'Xunzi's Systematic Critique of Mencius', *Philosophy East and West*, vol. 53 (2), 2003, pp. 215–33.

Christian, David, *Maps of Time: An Introduction to Big History* (Berkeley, CA: University of California Press, 2004).

Clark, Carol Lea, 'Aristotle and Averroes: The Influences of Aristotle's Arabic Commentator upon Western European and Arabic Rhetoric', in *Review of Communication*, vol. 7 (4), 2007, pp. 369–87.

Clooney SJ, Francis X., *Thinking Ritually: Rediscovering the Pūrva Mīmāṃsā of Jaimini* (Vienna: De Nobili, 1990).

Clooney SJ, Francis X., *Hindu God, Christian God: How Reason Helps Break Down the Barriers Between Religions* (Oxford: Oxford University Press, 2001).

Clooney SJ, Francis X., *Comparative Theology: Deep Learning across Religious Borders* (Oxford: Wiley-Blackwell, 2010).

Coakley, Sarah, *Powers and Submissions: Spirituality, Philosophy and Gender* (Oxford: Blackwell, 2002).

Coleman, W., *Biology in the Nineteenth Century: Problems of Form, Function and Transformation* (Cambridge: Cambridge University Press, 1977).

Coleridge, S.T., *Bibliographia Literaria*, ed. James Engell and W. Jackson Bate (Princeton, NJ: Princeton University Press, 1983).

Connolly, Peter, 'The Vitalistic Antecedents of the Ātman-Brahman Concept', in Peter Connolly and Sue Hamilton (eds.), *Indian Insights: Buddhism, Brahmanism, and Bhakti* (London: Luzac Oriental, 1997), pp. 21–38.

Connolly, William E., *The Fragility of Things: Self-Organizing Processes, Neoliberal Fantasies, and Democratic Activism* (Durham, NC: Duke University Press, 2013).

Cook, Scott, *The Bamboo Texts of Guodian: A Study and Complete Translation*, Part 1 (New York: East Asia Program Cornell University, 2012).

Coyne, Jerry A., *Why Evolution Is True* (Oxford: Oxford University Press, 2010).

Cross, Richard, 'Duns Scotus: Some Recent Research', *Journal of the History of Philosophy*, vol. 49 (3), 2011, pp. 271–95.

Crowe, Benjamin, *Heidegger's Phenomenology of Religion: Realism and Cultural Criticism* (Bloomington, IN: Indiana University Press, 2008).

Cruz, H. de and R. Nichols, *Advances in Religion, Cognitive Science, and Experimental Philosophy* (London: Bloomsbury, 2016).

Cummings, E.E., *Complete Poems 1904–1962*, ed. George James Firmage (New York: Norton and Co., 2016 [1973]).

Cummings, Vicki, 'What Lies beneath: Thinking about the Qualities and Essences of Stone and Wood in the Chambered Tomb Architecture of Neolithic Britain and Ireland', *Journal of Social Archaeology*, vol. 12 (1), 2011, pp. 29–50.

Daley, Brian, 'The Nouvelle Théologie and the Patristic Revival: Sources, Symbols and the Science of Theology', *International Journal of Systematic Theology*, vol. 7, 2005, pp. 362–82.

Dalferth, Ingolf U., 'The Idea of Transcendence', in Robert N. Bellah and Hans Joas (eds.), *The Axial Age and its Consequences* (Cambridge, MA: Harvard University Press, 2012), pp. 146–88.

Das, Rahul Peter, *The Origin of the Life of a Human Being: Conception and the Female according to Ancient Indian Medical and Sexological Literature* (New Delhi: MLBD, 2003).

Dasgupta, Surendranath, *History of Indian Philosophy*, 5 vols (Delhi: MLBD, 1975 [1922]).

Davidson, Herbert H., *Alfarabi, Avicenna, and Averroes on Intellect* (New York and Oxford: Oxford University Press, 1992).

Davies, Oliver, *God within: The Mystical Tradition of Northern Europe* (London: Darton, Longman and Todd, 1988).

Davies, Oliver, *Theology of Compassion: Metaphysics of Difference and the Renewal of Tradition* (London: SCM, 2010).

Davies, Oliver, *Theology of Transformation: Faith, Freedom, and Christian Act* (Oxford: Oxford University Press, 2013).

Davies, Oliver, 'Niche Construction, Social Cognition, and Language: Hypothesizing the Human as the Production of Place', *Culture and Brain*, vol. 3 (2), 2016, pp. 87–112.

Dawkins, Richard, *The Extended Phenotype: The Gene as a Unit of Selection* (San Francisco, CA: Freeman, 2002).

Dawkins, Richard, *The Selfish Gene*, 4th ed. (Oxford: Oxford University Press, 2016).

Dayan, Tamar, 'Early Domesticated Dogs of the Near East', *Journal of Archaeological Science*, vol. 21 (5), 1994, pp. 633–40.

Dean, Kenneth, 'Underworld Rising: The Fragmented Syncretic Ritual Field of Singapore', unpublished paper presented at 'Political Theologies and Development in Asia', Asia Research Institute, NUS, 21 February 2017.

Deane-Drummond, Celia and Agustin Fuentes (eds.), *The Evolution of Human Wisdom* (Lanham, MD: Lexington Books, 2017).

De Gelder, B. and R. Rouw, 'Beyond Localization: A Dynamic Dual Route Account of Face Recognition', *Acta Psychologica*, vol. 107, 2001, pp. 183–207.

Deghaye, Pierre, 'La théosophie de Jacob Boehme. Les trois mystères du livre *De la Signature des choses*', *Les Études philosophiques*, vol. 2, 1999, pp. 147–65.

Deleuze, Gilles, *Difference and Repetition*, trans. Paul Patton (London: The Athlone Press, 1994).

Deleuze, Gilles, 'Immanence: une vie . . . ' Philosophie 47, Sept. 1995, trans. Nick Millet, 'Immanence: A Life . . . ', *Theory, Culture, Society*, vol. 14 (2), 1997, pp. 3–7.

Deleuze, Gilles, *The Logic of Sense*, trans. Mark Lester (London: Bloomsbury, 2004 [1990]).

Deleuze, Gilles and Felix Guattari, *A Thousand Plateaus: Capitalism and Schizophrenia*, trans. Brian Massumi (Minneapolis, MN and London: University of Minnesota Press, 1987).

Derr, Mark, *How the Dog Became the Dog: From Wolves to our Best Friends* (New York: Overview Books, 2011).

Derrida, Jacques, *Edmund Husserl's 'Origin of Geometry': An Introduction*, trans. John P. Leavey (Lincoln, NE and London: University of Nebraska Press, 1978).

Detienne, Marcel and J.P. Vernant (eds.), *The Cuisine of Sacrifice among the Greeks* (Chicago, IL: Chicago University Press, 1989).

Di Paulo, Ezequiel and Hanne de Jaegher, 'The Interactive Brain Hypothesis', *Frontiers in Human Neuroscience*, vol. 6, June 2012, article 163, p. 1.

Diamond, Eli, *Mortal Imitations of Divine Life: The Nature of the Soul in Aristotle's* De Anima (Evanston, IL: Northwestern University Press, 2015).

Dillon, Michael, *China: A Modern History* (London: I.B. Tauris, 2010).

Dillon, Michael and Paul Fletcher (eds.), *Violence, Sacrifice, Desire, Cultural Values*, vol. 4 (2), 2000.

Donald, Merlin, *The Origins of the Modern Mind: Three Stages in the Evolution of Culture and Cognition* (Cambridge, MA: Harvard University Press, 1991).

Doniger, Wendy, *On Hinduism* (Oxford: Oxford University Press, 2014).

Doniger O'Flaherty, Wendy, 'Karma and Rebirth in the Vedas and Purāṇas', in Wendy Doniger O'Flaherty (ed.), *Karma and Rebirth in Classical Indian Traditions* (Berkeley, CA, London: University of California Press, 1980), pp. 1–37.

Dodds, E.R., *Greeks and the Irrational* (Berkeley, CA: University of California Press, 1951).

Downing, F.G., 'Cosmic Eschatology in the First Century: "Pagan", Jewish, and Christian', *L'Antiquité Classique*, vol. 64, 1995, pp. 99–109.

Dumont, Louis, *Homo Hierarchicus* (Chicago, IL: Chicago University Press, 1980 [1966]).

Dumoulin, H., *Zen Buddhism: A History: India and China*, trans. J.W. Heisig and P. Knitter (Bloomington, IN: World Wisdom, 2005).

Dunbar, Robin, *Grooming, Gossip, and the Evolution of Language* (London: Faber and Faber, 1996).

Dunbar, Robin, 'Why Are Humans Not Just Great Apes?', in Charles Pasternak (ed.), *What Makes Us Human* (Oxford: One World, 2007), pp. 37–8.

Dunbar, Robin, 'Gossip and the Social Origins of Language', in Kathleen R. Gibson and Maggie Tallerman (eds.), *The Oxford Handbook of Language Evolution* (Oxford: Oxford University Press, 2011), pp. 343–5.

Duncan, Ann W., 'Sacred Pregnancy in the Age of the "Nones"', *Journal of the American Academy of Religion*, vol. 85 (4), 2018, pp. 1089–115.

Dunn, James, review of Englberg-Pedersen, 'Cosmology and Self in the Apostle Paul', *Journal of Theological Studies*, vol. 61 (2), 2010, pp. 748–50.

Dupuche, John R., *Abhinavagupta: The Kula Ritual as Elaborated in Chapter 29 of the Tantraloka* (Delhi: MLBD, 2003).

Durkheim, Emile, *Elementary Forms of the Religious Life*, trans. Carol Cosman (Oxford: Oxford University Press, 2001).

Eckstrand, Nathan, 'Deleuze, Darwin and the Categorisation of Life', *Deleuze Studies*, vol. 8 (4), 2014, pp. 415–44.

Ekman, Paul, 'Universals and Cultural Differences in Facial Expressions of Emotion', in J. Cole (ed.), *Nebraska Symposium on Motivation 1971* vol. 19 (Lincoln, NE: University of Nebraska Press, 1972), pp. 207–83.

Ekman, Paul and W.V. Friesen, 'Constants across Cultures in the Face and Emotions', *Journal of Personality and Social Psychology*, vol. 17, 1971, pp. 124–9.

Ekman, Paul, W.V. Friesen, M. O'Sullivan, A. Chan et al., 'Universals and Cultural Differences in the Judgments of Facial Expression of Emotion', *Journal of Personality and Social Psychology*, vol. 53, 1989, pp. 112–17.

Elias, Norbert, *The Civilizing Process* (Oxford: Blackwell, 1978).

Eliot, T.S., *The Four Quartets* (London: Faber and Faber, 1995).

Emmeche, Clause, David Budtz Pedtesen, and Frederik Stjernfelt (eds.), *Mapping Frontier Research in the Humanities* (London: Bloomsbury, 2017).

Engberg-Pedersen, Troels, 'Stoicism and the Apostle Paul: A Philosophical Reading', in S.K. Strange and J. Zupko (eds.), *Stoicism: Traditions and Transformations* (Cambridge: Cambridge University Press, 2004), pp. 52–75.

Engberg-Pedersen, Troels, *Cosmology and Self in the Apostle Paul: The Material Spirit* (Oxford: Oxford University Press, 2011).

Esposito, Roberto, *Bios: Biopolitics and Philosophy*, trans. Timothy Campbell (Minneapolis, MN: University of Minnesota Press, 2008).

Esposito, Roberto, *Immunitas: The Protection and Negation of Life*, trans. Zakiya Hanafi (Cambridge: Polity Press, 2011).

Esposito, Roberto, Rhiannon Welch, and Vanes Lemm, *Terms of the Political: Community, Immunity, Biopolitics* (New York: Fordham University Press, 2012).

Evans, G.R., *Philosophy and Theology in the Middle Ages* (London: Routledge, 1993).

Evans-Pritchard, E.E., *Nuer Religion* (Oxford: Oxford University Press, 1956).

Ewing, A.H., 'The Hindu Conception of the Functions of Breath: A Study in Early Hindu Psycho-Physics', *Journal of the American Oriental Society*, vol. 22, 1901, pp. 249–308.

Falk, Dean, 'Prelinguistic Evolution in Hominins: Whence Motherese?', *Behavioural and Brain Sciences*, vol. 27, 2004, pp. 491–503.

Falk, Dean, 'Hominin Paleoneurology: Where Are We Now?', in M. Hoffman and D. Falk (eds.), *Progress in Brain Research*, vol. 195, 2012, pp. 255–72.

Falkenhausen, Lothar van, *Suspended Music: Chime Bells in the Culture of Bronze Age China* (Berkeley, CA: University of California Press, 1993).

Faure, Bernard, *The Rhetoric of Immediacy: A Cultural Critique of Chan/Zen Buddhism* (Princeton, NJ: Princeton University Press, 1991).

Faure, Bernard, *Double Exposure: Cutting across Buddhist and Western Discourse* (Palo Alto, CA: Stanford University Press, 2004).

Ferraris, Maurizio, *Introduction to New Realism*, trans. Sarah De Sanctis (London: Bloomsbury, 2015).

Fiddes, Paul S., *Participating in God: A Pastoral Doctrine of the Trinity* (London: Darton, Longman and Todd, 2000).

Filliozat, Jean, *The Classical Doctrine of Indian Medicine* (New Delhi: Munshiram Manoharlal, 1964).

Fitzgerald, Timothy, *Discourse on Civility and Barbarity: A Critical History of Religion and Related Categories* (Oxford: Oxford University Press, 2007).

Fleet, Barry, *Plotinus Ennead IV.8: On the Descent of the Soul into Bodies* (Las Vegas, NV, Zurich, Athens: Parmenides Publishing, 2012).

Flood, Gavin, *Consciousness Embodied: Body and Cosmology in Kashmir Śaivism* (San Francisco, CA: Mellen Press, 1993).

Flood, Gavin, 'The Purification of the Body in Tantric Ritual Representation', *Indo-Iranian Journal*, vol. 45, 2002, pp. 22–43.

Flood, Gavin (ed.), *The Blackwell Companion to Hinduism* (Oxford: Blackwell, 2003).

Flood, Gavin, *The Ascetic Self: Subjectivity, Memory and Tradition* (Cambridge: Cambridge University Press, 2004).

Flood, Gavin, *The Importance of Religion: Meaning and Action in our Strange World* (Oxford: Wiley-Blackwell, 2012).

Flood, Gavin, 'Sacrifice as Refusal', in Julia Meszaros and Johannes Zachhuber (eds.), *Sacrifice and Modern Thought* (Oxford: Oxford University Press, 2013), pp. 115–31.

Flood, Gavin, *The Truth Within: A History of Inwardness in Christianity, Hinduism, and Buddhism* (Oxford: Oxford University Press, 2013).

Foard, J., M. Solomon, and R.K. Payne (eds.), *The Pure Land Tradition: History and Development* (Berkeley, CA: Institute of Buddhist Studies, 1996).

Foley, Robert and Marta Mirazón Lahr, 'Mode 3 Technologies and the Evolution of Modern Humans', *Cambridge Archaeological Journal*, vol. 7 (1), 1997, pp. 3–36.

Fontana, Walter, 'Algorithmic Chemistry', in C.G. Langton, C. Taylor, J.D. Farmer, and S. Rasmussen (eds.), *Artificial Life II* (Redwood City, CA: Addison-Wesley, 1992), pp. 159–209.

Fontana, Walter and L.W. Buss, 'What Would Be Conserved If the Tape Were Played Twice?', *Proceedings of the National Academy of Science*, vol. 91, 1994, pp. 757–61.

Fontana, Walter and L.W. Buss, 'The Arrival of the Fittest: Toward a Theory of Biological Organization', *Bulletin of Mathematical Biology*, vol. 56, 1994, pp. 1–64.

Foucault, Michel, *Dits et ecrits* (Paris: Gallimard, 1994).

Fragaszy, Dorothy M. and Susan Perry (eds.), *The Biology of Traditions: Models and Evidence* (Cambridge: Cambridge University Press, 2003).

Frank, Manfred, *Ein Einführung in Schellings Philosophie* (Frankfurt: Suhrkamp, 1985).

Fraser, Kyle, 'Seriality and Demonstration in Aristotle's Ontology', in *Oxford Studies in Ancient Philosophy*, vol. 22, 2002, pp. 43–82.

Frauwallner, Eric, *Geschichte des indischen Philosophie* (Salzburg: Otto Mullet Verlag, 1953), trans. V.M. Bedekar, *History of Indian Philosophy*, 2 vols (Delhi: Motilal Banarsidass, 1973).

Frauwallner, Erich, 'Der ursprünglische Afgang der Vaiśeṣika-Sutrem', in *Nachgelassene Werke I* (Vienna: Verlag der Österreichischen Akademie der Wissenschaften, 1984), pp. 35–41.

Frazier, Jessica, *Hindu Worldviews: Theories of Self, Ritual, Reality* (London: Bloomsbury, 2016).

Freeman, Rich, 'The Dancing of the Teyyams', in Gavin Flood (ed.), *The Blackwell Companion to Hinduism* (Oxford: Blackwell, 2003), pp. 307–26.

Freud, Sigmund, *Civilization and its Discontents*, trans. Joan Riviere (London: Hogarth Press, 1930).

Freud, Sigmund, *Totem and Taboo*, trans. James Strachey, reprinted in *The Origins of Religion* (London: Penguin, 1985).

Freudenthal, Gad, 'The Medieval Astrologization of Aristotle's Biology: Averroes on the Role of Celestial Bodies in the Generation of Animate Beings', *Arabic Sciences and Philosophy*, vol. 12, 2002, pp. 111–37.

Frey, U. (ed.), *The Nature of God: Evolution and Religion* (Marburg: Kubitza, Heinz-Werner, Tectum Verlag, 2016).

Fridja, N.H. and A. Tcherkassof, 'Facial Expressions as Modes of Action Readiness', in J.A. Russell and J.A. Fernandez-Dols (eds.), *The Psychology of Facial Expression* (New York: Cambridge University Press, 1997), pp. 78–102.

Fridlund, A.J., *Human Facial Expression: An Evolutionary View* (San Diego, CA: Academic Press, 1994).

Frith, Chris D., *Making up the Mind: How the Brain Creates our Mental World* (Oxford: Blackwell, 2007).

Frost, Samantha, *Biocultural Creatures: Toward a New Theory of the Human* (Durham, NC: Duke University Press, 2016).

Fuentes, Agustin, 'Evolutionary Perspectives and Transdisciplinary Intersections: A Roadmap to Generative Areas of Overlap in Discussing Human Nature', *Theology and Science*, vol. 11 (2), 2013, pp. 106–29.

Fuentes, Agustin, 'Manipulating Materials, Bodies, and Signs: How the Ecology of Creative Problem Solving, Tool Manufacture, and Imaginative Sociality Set the Context for Language in the Later Pleistocene Human Niche', in Celia Deane-Drummond and Agustin Fuentes (eds.), *The Evolution of Human Wisdom* (Lanham, MD: Lexington Books, 2017), pp. 191–204.

Fuentes, Agustin, *The Creative Spark: How Imagination Made Humans Exceptional* (New York: Dutton, 2017).

Fung, Yu-lan, *History of Chinese Philosophy, vol. 2: The Period of Classical Learning from the Second Century* BC *to the Twentieth Century* AD, trans. Derek Bodde (Princeton, NJ: Princeton University Press, 1951–2).

Gandhi, M.K., *'Hind Swaraj' and Other Writings*, ed. Anthony J. Parel (Cambridge: Cambridge University Press, 1997).

Ganeri, Martin, 'Free-Will, Agency, and Selfhood in Rāmānuja', in Matthew R. Dasti and Edwin F. Bryant (eds.), *Free Will, Agency, and Selfhood in Indian Philosophy* (Oxford: Oxford University Press, 2014), pp. 232–54.

Garfield, Jay, *Engaging Buddhism: Why It Matters to Philosophy* (Oxford: Oxford University Press, 2015).

Gauchet, Marcel, *The Disenchantment of the World*, trans. Oscar Burge (Princeton, NJ: Princeton University Press, 1997).

Gay, Peter, *Freud: A Life for our Time* (London: J.M. Dent and Sons Ltd., 1988).

Gaziel, Ahuva, 'Questions of Methodology in Aristotle's Zoology: A Medieval Perspective', *Journal of the History of Biology*, vol. 45, 2012, pp. 329–52.

Gérard, Vallée, J.B. Lawson and C.G. Chapple (trans.), *The Spinoza Conversations between Lessing and Jacobi: Text with Excerpts from the Ensuing Controversy* (Lanham, MD and London: University Press of America, 1988).

Ge, Zhaoguang, *Taoism and Chinese Culture* (Shanghai: Shanghai People's Press, 1987).

Ge, Zhaoguang, *History of Chinese Thought*, 3 vols (Fudan: Fudan University Press, 2001–12).

Geaney, Jane, *On the Epistemology of the Senses in Early Chinese Thought* (Honolulu, HI: University of Hawaii Press, 1992).

Geertz, Armin W., 'Cognitive Approaches to the Study of Religion', in P. Antes, A.W. Geertz, and R.R. Warne (eds.), *New Approaches to the Study of Religion Volume 2: Textual, Comparative, Sociological, and Cognitive Approaches* (Berlin: Walter de Gruyter, 2004), pp. 347–99.

Geertz, Clifford, 'Religion as a Cultural System', in *The Interpretation of Cultures* (New York: Basic Books, 1973), pp. 87–125.

Genette, Gérard, *Métalepse: De la figure à la fiction* (Paris: Seuil, 2004).

Gennep, Arnold van, *The Rites of Passage*, trans. M.B. Vizedom and G.L. Cafee (Chicago, IL: University of Chicago Press, 1960 [1909]).

Gerbaut, P. et al., 'Evolution of Lactase Persistence: An Example of Human Niche Construction', *Philosophical Transactions of the Royal Society B: Biological Sciences*, vol. 366, 2011, pp. 863–77.

Gethin, Rupert, *The Foundations of Buddhism* (Oxford and New York: Oxford University Press, 1998).

Gethin, Rupert, *The Buddhist Path to Awakening* (Oxford: One World, 2001 [1992]).

Gibson, K.R. (eds.), *Evolutionary Anatomy of the Primate Cerebral Cortex* (Cambridge: Cambridge University Press, 2001).

Gibson, Kathleen R. and Maggie Tallerman (eds.), *The Oxford Handbook of Language Evolution* (Oxford: Oxford University Press, 2011).

Gilders, William K., *Blood Ritual in the Hebrew Bible: Meaning and Power* (Baltimore, MD: Johns Hopkins University Press, 2004).

Gill, A. Le, C. Leyrat, M.A. Janssen, G.C. Choblet, G. Tobie, O. Bourgeois, A. Lucas, C. Satin, C. Howett, R. Kirk, R.D. Lorenz, R.D. West, A. Stoizenbach, M. Massé, A.H. Hayes, L. Bonnefoy, G. Veyssière, and F. Paganelli, 'Thermally Anomalous Features in the Subsurface of Enceladus's South Polar Terrain', *Nature Astronomy*, vol. 1, 0063, 2017.

Gintis, Herbert, 'Gene-Culture Coevolution and the Nature of Human Sociality', *Philosophical Transactions of the Royal Society B: Biological Sciences*, vol. 366, 2011, pp. 878–88.

Girard, Rene, *Des choses cachées depuis la fondation du monde* (Paris: Le Livre de poche, 1983).

Girardot, Norman J., *The Victorian Translation of China: James Legge's Oriental Pilgrimage* (Berkeley, CA: University of California Press, 2003).

Glasner, Ruth, *Averroes' Physics: A Turning Point in Medieval Natural Philosophy* (Oxford: Oxford University Press, 2009).

Goetschei, Wili, *Spinoza's Modernity: Mendelssohn, Lessing, and Heine* (Madison, WI: University of Wisconsin Press, 2004).

Goodenough, Ursula, *The Sacred Depths of Nature* (Oxford: Oxford University Press, 1998).

Gourinat, Jean-Baptiste, 'The Stoics on Matter and Prime Matter', in R. Salles (ed.), *God and Cosmos in Stoicism* (Oxford: Oxford University Press, 2009), chapter 2.

Granet, Marcel, *La Pensée chinoise* (Paris: Albin Michel, 1934).

Green, N., *Sufism: A Global History* (Oxford: Wiley-Blackwell, 2012).

Gregory, Peter N., *Inquiry into the Origin of Humanity: An Annotated Translation of Tsung-mi's Yuan jen lun* (Honolulu, HI: University of Hawaii Press, 1995).

Griffiths, Paul J., *Problems of Religious Diversity* (Oxford: Blackwell, 2001).

Guéry, François and Didier Deleule, *The Productive Body*, trans. Philip Barnard and Stephen Shapiro (Winchester and Washington, DC: Zero Books, 2014).

Gurtler, Gary M., *Plotinus: The Experience of Unity* (New York: Peter Lang, 1988).

Gurtler, Gary M., *Enneads of Plotinus: Ennead 4.30–45 and IV.5: Problems concerning the Soul: Translation, with an Introduction and Commentary* (Las Vegas, NV, Zurich, Athens: Parmenides Publishing, 2015).

Hadot, Pierre, *Plotinus or the Simplicity of Vision*, trans. Michael Chase with an introduction by Arnold L. Davidson (Chicago, IL: University of Chicago Press, 1993 [1989]).

Haggard, Patrick, 'Sense of Agency in the Human Brain', *Nature Reviews: Neuroscience*, vol. 18, 2017, pp. 197–208.

Hahm, David E., *The Origins of Stoic Cosmology* (Columbia, OH: Ohio State University Press, 1977).

Halbfass, Wilhelm, *Tradition and Reflection* (Albany, NY: SUNY Press, 1988).

Halbfass, Wilhelm, *On Being and What There Is: Classical Vaiśeṣika and the History of Indian Ontology* (Albany, NY: SUNY Press, 1992).

Hamilton-Kelly, R.G. (ed.), *Violent Origins* (Stanford, CA: Stanford University Press, 1987).

Hanneder, J., *Abhinavagupta's Philosophy of Revelation* (Groningen: Egbert Forsten, 1998).

Hanvey, James, 'Tradition as Subversion', *International Journal of Systematic Theology*, vol. 6 (1), 2004, pp. 50–68.

Haraway, Donna, 'The Cyborg Manifesto', in *Simians, Cyborgs and Women: The Reinvention of Nature* (London: Free Association Books, 1991), pp. 149–81.

Harman, Graham, *Quentin Meillassoux: Philosophy in the Making* (Edinburgh: Edinburgh University Press, 2011).

Harvey, Graham, *Animism: Respecting the Living World* (New York: Columbia University Press, 2017).

Harvey, Peter, *Introduction to Buddhism*, 2nd ed. (Cambridge: Cambridge University Press, 2013 [1990]).

Hassin, Ran R., Kevin N. Ochsner, and Yaacov Trope (eds.), *Self Control in Society, Mind, and Brain* (Oxford: Oxford University Press, 2010).

Hausner, Sondra, *Wandering with Sadhus: Ascetics in the Hindu Himalayas* (Bloomington, IN: Indiana University Press, 2007).

Hausner, Sondra and David Gellner, 'Category and Practice as Two Aspects of Religion: The Case of Nepalis in Britain', *Journal of the American Academy of Religion*, vol. 80 (4), 2012, pp. 971–97.

Hayes, Glen A., 'Possible Selves, Body Schemas, and Sādhana: Using Cognitive Science and Neuroscience in the Study of Medieval Vaiṣṇava Sahajiyā Hindu Tantric Texts', *Religions*, vol. 5, 2014, pp. 684–99.

Hedley, Douglas, *Coleridge, Philosophy and Religion: Aids to Reflection and the Mirror of the Spirit* (Cambridge: Cambridge University Press, 2000).

Hedley, Douglas, *Sacrifice Imagined: Violence, Atonement, and the Sacred* (London: Continuum, 2011).

Heelas, Paul, *The New Age Movement* (Oxford: Blackwell, 1996).

Heelas, Paul, *Spiritualities of Life: New Age Romanticism and Consumptive Capitalism* (Oxford: Wiley-Blackwell, 2009).

Heelas, Paul and Linda Woodhead, *The Spiritual Revolution: Why Religion Is Giving Way to Spirituality* (Oxford: Blackwell, 2005).

Heesterman, J.C., *The Broken World of Sacrifice: An Essay in Ancient Indian Ritual* (Chicago, IL: University of Chicago Press, 1993).

Hegel, G.W.F., *Phenomenology of Spirit*, trans. A.V. Millar (Oxford: Oxford University Press, 1977); *Phänomenologie des Geistes* (Stuttgart: Reclam, 2009 [1807]).

Hegel, G.W.F., *Logic*, trans. William Wallace (Oxford: Clarendon Press, 1975 [1873]).

Hegel, G.W.F., *Lectures on the Philosophy of Religion*, 3 vols, ed. Peter C. Hodgson (Oxford: Oxford University Press, 2007); Walter Jaeschke (ed.), *Vorelsungen über die Philosophie der Religion* (Hamburg: Felix Meiner, 1985).

Heidegger, Martin, *An Introduction to Metaphysics*, trans. Ralph Mannheim (New Haven, CT and London: Yale University Press, 1959); *Einfühlung in die Metaphysik, Gesamtausgabe*, vol. 40 (Frankfurt: Vittorio Klostermann, 1983).

Heidegger, Martin, *Being and Time*, trans. John Macquarrie and Edward Robinson (Oxford: Blackwell, 1962); *Sein und Zeit, Gesamtausgabe*, vol. 2 (Frankfurt: Vittorio Klostermann, 1977).

Heidegger, Martin, *Phenomenological Interpretations of Aristotle*, trans. Richard Rojcewicz (Bloomington and Indianapolis, IN: Indiana University Press, 1985); *Phänomenologische Interpretationen zu Aristotles: Einführung in die Phänomenologische Forschung, Gesamtausgabe*, vol. 61 (Frankfurt: Vittorio Klostermann, 1985).

Heidegger, Martin, *The Basic Problems of Phenomenology*, trans. Albert Hofstadter (Bloomington and Indianapolis, IN: Indiana University Press, 1988).

Heidegger, Martin, *The Phenomenology of Religious Life*, trans. Matthias Fritsch and Jennifer Anna Gosetti-Ferecei (Bloomington and Indianapolis, IN: Indiana University Press, 2004); *Phänomenologie des Religiösen Lebens: Gesamtausgabe*, vol. 60 (Frankfurt: V. Klostermann, 1995).

Helmreich, S., *Alien Ocean: Anthropological Voyages in Microbial Seas* (Berkeley, CA: University of California Press, 2009).

Hervieu-Léger, Danielle, 'The Role of Religion in Establishing Social Cohesion', in M. Michalski (ed.), *Religion in the New Europe* (New York: Central European University Press, 2006), pp. 45–63.

Hessayon, Ariel and Sarah L.T. Apertrei (eds.), *An Introduction to Jacob Boehme: Four Centuries of Thought and Reception* (London: Routledge, 2014).

Hird, Myra, *The Origins of Sociable Life: Evolution after Science Studies* (New York: Palgrave Macmillan, 2009).

Hirstenstein, Stephen, *Unlimited Mercifier* (Oxford: Aqua Publishing, 1999).

Hiscock, Peter, 'Learning in Lithic Landscapes: A Reconsideration of the Hominid "Toolmaking" Niche', *Biological Theory*, vol. 9, 2014, pp. 27–41.

Hobbes, Thomas, *Leviathan*, edited with notes by Edwin Curley (Indianapolis, IN and Cambridge: Hackett, 1994).

Hoffmeyer, Jesper, 'Some Semiotic Aspects of the Psycho-Physical Relation: The Endo-Exosemiotic Boundary', in Thomas A. Sebeok and Jean Umiker-Sebeok (eds.), *Biosemiotics: The Semiotic Web 1991* (Berlin and New York: Mouton de Gruyter, 1992), pp. 101–23.

Hoffmeyer, Jesper, *Biosemiotics: An Examination into the Signs of Life and the Life of Signs*, trans. Jesper Hoffmeyer and Donald Favaresceau (Scranton, PA and London: University of Scranton Press, 2008).

Hollon, Bryan C., *Everything Is Sacred: Spiritual Exegesis in the Political Theology of Henry de Lubac* (Eugene, OR: Cascade Books, 2009).

Hopcroft, Rosemary L. (ed.), *The Oxford Handbook of Evolution, Biology, and Society* (Oxford: Oxford University Press, 2018).

Horn, Friedrich Wilhem, *Das Angel des Geistes: Studien zur paulischen Pneumatologie*, Forschungen zur Religion und Literatur des Alten und Neuen Testaments 154 (Göttingen: Vandenhoeck and Ruprecht, 1992).

Houben, Jan, 'Liberation and Natural Philosophy in Early Vaiśeṣika: Some Methodological Problems', *Asiatische Studien: Zeitschrift der Schwerizeerischen Asiengesellschaft*, vol. 2, 1994, pp. 711–48.

Huang, Tsung-hsi, *Waiting for the Dawn, a Plan for the Prince (Ming-I tai-fang lu)*, trans. Wm Theodore de Bary (New York: Columbia University Press, 1993).

Hubert, Henri and Marcel Mauss, *Sacrifice: Its Nature and Function*, trans. W.D. Halls (London: Cohen and West, 1964). First published as 'Essai sur la nature et la function du sacrifice' in *L'année sociologique*, vol. 2, 1897, pp. 29–138.

Hucker, Charles, *The Ming Dynasty: Its Origins and Evolving Institutions* (Ann Arbor, MI: University of Michigan Papers in Chinese Studies, 1978).

Hugenberg, Kurt and John Paul Wilson, 'Faces Are Central to Social Cognition', in Donal E. Carlston (ed.), *The Oxford Handbook of Social Cognition* (Oxford: Oxford University Press, 2013), pp. 167–93.

Huntington, Samuel P., *The Clash of Civilizations and the Remaking of our World Order* (London: Simon and Schuster, 1996).

Hurtado, Larry W., *Destroyer of the Gods: Early Christian Distinctiveness in the Roman World* (Baltimore, MD: Baylor University Press, 2016).

Husserl, Edmund, *Crisis in European Sciences and Transcendental Phenomenology*, trans. David Carr (Evanston, IL: Northwestern University Press, 1970).

Hyers, M. Conrad, *Zen and the Comic Spirit* (London: Rider, 1974).

Ingalls, Daniel H.H., 'Cynics and Pāśupatas: Seeking Dishonor', *Harvard Theological Review*, vol. 55 (4), 1962, pp. 281–98.

Ingold, Tim, *Being Alive: Essays on Movement, Knowledge and Description* (London and New York: Routledge, 2011).

Ingold, Tim, 'Prospect', in Tim Ingold and Gisli Palsson (eds.), *Biosocial Becomings: Integrating Social and Biological Anthropology* (Cambridge: Cambridge University Press, 2013), pp. 1–21.

Ingold, Tim and Gisli Palsson (eds.), *Biosocial Becomings: Integrating Biological and Social Anthropology* (Cambridge: Cambridge University Press, 2013).

Inwood, Michael, *Hegel* (London and New York: Routledge, 1983).

Irigaray, Luce, *Between East and West: From Singularity to Community*, trans. Stephen Pluhácek (New York: Columbia University Press, 2003).

Irigaray, Luce, 'The Age of Breath', *Key Writings* (London: Continuum, 2004), pp. 165–70.

Isaacson, Harunaga, 'Notes on the Manuscript Transmission of the Vaiśeṣikasūtra and its Earliest Commentaries', *Asiatische Studien: Zeitschrift der Schweizerischen Asiengesellschaft*, vol. 2, 1994, pp. 749–79.

Isayeva, Natalia, *Shankara and Indian Philosophy* (Albany, NY: SUNY Press, 1993).

Israel, J., *Radical Enlightenment: Philosophy and the Making of Modernity 1650–1750* (Oxford: Oxford University Press, 2001).

Ivanhoe, Philip J. and Bryan W. van Norden, *Readings in Classical Chinese Philosophy* (New York: Hackett, 2001).

Jack, Anthony I., Jared Parker Friedman, Richard Eleftherios Boyatzis, and Scott Nolan Taylor, 'Why Do You Believe in God? Relationships between Religious Belief, Analytic Thinking, Mentalizing and Moral Concern', *Plos One*, vol. 11 (3), 2016, doi: 10.1371/journal.pone.0149989.

Jack, Anthony I., Abigail J. Dawson, Katelyn L. Begany, Regina L. Leckie, Kevin P. Barry, Angela C. Ciccia, and Abraham Z. Snyder, 'fMRI Reveals Reciprocal Inhibition between Social and Physical Cognitive Domains', *Neuroimage*, vol. 66, 2013, pp. 385–401.

Jack, R.E., Oliver G.B. Garrod, Hui Yu, Roberto Caldera, and Philippe G. Schyns, 'Facial Expressions of Emotion Are Not Culturally Universal', *Proceedings of the National Academy of Sciences of the United States of America*, vol. 109 (19), 2012, pp. 7241–4.

Jaffa, Harry V., *Thomism and Aristotelianism* (Chicago, IL: University of Chicago Press, 1952).

Jankélévitch, Vladimir, *Henri Bergson*, trans. Nils V. Schott (Durham, NC and London: Duke University Press, 2015 [1959]).

Jeanrond, Werner, *A Theology of Love* (London: T and T Clark, 2010).

Jonas, Hans, *The Phenomenon of Life: Toward a Philosophical Biology*, trans. L. Vogel (Evanston, IL: Northwestern University Press, 2001 [1966]).

Jones, James W., *Can Science Explain Religion? The Cognitive Science Debate* (Oxford: Oxford University Press, 2016).

Juarrero, Alicia, *Dynamics in Action: Intentional Behaviour as a Complex System* (Cambridge, MA: MIT Press, 1999).

Jullien, François, *Vital Nourishment: Departing from Happiness*, trans. Arthur Goldhammer (New York: Zone Books, 2007).

Jullien, François, *The Philosophy of Living*, trans. Michael Richardson and Kryzysztof Fijalkowski (London, New York, Calcutta: Seagull Books, 2016).

Jung, Carl Gustav, *Symbols of Transformation: An Analysis of the Prelude to a Case of Schizophrenia*, trans. R.E.C. Hull (London and New York: Routledge, 1956).

Jung-Yeup Kim, *Zhang Zai's Philosophy of Qi: A Practical Understanding* (Lanham, MD, Boulder, CO, New York, London: Lexington Books, 2015).

Junyi, Tang, 'Chang Tsai's Theory of Mind and its Metaphysical Basis', *Philosophy East and West*, vol. 6, 1956, pp. 113–36.

Kafka, Franz, 'In the Penal Colony', trans. Stanley Corngold, in *Kafka's Selected Stories*, Norton Critical Edition (New York: Norton, 2007), pp. 35–59.

Kant, Immanuel, 'Anthropology from a Pragmatic Point of View', in Günther Zöller and Robert B. Louden (eds.), *Anthropology, History, and Education* (Cambridge: Cambridge University Press, 2007).

Karfik, Filip and Euree Song (eds.), *Plato Revived: Essays on Ancient Platonism in Honour of Dominic O'Meara* (Boston, MA and Berlin: Walter de Gruyter, 2013).

Katzenstein, Peter, 'Civilizational States, Secularisms and Religions', in Craig Calhoun, Mark Juergensmeyer, and Jonathan Van Antwerpen (eds.), *Rethinking Secularism* (New York: Oxford University Press, 2011), pp. 145–65.

Kelly, Anthony J., '"The Body of Christ Amen!" The Expanding Incarnation', *Theological Studies*, vol. 71, 2010, pp. 792–816.

Kendal, Jeremy, Jamshid J. Tehrani, and John Odling-Smee, 'Introduction: Human Niche-Construction in Interdisciplinary Focus', *Philosophical Transactions of the Royal Society B: Biological Sciences*, vol. 366, 2011, pp. 785–92.

Keynton, K.M. and T.A. Holland (eds.), *Excavation at Jericho: The Architecture and Stratigraphy of the Tel* (London: British School of Archaeology at Jerusalem, 1981).

King, Sallie B., *Buddha Nature* (Albany, NY: SUNY Press, 1991).

Kirschner, Marc W. and John C. Gerhart, *The Plausibility of Life: Resolving Darwin's Dilemma* (New Haven, CT and London: Yale University Press, 2005).

Kisiel, Theodore, *The Genesis of Heidegger's* Being and Time (Berkeley, CA: University of California Press, 1993).

Kleeman, Terry F., *Celestial Masters: History and Ritual in Early Daoist Communities* (Cambridge, MA: Harvard University Press, 2016).

Knipe, David M., 'Sapiṇḍakaraṇa: The Hindu Rite of Entry into Heaven', in Frank Reynolds and Earle H. Waugh (eds.) *Religious Encounters with Death: Insights from the History and Anthropology of Religions* (Philadelphia, PA: Pennsylvania State University Press, 1977), pp. 111–24.

Knutson, B., 'Facial Expressions of Emotion Influence Interpersonal Trait Inferences', *Journal of Non-Verbal Behaviour*, vol. 20, 1996, pp. 165–82.

Kobayashi, H. and S. Kohshim, 'Unique Morphology of the Human Eye and its Adaptive Meaning: Comparative Studies on External Morphology of the Primate Eye', *Journal of Human Evolution*, vol. 40, 2001, pp. 419–35.

Koca, Basak, Mehmet Sagir, and Ismail Özer, 'Secular Changes in the Height of the Inhabitants of Anatolia (Turkey) from the 10th Millennium B.C. to the 20th Century A.D.', *Economics and Human Biology*, vol. 9, 2011, pp. 211–19.

Kojève, A., *Introduction to the Reading of Hegel*, trans. James H. Nichols (Ithaca, NY: Cornell University Press, 1980).

Krell, David Farell, *The Tragic Absolute: German Idealism and the Languishing of God* (Bloomington and Indianapolis, IN: Indiana University Press, 2005).

Kross, Ethan and Kevin K. Ochsner, 'Integrating Research on Self-Control across Multiple Levels of Analysis: Insights from Social Cognitive and Affective Neuroscience', in Ran R. Hassin, Kevin N. Ochsner, and Yaacov Trope (eds.), *Self Control in Society, Mind, and Brain* (Oxford: Oxford University Press, 2010), pp. 76–92.

Kühl, H.S. et al., 'Chimpanzee Accumulative Stone Throwing', *Scientific Reports*, vol. 6, 2016, p. 22219.

Kull, Kalevi, Claus Emmeche, and Jasper Hoffmeyer, 'Why Biosemiotics? An Introduction to our View on the Biology of Life Itself', in Kalevi Kull and Claus Emmeche (eds.), *Towards a Semiotic Biology: Life Is the Action of Signs* (London: Imperial College Press, 2014), pp. 1–21.

Kuriyama, Shigehisa, *The Expressiveness of the Body and the Divergence of Greek and Chinese Medicine* (New York: Zone Books, 2002).

Laidlaw, James, *Riches and Renunciation: Religion, Economy, and Society among the Jains* (Oxford: Clarendon Press, 1995).

Lancy, David F., 'Homo faber juvenalis: A Multidisciplinary Survey of Children as Tool Makers/Users', *Childhood in the Past*, vol. 10 (1), 2017, pp. 72–90.

Lapidge, Michael, 'Stoic Cosmology', in J.M. Rist (ed.), *The Stoics* (Berkeley and Los Angeles, CA: University of California Press, 1978), pp. 161–85.

Latour, Bruno, *We Have Never Been Modern*, trans. Catherine Porter (Cambridge, MA: Harvard University Press, 1993).

Latour, Bruno, *Reassembling the Social: An Introduction to Actor Network Theory* (Oxford: Oxford University Press, 2005).

Latour, Bruno, *Modes of Existence: An Anthropology of the Moderns* (Cambridge, MA: Harvard University Press, 2013).

Law, John, 'The Actor Network Resource', http://www.lancaster.ac.uk/fass/centres/css/ant/ant.htm, accessed 27 August 2016.

Leach, Edmund, *Culture and Communication: The Logic by Which Symbols Are Connected* (Cambridge: Cambridge University Press, 1976).

Leinhardt, Godfrey, *Divinity and Experience: Religion among the Dinka* (Oxford: The Clarendon Press, 1987).

Leslie, C.R., *Memoirs of John Constable Composed Chiefly of his Letters* (London: Phaidon Press, 1951 [1845]).

Levinson, Stephen, *Pragmatics* (Cambridge: Cambridge University Press, 1983).

Lévi-Strauss, Claude, 'The Sorcerer and his Magic', in *Structural Anthropology* vol. 1, trans. Claire Jackobsen and Brooke G. Schoepf (London: Penguin, 1963), pp. 167–85.

Lewis, C.S., *The Weight of Glory* (London: Harper Collins, 1980 [1941]).

Lewis, Geraint F. and Luke A. Barnes, *A Fortunate Universe: Life in a Finely Tuned Cosmos* (Cambridge: Cambridge University Press, 2016).

Lewis-Williams, David, *The Mind in the Cave* (London: Thames and Hudson, 2002).

Lewontin, R.C., 'Gene, Organism, and Environment', in D.S. Bendall (ed.), *Evolution from Molecules to Men* (Cambridge: Cambridge University Press, 1983), pp. 273–85.

Lewontin, Richard, 'Foreword' to Susan Oyama, *The Ontogeny of Information: Developmental Systems and Evolution*, 2nd ed. (Durham, NC: Duke University Press, 2000), pp. xii–xiii.

Leyser, Ottoline and Stephen Day, *Mechanisms in Plant Development* (Oxford: John Wiley and Sons, 2009).

Lipner, Julius J., *The Face of Truth: A Study of Meaning and Metaphysics in the Vedantic Theology of Ramanuja* (London: Macmillan, 1986).

Loewe, Michael, *Ways to Paradise: The Chinese Quest for Immortality* (London: George Allen & Unwin, 1979).

Löw, Reinhold, *Philosophie des Lebendigen. Der Begriff des Organischen bei Kant, sein Grund und seine Aktualitat* (Munich: Suhrkamp Verlag, 1980).

Lubac, Henri de, *Surnaturel: Études historiques* (Paris: Aubier, 1946).

Lubac, Henri de, *Teilhard de Chardin, the Man and his Meaning*, trans. Réne Hague (New York: Hawthorn Press, 1964).

Lubac, Henri de, *A Brief Catechesis on Nature and Grace*, trans. Richard Arnandez (San Francisco, CA: Ignatius, 1984).

Luhmann, Niklas, *A Systems Theory of Religion*, trans. David Brenner (Stanford, CA: Stanford University Press, 2000).

Luhmann, Niklas, *Theory of Society*, 2 vols, trans. Rhodes Barrett (Stanford, CA: Stanford University Press, 2012).

Luhmann, Niklas, *Introduction to Systems Theory*, trans. Peter Gilgen (Cambridge: Polity Press, 2013).

Luhrmann, Tanya M., *Persuasions of the Witches Craft: Ritual Magic in Contemporary England* (London: Picador, 1994).

Luper, Steven (ed.), *Life and Death* (Cambridge: Cambridge University Press, 2014).

MacLarnon, Anne, 'The Anatomical and Physiological Basis of Human Speech Production: Adaptations and Exaptions', in K. Gibson and M. Tallerman (eds.), *The Oxford Handbook of Language Evolution* (Oxford: University Press, 2011), pp. 224–35.

Madeley, John, 'Religion and the State', in Jeffrey Haynes (ed.), *Religion and Politics* (London: Routledge, 2009), pp. 174–91.

Malabou, Catherine, *The Ontology of the Accident: An Essay on Destructive Plasticity* (Cambridge: Polity, 2012).

Malamoud, Charles, *Cooking the World: Ritual and Thought in Ancient India* (Delhi: Oxford University Press, 1996).

Malamoud, Charles, 'Semantics and Rhetoric in the Hindu Hierarchy of the "Aims of Man"', in *Cooking the World: Rituals and Thought in Ancient India* (Delhi: Oxford University Press, 1996), pp. 109–29.

Malinar, Angelika, *The Bhagavadgītā: Doctrines and Contexts* (Cambridge: Cambridge University Press, 2007).

Marder, Michael, *Plant Thinking: A Philosophy of Vegetal Life* (New York: Columbia University Press, 2013).

Margulis, Lyne, *Symbiosis in Cell Evolution: Life and its Environment on the Early Earth* (San Francisco, CA: Freeman, 1981).

Martin, Craig, *Subverting Aristotle: Religion, History, and Philosophy in Early Modern Science* (Baltimore, MD: Johns Hopkins University Press, 2014).

Martin, Lucinda, review of Andrew Weeks' translation of Boehme's Aurora, *Aries: A Journal for the Study of Western Esotericism*, vol. 16 (2), 2016, pp. 241–5.

Martin, William, 'Woe Is the Tree of Life', in Jan Sapp (ed.), *Microbial Phylogeny and Evolution: Concepts and Controversies* (Oxford: Oxford University Press, 2005).

Maryanski, Alexandra and Jonathan H. Turner, 'The Neurology of Religion: An Explanation from Evolutionary Sociology', in Rosemary L. Hopcroft (ed.), *The Oxford Handbook of Evolution, Biology, and Society* (Oxford: Oxford University Press, 2018), pp. 113–42.

Maspero, Henri, *Daoism and Chinese Religion*, 2nd ed., trans. Frank A. Kierman (Melbourne and Basel: Quirin Press, 2014).

Masuzawa, T., *The Invention of World Religions or How European Universalism Was Preserved in the Language of Pluralism* (Chicago, IL: Chicago University Press, 2005).

Matar, Nabil, *Islam in Britain 1558–1685* (Cambridge: Cambridge University Press, 1998).

Matthews, Bruce, *Schelling's Organic Form of Philosophy: Life as a Schema of Freedom* (Albany, NY: SUNY Press, 2011).

Matthews, Gareth B., 'De Anima 2.2–4 and the Meaning of Life', in Martha C. Nussbaum and Amélie Oksenberg Rorty (eds.), *Essays on Aristotle's* De Anima (Oxford: Clarendon Press, 1992), pp. 185–93.

Maturana Romesin, Humberto, 'Autopoiesis, Structural Coupling and Cognition: A History of These and Other Notions in the Biology of Cognition', *Cybernetic and Human Knowing*, vol. 9 (3–4), 2002, pp. 5–34.

Mauss, Marcel, 'The Category of the Person', in Michael Carrithers, Steven Collins, and Steven Lukes (eds.), *The Category of the Person: Anthropology, Philosophy, History* (Cambridge: Cambridge University Press, 1985), pp. 1–25.

Mauss, Marcel, *On Prayer*, trans. Susan Leslie (New York and Oxford: Berghahn Press, 2003).

McCauley, Robert N., *Why Religion Is Natural and Science Is Not* (Oxford: Oxford University Press, 2011).

McCauley, Robert N. and E.T. Lawson, *Bringing Ritual to Mind: Psychological Foundations of Cultural Forms* (Cambridge: Cambridge University Press, 2002).

McDougall, Lorna, 'Symbols and Somatic Structures', in John Blacking (ed.), *The Anthropology of the Body* (London and New York: Academic Press, 1977), pp. 391–401.

McGinn, Bernard, *The Presence of God: A History of Western Christian Mysticism, vol. 3: The Flowering of Mysticism* (Minneapolis, MN: Crossroad Press, 1998).

McGrath, Alistair, *Inventing the Universe: Why We Can't Stop Talking about Science, Faith and God* (London: Hodder and Stoughton, 2015).

Meillassoux, Quentin, *After Finitude: An Essay on the Necessity of Contingency*, trans. Roy Brassier (London and New York: Bloomsbury, 2009).

Mele, Alfred, *Effective Intentions: The Power of Conscious Will* (New York: Oxford University Press, 2009).

Mellars, Paul and Chris Stringer (eds.), *The Human Revolution: Behavioural and Biological Perspectives on the Origins of Modern Humans* (Edinburgh: Edinburgh University Press, 1989).

Merleau-Ponty, Maurice, *The Phenomenology of Perception*, trans. Colin Smith (London and New York: Routledge, 1958).

Meszaros, Julia and Johannes Zachhuber (eds.), *Sacrifice and Modern Thought* (Oxford: Oxford University Press, 2013).

Michot, Yahya, 'La pandémie avicennienne', *Arabica (Paris)*, vol. 40, 1993, pp. 287–344.

Milbank, John, *The Word Made Strange* (Oxford: Blackwell, 1997).

Miles, Jack (general ed.) and Wendy Doniger, Donald S. Lopez, and James Robson (eds.), *The Norton Anthology of World Religions*, vol. 1 (New York and London: W.W. Norton and Company, 2015).

Mithen, Steven, *The Singing Neanderthals: The Origin of Language, Music, Mind and Body* (London: Weidenfeld and Nicolson, 2005).

Moise, Ionut, 'The Nature and Function of Vaiśeṣika Soteriology with Particular Reference to Candrānanda's Vṛtti', unpublished DPhil. thesis, Oxford University, 2018.

Moosa, Ebrahim, *Ghazzali and the Poetics of Imagination* (Oxford: Oxford University Press, 2005).

Muller, F. Max, *The Six Systems of Indian Philosophy* (London: Longman, Greens and Co., 1899).

Mumme, Patricia, 'Haunted by Śaṅkara's Ghost', in Jeffrey R. Timm (ed.), *Texts in Context: Traditional Hermeneutics in South Asia* (Albany, NY: SUNY Press, 1992), pp. 69–84.

Nadler, Steven, *Spinoza: A Life* (Cambridge: Cambridge University Press, 2001).

Nehru, Jawaharlal, *The Discovery of India* (Delhi: Oxford University Press, 1989 [1946]).

Nicholson, Andrew J., *Unifying Hinduism: Philosophy and Identity in Indian Intellectual History* (New York: Columbia University Press, 2010).

Noe, Alva, *Action in Perception* (Cambridge, MA: MIT Press, 2006).

Nonaka, Tetsushi, Blandine Bril, and Robert Rein, 'How Do Stone Knappers Predict and Control the Outcome of Flaking? Implications for Understanding Early Stone Tool Technology', *Journal of Human Evolution*, vol. 59, 2010, pp. 155–67.

Norden, Bryan W. van (ed.), *Confucius and the Analects: New Essays* (New York: Oxford University Press, 2001).

Nowotny, Helga and Giuseppe Testa, *Naked Genes: Reinventing the Human in the Molecular Age* (Cambridge, MA and London: MIT Press, 2010).

Nummenmaa, J. Hyönä and J.K. Hietanen, 'I'll Walk This Way: Eyes Reveal the Direction of Locomotion and Make Passersby Look and Go the Other Way', *Psychological Science*, vol. 20, 2009, pp. 1454–8.

Nussbaum, Martha, 'Compassion, the Basic Social Emotion', *Social Philosophy and Policy*, vol. 13 (1), 1996, pp. 27–58.

Nussbaum, Martha C. and Amélie Oksenberg Rorty (eds.), *Essays on Aristotle's* De Anima (Oxford: Clarendon Press, 1992).

Oberlies, T., 'Die Śvetāśvatara-Upaniṣad: Einleitung-Edition und Übersetzung von Adhyāya I', *Weiner Zeitschrift für die Kunde Südasiens*, vol. 39, 1995, pp. 61–102.

Odling-Smee, F. John, 'Niche Construction Phenotypes', in H.C. Plotkin (ed.), *The Role of Behaviour in Evolution* (Cambridge, MA: MIT Press, 1988).

Odling-Smee, F. John, Kevin N. Laland, and Marcus W. Feldman, *Niche Construction: A Neglected Process in Evolution* (Princeton, NJ: Princeton University Press, 2003).

O'Hear, Anthony (ed.), *Philosophy, Biology, and Life* (Cambridge: Cambridge University Press, 2015).

Olivelle, Patrick, *The Āśrama System: The History and Hermeneutics of a Religious Institution* (Oxford: Oxford University Press, 1993).

Onians, R.B., *The Origins of European Thought about the Body, the Mind, the Soul, the World, Time and Fate* (Cambridge: Cambridge University Press, 1951).

Oyama, Susan, *The Ontogeny of Information: Developmental Systems and Evolution*, 2nd ed. (Durham, NC: Duke University Press, 2000).

Padoux, André, *Comprendre le Tantrisme: les sources hindoues* (Paris: Albin Michel, 2010).

Padoux, André, *The Hindu Tantric World: An Overview* (Chicago, IL: Chicago University Press, 2017).

Pan, Dawei, 'Is Chinese Culture Dualist? An Answer to Edward Slingerland from a Medical Philosophical Viewpoint', *Journal of the American Academy of Religion*, vol. 85 (4), 2018, pp. 1017–31.

Panikkar, Raimundo, *The Vedic Experience: Mantramañjari: An Anthology of the Vedas for Modern Man and Contemporary Celebration* (London: Darton, Longman and Todd, 1977).

Parker, Robert, *On Greek Religion* (Ithaca, NY and London: Cornell University Press, 2011).

Parkinson, B., 'Do Facial Movements Express Emotions or Communicate Motives?', *Personality and Social Psychology Review*, vol. 9, 2005, pp. 278–311.

Passingham, R.E. and S.P. Wise, *The Neurobiology* of the Prefrontal Cortex: Anatomy, Evolution, and the Origin of Insight (Oxford: Oxford University Press, 2014).

Pearson, Keith Ansell, *Germinal Life: The Difference and Repetition of Deleuze* (London: Routledge, 1999).

Pearson, Keith Ansell, *Henri Bergson: An Introduction* (London: Routledge, 2011).

Pellegrini, Maura, John Pouncett, M. Jay et al., 'Tooth Enamel Oxygen "Isoscapes" Show a High Degree of Human Mobility in Prehistoric Britain', *Scientific Reports*, vol. 6 (34986), 2016, pp. 1–9.

Pessin, Sarah, 'Matter, Form and the Corporeal World', in Tamar Rudavsky and Steven Nadler (eds.), *The Cambridge History of Jewish Philosophy: From Antiquity to the Seventeenth Century* (Cambridge: Cambridge University Press, 2009), pp. 269–301.

Pessin, Sarah, 'Solomon Ibn Gabirol [Avicebron]', *The Stanford Encyclopedia of Philosophy* (Summer 2014 Edition), Edward N. Zalta (ed.), http://plato.stanford.edu/archives/sum2014/entries/ibn-gabirol.

Petropoulou, Maria-Zoe, *Animal Sacrifice in Greek Religion, Judaism, and Christianity, 100 BC to AD 200* (Oxford: Oxford University Press, 2008).

Pinkard, Terry, *Hegel's Phenomenology: The Sociality of Reason* (Cambridge: Cambridge University Press, 2014).

Pinker, Steven, *Enlightenment Now: The Case for Reason, Science, Humanism and Progress* (London: Allen Lane, 2018).

Plotkin, H.C. (ed.), *The Role of Behaviour in Evolution* (Cambridge, MA: MIT Press, 1988).

Pollard, David, 'Ch'i in Chinese Literary Theory', in Adele Austin Rickett (ed.), *Chinese Approaches to Literature from Confucius to Liang Chi-chao* (Princeton, NJ: Princeton University Press, 1978), pp. 43–66.

Praet, Istvan, 'Humanity and Life as the Perpetual Maintenance of Specific Efforts: A Reappraisal of Animism', in Tim Ingold and Gisli Palsson (eds.), *Biosocial Becomings: Integrating Biological and Social Anthropology* (Cambridge: Cambridge University Press, 2013), pp. 191–210.

Prayer Book for Orthodox Christians (Boston, MA: Holy Transfiguration Monastery, 1987).

Pregasio, Fabrizio, 'Which Is the Daoist Immortal Body?', *Micrologus*, vol. 26, 2018, pp. 385–407.

Pregasio, Fabrizio, 'The Alchemical Body in Daoism', in Manuel Vasquez and Vasudha Narayana (eds.), *The Wiley-Blackwell Companion to Material Religion* (Oxford: Wiley-Blackwell, 2019).

Pyysiäinen, I., *How Religion Works: Towards a New Cognitive Science of Religion* (Leiden: Brill, 2001).

Pyysiäinen, Ilkka and Marc Hauser, 'The Origins of Religion: Evolved Adaptation or By-Product?', *Trends in Cognitivist Science*, vol. 14 (3), 2009, pp. 104–9.

Rabinow, Paul (ed.), *Essays on the Anthropology of Reason* (Princeton, NJ: Princeton University Press, 1996).

Radhakrishnan, S., *The Hindu View of Life* (London: George Allen and Unwin, 1927).

Rahman, Shahid, Tony Street and Hassan Tahiri (eds.), *The Unity of Science in the Arabic Tradition: Science, Logic, Epistemology and their Interactions* (Dordrecht, Boston, MA, and London: Springer, 2008).

Rahner, Karl, *Theological Investigations vol. 2: Man in the Church*, trans. Karl H. Kruger (London: Darton, Longman and Todd, 1963).

Raichle, M.E. et al., 'A Default Mode of Brain Function', *Proceedings of the National Academy of Science USA*, vol. 98 (2), 2001, pp. 676–82.

Ram-Prasad, C., *Divine Self, Human Self: the Philosophy of Being in Two Gita Commentaries* (London: Bloomsbury, 2013).

Rappaport, Roy, *Ritual and Religion in the Making of Humanity* (Cambridge: Cambridge University Press, 1999).

Reilly, Kevin (ed.), *Readings in World Civilizations*, vol. 1 (New York: St Martin's Press, 1994).

Reisebrodt, Martin, *The Promise of Salvation: A Theory of Religion* (Chicago, IL: University of Chicago Press, 2014).

Richards, Robert J., *The Romantic Conception of Life: Science and Philosophy in the Age of Goethe* (Chicago, IL: Chicago University Press, 2002).

Ricoeur, Paul, 'Phenomenology and Hermeneutics', in John B. Thompson (ed. and trans.), *Paul Ricoeur: Hermeneutics and the Social Sciences* (Cambridge: Cambridge University Press, 1981), pp. 101–28.

Ricoeur, Paul, 'The Model of the Text: Meaningful Action Considered as a Text', in *Hermeneutics and the Human Sciences*, trans. John B. Thompson (Cambridge: Cambridge University Press, 1981), pp. 197–221.

Ricoeur, Paul, 'Qu'est-ce qu'un texte?', in *Du texte a l'action. Essais hermeneutique II* (Paris: Editions de du Seul, 1986), pp. 153–78.

Ricoeur, Paul, *Oneself as Another*, trans. Kathleen Blamey (Chicago, IL: Chicago University Press, 1992).

Ricoeur, Paul, 'Hermeneutics of the Idea of Revelation', in *Hermeneutics*, trans. David Pellauer (Cambridge: Polity Press, 2013), pp. 111–70.

Riel, Gerd van and Pierre Destrée (eds.), *Ancient Perspectives on Aristotle's* De Anima, Ancient and Medieval Philosophy Series I 41 (Leuven: Leuven University Press, 2009).

Rist, John M., *Augustine: Ancient Thought Baptized* (Cambridge: Cambridge University Press, 1994).

Roberts, Richard, *Religion, Theology and the Human Sciences* (Cambridge: Cambridge University Press, 2002).

Robinet, I., *Taoist Meditation: The Mao-shan Tradition of Great Purity* (Albany, NY: SUNY Press, 1993 [1979]).

Roetz, Heiner, *Confucian Ethics of the Axial Age* (Albany, NY: SUNY Press, 1993).

Romano, Claude, *At the Heart of Reason*, trans. Michael B. Smith and Claude Romano (Evanston, IL: Northwestern University Press, 2015).

Rose, Nikolas, *The Politics of Life Itself: Biomedicine, Power, and Subjectivity in the Twenty-First Century* (Princeton, NJ and Oxford: Princeton University Press, 2007).

Rose, Nikolas, 'The Human Sciences in a Biological Age', *Theory, Culture and Society*, vol. 30 (1), 2013, pp. 3–34.

Rosen, Robert, *Life Itself: A Comprehensive Inquiry into the Nature, Origin, and Fabrication of Life* (New York: Columbia University Press, 1991).

Rousseau, Jean-Jacques, *Discourse on the Origin of Inequality*, trans. Franklin Philip (Oxford: Oxford University Press, 1994).

Rowland, Christopher, *The Open Heaven: A Study of Apocalyptic Judaism and Early Christianity* (London: SPCK, 1982).

Rowland, Christopher and C.R.A. Morray-Jones, *The Mystery of God: Early Jewish Mysticism and the New Testament* (Leiden: Brill, 2009).

Ruse, Michael, 'Evo-Devo: A New Evolutionary Paradigm?', in Anthony O'Hear (ed.), *Philosophy, Biology, and Life* (Cambridge: Cambridge University Press, 2015), pp. 105–24.

Russell, Bertrand, 'The Philosophy of Bergson', *The Monist*, vol. 22, 1912, pp. 321–47.

Salguero, C. Pierce, *Translating Buddhist Medicine in Medieval China* (Philadelphia, PA: University of Pennsylvania Press, 2014).

Sanderson, Alexis, 'Purity and Power among the Brahmans of Kashmir', in Michael Carrithers, Steven Collins, and Steven Lukes (eds.), *The Category of the Person: Anthropology, Philosophy, History* (Cambridge: Cambridge University Press, 1985), pp. 190–216.

Sanderson, Alexis, 'Meaning in Tantric Ritual', in A.-M. Blondeau and K. Schipper (eds.), *Essais sur le Rituel III: Colloque du Centenaire de la Section des Sciences religieuses de l'École Pratique des Hautes Études*, Bibliothèque de l'École des Hautes Études, Sciences Religieuses, Volume CII (Louvain-Paris: Peeters, 1995), pp. 15–95.

Sanderson, Alexis, 'Levels of Initiation and Practice in the Śaivism of Abhinavagupta' (1995 unpublished MS); commentary on the first verse of the *Tantrasāra* (Handout, Trinity Term Oxford, 2003).

Sanderson, Alexis, 'The Śaiva Age: The Rise and Dominance of Śaivism during the Early Medieval Period', in Shingo Einoo (ed.), *Genesis and Development of Tantrism* (Tokyo: Institute of Oriental Culture, University of Tokyo, 2009), pp. 41–350.

Sanderson, Alexis, 'Śaiva Literature', *Journal of Indological Studies (Kyoto)*, vols 24 and 25, 2014, pp. 1–113.

Sapp, Jan (ed.), *Microbial Phylogeny and Evolution: Concepts and Controversies* (Oxford: Oxford University Press, 2005).

Schaefer, Donovan O., *Religious Affects: Animality, Evolution, and Power* (Durham, NC and London: Duke University Press, 2015).

Schelling, F.W., *System of Transcendental Idealism*, trans. Peter Heath (Charlottesville, VA: University Press of Virginia, 1978).

Schelling, F.W.J., *First Outline of a System of Philosophy of Nature*, trans. Keith R. Peterson (Albany, NY: SUNY Press, 2004).

Schilbach, Leonard, Bert Timmermans, Vasudevi Reddy, Alan Costall, Gary Bente, Tobias Schlicht, and Kai Vogeley, 'Towards a Second Person Neuroscience', *Behavioural and Brain Sciences*, vol. 36 (4), 2013, pp. 393–414.

Schipper, Kristofer, *The Taoist Body*, trans. Karen C. Duval (Berkeley, CA: University of California Press, 1993).

Schleiermacher, Friedrich, *On Religion: Speeches to its Cultured Despisers*, trans. Richard Crouter (Cambridge: Cambridge University Press, 1988).

Schmidt, Klaus, *Göbekli Tepe: A Stone Age Sanctuary in South-Eastern Anatolia* (Berlin: Ex Oriente e.V., 2012).

Schmitt, Carl, *Political Theology: Four Chapters on the Concept of Sovereignty* (Chicago, IL: University of Chicago Press, 2005).

Schmitt, Carl, *The Nomos of the Earth in the International Law of the Jus Publicum Europaeum*, trans. G.L. Ulman (New York: Telos Press, 2006).

Schrödinger, Edwin, *What Is Life?* (Cambridge: Cambridge University Press, 1992).

Schutz, Alfred, *The Structure of the Life-World*, vol. 1, trans. Richard M. Zaner and Tristram Engelhardt Jr. (Evanston, IL: Northwestern University Press, 1973).

Scott, Michael, 'Steps to a Methodological Nondualism', in *The Group for Debates in Anthropological Theory (GDAT), The University of Manchester: The 2011 Annual Debate—Nondualism Is Philosophy Not Ethnography*, ed. Soumhya Venkatesan et al., pp. 303–8, 356; *Critique of Anthropology*, vol. 33 (3), pp. 300–60.

Sebeok, Thomas A. and Jean Umiker-Sebeok (eds.), *Biosemiotics: The Semiotic Web 1991* (Berlin and New York: Mouton de Gruyter, 1992).

Selby, Martha, 'Narratives of Conception, Gestation, and Labour in Sanskrit Ayurvedic Texts', *Asian Medicine*, vol. 1 (2), 2005, pp. 254–75.

Sergeant, Bernard, *Genése de l'Inde* (Paris: Payot, 1997).

Seshami, S., A.I. Blazejewska, S. Mckown, J. Caucutt, M. Dighe, C. Gatenby, and C. Studholme, 'Detecting Default Mode Networks in Utero by Integrated 4D fMRI Reconstruction and Analysis', *Hum Brain Mapp*, vol. 37 (11), 2016, pp. 4158–78.

Shapin, Steven, *A Social History of Truth: Civility and Science in Seventeenth-Century England* (Chicago, IL: University of Chicago Press, 1995).

Sheets-Johnstone, Maxine, *The Corporeal Turn: An Interdisciplinary Reader* (Charlottesville, VA: Imprint-Academic-Dotcom, 2009).

Silburn, Lilian, *Kuṇḍalinī: The Energy from the Depths*, trans. Jacques Gontier (Albany, NY: SUNY Press, 1990).

Sinha, Chris, 'Language and Other Artifacts: Socio-Cultural Dynamics of Niche Construction', *Frontiers in Psychology*, vol. 6, 2015, p. 1601.

Skof, Lenart, *Breath of Proximity: Intersubjectivity, Ethics, and Peace* (Dordrecht: Springer, 2015).

Slavich, George M. and Steven W. Cole, 'The Emerging Field of Human Social Genomics', *Clinical Psychological Science*, vol. 1 (3), 2013, pp. 331–48.

Slingerland, Edward, *What Science Offers the Humanities: Integrating Body and Culture* (Cambridge: Cambridge University Press, 2008).

Slingerland, Edward, 'Body and Mind in Early China: An Integrated Humanities–Science Approach', *Journal of the American Academy of Religion*, vol. 81 (1), 2013, pp. 6–55.

Slingerland, Edward, Ryan Nichols, Kristoffer Neilbo, and Carson Logan, 'The Distant Reading of Religious Texts: A "Big Data" Approach to Mind–Body Concepts in Early China', *Journal of the American Academy of Religion*, vol. 85 (4), 2018, pp. 985–1016.

Slone, D.J. and J. Van Slyke, *The Attraction of Religion* (London: Bloomsbury Academic Press, 2015).

Sloterdjik, Peter, *Bubbles*, trans. Wieland Hoban (Los Angeles, CA: Semiotext, 2011).

Sloterdijk, Peter, *You Must Change Your Life*, trans. Wieland Hoban (Cambridge: Polity Press, 2013).

Smail, David Lord, *On Deep History and the Brain* (Berkeley and Los Angeles, CA: University of California Press, 2008).

Smith, Bruce D., *The Emergence of Agriculture* (New York: Scientific American, 1995).

Smith, Daniel W., 'The Doctrine of Univocity: Deleuze's Ontology of Immanence', in Mary Bryden (ed.), *Deleuze and Religion* (London: Routledge, 2001), pp. 167–83.

Smith, John Maynard and Eörs Szathmáry, *The Origins of Life: From the Birth of Life to the Origins of Language* (Oxford: Oxford University Press, 1999).

Smith, Jonathan Z., *Imagining Religion, from Babylon to Jonestown* (Chicago, IL: University of Chicago Press, 1982).

Smith, Jonathan Z., 'The Domestication of Sacrifice', in R.G. Hamilton-Kelly (ed.), *Violent Origins* (Stanford, CA: Stanford University Press, 1987), pp. 278–304.

Smith, M.L., F. Cottrell, F. Gosselin, and P.G. Schyns, 'Transmitting and Decoding Facial Expressions', *Psychological Science*, vol. 16, 2005, pp. 184–9.

Smith, Steven B., *Spinoza's Book of Life: Freedom and Redemption in Ethics* (New Haven, CT: Yale University Press, 2003).

Sorabji, Richard, 'Body and Soul in Aristotle', *Philosophy*, vol. 29, 1974, pp. 3–89.

Sørensen, J., 'Religion, Evolution, and an Immunology of Cultural Systems', *Evolution and Cognition*, vol. 10 (1), 2004, pp. 61–73.

Spaemann, Walter, *Persons: The Difference between 'Someone' and 'Something'*, trans. Oliver O'Donovan (Oxford: Oxford University Press, 2012).

Sperber, Dan, *Rethinking Symbolism* (Cambridge: Cambridge University Press, 1975).

Spinoza, *Ethics*, trans. R.H.M. Elwes, *Benedict de Spinoza: On the Improvement of the Understanding, The Ethics, Correspondence* (New York: Dover Publications, 1955 [1883]).

Staal, Frits, *Advaita and Neoplatonism: A Critical Study in Comparative Philosophy* (Madras: University of Madras, 1961).

Staal, Frits, *Rules without Meaning: Ritual, Mantras, and the Human Sciences* (New York: Peter Lang, 1989).

Staal, Frits, 'The Indian Sciences', in Gavin Flood (ed.), *The Blackwell Companion to Hinduism* (Oxford: Blackwell, 2003), pp. 348–409.

Sterelney, Kim, *The Evolved Apprentice: How Evolution Made Humans Unique* (Cambridge, MA: MIT Press, 2012).

Stollberg, Gunnar, 'Vitalism and Vital Force in Life Sciences: The Demise and Life of a Scientific Conception', unpublished manuscript, http://www.uni-bielefeld.de/soz/pdf/Vitalism.pdf, accessed 10 July 2017.

Stroumsa, Guy G., *La fin du sacrifice: les mutations religieuses de l'antiquité tardive* (Paris: Odile Jacob, 2005).

Stroumsa, Guy G., *A New Science: The Discovery of Religion in the Age of Reason* (Cambridge, MA: Harvard University Press, 2010).

Stroumsa, Guy, *The Making of the Abrahamic Religions in Late Antiquity* (Oxford: Oxford University Press, 2015).

Swain, James E. and S. Shaun Ho, 'Baby Smile Response Circuits of the Parental Brain', *Behavioral and Brain Sciences*, vol. 33 (6), 2010, pp. 460–1.

Takasaki, Jikido, *A Study of the Ratnagotravibhāga: Being a Treatise on the Tathāgatagarbha Theory of Mahāyāna Buddhism* (Rome: Is.M.E.O., 1964).

Taubes, Jacob, *The Political Theology of Paul* (Palo Alto, CA: Stanford University Press, 2004).

Taylor, Charles, *Hegel* (Cambridge: Cambridge University Press, 1975).

Taylor, Charles, 'Interpretation and the Sciences of Man', in *Philosophy and the Human Sciences: Philosophical Papers*, vol. 2 (Cambridge: Cambridge University Press, 1985), pp. 15–57.

Taylor, Charles, *Sources of the Self: The Making of Modern Identity* (Cambridge: Cambridge University Press, 1989).

Taylor, Charles, 'Overcoming Epistemology', in *Philosophical Arguments* (Cambridge, MA: Harvard University Press, 1995), pp. 1–19.

Taylor, Charles, 'What Was the Axial Revolution?', in Robert N. Bellah and Hans Joas (eds.), *The Axial Age and Its Consequences* (Cambridge, MA: Harvard University Press, 2012), pp. 30–46.

Taylor, Charles, *The Language Animal: The Full Shape of the Human Linguistic Capacity* (Cambridge, MA: Belknap Press, 2016).

Taylor, Richard C., 'Averroes on the Ontology of the Human Soul', *The Muslim World*, vol. 102 (2–4), 2012, pp. 580–96.

Tebbich, S., M. Taborsky, B. Fessl, and D. Blomqvist, 'Do Woodpecker Finches Acquire Tool-Use by Social Learning?', *Proceedings of the Royal Society B: Biological Sciences*, vol. 268, 2001, pp. 2189–93.

Tennie, Claudio, David R. Braun, and Shannon P. McPherron, 'The Island Test for Cumulative Culture in the Paleolithic', in Miriam N. Haidle, Nicholas J. Conrad, and Michael Bolus (eds.), *The Nature of Culture* (Dordrecht: Springer, 2016), pp. 121–33.

Tennie, Claudio, Joseph Call, and Michael Tomasello, 'Push or Pull: Imitation vs Emulation in Great Apes and Human Children', *Ethology*, vol. 112, 2006, pp. 1159–69.

Thacker, Eugene, *After Life* (Chicago, IL: University of Chicago Press, 2010).

Thapar, Romila, *Interpreting Early India* (Oxford: Oxford University Press, 1992).

Thomas, Julian, 'Ritual and Religion in the Neolithic', in Timothy Insoll (ed.), *The Oxford Handbook of the Archaeology of Ritual and Religion* (Oxford: Oxford University Press, 2011), pp. 371–86.

Thomson, George, *Fire in the Mind: Science, Faith, and the Search for Order* (London: Penguin, 1995).

Timalsena, Sthanshwar, *Tantric Visual Culture: A Cognitive Approach* (London: Routledge, 2015).

Todd, Robert B., 'The Stoics and their Cosmology in the First and Second Centuries AD', in W. Haase (ed.), *Aufstieg und Niedergang der Römischen Welt* (Berlin: de Gruyter, 1989), pp. 1365–78.

Tomasello, Michael, *Why We Cooperate* (Cambridge, MA: MIT Press, 2009).

Tomasello, Michael, *A Natural History of Human Thinking* (Cambridge, MA and London: Harvard University Press, 2014).

Tonino, Giulio, Melanie Boly, Marcello Massimini, and Kristof Koch, 'Integrated Information Theory: From Consciousness to its Physical Substrate', *Nature Reviews Neuroscience*, vol. 17 (7), 2016, pp. 450–61.

Tsing, Anna Lowenhaupt, *The Mushroom at the End of the World: On the Possibility of Life in Capitalist Ruins* (Princeton, NJ: Princeton University Press, 2015).

Tu, Youguang, 'Daoism Stresses Individual Objects', *Contemporary Chinese Thought*, vol. 30 (1), 1998, pp. 45–57.

Turner, Denys, *The Darkness of God: Negativity and Christian Mysticism* (Cambridge: Cambridge University Press, 1995).

Turner, Denys, *Faith, Reason, and the Existence of God* (Cambridge: Cambridge University Press, 2004).

Turner, Denys, *Thomas Aquinas: A Portrait* (New Haven, CT: Yale University Press, 2013).

Turner, J.H., A. Maryanski, A. K. Petersen, and A.W. Geertz, *The Emergence and Evolution of Religion: By Means of Natural Selection* (London: Routledge, 2017).

Turner, Victor, *The Forest of Symbols: Aspects of Ndembu Ritual* (Ithaca, NY: Cornell University Press, 1967).

Tweedale, Martin M., *Scotus vs. Ockham: A Medieval Dispute Over Universals [Scotus vs. Ockham]*, 2 vols (Lampeter: Edwin Mellen Press, 1999).

Twitchett, Denis and Michael Loewe, *The Cambridge History of China vol. 1: The Chin and Han Empires* (Cambridge: Cambridge University Press, 1986).

Tymieniecka, A.T. (ed.), *Timing and Temporality in Islamic Philosophy and Phenomenology of Life* (Dordrecht: Springer, 2007).

Unschuld, Paul U., *Medicine in China: A History of Ideas* (Berkeley, CA: University of California Press, 1985).

Urban, Greg, 'The I of Discourse', in Benjamin Lee and Greg Urban (eds.), *Semiotics, Self and Society* (Berlin and New York: Mouton de Gruyter, 1989), pp. 27–51.

Urgesi, C., S.M. Aggioti, M. Skrap, and F. Fabbro, 'The Spiritual Brain: Selective Cortical Lesions Modulate Human Self-Transcendence', *Neuron*, vol. 65, 2010, pp. 509–19.

van der Veer, Peter, *The Value of Comparison* (Durham, NC: Duke University Press, 2016).

Vasquez, Manuel A., *More Than Belief: A Materialist Theory of Religion* (Oxford: Oxford University Press, 2011).

Vasudeva, Somadeva, *The Yoga of the Mālinīvijayottaratantra* (Pondicherry: IFP, EFDE, 2004).

Varela, Francisco J., Eleanor Rosch, and Evan Thompson, *The Embodied Mind: Cognitive Science and Human Experience* (Cambridge, MA: MIT Press, 1992).

Venter, Craig, *Life at the Speed of Light: From the Double Helix to the Dawn of Digital Life* (London: Little, Brown, 2013).

Vetö, Miklos, 'Jacob Boehme et l'idéalisme postkantien', *Les Études philosophiques*, vol. 2, 1999, pp. 167–80.

Vieillard-Baron, Jean-Louis, 'Schelling et Jacob Boehme: Les Recherches de 1809 et la lecture "la Lettre pastorale"', *Les Études philosophiques*, vol. 2, 1999, pp. 223–42.

Vijyer, G. Van de (ed.), *Evolutionary Systems: Biological and Epistemological Perspectives in Selection and Self Organization* (Dordrecht: Kluwer, 1998).

Vries, Hent de, *Philosophy and the Turn to Religion* (Baltimore, MD and London: Johns Hopkins University Press, 1999).

Waal, Frans de, *The Age of Empathy: Nature's Lessons for a Kinder Society* (London: Souvenir Press, 2009).

Walker, Matthew D., *Aristotle on the Uses of Contemplation* (Cambridge: Cambridge University Press, 2018).

Walker, P.L. and J.T. Eng, 'Long Bone Dimensions as an Index of Socio-Economic Change in Ancient Asian Population', *American Journal of Physical Anthropology*, supplement 42, 2007, p. 241.

Walsh, David, *The Mysticism of Innerworldly Fulfillment: A Study of Jacob Boehme* (Gainesville, FL: University Press of Florida, 1983).

Wang, Robin, 'Zhou Dunyi's Diagram of the Supreme Ultimate Explained (Taiji Shuo): A Construction of the Confucian Metaphysics', *Journal of the History of Ideas*, vol. 66 (3), 2005, pp. 307–23.

Wang, Robin R., *Yinyang: The Way of Heaven and Earth in Chinese Thought and Culture* (Cambridge: Cambridge University Press, 2012).

Ward, Graham, *How the Light Gets in: Ethical Life I* (Oxford: Oxford University Press, 2016).

Ward, Keith, *The Big Questions in Science and Religion* (West Conshohocken, PA: Templeton Foundation Press, 2008).

Warneken, Felix and Michael Tomasello, 'Altruistic Helping in Human Infants and Young Chimpanzees', *Science*, vol. 311, 2006, pp. 1301–3.

Warneken, Felix and Michael Tomasello, 'Helping and Cooperation at 14 Months of Age', *Infancy*, vol. 11 (3), 2007, pp. 271–94.

Watkin, Christopher, *Difficult Atheism: Post-Theological Thinking in Alain Badiou, Jean-Luc Nancy and Quentin Meillassoux* (Edinburgh: Edinburgh University Press, 2011).

Watts, F. and L.P. Turner (eds.), *Evolution, Religion, and Cognitive Science: Critical and Constructive Essays* (Oxford: Oxford University Press, 2014).

Weeks, Andrew, *Boehme: An Intellectual Biography of the Seventeenth-Century Philosopher and Mystic* (Albany, NY: SUNY Press, 1991).

Weigel, Peter, *Aquinas on Simplicity: An Investigation into the Foundations of his Philosophical Theology* (Oxford: Peter Lang, 2008).

Welton, Donn, *The Other Husserl: The Horizons of Transcendental Phenomenology* (Bloomington and Indianapolis, IN: Indiana University Press, 2000).

Wernicke-Olesen, B. and S. L. Einarsen, 'Übungswissen in Yoga, Tantra und Asketismus,' in Almut-Barbara Renger and Alexandra Stellmacher (eds.), *Übungswissen in Religion und Philosophie: Produktion, Weitergabe, Wandel* (Berlin: Lit Verlag, 2018), pp. 241–57.

Wheeler, Wendy, *Expecting the Earth: Life, Culture, Biosemiotics* (London: Lawrence and Wishart, 2016).

Whitehouse, Harvey and Robert N. McCauley, *Mind and Religion: Psychological and Cognitive Foundations of Religiosity* (Walnut Creek, CA and Oxford: Altamira Press, 2005).

Whitehouse, Harvey and James Laidlaw, *Religion, Anthropology, and Cognitive Science* (Durham, NC: Carolina Academic Press, 2007).

Wilken, Robert Louis, *The Spirit of Early Christian Thought* (New Haven, CT: Yale University Press, 2003).

Williams, Paul, *Buddhist Thought: A Complete Introduction to the Indian Tradition* (London: Routledge, 2000).

Williams, Paul, *Mahāyāna Buddhism* (London: Routledge, 2009 [1989]).

Wilson, David Sloane, *Darwin's Cathedral: Evolution, Religion, and the Nature of Society* (Chicago, IL: Chicago University Press, 2003).

Wilson, Edward O., *The Insect Societies* (Cambridge, MA: Harvard University Press, 1971).

Wilson, Edward O., *On Human Nature* (Cambridge, MA: Harvard University Press, 1978).

Wisnovsky, Robert, *Avicenna's Metaphysics in Context* (London: Duckworth, 2003).

Wittgenstein, Ludwig, *Remarks on Frazer's Golden Bough*, trans. A.C. Miles, rev. and ed. Rush Rhees (Nottingham: Brynmill Press Ltd., 1979).

Witzel, E.J. Michael, 'Vedas and Upaniṣads', in Gavin Flood (ed.), *The Blackwell Companion to Hinduism* (Oxford: Blackwell, 2003), pp. 68–101.

Witzel, E.J. Michael, *The Origins of the World's Mythologies* (Oxford: Oxford University Press, 2012).

Wordsworth, William and S.T. Coleridge, *Lyrical Ballads 1798 and 1802* (Oxford: Oxford University Press, 2013).

Wright, M.W., *Cosmology in Antiquity* (London and New York: Routledge, 1995).

Wujastyk, Dominik, *The Roots of Ayruveda* (London: Penguin, 1998).

Wujastyk, Dominik, 'Interpréter l'image du corps humain dans l'Inde pré-moderne', in Véronique Boullier and Gilles Tarabout (eds.), *Images du corps dans le monde hindou* (Paris: CNRS, 2002), pp. 71–99.

Wujastyk, Dominik, 'The Science of Medicine', in Gavin Flood (ed.), *The Blackwell Companion to Hinduism* (Oxford: Blackwell, 2003), pp. 393–409.

Wunn, Ina, *Die Religionen in vorgeschictlicher Zeit* (Stuttgart: Kohlhammer, 2005).

Yao, Xinzhong, *Introduction to Confucianism* (Cambridge: Cambridge University Press, 2000).

Yates, Frances, *Giordano Bruno and the Hermetic Tradition* (London: Routledge, 1964).

Yü Yingshih, *Chinese History and Culture: Sixth Century BCE to Seventeenth Century* (New York: Columbia University Press, 2016).

Zachhuber, Johannes, 'Transcendence and Immanence', in Daniel Whistler (ed.), *The Edinburgh Critical History of Nineteenth-Century Christian Theology* (Edinburgh: Edinburgh University Press, 2017), chapter 9.

Zaehner, R.C., *The Bhagavad Gita* (Oxford: Oxford University Press, 1969).

Zeman, Adam and Oliver Davies, 'A Radical New Way to Understand the Brain', *Standpoint*, Sept. 2013, pp. 53–5.

Zuber, Mike A., 'Copernican Cosmo-Theism: Johann Jacob Zimmermann and the Mystical Light', *Aries: Journal for the Study of Western Esotericism*, vol. 14, 2014, pp. 215–45.

Zürcher, Erik and Jonathan Silk (eds.), *Buddhism in China: Collected Papers of Erik Zürcher* (Leiden: Brill, 2013).

Zydenbos, Robert J., *Mokṣa in Jainism, according to Umāsvāti* (Wiesbaden: Franz Steiner, 1983).

Zysk, Kenneth, *Asceticism and Healing in Ancient India: Medicine in the Buddhist Monastery* (New Delhi: MLBD, 1991).

Zysk, K.G., 'The Science of Respiration and the Doctrine of Bodily Winds in Ancient India', *Journal of the American Oriental Society*, vol. 113, 1993, pp. 198–213.

Index

abduction / abductive argument 4–5, 304–5, 367
Abhinavagupta 32, 167–73, 308–9, 315–17
Abrahamic Religions 2, 33, 115–16, 218, 228–9, 245, 271, 363
Acheulian hand axe 80
actor network theory / ANT 312–13
actuality / the actual 225–7, 235–6, 249–51, 264–5, 303–4, 308–10, 320–1
affect theory 76–7
After Finitude (Meillasoux) 335–6
Agamben, Giorgio 344–6, 349–54, 356–7
agency 30, 50, 80, 141, 300, 312–13, 317–18, 320–1, 339–41
  distributive 320–2, 339
agriculture 13–15
Al-Futūḥāt al-makkiyya (Ibn Arabi) 261
Al-Qanun fi 't-Tibb (Avicenna) 264–5
Al Ghazzālī 258–60, 264–5
alchemy 42–3
  Chinese 196–200
Altamira 86–9
Alter, Joe 156–7
altruism 69, 71–4, 76, 371–2
anabasis 96–7
Analects (*Lunyu*) 179–80
ancestors 13–16, 85, 112, 184, 188, 197
ancestrality 335–6
angels 235–6, 254–5, 259–62, 316–17, 344, 373–4
anima 26, 225, 249–53, 257
animism 276, 285–6
anthropic cosmological argument 52–3
anthropocentrism 341–2, 357
anthropology 30–1, 35–6, 40–1
  evolutionary 67–9, 90–1
  Christian 356–9
  non-dualist 314
Anthropology from a Pragmatic Point of View (Kant) 40–1
Apollo 310–11
Aquinas, Thomas 247–54, 257–8, 264–5, 267–8, 275–7, 290, 354–6, 360–1
archaeology 91, 184
arche fossil 335–6, 339–40
Arendt, Hannah 331, 353–4
Aristotle 26, 42–5, 230–2, 245, 247–55, 258–9, 263–71, 290, 302–3, 315–16, 326–7, 392–3
  Heidegger's lectures on 324–5
  on the soul 224–8
  on types of life 228, 345–6
archaea 54–5
art 2, 12–13, 81, 91–2, 295–8, 385–6
Artemis 223
ascension 345–6, 360–1
ascentional psychism 310
asceticism 5–7, 30–1, 61, 94–5, 104–5, 112, 119–20, 149–50, 206, 224–5, 310
  inner worldly 244–5
āśrama system 124–5
ātman 111, 134–6, 142, 144–9
atheism 46–8, 279–80, 340
Ati Mārga 165–6
Aufhebung 22–3, 289, 295–6
Augustine 243–7, 254
Aurora, The (Boehme) 277–8
australopithecenes 79–81
autocracy 392
autopoiesis 377–80
Avataṃsaka-sūtra 207–8
Avebury 15
Averroës / Ibn Rushd 252, 257–8, 263–7
Avicenna 251–2, 256–7, 315–17
Ayurveda 138, 141–2, 150, 154–7, 174
Axial Age/ Axial Shift 16–18, 34, 120–1, 124, 131–2, 134, 148, 149, 174

Baader, Franz von 276
bacteria 54–5
Badiou, Alain 333–4
bare life 113, 306–7, 310–11, 344–6, 348, 350–4, 356, 362
Barnard, Alan 77, 81–2
Barnes, Luke A. 52–3
Barth, Karl 344–5
Bartley, Chris 144
barzakh 261–2
Bataille, Georges 98, 100–6, 110–14, 124, 126–30, 302–3, 352–3
beatific vision 355–6
beauty / the beautiful 231–4, 246–7
Bedau, Mark A. 53–4
Beethoven, Ludwig van 28
being-for-itself 129, 288
being-in-itself 119–20, 288
being-in-and-for-itself 129, 288

being-in-the-world 290, 326–8
Bellah, Robert 19–21, 24–5, 66–7
Benjamin, Walter 358–9
Bennett, Jane 317–19, 339–42
Benveniste, Émile 66
Berdyaev, Nicolai 24–5, 361–2
Bergson, Henri 20–1, 26–7, 236–7, 275–6, 299–305, 310, 315–16, 320–1, 323, 339, 341–2
Bernard, Claude 303–4
Bhagavad-gītā 138–9
bhakti 164–5
Biardeau, Madeleine 138–9
Bickerton, Derek 79–81
bio-energy 50–1, 61, 66–7, 76–7, 372–4, 378, 382, 385
biopolitics 347–8, 351, 353–4
bio-semiotics 53–4
bio-sociology 4–5, 217, 312, 320–1, 367, 372–3, 380–1, 383
  of transformation 65
  religion as 380
biology 2–3, 6–7, 29, 34–5, 50–62, 157, 272, 326–7, 385–6, 393
  founded on Dasein 328–9
  holistic 48
  human 2–3, 69–70, 383
  molecular 340–2, 388–9
bios 26, 227–8, 235–6, 345–6, 350, 356
Blake, William 24–5, 272, 277–8, 361–2
Blombos cave 81–2, 86–7
Blondel, Maurice 355–6
Blumenberg, Hans 110
Bodhidharma 207–8, 210–11
body politic 346–7
Boehme, Jacob 24–5, 272, 277–9, 281, 297–8, 361–2
Boethius 254–5
Bolland, Mechtilde 135–6
Bonaventure 245
Bonhoeffer, Dietrich 357
Book of Animals (Aristotle) 265–6
Bowie, Andrew 284–7
Bowker, John 67, 90
Boyle, Robert 40
Brahma-sūtra 142–3
brahman 111, 119–20, 122–3, 134–7, 142, 146, 148–9, 155–6
Brahmanism 120–1, 149–50, 164–6, 172–3
brain 22–3, 29, 69–70, 339–40, 370–2, 374, 379–80, 382
  plasticity of 373
  quantum 340
  social 5–6
breath 26–7, 134–7, 148, 219–20, 222, 232, 238–9, 243–4, 247
  of God 26, 221–2, 261–2
  of life 262–3
breathing 219–20
Bretano, Franz 228–9
Bṛhadaranyaka-upaniṣad 131–2, 140, 145–6
Bronkhorst, Johannes 21, 149–50, 152–3, 174
Brown, Norman 24–5, 272, 277–8, 361–2
Bruno, Giordano 272, 276–7, 297–8
Buddha 149–50, 154–5, 200, 207–8, 211–12, 299
Buddha nature (*foxing*) 206–10
Buddhahood 208
Buddhism 1–2, 4, 17–20, 32, 43–4, 96–7, 115–16, 149–53, 174, 188, 214, 216
  Chinese 176, 188, 197–8, 202–8, 213
  Mahāyāna 125
Burkert, Walter 96–7, 99–100, 126–7
Butler, Judith 299–300

Campbell, Scott M. 323
Canon of Medicine see *Al-Qanun fi 't-Tibb*
capitalism 33–4, 244–5, 348–9, 392–3
Caraka-saṃhitā 154–5
care 324–5, 327–8
catharsis 98–100
Catholicism 39–40
causation 46–8, 317–18, 320–1, 338–40, 379–80
celibacy 130
centralization 392–3
Certeau, Michel de 272, 276–7
Chan Buddhism (*see also* Buddhism, Chinese) 207–8
Chāndogya-upaniṣad 145–6
Chen Duxiu 215–16
Chengzi 184–5
Childe, Gordon 13
Chinese 175–6
Chittick, William 260
Christ, Jesus 241–2, 311–12, 344–5, 357–61
  resurrection of 360
Christian, David 14–15
Christianity 17–18, 32–3, 41, 96–7, 100, 106, 218–19, 237–9, 241–8, 254, 257–8, 265–8, 275, 293–5, 330, 344, 358–60, 362–4, 382, 388–9
Chrysostom, John 359–60
Church Fathers 245–6, 355
Cicero 238–41, 247
citizen 352–3
civilization 1–8, 30–1, 34–5, 39, 119, 177–80, 182–3, 186–7, 190–1, 217, 342–3, 388
  as secular sacred 361–2
  as self-repair 362, 365, 393
  emergence of 11–18

global 392–3
religion and 4–5, 18–26
Clooney SJ, Francis X. 35–6
cognitive science 67–8
Coleridge, Samuel Taylor 277–8, 280–1, 295–7
commensality 22–3
communication 369–73, 375–6, 378, 388
function of 379
religion as 380
systems of 372–4, 378–80
comparative theology 320
compassion 7, 71–2, 76, 236–7, 372–3
complexity 30–1, 379–80
reduction of 381–2
Confucius *see* Kongzi
Confucianism 176, 179–83, 188, 190–1, 200–1, 207–8, 214–17
Neo- 184, 188, 201–6, 216–17
consciousness 21, 91, 306–9, 315, 370–1, 384–6
advanced linguistic 4–7, 11–13, 22–5, 28, 34, 211, 367, 388–9
altered states of 70
Hegel and 292–5
rights and 215–16
Śaiva understanding of 166–9, 171–3
Constable, John 296–7
constraint 30–1, 46–8, 67, 317–18
contemplation 186, 236–7
Cook, Scott 185
Corbin, Henry 260
correlation 337–8
correlationism 335–7, 340
cosmic man 127–8
cosmology 42–3, 157, 272, 276–7, 357
hierarchical 245, 261–2
retreat of religion from 358–9
Courbet, Gustave 296–7
Coyne, Jerry A. 46–8
Creative Evolution (Bergson) 301–2, 304–5
Critique of Judgement (Kant) 329–30
culture 4–5, 49, 59–61, 369–70, 382–3, 388
Cummings, E.E. 43–4
cyborg feminism 301
Cynics 310–11

Dalai Lama 393
dao 177–8, 188, 190–2, 194–8, 200–2, 208–9, 211, 216–17
Daodejing (Laozi) 189–91, 197
Daofa huiyaun 198
Daoism 20–1, 30, 32, 175–8, 188–91, 196–7, 200–3, 208, 211, 214, 216–17
Daoist possession cults 362–3
Darwin, Charles 27–8, 285–6
Dasein 23–4, 323, 326–31, 333, 338–9
Davidson, Herbert H. 265–6
Davies, Oliver 9, 360–1
Dawkins, Richard 39
De Anima (Aristotle) 225, 228–9, 249–50, 264–5
De Natura Deorum (Cicero) 238–9
De Substantis Separatis (Aquinas) 251–2
De Trinitate (Augustine) 247
death 1, 24–5, 28, 43–5, 66, 94–8, 127–8, 137, 144–5, 216, 251, 295, 331–2, 348, 353, 361–2
acceptance of 361–2
being towards 327–8
desire for 33–4, 382
laudation of 349–50
meanings of 381
personified 132–4
transcending 81, 94–5
definite descriptions 377, 379
Deleule, Didier 348–9
Deleuze, Gilles 26–7, 32, 254, 257–8, 275–6, 299, 301, 308–22, 326–7, 339, 341–2
democracy 217, 346–8, 352–3, 356–7, 392
Derrida, Jacques 7–8, 70, 393
Descartes, René 40, 247, 278–81, 283–4
Description of Greece (Pausanias) 223
desire 126–7, 201, 227–8
repression of 23–4
detachment 138–40, 190, 216, 241
Detienne, Marcel 96–7
deus absconditus 333
deus sive natura 279–80
dharma 12–13, 120–1
Dharmaśāstras 147–50
Dhvanyalocana (Ānandavardhana) 167
Diagram of the Great Ultimate see *Taiping jing*
dictatorship 347–8
Difference and Repetition (Deleuze) 308
dignity 224–5, 355–6
Dionysius 310–11
Dionysius the Pseudo-Areopagite 245, 276–7
Discarded Image, The (C.S. Lewis) 41
Divine Hierarchy (Pseudo Denys) 245–6
Divine Right of Kings 346
Dogen 210–11
domestication 81–2
Driesch, Hans 276
drugs 21
dualism 32, 176–7, 253, 279–82, 331–2
Kantian 340
dualistic ontology 355
Dumont, Louis 122–5
Dunbar, Robin 77

Duns Scotus, John 32, 196, 247–8, 253–8, 275–7, 307–8, 315–16, 319, 326–7
Durkheim, Emile 19–21, 66–7, 345–6, 383

Eckhart, Meister 272, 276–7
Eckman, Paul 369–70, 375
ecosystem 333
ecstasy 169, 286–7
education 187
élan vital see vital principle
Elias, Norbert 12–13
emotions 180–1, 185
  cross-cultural 369–70
  pro-social 8–9, 19, 64–5, 68–70, 75–7, 83–4, 89, 214
empathy 46–8, 64–5, 69–72, 76, 91, 372–3
Empedocles 310
emptiness 125
Encyclopedia of the Philosophical Sciences (Hegel) 289–90
Engberg-Pedersen, Troels 241–3
entelecheia 326–7
entelechy 326–7
Ethics (Spinoza) 279
environment 29, 327, 357, 367, 378–80
  distinct from religion 380
environmentalism 362
Eros and Thanatos 23–4, 361–2
eschatology 6–7, 251
  Christian 353–4
esotericism 278–9
Esposito, Roberto 66, 346–8
ethics 17–18, 26, 70, 190–1, 200–1, 203, 332
ethology 68–9
ethopolitics 391
eukaryote 54–7
Evans-Pritchard, E.E. 106
evolution 11, 48–50, 53–4, 69, 301–2, 313–14, 320–1, 342–3, 369–70, 372
  co- 8, 11, 29–30, 33–4, 49, 59–60, 319
  human 2–3, 46–8, 76
exception, state of 345–6, 349–52
existentialism 332–3
extended evolutionary synthesis 48, 68–9, 319
extrinsicism 355–6, 359–60

face, the 311–12
face-to-face encounter 71–2, 76–7, 368–76, 379, 382–6
faciality 311–12
facticity 324, 326–7, 337–8
Falk, Dean 79
family, the 382–3
fasting 384–5, 390–1
Faure, Bernard 211–12
Feldman, Marcus W. 58–9
Ferraris, Maurizio 334–5
Fertile Crescent 13–14
Feuerbach, Ludwig 67
Fichte, Johann Gottlieb 282
Ficino, Marsilio 236–7, 272, 276–7, 297–8
filial piety 175, 179–80, 182–3, 204–7, 214, 216–17
Filliozat, Jean 154–5
fine-tuned universe / fine-tuning 30–1, 52–4
flint knapping 74–5
flow 216, 319, 326–7
Fons Vitae (Avicebron) 267–8
form/s 225–33, 235–6, 249–51, 253, 255–6, 268–71, 289
formal distinction 254–7
formal indication 63–4, 114–15, 324–5, 330
Foucault, Michel 348–9, 353–4
Frankfurt School 344
Frauwallner, Eric 152–3
freedom 2–3, 5–6, 25, 29–57, 62, 281–7, 289, 293–4, 296–7, 299, 320–1, 381, 388
Freeman, Rich 109–10
French Revolution 293–4
Freud, Sigmund 23–5, 67, 99–100, 361–2
Freudenthal, Gad 265–6
Fuentes, Agustin 48, 79
Fung, Yu-lan 196–7

Gadamer, Hans Georg 331, 334–5
Galapagos finches 58–9, 61
Galen 265–6
Galilean Mathematical Model (GMM) 40–1, 44–5, 62, 119, 218–19, 272, 275, 281, 388–9
Galileo Galilei 40, 62
Gandhi, Mohandas 12–13
Garfield, Jay 161
Gauchet, Marcel 388–9, 391
Geertz, Clifford 374
Geist 242–4, 287–8, 344
Gennep, Arnold van 85–6
Gerhart, John C. 56–7
Girard, René 99–100, 103, 105–6, 113–14
Gītagovinda (Jayadeva) 164–5
Gitchel, Johann Georg 272
Gnostic narrative 32–3, 351–4, 361
Gnosticism 4–5, 32, 236–7, 245–7, 331–3
Göbekli Tepe 15
Goethe, Johann Wolfgang 236, 295–6, 361–2
golden rule 7
Good, the 231–6, 245, 251–2, 292
governance 187
grace 262–3, 271–2, 354–9
great outdoors 338, 340
Guan Wuliangshou Jing 212
Guattari, Felix 311, 317, 339

Guéry, François 348–9
Guide to the Perplexed (Maimonides) 267–8
Guodian 184

habeas corpus 351
Hades 81
Hadot, Pierre 229–32, 236
haecceitas / haeccity 254–7, 308–9, 314–16
Halbfass, Wilhelm 151–2
Hammurabi 14–15
Haraway, Donna 301, 357, 388–9
harmony 175, 190, 203, 217, 231–2
health 66, 266–7, 351–2
  of social body 346–7
heaven 175, 177–8, 185, 190–1, 197, 201, 360
  mandate of 186
Hedley, Douglas 96–7
hedonism 330
Heelas, Paul 33–4, 391–2
Heesterman, Jan 94–8, 114–15, 126–9
Hegel, G.W.F. 24–5, 242–4, 277–81, 286–96, 299, 308, 344, 361–2, 385–6
Heidegger, Martin 23–4, 27–8, 62–4, 91, 129, 299–300, 322–31, 333, 338, 341–2, 383–5
  critique by speculative materialism 335, 337–8
hermeneutical phenomenology 35–6, 50, 90, 383
hermeneutics 62–3, 90–1
Hilsum, Etty 357
Hinduism 1–2, 19–20, 43–4, 96–7, 164–5
historicism 319, 355–6, 359
history 358–61, 363–4, 393
  of salvation 359–60
Hobbes, Thomas 214–15, 346–51, 354–5
Hobbesian tradition 346
Hoffmeyer, Jesper 29–30
Hölderlin, Friedrich 280–1, 295–6
holiness 342–3, 365, 388–9, 393
holism 176–7
Holy Spirit 243–4, 257
homa 96
hominins 34
Homer 26, 222
Homo heidelbergensis 76, 81–2
*homo sace*r 345–6, 350–4, 358–9, 362
Homo sapiens 5–6, 13–14, 43–4, 60–1, 72–4, 78–82, 86, 94, 367, 372–3
Honen 213
Hongzan Fahua Zhuan 208
hope 9, 129, 356
  eschatological 7, 17–18, 25–6, 124–5, 131–2, 134, 141–4, 146–54, 158–9, 164, 173–4, 218, 228, 243–7, 253, 360–1
  soteriological 107–8
Hopkins, Gerard Manley 254, 326–7, 361–2
Horn, Friedrich Wilhelm 242–3
Houben, Jan 152–3
Huainanzi 192–3
Huang Tsung-hsi 178–9, 214–15
Hubert, Henri 96, 98–9, 126
Hui-neng 209–11
human condition 1–2, 104–5, 388
human nature 180–1, 185–7, 201, 215–16, 346–7, 392
  as conflictual 347
human rights 33–4, 215–16, 362, 392–3
humanism 281, 388
  atheist 355–6
humanities 4–5
humanity 175, 182–3, 187, 201, 355–6, 393
  in new creation 360
  philosophy of 383
Hume, David 338
Husserl, Edmund 23–4, 27, 40, 62, 322–3, 338, 341–2, 384–5
  critique by speculative materialism 335
hylomorphism 49, 228, 267–71
hylozoism 281

Ibn al-Arabī 259–60
Ibn Gabirol / Avicebron 267–8
idealism 162, 299–300, 331–2
  German 299–300, 334–5
identity, personal 373–4
  idem 373–4
  ipse 373–4
  narrative 373–4
ignorance 1–2
Illiad (Homer) 222
imagination/ imaginaire 6–7, 50, 64–5, 124, 126, 175–7, 261–2, 280–1, 296–8, 362, 381–2, 388–9
immanence 134, 144–5, 168–70, 172, 236, 272, 276, 280, 305–8, 319
  field of 276–9, 306–22, 341–2, 384–5
immortality 81, 111–12, 177–8, 196–8
immunity / immune system 66, 346–7
Incoherence of the Incoherence see *Tahafur al-tanafut*
Incoherence of the Philosophers see *Tahafut al-falasifa*
index 372–6
indexicality/indexicals 374, 376–7, 379, 381–2, 385
indexical-I 376, 385
individualism 110, 122–3, 217, 320–1
individuation 254–5
Indo-European Languages 66
Indus Valley Civilization 12–15
Industrial Revolution 11

Ingalls, Dan 310
Ingold, Tim 49, 299–301, 313, 318
injunction 111
intentionality 339
  collective 76, 377
  individual 74–5
  joint 74–6, 371–2
  shared 369–72, 374–5
interactive brain hypothesis (IBH) 370–1
intimacy 102–5, 107, 111–12, 126–8, 141, 352–3
intuition 287–8
inwardness /interiority 186, 236–7, 262–3, 358–9
Irigaray, Luce 26–7
Islam 17–19, 33–4, 46, 218, 247–8, 254, 257–9, 265–8, 363, 366, 382, 388–9
Īśvarapratyabhijñā-vimārśinī 167

Jacobi, Friedrich Heinrich 279–80
Jaegher, Hanne de 370–1
Jainism 19–20, 32, 115–16, 120–1, 149–50, 152–3, 170
James, William 20–1
Jankélévitch, Vladimir 303–4
Jaspers, Karl 16–17, 317, 329
Jericho 14–15, 311–12
Jhering, Rudoph von 215–16
jīva 26, 33, 145–7
jīvana 26
jñāna 112
Jonas, Hans 32, 49–50, 62–3, 299–300, 330–3
Josephus 220
Judaism 17–18, 32, 81, 96–7, 218, 220, 237–8, 247–8, 254, 267–8, 363, 382
Jullien, François 27–8
Jung, Carl Gustav 21, 199–200

Kafka, Franz 352–3
Kant, Immanuel 40–1, 50–1, 62–3, 280–2, 284–5, 297–8, 329–30, 335–6
Kantian Humanist Model (KHM) 40–1, 44–5
karma 144–5, 149–50, 235
Katzenstein, Peter 12–13
Kauṣītaki-upaniṣad 137–8
Keter Malkhut (Avicebron) 270
Kierkegaard, Soren 299
Kim, Jung-Jeup 203
Kirschner, Marc W. 56–7
Kojève, Alexandre 29, 293–4
Kongzi/Confucius 179–80, 184
Kropotkin, Peter 392
Kula 171–2
Kumārajīva 206–8

Laland, Kevin N. 58–9
language 5–7, 12–15, 21, 50, 61, 69, 76, 147–8, 175–6, 345–6, 371–4, 376–8, 381
  as *langue* and *parole* 385
  dominant 12–13
  origin of 68–9, 77–83
  sacred 15
  signification and 377
  structural coupling and 379
Laṅkāvatāra-sūtra 162–3
Laozi/ *Laozi* 17–18, 184, 189–90, 197, 199–200
Lascaux 88–9
Latour, Bruno 299–300, 316–18
law 14–15, 22–3, 214–15, 345–7, 350–1, 362, 366, 372–3, 379, 382, 392
  God's 363
  Islamic 259
  Roman 350
  sacred 43–4
Leach, Edmund 374–5
Lefebvre, Henri 313
Legge, James 192
legislature / legislative system 392
Leinhardt, Godfrey 18
Lessing, Gotthold Ephraim 279–80
Levallois flint technology 89, 91–2
Leviathan (Hobbes) 351
Levi-Strauss, Claude 86–8
Lewis, C.S. 41
Lewis, Geraint F. 52–3
Lewis-Williams, David 87–8, 91
Lewontin, Richard C. 57–9
Leyser, Ottoline 46–8
Liang Qichao 215–16
liberation 119–20, 123–4, 144, 147–8, 153–4, 174
liberty 351
life, biology of 49–50
  definition of 53–4
  desire for 2, 4, 6–8, 13, 17–18, 23–5, 31–4, 113, 123–4, 388–9, 393
  good 46
  holy 261
  images of 191–2
  meaning of 1, 51, 327–8, 391
  narratives of 3
  ontology of 49–50
  practice of 4–5, 25
  politicization of 346
  sacrality of 254, 354
  sanctification of 356–7
life force (see also *qi*) 44–5, 50–1, 177–8, 180, 182–3, 187, 192–3, 201–2, 204–5, 214–16, 238–9, 351
  virtue and 214

lifeworld 27–8, 44–5, 322–3, 327–8, 333–4, 340
light 233
Lingbao 197
literature 12–13
Liuzi-tan-jing (Hui-neng) 209
Loach, Ken 357
Locke, John 18, 214–15, 275, 354–5
logos 229–30, 235, 238–9, 291, 327
Lombard, Peter 254–5
long life 196–8
Lotus-sūtra 207–8
love 7–8, 76, 180, 357
politics of 7–8, 393
Lovejoy, Arthur O. 42–3
Löw, Reinhold 286
Löwith, Karl 351
Lubac, Henri de 299–300, 344–5, 354–9
Lucian 223
Luhmann, Niklas 300, 312, 376–83
Luther, Martin 361–2
Lyrical Ballads (Wordsworth) 297–8

Magna Carta 351
Mahābhārata 138–9
Maimonides 146–7
Mālinīvijayottara-tantra 167
Mandanamiśra 142–3
mantra 21, 78–9
Mantra Mārga 165–6
Mao tse tung 215–17
Maoism 203
Marcion 32
Martens, Wil 378
Marx, Karl 294–5
Marxism 217, 322
materialism 241, 243–5, 251, 331–2, 342–3, 383–6
materiality 253, 385
matter 126, 229–30, 235, 249–50, 269–70, 281–2, 284–8, 331–2, 340
Matthews, Gareth 225
Maturana Romesin, Humberto 377–8
Mauss, Marcel 96, 98–9, 122–3, 126
māyā 142
Mayr, Ernst 58–9
meaning 5–6, 44–5, 50–1, 64–5, 90–1, 96–7, 324–5, 339–40, 342–3, 377–9, 381
life charged with 326–7
medicine 156–7, 224–5, 272
Chinese 176–7
medieval exegesis 354–5
meditation 6–7, 21, 30–1, 119–21, 149–50, 203, 385–6
Meillassoux, Quentin 335–40
memory 373
Mendelsohn, Moses 279–80
Mengzi/ Mencius 42–3, 175, 180–5, 192, 201–2, 214–15
mentalizing system 370, 372
Menzel, Emil and Charles 73–4
Merleau-Ponty, Maurice 370–1, 383–5
meshwork 313–14
Messiah 85–6, 237–8
metaphor 374–5, 377–8
Metaphysical Foundations of Natural Science (Kant) 372
metonymy 374–5
Meuli 223
Milbank, John 39–40, 358–9
mimetic desire 346–7
Mīmāṃsā 129–30
Ming-I tai-fang lu (Huang Tsung-hsi) 214–15
Mingzhen zhai 197
mirror neuron system 370, 372
Mishnah 220
modernity 3, 42–3, 110, 120–1, 215–16, 272, 275–6, 298, 317, 323, 331–2, 347–8, 358–60, 385–6
bare life and 345–6
desire for death in 362
Mohammad 299
mokṣa 148
monasticism 124–5, 151–2
monos 230, 232
monotheism 219–20
Moore, Adrian 287–8, 308, 326–7
moral action 1
moral ontology 2–5, 45–6
More, Henry 40
Moses 242–3, 322
Mozi 185
Mrs. Dalloway (Woolf) 306–7
music 2, 12–13, 21, 78–9, 184–7, 385–6
mutualism 371–2
mythology 91
mystical ascent 236–7, 246–7, 358–9
mystical experience/state 19–20, 22
mystical theology 257–9, 276–7
mysticism 257–8, 272, 276–7

Napoleon 242–3, 293–4
Nārāyaṇa 144–5
Narcissus 236
National Socialism 331, 333, 344–5, 349, 351–2
nationalism 365–6
Natufian Culture 13, 82
naturalism 46–8, 67–8, 153–4, 165, 265–6
nature 4–5, 42–3, 49, 204, 238–9, 271–2, 275, 278–81, 284–9, 296–7, 317, 333–5
affirmation of 357

nature (*cont.*)
as God 279–80
external 291
grace and 354–7
philosophy of 152–3, 282, 286–7, 297–8, 334–5
pure 355–6
state of 346–7
Ndembu 85–6
Neanderthals 77–9, 81–2, 86
Nehru, Jawaharlal 12–13
Neiye 197
neo-cortex 371–4, 379–80, 382, 388–9
Neolithic 13–14, 16
Neolithic Revolution 13, 94
neurobiology 384–5
neuroscience 65, 67, 320–1, 367
second person 370–2
New Age 276, 391–2
new realism 333–5
Newton, Isaac 40
niche construction 2–3, 5–7, 25–6, 33–4, 48–9, 58–61, 64, 69–70, 73, 313, 319, 342–3, 374, 378, 388–9
Nicomachean Ethics (Aristotle) 248–9
Nietzsche, Friedrich 279–80, 299–300, 308, 310, 332–3, 348–52
nihilism 293–4, 332–3, 344–5, 348, 355–6, 361–2
Nomos of the Earth, The (Schmitt) 363–4
non-dualism / monism 32, 166, 170, 276–7, 279, 308–9, 314, 319–21, 332, 340–2
non-dualist narrative 32
nothing / nothingness 94–7, 105, 115–16, 132, 261–2, 286–7, 308, 336–7, 348
creation from 245–7
nous 228, 230, 232, 238–41, 269–70, 291
nouvelle théologie 355
Novalis 24–5, 278–9, 361–2
Nowotny, Helga 43–4, 341–2, 388–91
Nuer 106
Nussbaum, Martha 228–9

Oddling-Smee, F. John 58–60, 313
Odysseus 81, 222, 232–3, 236
Odyssey (Homer) 81, 222
Olivelle, Patrick 131–5, 137
omega point 360
On Human Nature (Wilson) 67–8
On the History of Modern Philosophy (Schelling) 282–3
One, the 232–7, 246–7, 270
ontogeny 57, 73, 302–3
ontology 329–30, 362
confusion with epistemology 334–5
of death 331–2
participative 358–9
realist 383
onto-theology 333–4
Organon (Aristotle) 245–6
Origen 246–7
Otto, Rudolf 351
Our Mutual Friend (Dickens) 306–7
Oyama, Susan 57

Padoux, André 165
Paleolithic 33–4, 60–1, 74–5, 81–2, 94
panentheism 144
panpsychism 334
pantheism 279, 281, 299
pantheism controversy 279–80
panmechanism 331–2
panvitalism 331–2
Panikkar, Raimundo 129–30
Parker, Robert 223–4
Part maudite, La (Bataille) 101
participation 111–12, 144, 228, 231–2, 353–6, 359–60, 364
in Christ 358–9
in spirituality 222
particularity 143–4, 326–7, 151–4, 196, 254–5, 257, 275, 300, 306–9, 312–13, 325–7
Pāśupatas 165–6, 310–11
patriarchy 271–2
pattern (*li*) 204–5, 214
Paul, Saint 63–4, 237–8, 241–2, 245, 344–6, 353–4, 361
Pauli, Ezequiel Di 370–1
Pausanias 223
peace 41
Peirce, Charles Sanders 376
person 30–1, 312, 320–2, 327, 341–3, 352–3, 356–7, 373–4, 376, 383
as mind and body 384–5
as psychic system 378
sanctity of 357
social system and 380–1, 383–4
personalism 358–9
Pessim, Sarah 267–8
Petrapoulou, Maria-Zoe 222
Phenomenology 35–6, 62–4, 275–6, 317–18, 327–8, 330–1, 333–4, 338–40, 380–1, 383
not worldview 329–30
of life 333–4
Philo 220–3, 237–8
philology 35–6, 64–5
phrenes 222
phylogeny 73, 302–3
physics 385–6
Physics (Aristotle) 264–5
Pinkard, Terry 385–6

Platform-sūtra see *Liuzi-tan-jing*
Plato 32, 224–5, 229–33, 245–6, 258–9, 263–4, 271, 291, 310
Platonism 228–30, 237–9, 245–7, 299–300
  Christian 285–6
  Neo- 224–5, 229–30, 267–9, 276–7
play 24–5, 361–2
pleasure 393
Plotinus 32, 229–37, 245–7, 270
pluralism 392–3
Plutarch 223
pneuma 26, 39–40, 154–5, 192, 217, 232, 238–44, 257, 344, 359
pneumatikos 356
poetry 272, 275–6, 295–8, 385
pointing 73–4, 77–8, 376
polis 227–8, 242–3, 344–6
political eschatology 174
political philosophy 217, 339, 346–7, 354–5
political theology 174, 342–5, 354, 357–9, 361–4
politics 347, 351–2
  identity 365
polytheism 219–20
Pompeius Festus 350
Pope, the 393
Porphyry 235, 246–7
postmodernity 385–6
power 347, 361
  axes of 392
  state 346
  will to 348
power scavenging 80
practice 385–6
pragmatics 385
prāṇa 26, 39–40, 134–8, 144–5, 148–9, 154–5, 192, 217
Pratyabhijñā 167
prayer 4–6, 30–1, 61, 261, 384–5, 390–1
present-at-hand (*vorhanden*) 327–8
primates 4–6, 73–5
pro-sociality 4–6, 23–4, 50, 61, 70, 91, 214, 217, 236–7
Productive Body, The (Guéry and Deleule) 348–9
prokaryote 54–7
proper names 377, 379
Protestantism 361–2
Proust, Marcel 27–8
psuchē 26, 217, 220–1, 225, 227, 230–2, 238–41, 249–50, 253
psychoanalysis 24–5, 272, 277–8
Psychology of Worldview (Jaspers) 329–30
Pure Land Buddhism 212–13
purification 99–100, 108–10, 119–20, 169–70, 220–2, 233–4
  ascetic 129–30
  of the body 221–2
  of the tribe 100
purity 121–2, 165–6, 171–2, 283, 351–2
purposes of life 148
Putnam, Hilary 228–9, 333–4

qi (*see also* life-force) 26, 39–40, 180, 185, 191–6, 198–205, 207–8, 214, 216–17
Qur'an 258–9, 264–5, 271

racism 311
Rahner, Karl 344–5, 359–60
Rappaport, Roy 34–5, 83–4
Rāmānuja 33, 141–7, 174
rationalism 218–19, 266–7, 279–80, 354, 364
rationality 279–80, 372–3
Ratnagotravibhāga 162–3
ready-to-hand (*zuhanden*) 327–8
realism 336–7, 385–6
redemption 28, 33, 300, 353, 358–60
reductionism 39–40, 46–8, 69–70
Reformation 244–5
reincarnation 149–50, 206–7, 235
Reisebrodt, Martin 6–7
relativism 334–5
religion 2–3, 5–6, *passim*
  as autopoietic system 86, 115–16, 320–1
  as narrative, law and practice 380–1
  as personal experience 320–1, 383
  as process 383–4
  as transformation of bio-energy 383–4
  definition 66
  of morality 385–6
  origins of 68–70, 91, 379–80
  proto- 13
Religious Cosmological Model (RCM) 41–3, 119, 172, 218–19, 272, 281, 388–9
religious studies 33–4, 366–7, 382–3, 387
Renaissance 41, 265–6, 276–7, 331–2
renunciation 94–5, 123–6, 147
representation 385–6
repression 362
ressourcement 354
resurrection 24–5, 243–4, 251, 353, 356–7, 359–62
Resurrection, Cookham, The (Spencer) 359–60
Ṛg-veda 132
Ricoeur, Paul 50–1, 62–5, 114–15, 373, 376–7, 379, 383–4
Rilke, Rainer Maria 272, 361–2
ritual 1, 4–6, 16, 19–20, 34–5, 61, 70, 119–20 78–9, 83–6, 179–80, 184–7, 203, 382–3
  as repair 35–6
  Axial shift from 17–18

ritual (*cont.*)
centres of 15
proclivity to 12–13
propriety 16, 187, 203–4
purification 221–2
Romano, Claude 28, 327–8
Romanticism 282–8
Rose, Nikolas 391–3
Rousseau, Jean Jacques 45–6
ruach 26, 243–4

sacrality 67, 350–4, 361
of life 254, 351–4
secular creation of 364
sacred, the / sacredness 66, 345–6, 353–4, 358–9
secular 361–2
sacred man see *homo sacer*
sacrifice 34–5, 94–116, 124–6, 179–80, 187, 219–20, 263, 352–3
as refusal of death 94–9, 104–6, 110–12, 114
capital punishment as 350–1
Greek 220–5
of Christ 356
theme of 138–42
Vedic 120, 123–4, 126–31
sacrificial imaginary 119–22, 125–6, 137–8, 142–3, 148-50, 154–5, 160–1, 163–6, 170–1, 174, 218–19
Śaivism 165–6, 172, 310
Śaiva Siddhānta 165–6, 169–70
śakti 26, 39–40, 167, 170, 308–9
salvation 6–7, 124–5, 177–8, 356
Christian 356
Sāṃkhya 125, 150, 152–3
Sanderson, Alexis 164–6, 171–2
Śaṅkara 141–2, 174
Sanskrit 120–1, 175–6
Saumarez, Richard 297–8
Saussure, Ferdinand de 376
Schelling, Friedrich Wilhelm Joseph 277–9, 282–8, 295–8, 341–2
Schilbach, Leonard 370
Schipper, Kristofer 189, 194
Schleiermacher, Friedrich Daniel Ernst 296–7
Schmitt, Carl 345–8, 353–4, 363–4
Scholasticism 247, 263–5, 267–8, 271
Neo- 354–5
Schrödinger, Edwin 58–9
Schutz, Alfred 20–1
science 7, 33–4, 39–44, 46–8, 51, 63, 119, 157, 224–5, 264–7, 272, 275, 296–7, 319, 334–5, 340–2, 366–7, 383–4, 388, 390–1
biological 39–40, 44–5, 49, 301–2, 333, 380–1
cognitive 2–3, 67–8, 372–3
human 51, 63–5, 67, 114, 385–6
integration with humanities 2–3, 30
medical 122, 150
molecular 341
natural in contrast to phenomenological 62
of life 27–8, 63, 119, 137–8, 157, 275–6, 285–6, 328–30
religion and 42–3
social 99–100, 114–15
technology and 389–90
truths of 28
scientism 62
Scott, Michael 314
Secret of the Golden Flower see *Taiyi Jinhua zonzhi*
secular narrative 33–4, 354
secularism 33–4, 354, 356–7
secularization 351–2, 356, 365–6
self 122–3, 125
self-actualization 391–2
self-cultivation 178–9
self-realization 325–6
self-repair 2, 6–7, 34–5, 69–70, 90–3, 105, 113–14, 119–21, 123–4, 177–8, 342–3, 362, 374, 388–91, 393
semantic density 385–6
Sentences (Lombard) 254–5
sequestration 122–3, 128
sexuality 22–3, 193–4
Shamanism 87–8, 189
shen 194
Shishu (Zhu Xi) 204–5
Shinran 213
sign 367, 372–6, 379, 381
face as 369–70
Signature of All Things (Boehme) 277–8
Simmel, Georg 299–300
sin 1–2, 33, 108–9, 220–1, 246–7, 347–8, 355–6
Śiva 119–20, 164–5, 167, 169–70, 352–3, 355–6
Śivadṛṣṭi (Somānanda) 167
Slingerland, Edward 176–7
Sloterdijk, Peter 90, 92, 320, 369–70, 372–3
Smail, David Lord 7–8
Smith, Jonathan Z. 96–7
social contract 346–7
social cognition 4–7, 22–3, 50, 61, 64–5, 67–70, 89–91, 126–7, 320–1, 367–75, 377–8, 382–3, 385–6, 388–9, 393
inclusivity of 393
prior to language 377–8
transcending 379
social justice 217
social system 119, 312–13, 378–81

sociality 368–9, 376, 378, 382
sociology 64–5, 322–3, 380–1, 383
sociobiology 30–1, 49, 59–60, 68–9
solipsism 293–4
soteriology 6–7, 152–3
soul 26, 32, 144–5, 218–19, 225–7, 232–6, 238–9, 251–3, 263–5, 270, 288, 332–3
  animating force of body 249–50
  as life 290
  inner ascent of 246–7
  universal 269–70
sovereign power 344–7, 349–51, 357, 362
sovereignty 49, 346–7
Spaemann, Walter 358–9
speculative materialism 333–4
speech 219–20
Spencer, Stanley 359–60
Spinoza, Baruch 275–6, 278–81, 283–4, 297–9, 301, 308–9, 317, 320–1, 339
spirituality 365, 388–9, 391–2
spiritus 243–4
Śramanas 120–1, 149
Staal, Frits 19–20, 77–9, 83–6, 126
state, sovereign 363–4
Stahl, Georg Ernst 276
Sterelney, Kim 60–1
Stoicism 238–9, 241–2, 293–4
Stonehenge 15
Stravinski, Igor 298, 364
subjectivity 30, 90–1, 104–5, 280–3, 289, 292, 338–9, 344, 376, 388–9
  absolute 286–7, 295–6
Sukhāvativyūha 207–8, 213
Summa Theologia (Aquinas) 248–9
Śunaḥśepa 128–9
supernatural, the 354–6
Surnaturel: Études historiques (de Lubac) 354–5
survival 393
symbol 367, 372–5, 379, 381–2
System of Transcendental Idealism (Schelling) 282–3

taboo 351
Tahafut al-falasifa (Al-Ghazzālī) 264–5
Tahafut al-tahafut (Averroës) 264–5
Taijitsu shuo 201–2
Taiping jing 177–8
Taittirīya-upaniṣad 154–5
Taiyi Jinhua zonzhi 199–200
Tannisho (Shinran) 213
Tantrāloka (Abhinavagupta) 167
Tantras 108–9, 164–5, 167
Tantrasāra (Abhinavagupta) 132, 167
Tantrism 164–5
Tattvārtha-sūtra 158–9
Taubes, Jacob 242–3, 344, 353–4
Taylor, Charles 2–3, 16–18, 30–1, 33–4, 45–6, 90–1, 124, 316–17, 365–6, 388–9
technē 348
technology 215–16, 342–3, 349–50, 388–9
  assisted reproductive 341
  bio- 392–3
Teilhard de Chardin 247, 257–8, 299–300, 344–5, 356, 358–60
telos 287–8, 292, 303–4, 326–7
tertium quid 290–2
teyyam 85–6, 109–10
Thacker, Eugene 225, 249–50
theism 33, 279
theistic narrative 33, 351–2
theocracy 392
theory of mind (ToM) 370–1
Thomism 254
Thousand Plateaus, A (Deleuze) 306–7
thrownness 326–7
time 175, 193, 259, 263, 297–9, 301–2, 304, 308, 329–30, 341–2
  sacred 85–6
  social 85–6
Tomasello, Michael 45–6, 72–7, 80–1, 89, 369–70, 376
totalitarianism 351–3, 356–7
Totemism 99–100
tradition 385
tragedy 282–3
tragic, the 1–2
tragic sense of life 23–4
trance 6–7
transcendence 42–3, 70, 120–6, 143–5, 147–8, 157, 231, 236–8, 271–2, 320–1, 330, 353–6, 373–4, 381
  absence of in China 175, 186
Trinity, Holy 243–5, 257
Turner, Denys 225, 249, 251
Turner, Victor 85–6
Tyron, Thomas 266–7
Tweedale, Martin M. 256–7

Uexküll, Jakob von 299–300
Umwelt 29, 327, 357
unhappy consciousness 385–6
United Nations 362
universalism 7
univocity 196, 254, 257, 270, 312–15
unrepeatability 306–9, 313–14, 326–7
Upaniṣads 17–18, 44–5, 111, 120–1, 126–8, 131–2, 140–3, 145–50, 152–3, 164, 170, 174
Urban, Greg 385

Vaiśeṣika 157, 174
realism 150–4
Vaiśeṣika-sūtra 153, 155–8
Varella, Francisco 370–1, 377–8, 383–5
Vasquez, Manuel 19–20
Veda/s 108, 112, 120–1, 130, 142, 171–4
Vedānta 120–1, 125–6, 129–30, 142, 149–50, 167
Advaita 125, 142–3
Veer, Peter van der 11
vegetarianism 266–7, 382
Venus of Willendorf 81–2
verticality/ vertical ascent 2–3, 91–2, 113–16, 122–3, 134, 153, 156–7, 228–9, 233–4, 237–8, 261–2, 265–6, 309–10, 320
violence 1–2, 11, 99–100, 103, 129, 347, 349–50
of history 358–9
virtual, the / virtuality 303–4, 308–10, 320–1
virtue 182–3, 185–7, 190–1, 196, 214, 233–4, 309
vital materialism 334, 339–42, 344
vital principle /impulse/ *élan vital* 301–2
vitalism 26–7, 120, 125, 154–5, 174, 178–9, 203, 247–8, 251–2, 257–8, 263, 276–7, 284–5, 287–8, 299, 303–4, 320–1, 330, 333–4, 344–5, 351–2, 357, 362–3, 383
dark 33–4, 344–5, 349–53
German 363
in contrast to mechanism 389–90
new 334, 391
Viṣṇu 142, 144–5
Vogelherd cave 86–7
Vries, Hent de 96

Waal, Frans de 72–4
Wang, Robin 191–2
Wang Yangming 205–6
Ward, Graham 358–9
Warneken, Felix 72
Weber, Max 46–8, 122–3, 244–5
Whitehead, Alfred North 333–4
Whitman, Walt 272
Wilhelm, Richard 199–200
Wilson, Edward 68–9, 90–1
wisdom 216, 263
Wittgenstein, Ludwig 18–19, 337–8
Witzel, Michael 34, 86–9, 96–7
Wordsworth, William 280–1, 296–7
world spirit /world soul 40, 242–4, 291
worship 385–6
Wujastyk, Dominik 154–5
wujūd 261
Wuzhen pian (Zhang Boduan) 198

Xici 194–5

Yajña 96
yang and *yin* / *yinyang* 177–8, 192–5, 201–3, 216
Yang Zhu 180–1
Yijing (*Book of Changes*) 187, 194–5, 197
yoga 19–20, 120–2, 150–1, 199
Yogaśāstra (Hemacandra) 158
yogic suicide 108–9
Youguang, Tu 196
Yuan Ren Lun 209

Zaehner, R.C. 140
Zhang Zai 202–4
Zhengmeng (Zhang Zai) 203
Zhouyi cantong qi 197
Zhu Xi 201–2, 204–6, 214–15
Zhuangzi / *Zhuangzi* 177–8, 195–7, 211
Zimmermann, Johann Jacob 277–8
zoē 26, 225, 227–8, 235–6, 315–16, 345–6, 350
Zuozhuan 192–3